Fundamentals of Microelectromechanical Systems (MEMS)

About the Author

Eun Sok Kim received the B.S., M.S., and Ph.D. degrees from the University of California, Berkeley, CA, USA, in 1982, 1987, and 1990, respectively, all in electrical engineering.

He joined the University of Southern California, Los Angeles, in Fall 1999, where he is currently a Professor in the Department of Electrical and Computer Engineering–Electrophysics. Since July 2009, he has chaired the electrophysics division of the department, and oversaw a net tenure-track-faculty growth of 2.5 (from 15.25 to 17.75) with 6.5 new tenure-track-faculty hires. Over those nine years, US News' ranking raw score on USC EE's Graduate Program rose from 3.9 to 4.2 (out of 5.0 max).

From Spring 1991 to Fall 1999, he worked at the University of Hawaii at Manoa as a faculty member. Previously, he worked at IBM Research Laboratory, San Jose, CA, NCR Corp., San Diego, CA, and Xicor Inc., Milpitas, CA as a co-op student, design engineer, and summer-student engineer, respectively.

Professor Kim is an expert in acoustic, piezoelectric, and vibration-energy-harvesting MEMS, having published about 250 refereed papers, 16 issued US patents, and 7 pending US patents in the field. He is a Fellow of the Institute of Electrical and Electronics Engineers (IEEE) and the Institute of Physics (IOP). He serves as an editor for IEEE/ASME Journal of Microelectromechanical Systems. He has been awarded a Research Initiation Award (1991–1993) and a Faculty Early Career Development (CAREER) Award (1995–1999) by the National Science Foundation. He received the Outstanding EE Faculty of the Year Award at the University of Hawaii in May 1996 as well as the IEEE Transactions on Automation Science and Engineering 2006 Best New Application Paper Award.

Fundamentals of Microelectromechanical Systems (MEMS)

Eun Sok Kim

New York Chicago San Francisco
Athens London Madrid
Mexico City Milan New Delhi
Singapore Sydney Toronto

Cataloging-in-Publication Data is on file with the Library of Congress.

McGraw Hill books are available at special quantity discounts to use as premiums and sales promotions or for use in corporate training programs. To contact a representative, please visit the Contact Us page at www.mhprofessional.com.

Fundamentals of Microelectromechanical Systems (MEMS)

1 2 3 4 5 6 7 8 9 CCD 26 25 24 23 22 21

ISBN 978-1-264-25758-4
MHID 1-264-25758-9

This book is printed on acid-free paper.

Sponsoring Editor
Lara Zoble

Editorial Supervisor
Donna M. Martone

Production Supervisor
Lynn M. Messina

Acquisitions Coordinator
Elizabeth M. Houde

Project Manager
Revathi Viswanthan,
KnowledgeWorks Global Ltd.

Copy Editor
Namita Panda

Proofreader
Upendra Prasad

Indexer
Alexandra Nickerson

Art Director, Cover
Jeff Weeks

Composition
KnowledgeWorks Global Ltd.

To Joanne

Contents

Preface

With burgeoning internet-of-things (IoT) and artificial intelligence (AI), trillion sensors and transducers are on the horizon with microelectromechanical systems (MEMS) technology poised to be the only one that will make trillion transducers possible. Consequently, a comprehensive textbook on fundamentals of MEMS with accompanying in-depth questions and problems can contribute to making the world be connected through trillion sensors and transducers, as the textbook educates workforce in MEMS industry and also promotes MEMS research and development.

This textbook is intended for undergraduate seniors and graduate students in electrical, mechanical, and biomedical engineering as well as professional engineers who want to learn MEMS technology. This textbook is aimed to teach design, fabrication and testing of MEMS and contains many questions and problems for the students' homework and exams for the sake of in-depth understanding of MEMS as well as problem-solving and/or design skills in MEMS analysis and design. Also, it covers some commercially-significant topics (such as radio-frequency [RF] front-end filters and oscillators, acoustic transducers, vibration energy harvesting, etc.) that have not been covered in other MEMS textbooks.

The topics for this textbook are chosen along the line of the most commercially successful MEMS: pressure sensors, microphones, inertial sensors, RF front-end duplexers, micromirror array, etc. Many of the topics are covered in depth to the level of allowing the students to be able to design microfabrication process, MEMS structures, commercially successful MEMS, etc.

This textbook is divided into 10 chapters covering the microfabrication and MEMS processing technologies in the first two chapters, transduction principles in Chap. 3, RF and optical MEMS in Chaps. 4 and 5, mechanics and inertial sensors in Chap. 6, film properties and issues as well as silicon pressure sensors and MEMS microphone in Chap. 7, microfluidic systems and bio-MEMS in Chap. 8, vibrational energy harvesting in Chap. 9, and interface analog and nonlinear circuits for MEMS in Chap. 10. Most of the contents in this textbook can be taught over a semester of 15 weeks with 4 hours of lecture per week (i.e., for a four-unit course) to undergraduate seniors or graduate students in engineering without any particular prerequisite. PowerPoint presentation files of the figures as well as the solutions to the questions and problems in this textbook will be available to the faculty who adopt this book as a textbook for a course.

This textbook is a result of my teaching MEMS over about three decades, which has been greatly enriched through the lessons that I received from my alma mater, colleagues, and students. I would like to particularly acknowledge the following professors, some of whose course materials I have incorporated in the topics noted in parentheses: the late Nathan W. Cheung (basic microfabrication and vacuum technology in Chap. 1); Roger T. Howe (basic surface micromachining in Chap. 2 and energy method and accelerometers in Chap. 6); Paul R. Gray (electronic noise in Chap. 10); and Robert G. Meyer (oscillators in Chap. 10).

Finally, this textbook has been greatly enriched through my doing research on MEMS as a PhD student and then leading an active research program on MEMS for more than three decades. Thus, I would like to acknowledge the invaluable advice and encouragement from my PhD advisor, Prof. Richard S. Muller, and the extremely fruitful research endeavors of the students and post docs who have worked in my research group, whose names are listed in my research group's web site. As I read through this book, I pretty much know who have contributed on which topics, and there are many, a bit too many for me to list them here without potentially missing out some. I am very grateful that so many excellent students and post docs have worked with me in my research group.

E.S. Kim
Los Angeles, California
March 2021

Introduction

Commercial applications of microelectromechanical systems (MEMS) have ever been growing for many decades, starting with silicon pressure sensors. Some of the commercially successful MEMS include accelerometers, gyroscopes, digital mirror array of 0.3–1 million movable mirrors, microbolometers for room temperature infrared (IR) sensing or for night vision. The nighttime vision without ultra low temperature for a camera is possible because of excellent thermal isolation made possible through MEMS structure, while the large number of movable mirrors is possible due to the unique features of MEMS fabrication process. There is currently no other technology that will allow an array of 0.3–1 million movable micromirrors on a single chip.

More recently, MEMS microphones have been replacing conventional electret condenser microphones (ECMs), largely due to the MEMS microphones' ability to withstand the soldering temperature of about 250°C which allows them to be mounted on printed circuit boards (PCBs) through robotic pick-and-placement. Another MEMS commercialization success has lately been film bulk acoustic-wave resonators (FBARs) for RF front-end filters for cell phones, as FBARs offer high-quality resonant characteristics due to air-backing on top and bottom of a piezoelectric film, made possible by MEMS fabrication processing. The growth of MEMS commercialization has lately been 30% annually, and as of 2020, MEMS market is about a tenth of integrated circuits (IC) market.

Next Big Thing

Over the past seven decades, there have been some major technology advancements, starting with mainframe computer and then minicomputer, followed by notebook tablet and smartphones. These are the unique devices that have made the economy grow and also impacted people's lives greatly. So, the question in 2021 is what is the next big thing. Internet of things (IoT) has been hailed as the potentially next big thing, as smart wearables such as smart watches, glasses, and clothes can monitor or track people's health and activity. Smart home with smart appliances, smart car with collision avoidance or driving assistance, and smart city with smart parking are the areas where IoT will be needed. In all these cases, though, what are critically needed is the ability to sense and/or actuate and to compute at the sensor/actuator nodes (or the edges of cloud computing).

One notable example of IoT is a wearable activity tracker, such as FitBit. It contains a microcontroller, MEMS inertial sensors to sense the wearer's motion, light-emitting diode (LED) and photodetector, analog-to-digital converter, RF communication circuits and antenna, and battery. It is an intelligent edge device that processes sensed signal and communicate the data wirelessly to the cloud.

Edge devices may include actuators and/or biomedical platform, in addition to sensors. The number of the expected edge devices containing sensors and actuators is astronomically large. In comparison, the number of personal computers or laptops is limited, while the number of smartphones is 10 times more. With edge devices, though, the number quickly reaches hundreds of billions or even trillion. Thus, the manufacturing costs for edge devices must be very low, likely below $1 per unit. There is currently no technology, other than MEMS technology, that will allow such a massive number of edge devices to be manufactured at such a low cost.

Current MEMS as Edge Devices

Already there are many edge sensors. For example, a cell phone has a microphone, temperature sensor, accelerometer, gyroscope, absolute pressure sensor for measuring ambient pressure at different altitudes, and magnetometer, most of which are MEMS sensors. Humidity sensor, CO_2 sensors, gesture sensor based on ultrasonic sensors, proximity sensor, blood glucose sensor, blood pressure sensor, autofocusing based on MEMS, etc. (again most of which are MEMS devices) are already used in smartphones or will likely be in smartphones in near future.

Emerging MEMS Applications

In addition to IoT, emerging applications of MEMS include sensors and actuators for wearable technology, augmented reality, or virtual reality; MEMS microspeakers which are now in the market; microfluidic systems and bio-MEMS; and MEMS RF switches for antenna tuning for 5G communication. In case of 5G communication that uses so many different frequency bands, antenna will have to perform well for all those different frequencies, and an array of antennas is used for 5G wireless transceivers. Thus, impedance matching to various antennas is needed. One way to present different capacitances to antennas for impedance matching is to use a MEMS tunable capacitor or MEMS RF switches in combination with a bank of capacitors to present different values of capacitance through turning on/off the switches. Another emerging applications of MEMS are in automobiles, as the auto industry is transforming with driving assistance or self-driving features. There are already many MEMS sensors in cars such as pressure sensors, accelerometers, etc. However, radar and imaging sensors aided by or with MEMS technology are likely of high demand in future cars.

MEMS Industry

Among various MEMS manufacturers, Broadcom became the number one in 2017, over Bosch which sells many different kinds of MEMS sensors, due to the exploding number of RF front-end filters for smartphones. In 2020, Broadcom is still the largest MEMS manufacturer in revenue, while Bosch is closely the second. There are many other MEMS companies, almost all of which are growing in revenue and the number of the MEMS units that they are selling. Most of the current MEMS companies make their MEMS at their facilities. However, there are also many MEMS foundries which fabricate MEMS based on the designs done by MEMS companies, similar to silicon foundries. However, unlike silicon foundries (the number of which is rapidly shrinking), MEMS foundries are many in numbers, because vast majority of MEMS requires quite different fabrication processes. There is no one standard fabrication process that fits all kinds of MEMS. For example, some MEMS use electrostatic transduction, while others use electromagnetic or piezoelectric transductions that require quite a different set of materials and fabrication processes. Some MEMS are manufactured through wet micromachining and some others through dry micromachining. Thus, it is unlikely that a few foundries can dominate, but rather, there will continue to be many foundries for MEMS design houses.

MEMS Advantages

Microfabrication process of MEMS is based on silicon microfabrication technology that processes a batch of wafers in a fabrication run. In a fabrication of one batch of say, about 25 wafers, hundred or thousand sensors on each wafer can be processed for the 25 wafers. Consequently, at the completion of such a batch processing, there will be 2,500–25,000 sensors. This mass manufacturability due to batch processing is one of the key advantages of MEMS, because the manufacturing costs can be reduced tremendously. For IoT, trillion sensors and transducers are expected to be deployed around the world. Trillion devices will be possible, only if each device costs very little, and MEMS technology is currently the only one that can offer such possibility.

Another advantage of MEMS is that sensor or actuator can be integrated with IC on a single chip. By having a sensor and IC on a single chip, parasitic capacitance between the sensor and the IC can precisely be characterized, and the sensor's minimum detectable signal level can be improved greatly. Also, with IC already integrated on a single chip with a sensor, the integrated sensor can be small and inexpensive. In other words, through integration of MEMS with complementary metal–oxide semiconductor (CMOS) on a single chip or in a single package, the size, weight, power, and cost (SWaP-C) can be reduced. One of the commercially successful MEMS is an accelerometer integrated with signal-processing and temperature-compensating circuits on a single chip.

Learning MEMS

Though MEMS is an acronym for a microelectromechanical systems, it covers also microelectrothermal, microelectrochemical, microfluidic systems, and bio-MEMS. Thus, there are much more than just electromechanical expertise that is needed for an engineer to thrive in MEMS. Multidisciplinary knowledge is needed, and that is good for an engineer to take a lifetime to learn and keep on innovating. I myself majored in electrical engineering during my undergraduate years, and when I started my PhD research in MEMS, more specifically a resonant silicon pressure sensor (as MEMS had not yet been coined in mid-80s when I started the research), I had pretty good knowledge on microfabrication and circuits but quite limited knowledge on mechanics. Thus, I had to learn classical mechanics, particularly the statics and dynamics of plates. Ever since then, I have kept on learning new knowledge and technologies on many other areas such as acoustic, piezoelectric, thermal, and biological principles. This kind of multidisciplinary knowledge is what makes the job interesting for many decades and also what provides engineers with unique capability to innovate. With MEMS, it is not only depth but also breadth of knowledge and skills. As we keep learning new scientific knowledge, new technologies, new scientific disciplines, etc., we see more and further, allowing us to contribute to humanity through innovative and useful MEMS.

Another aspect of MEMS that excites me is that it is open-ended, with so much application and innovation wide open for well-educated and motivated MEMS engineers. Engineers and scientists with multidisciplinary knowledge and good ideas will be able to benefit the world greatly through many new MEMS-based devices and systems.

Impact of MEMS

Impact of MEMS has been already great, but will be much more than ever, since MEMS technology not only enables product differentiation but also allows new product capability for transducers. Already many conventional transducers such as electret condenser microphone, acceleration switch, cooled IR camera, etc. have been replaced by MEMS counterparts, and many more conventional transducers will likely be replaced by MEMS counterparts. More excitingly, though, many brand new transducers or systems based on MEMS (e.g., single-cell gene sequencing) will be invented and used.

It is likely that MEMS will be the key technology for the "next big thing," but how big the impact of MEMS will depend on the people who work on MEMS, and create MEMS devices and their applications. In other words, innovative engineers and scientists will be the key on how strong MEMS' future impact on human lives will be.

So, let us delve into learning microfabrication and MEMS processing technologies, transduction principles, RF and optical MEMS, mechanics and inertial sensors, film properties and issues as well as silicon pressure sensors and MEMS microphone, microfluidic systems and bio-MEMS, vibrational energy harvesting, and interface analog and nonlinear circuits for MEMS.

CHAPTER 1

Basic Microfabrication

1.1 Introduction

Fabrication process of microelectromechanical systems (MEMS) is based on the microfabrication process that was developed for manufacturing integrated circuits on semiconductor wafers and inherits most of the benefits that the microfabrication process offers, particularly in device miniaturization and mass manufacturability. One key technique of microfabrication process is photolithography that allows miniaturization (down to <10 nm in length) and almost unlimited repetition of same patterns over a whole wafer (that can allow extremely large number of same elements per device, e.g., one million micromirrors on a $1 \times 1\ cm^2$ silicon chip). Photolithography and various thin-film growth/deposition and etching techniques allow a batch of tens of wafers to be processed simultaneously at each of the processing steps, resulting in hundreds to thousands of same devices, per wafer, when one batch of wafers is completed of a microfabrication process. In this chapter, we will study some basic microfabrication processes that are commonly used in MEMS fabrication.

1.2 Photolithography

Photolithography involves the following: (1) a photomask that contains patterns to be transferred onto a wafer, (2) a wafer coated with photosensitive resin (commonly called photoresist [PR]), and (3) optical (or x-ray) pattern transfer system [1]. The photomask is produced by a so-called "pattern generator" that uses a computer file to produce opaque patterns on a mask (typically, a glass substrate for optical pattern transfer system). Photoresist is coated onto a wafer by dispensing a few droplets of photoresist and spinning the wafer at several thousand rpm (revolutions per minute). A pattern transfer system (either a contact printer or projection printer) takes a photomask and a photoresist-coated wafer and shines light onto the wafer through the photomask so that the patterns on the mask may be imprinted (through latent image) on the photoresist on the wafer, as illustrated in Fig. 1.1b. When the photoresist is developed in a photoresist developer, the latent image shows up, and a replica of the mask patterns now appears on the photoresist. Thus, patterned photoresist can now act as a masking layer for delineating the layer(s) underneath the photoresist. The delineation of the layer(s) is usually accomplished through wet or dry etching, after which the photoresist can be removed by a photoresist remover or oxygen plasma.

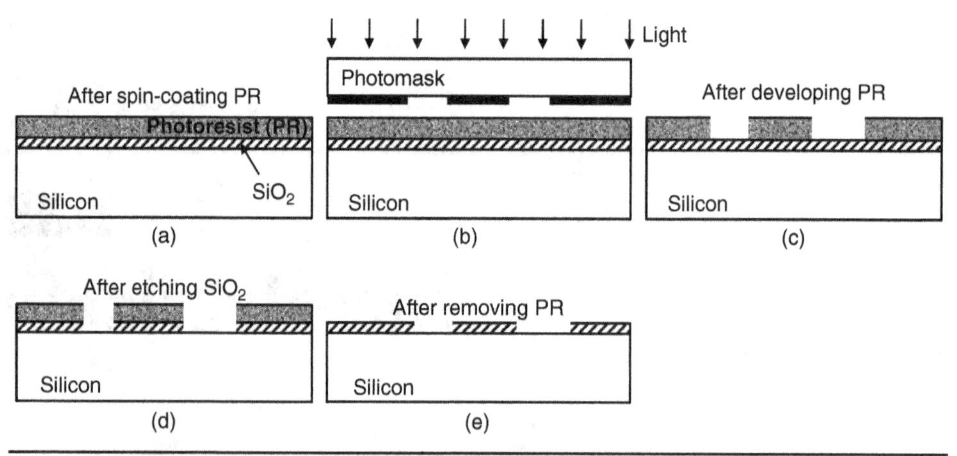

FIGURE 1.1 Pattern transfer onto a wafer through photolithography: photoresist is spin-coated on a wafer (a) and is exposed to ultraviolet light through a photomask (that contains opaque patterns blocking light passage), via a contact or projection printing (b). When the exposed photoresist is developed, the developed photoresist is a replica of the pattern on the photomask (c), and can be used as an etch mask for etching the underlying layer (e.g., oxide) (d).

1.2.1 Aerial Image of Contact/Proximity Printing

In transferring pattern from a photomask to a wafer coated with a photoresist, with light, the photomask and the wafer can be in contact or in proximity. In case of proximity printing, the minimum linewidth L that can be printed on the photoresist is about $\sqrt{1.6\lambda z}$, where λ and z are the wavelength of the light and the gap between the photomask and the wafer, respectively. For $\lambda = 0.4$ μm and $z = 25$ μm, $L = 4$ μm. In contact printing where $z = 0$, the L can be 0.5–1 μm.

1.2.2 Aerial Image of Projection Printing

In case of a projection printing, the minimum linewidth L that can be printed on a photoresist is about $\frac{m\lambda}{\text{Numerical Aperture (NA)}}$ with m being 0.25–1, depending on the quality of optics, photoresist, etc. For $\lambda = 0.4$ μm, NA = 0.28, and $m = 0.8$, $L = 1.1$ μm. In both projection and contact/proximity printings, the wavelength of the light is directly related to the minimum linewidth and has been pushed down over decades, down to 13.5 nm for extreme ultraviolet (EUV) lithography.

1.2.3 Photoresist

Photoresist (PR), a polymeric material with photoactive component, is made into a positive PR (which becomes more soluble when exposed to light) or a negative PR (which becomes less soluble when exposed to light). After PR is exposed to light, it is then developed in a PR developer and results in any of the two shapes illustrated in Fig. 1.2. The amount of light is what determines whether the PR becomes completely soluble (or nonsoluble in case of negative PR), and the so-called exposure energy, which is equal to (Exposure Time) × (Intensity), is what needs to be properly controlled, as shown in Fig. 1.2. The lower the ratio between E_f and E_i is, the sharper (in the thickness direction) the edge of the developed PR will be.

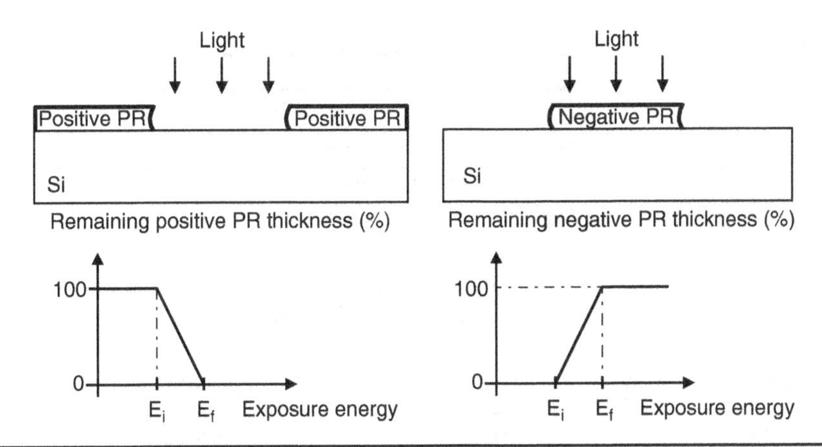

FIGURE 1.2 (Top) Cross-sectional views of photoresists after being exposed to light and then being developed. (Bottom) Fractions of photoresist thicknesses remaining versus exposure energy which is equal to (Exposure Time) × (Intensity).

Photoresist (PR) can be removed with acetone (in several minutes for 1–2 μm thick PR), which is then cleansed with methanol followed by cleansing with deionized (DI) water. Acetone does not attack dielectrics and metals, and it is usually a good choice to remove PR. However, it sometimes does not remove PR thoroughly and also damages some polymers. A photoresist remover at room temperature or an elevated temperature removes PR well, especially if the remover has been made for the particular PR. Some PR removers, though, may contain a small amount of acid(s) that damages metal. Plasma ashing with O_2 plasma at 300 W for 20 minutes does a very good job of removing 1–2-μm-thick PR (even some PR hardened due to a process step beyond 120°C or a process with carbon-containing plasma) without damaging any dielectric or metallic layers. If even O_2 plasma at 300 W for 20 minutes does not remove PR thoroughly, one can consider using Piranha (H_2SO_4:H_2O_2 = 5:1), if no metal (or any material that Piranha attacks) will be exposed to Piranha solution. To make Piranha, add five parts of H_2SO_4 to one part of H_2O_2, not the other way (i.e., H_2O_2 to H_2SO_4), to avoid splashing of H_2O_2 due to violent exothermic reaction between H_2O_2 and H_2SO_4.

1.3 Thermal Oxidation of Silicon

Silicon can be thermally oxidized with dry oxygen (O_2) or steam (H_2O) at 800–1200°C, with one silicon atom (Si) being consumed to produce one silicon oxide (SiO_2) molecule. To form SiO_2 with thickness of t_{ox}, the thickness of silicon consumed t_{si} is equal to $t_{ox}(N_{ox}/N_{si})$, where N_{ox} and N_{si} are the volume density of SiO_2 molecules and that of Si atoms, respectively. Thus,

$$t_{si} = t_{ox}\frac{2.3\times 10^{22}/\text{cm}^3}{5\times 10^{22}/\text{cm}^3} = 0.46\, t_{ox} \qquad (1.1)$$

Or 1-μm-thick layer of Si consumed gives 2.17-μm-thick SiO_2, as illustrated in Fig. 1.3. This property has been used in some MEMS fabrication processes to enclose an air gap, produce a sub-micron air gap, or produce a sharp tip. However, because a single SiO_2

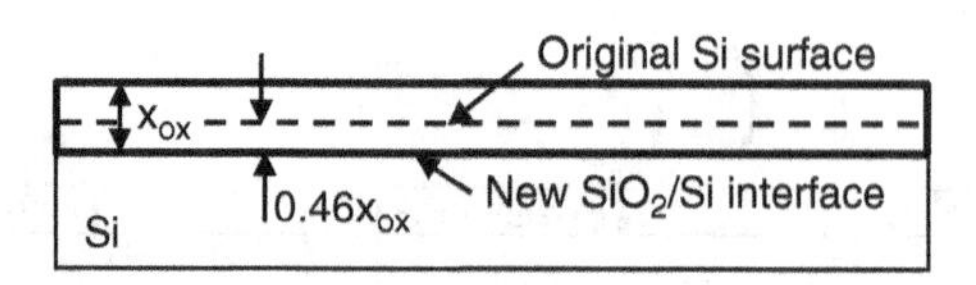

FIGURE 1.3 Cross-sectional view of thermal oxidation process of silicon where a Si atom is consumed to produce a SiO_2 molecule.

molecule occupies more space than a single Si atom (by a factor of about 2), thermally grown SiO_2 is under a substantial amount of compressive residual stress.

Question: A vertical deep groove (1 μm wide) shown below is etched in a Si substrate with reactive ion etching (RIE). The Si surface is covered with Si_3N_4, which acts as an oxidation mask. The structure is then oxidized in steam at 1,100°C. What will be the SiO_2 thickness when it is fully grown to fill up the groove? [Hint: Growing x thickness of SiO_2 requires $0.46x$ thickness of Si.]

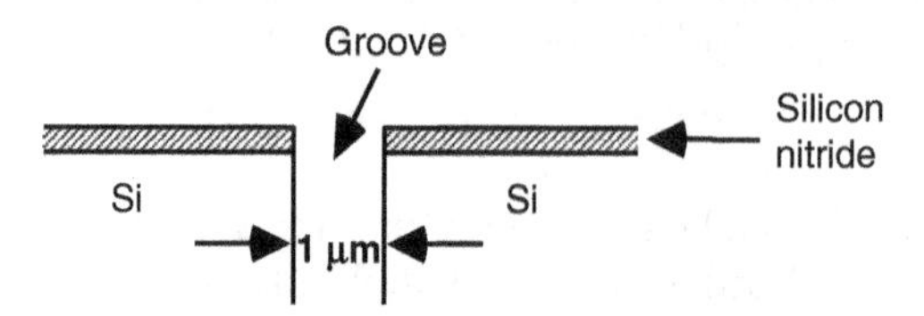

Answer: From the left side of the groove, $x_{ox} = 0.5\ \mu m + 0.46x_{ox} \rightarrow x_{ox} = 0.926\ \mu m$. Thus, the total SiO_2 thickness will be 1.85 μm.

1.3.1 Local Oxidation of Silicon

Using the fact that oxidation rate of silicon nitride (Si_3N_4) at 1,150°C with steam is about 25 times less than that of Si, one can locally mask Si oxidation with a patterned Si_3N_4 in a process called local oxidation of silicon (LOCOS). The LOCOS has been used for self-aligned channel stop, isoplanar dielectric (SiO_2) isolation, etc.

1.4 Silicon Doping

Silicon is doped with dopants to make it n-type (where the majority carriers are electrons) or p-type (where the majority carriers are holes) for resistors, diodes, transistors, etc. Doping level is typically from one dopant atom per 10^8 silicon atoms to one per 10^3 silicon atoms (i.e., $5 \times 10^{14} - 5 \times 10^{19}$ dopants/cm^3, since density of silicon atoms in a crystalline silicon is 5×10^{22} cm^{-3}). In bulk silicon micromachining with anisotropic etchant such as KOH, heavy boron doping and p-n junction are used for an etch stop in standard etching and electrochemical etching, respectively, as explained in Chap. 2. Also, doped silicon has been used to control the temperature coefficients of frequency (TCFs) of silicon microresonators [2]. Boron works as acceptor and makes boron-doped silicon p-type, while phosphorous, arsenic, and antimony work as donors and make the silicon (when doped with any of those) n-type.

Doping process typically involves the following two steps, both of which are governed by diffusion: pre-deposition of dopants (from gas, liquid, or solid source) and then drive-in of the pre-deposited dopants at a very high temperature (e.g., 900°C in N_2), as illustrated in Fig. 1.4. Note that dopants diffuse in all directions (including the horizontal direction under a diffusion mask) in both the pre-deposition and drive-in

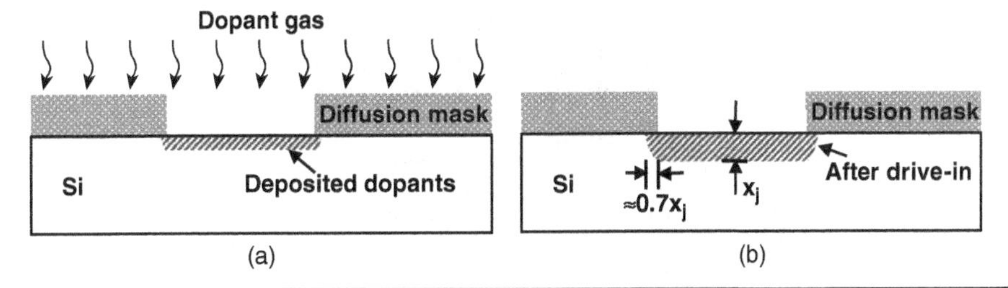

FIGURE 1.4 Typical doping process with (a) pre-deposition of dopants through diffusion from gas, liquid, or solid source, followed by (b) high-temperature drive-in step to spread the predeposited dopants into silicon substrate.

steps. The pre-deposition step can be carried out with (1) gas source such as AsH_3, PH_3, and B_2H_6, (2) solid source such as BN wafer placed next to a silicon substrate face-to-face, (3) spin-on-glass containing dopant oxide, or (4) liquid source such as boron bromide BBr_3 or phosphoryl chloride $POCl_3$. The drive-in step is typically done at an elevated temperature T to exponentially increase the diffusion coefficient $D = D_o \exp\left(\frac{-E_a}{k_B T}\right)$, where E_a and K_B are activation energy and Boltzman constant, respectively. The D_o is usually taken to be a constant that depends on dopant and substrate materials, but can be enhanced by built-in electrical field due to diffusing species, charged point defects, and under a growing oxide (for B and P). Consequently, dopants diffuse faster where the dopant concentration is higher, resulting in "box-like" diffusion profiles as shown in Fig. 1.5, where n and n_i are electron (or hole in case of p-type dopants) concentration and intrinsic carrier concentration ($n_i = 3.9 \times 10^{16} T^{3/2} \exp\left(\frac{-0.605 \text{ eV}}{k_B T}\right)$ in cm^{-3} for silicon), respectively.

For pre-deposition of dopants, ion implantation with dopant ions such as B^+, P^+, As^+, etc. can be used, instead of the diffusion, for accurate control of dopant dose and depth profile (as well as excellent dose uniformity over a large wafer area). The dopant profile with ion implantation is a little different from the one with diffusion, as shown in Fig. 1.6, but will be "box-like" after a drive-in. Ion implantation is done at room

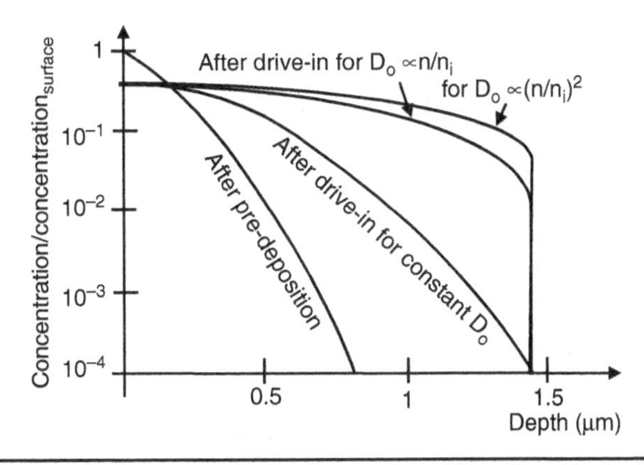

FIGURE 1.5 Dopant concentration versus depth after pre-deposition, after drive-in step (if the diffusion coefficient D_o is independent of dopant concentration), after drive-in step (if D_o is proportional to dopant concentration), and after drive-in step (if D_o is proportional to the square of dopant concentration).

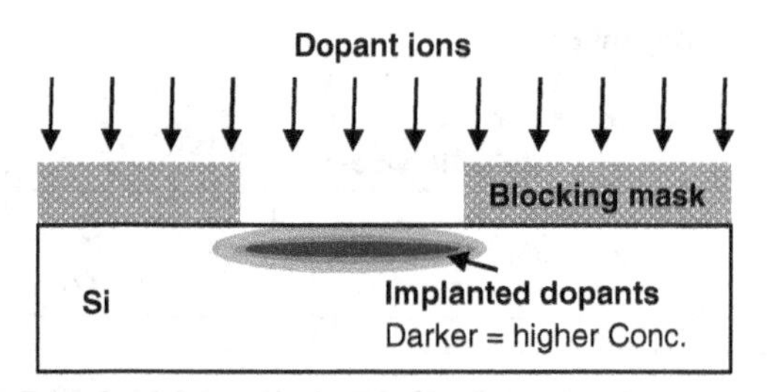

FIGURE 1.6 Implanted dopant profile after ion implantation.

temperature, and photoresist can be used as a blocking (or masking) material. However, the implanted ions need a post-implant annealing step (typically at ≈ 900°C for 30 minutes) to (1) repair the defects caused by the ion implantation (and restore silicon crystallinity) and (2) activate the dopants for electrical conduction.

1.5 Vacuum Basics

Before studying various thin-film deposition techniques, let us lay down some basics on gas property, flow, and transport. In a nutshell, we will see that temperature and pressure play very important roles in thin-film deposition.

1.5.1 Mean Free Path, Impingement Rate, and Pumping Speed

From Maxwell–Boltzmann distribution ($mv_{rms}^2/2 = 3kT/2$), the relative velocity distribution is $f_v = \frac{1}{N}\frac{dn}{dv} = \frac{4}{\sqrt{\pi}}\left(\frac{m}{2kT}\right)^{3/2} v^2 e^{-mv^2/2kT}$, and the average velocity of a molecule is $\bar{v} = \sqrt{8kT/\pi m}$; with k = Boltzmann constant, T = temperature, m = molecular weight, and $n = N/V$ with N and V being number of molecules and volume, respectively. The mean free path, λ, of a molecule before colliding with another molecule is equal to $\frac{1}{\pi n d_o^2 \sqrt{2}}$, where d_o = molecular diameter. As the ideal gas law is $PV = NkT$ with P being pressure, the mean free path depends on pressure. The impingement rate, Φ, (i.e., the number of molecules striking on unit surface per unit time) is $\frac{n\bar{v}}{4}$, and is equal to $2.64 \times 10^{20} \frac{P}{\sqrt{mT}}$ [$cm^{-2}s^{-1}$] with P in Pa and m in amu.

Example For air (amu ≈ 29) at 300 K, $\lambda = \frac{0.05}{P}$ [mm] and $\Phi = 2.83 \times 10^{21} P$ [$cm^{-2}s^{-1}$] with P in Torr. At $P = 10^{-6}$ Torr, $\Phi = 3 \times 10^{15}/cm^2s$. → If each striking molecule sticks on surface, it takes ≈ 0.3 seconds to form a monolayer on solid surface (since 1 monolayer of solid-surface atoms ≈ $10^{15}/cm^2$).

Pumping Speed (S) and Throughput (Q)

Pumping speed S is defined to be volume of a specific gas (e.g., N_2) passing a plane per unit time and is equal to dV/dt. Throughput Q, though, is defined to be $S{\cdot}P = (dV/dt){\cdot}P = kTdN/dt$ (from ideal gas law, $PV = NkT$) and is proportional to dN/dt (number of molecules removed per unit time).

For isothermal system at steady state, Q (= $kTdN/dt$) is constant across any plane of the tube between chamber and pump. However, P varies along the tube, and pumping speed $S = dV/dt$ ($= Q/P$) varies along the tube. Conductance C of a channel (held at constant temperature) connecting two parts of a vacuum system (e.g., orifice, pipe) is defined as $Q/\Delta P$ where ΔP is the pressure drop across the channel.

Example Let us consider a pump with pumping speed S_p and inlet pressure P_p that is pumping a chamber at a net pumping speed S_1, resulting in chamber pressure P_1. For isothermal system at steady state, Q is constant, and we have $P_1 = Q/S_1$, $P_p = Q/S_p$. Consequently, the conductance is $C = \frac{Q}{\Delta P} = \frac{Q}{P_1 - P_p} = \frac{Q}{Q/S_1 - Q/S_p} \rightarrow \frac{1}{S_1} = \frac{1}{S_p} + \frac{1}{C}$. In other words, pumping speed at chamber is reduced by C of the tube.

1.5.2 Regions of Gas Flow

Gas flow falls into one of several categories, depending on Knudsen number $K_n \equiv \lambda/d$, where d = characteristic dimension of vacuum component (e.g., diameter of tube).

Molecular Flow (When K_n >1)

When $\lambda > d$ (or $K_n > 1$) that happens in vacuum pressure, collisions between molecules and wall (rather than collisions among molecules) dominate. According to the momentum transfer method of Knudsen, conductance of a long tube (i.e., length l >20·diameter d) is $C = \frac{\pi \bar{v} d^3}{12 l}$. For air at 300 K, $C \approx 12.2 d^3/l$ (L/s) with d and l in cm.

In case of orifice (Fig. 1.7b) in molecular flow, $Q_1 = kT\frac{dN}{dt} = kT\left(\frac{n\bar{v}}{4}\right)A = P_1 A\sqrt{\frac{kT}{2\pi m}}$ and $Q_2 = P_2 A\sqrt{\frac{kT}{2\pi m}}$. Net flow in downstream, $Q_{\text{net}} = A(P_1 - P_2)\sqrt{\frac{kT}{2\pi m}}$. Thus, $C = \frac{Q_{\text{net}}}{\Delta P} = A\sqrt{\frac{kT}{2\pi m}} = \frac{A\bar{v}}{4}$. For air at 300 K, $C \approx 11.7A$ [L/s] with A in cm². For a short tube with length l <20·diameter d, the conductance can be obtained from $\frac{1}{C_{\text{short\_tube}}} = \frac{1}{C_{\text{long\_tube}}} + \frac{1}{C_{\text{orifice}}}$.

Viscous Flow (When K_n <0.01)

When $\lambda \ll d$ or K_n <0.01, molecules interact with each other as in a fluid (Fig. 1.8), and collisions between molecules dominate. In a long, straight tube of circular cross-section

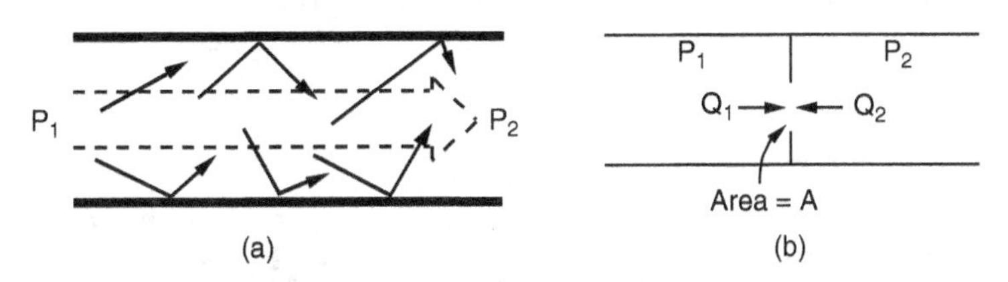

FIGURE 1.7 Molecular flow: (a) molecules colliding with the walls more often than with each other and (b) orifice in molecular flow.

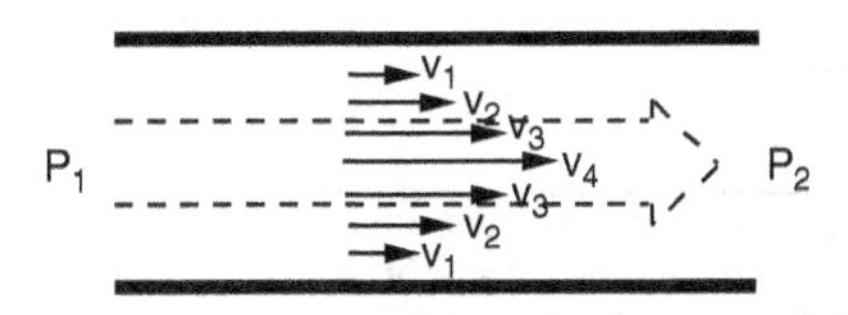

FIGURE 1.8 Viscous flow with the varying levels of velocity indicated with arrows.

with length l, Hagen–Poiseuille equation is $Q = C(P_1 - P_2) = \frac{\pi d^4}{128\eta l}\frac{P_1 + P_2}{2}(P_1 - P_2)$, where η is viscosity of gas and $\eta = 0.5\lambda n m \bar{v} = \frac{0.5\sqrt{4mkT}}{\pi^{3/2} d_o^2}$ [Pa·s = 10 poise]. For N_2 at 20°C, $C = 94d^4(P_1 + P_2)/l$ [L/s] with d and l in cm, while P_1 and P_2 are in Torr. Note that C depends on pressure.

Transition Flow (When 0.01< K_n <1)

For a long circular pipe, semi-empirically, $C = C_{viscous} + ZC_{molecular}$, where $Z = \frac{1 + 1.25d/\lambda}{1 + 1.55d/\lambda}$. But

$$C_{viscous} = \frac{\pi d^4}{128\eta l}\frac{P_1 + P_2}{2} = 0.0736\frac{d}{\lambda}C_{molecular}. \rightarrow C = C_{molecular}\left(0.0736\frac{d}{\lambda} + \frac{1 + 1.25d/\lambda}{1 + 1.55d/\lambda}\right).$$

Turbulent Flow

At high velocities and pressures, the viscous flow changes to turbulent flow. For air at 300 K, the flow is turbulent when $Q > 200 \cdot d$ [Torr·L/s] with tube diameter d in cm. It is encountered briefly in the throttle line during rough pumping, but rarely inside vacuum chamber.

1.6 Thin-Film Depositions

1.6.1 Deposition by Evaporation

Thin films can be deposited over a wafer by vaporizing atoms through heating a material in high vacuum (<10^{-5} Torr). There are two common practices for the heating: (1) resistive heating with a heating filament and (2) e-beam heating through bombardment with high-energy electrons (electron energy of several keV). Film deposition by evaporation is done under a high vacuum so that the evaporated atoms may travel (from the evaporation source) to the wafer with minimal number of collisions with gas molecules, since the collisions make the evaporated atoms lose their mechanical momentum to travel.

In a high-vacuum system, the flux of the evaporated atoms arrives at the wafer with directional energy and can miss to cover a step, as illustrated in Fig. 1.9a. Consequently, there may be discontinuity in the evaporation-deposited film over a step (Fig 1.9b); the taller the step, the more likely the discontinuity can happen. To solve this problem, one

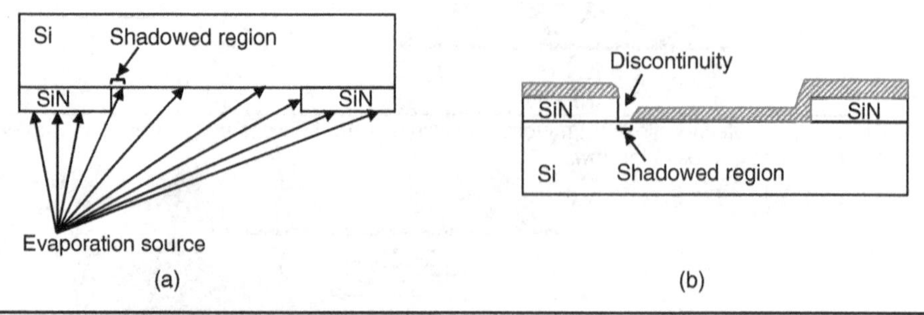

FIGURE 1.9 (a) Diagram to illustrate how a step can be missed in an evaporation process. (b) With directional flux of evaporated atoms, steps on a "wrong" side are not covered with the evaporated atoms, and discontinuity may occur on the evaporation-deposited film.

may have to rotate a wafer in the middle of evaporation process to make all the steps (in the wafer) be exposed to the incoming flux in the face directly at a certain percentage of the evaporation time. Or one can make the step's edge be tapered or sloped, as illustrated in Fig. 2.30 (Chap. 2).

1.6.2 Deposition by Sputtering

With noble gases such as Ar, Kr, and Xe, almost any material can be sputtered through physical impact by ionized gas molecules (drifted by a high electrical field from a plasm created between anode and cathode), as shown in Fig. 1.10. A sputter target (e.g., aluminum nitride [AlN]) is placed on the cathode (negatively biased with respect to the anode), while wafers are placed on the anode (which is typically grounded). The wafers receive molecules sputtered off from the sputter target, and a thin film of a sputter target material can be deposited on a wafer. Typical sputtering system uses gas pressure of 10–100 mTorr and a cathode voltage of –500 to –5,000 V (with anode grounded) to create a plasma. A DC sputtering system has a current density of 0.1–2.0 mA/cm^2 between anode and cathode and is good for sputtering electrically conductive materials. For non-conductive materials, an AC or RF sputtering system should be used. Due to government regulation, RF sputtering system uses either 13.56 MHz or 100 kHz. Deposition by sputtering covers steps better than deposition by evaporation (Fig. 1.9), because the sputtered molecules or atoms arrive at the wafers with little directional energy after having gone through many collisions (with gas molecules) in transit from the sputter target to the wafer.

Question: An Al-2%Si sputtering target is bombarded with 1 keV Ar ions. Estimate the surface composition of the target under *steady state* sputtering condition. Assume the sputter yields (≡ number of target atoms ejected ÷ incident particle) for Al and Si to be 1.5 and 0.6, respectively.

Answer: The sputter flux has $\frac{[\mathrm{Al}]}{[\mathrm{Si}]} = \frac{98}{2} = \frac{1.5}{0.6}\frac{[\mathrm{Al}]_{\mathrm{target}}}{[\mathrm{Si}]_{\mathrm{target}}}. \rightarrow \frac{[\mathrm{Al}]_{\mathrm{target}}}{[\mathrm{Si}]_{\mathrm{target}}} = \frac{98}{2}\cdot\frac{0.6}{1.5} = 19.6 \rightarrow 4.85\%$ Si and 95.1% Al.

1.6.3 Chemical Vapor Deposition

Films of various materials can be deposited on a silicon wafer with proper gases at an elevated temperature through a process called chemical vapor deposition (CVD). For example, polysilicon, amorphous silicon, or epitaxial silicon can be deposited through

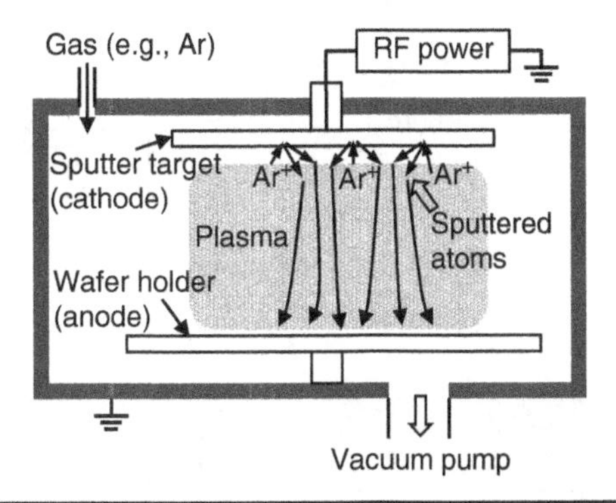

Figure 1.10 RF sputtering system: sputtering with ionized gas molecules drawn from the plasma between the target (on the cathode) and the substrate holder (the anode that is grounded).

$SiH_4 \rightarrow Si + 2H_2$ or $SiCl_4 + 2H_2 \rightarrow Si + 4HCl$ at 400–1,250°C (the higher temperature yields larger grains with crystalline silicon in each of the grains). For SiO_2 film, one can use $SiH_4 + O_2 \rightarrow SiO_2 + 2H_2$; and for Si_3N_4, $3SiH_4 + 4NH_3 \rightarrow Si_3N_4 + 12H_2$.

One very important feature of CVD is that the deposition is conformal over all kinds of surface topography, because reactant gases are transported through diffusion, and has no problem of covering any kind of step, unlike evaporation or sputtering. In fact, CVD is so conformal that it even deposits on two parallel surfaces of which the normal directions are opposite, as exploited in surface-micromachined pin joints (Fig. 2.27). The CVD's conformality is a key feature needed in many surface micromachining processes.

Kinetics of CVD

For CVD, gases are brought close to wafers, either through pumping (as shown in Fig. 1.11) in low-pressure CVD (LPCVD) or through a carrier gas in atmospheric pressure CVD (APCVD). The amounts of gas flows are precisely controlled by mass flow controllers (MFCs) such as the one shown in Fig. 1.20b.

With gas flow over a solid surface, a boundary layer or stagnant layer is formed over the surface, as illustrated in Fig. 1.12, and plays a major role in CVD process, as diffusion of gas molecules occurs through the boundary layer. In Fig. 1.12, $\delta(x)$ and L are the boundary layer thickness and the length of a plate (e.g., a substrate or a wall of CVD reactor), respectively. The gas velocity u, which is a function of x and y, is zero at the surface of the plate, but equal to U in the free gas stream. Friction force per unit area along the x direction is equal to $\eta \frac{\partial u}{\partial y}$, where $\eta \equiv \text{viscosity} = 0.5\lambda n m \bar{v} = \frac{0.5\sqrt{4mkT}}{\pi^{3/2} d_o^2}$.

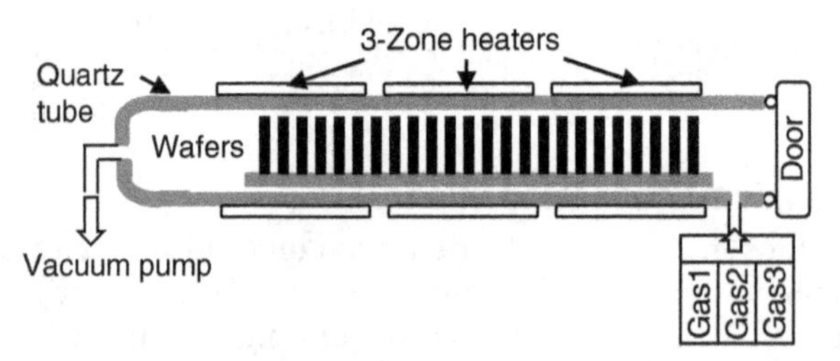

FIGURE 1.11 Schematic of a low-pressure chemical vapor deposition (LPCVD) system.

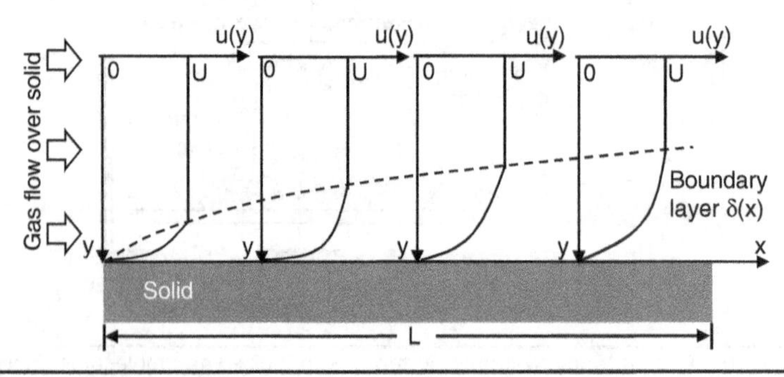

FIGURE 1.12 Profile of flow velocity over a solid surface.

(Because the gas stream velocity increases in y direction, away from the solid surface, within the boundary layer, molecules crossing a plane parallel to x and z directions from below the plane will transport less momentum than those crossing from above. Consequently, the gas undergoes shear, and thus a viscous force is present.) And total friction force on a differential volume element of unit depth (into the paper), height $\delta(x)$, and width dx is equal to $\eta\frac{\partial u}{\partial y}(1\cdot dx)=\eta\frac{\partial u}{\partial y}dx$ = decelerating force. On the other hand, total accelerating force on the element is equal to $\rho\delta(x)dx\frac{du}{dt}=\rho\delta(x)dx\frac{\partial u}{\partial x}\frac{dx}{dt}=\rho\delta(x)dx\frac{du}{dx}u$, where ρ is gas mass density. Balancing the forces leads to

$$\eta\frac{\partial u}{\partial y}=\rho\delta(x)u\frac{\partial u}{\partial x}, \tag{1.2}$$

from which $u(x,y)$ can be solved exactly.

Approximate Analysis of Boundary Layer and Diffusion through Boundary Layer

If we approximate $\frac{\partial u}{\partial y}\approx\frac{U}{\delta(x)}$, $\frac{\partial u}{\partial x}\approx\frac{U}{x}$, and $u\approx U$ in Eq. (1.2), then $\delta(x)\approx\sqrt{\frac{\eta x}{\rho U}}$, which is close to the dotted line ("parabolic dependence) in Fig. 1.12. Thus, the "average" boundary layer thickness is $\bar{\delta}=\frac{1}{L}\int_0^L\delta(x)dx=\frac{2}{3}\frac{L}{\sqrt{\rho UL/\eta}}=\frac{2}{3}\frac{L}{\sqrt{R_{eL}}}$, where R_{eL} ($\equiv\rho UL/\eta$) is called Reynold's number of the reactor. Flow is laminar viscous when R_{eL} <1,200, but turbulent when R_{eL} >2,200.

In the Grove model, it is assumed that mass transport across the boundary layer is carried by diffusion with diffusion flux $F_1\approx D_G\frac{C_G-C_S}{\delta}$, where D_G, C_G, and C_S are gas diffusivity, concentration at gas stream, and concentration at substrate, respectively. The average flux over the substrate length L is $\bar{F_1}\approx D_G\frac{C_G}{\bar{\delta}}$ for C_S being close to zero.

Film Deposition Rate in CVD

Reactant gas, when transported to a substrate surface, reacts at the surface and deposits a film through CVD. Surface reaction depends mainly on C_s (reactant concentration at substrate) and temperature T, and is proportional to $F_2=k_sC_s$, where k_s is surface reaction constant, which is proportional to exp $(-E_a/kT)$ with E_a being the activation energy. Thus, film deposition rate in CVD is determined by surface reaction and mass transport through diffusion across the boundary layer, and is proportional to $F=\frac{k_SC_G}{1+k_S\bar{\delta}/D_G}$. If $k_s\bar{\delta}/D_G$ is much larger than 1, the deposition rate is mass-transfer limited and is proportional to $\frac{D_GC_G}{\bar{\delta}}\propto 1/\bar{\delta}\propto\sqrt{U}\propto\sqrt{\text{gas flow rate}}$.

If $k_s\bar{\delta}/D_G$ is much less than 1 (e.g., due to a very high gas flow rate), the deposition rate is surface reaction limited and is equal to k_SC_G/ρ (which is mostly determined by temperature), where ρ is the number of molecules or atoms per unit volume for a deposited film (e.g., $\rho_{Si}=5\times10^{22}/\text{cm}^3$). The two distinct regions of deposition rate can be seen graphically in Fig. 1.13 that shows an exemplary deposition rate at 1,000°C versus gas flow rate.

At APCVD, the rate of mass transfer (reactants and products) is kept close to the rate of surface reaction. But the mass transfer depends on reactor configuration, flow

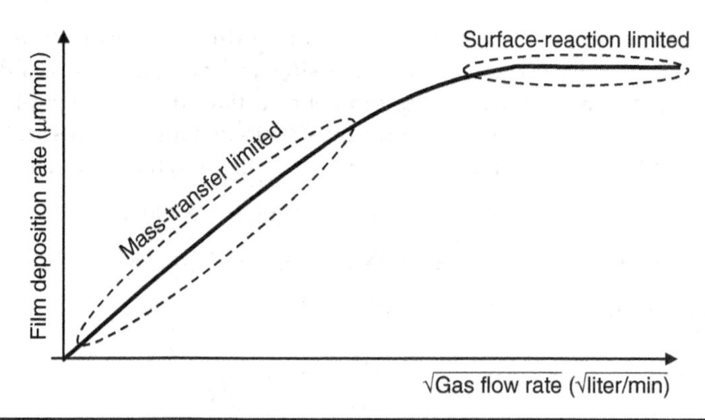

FIGURE 1.13 Exemplary CVD deposition rate versus gas flow rate at an elevated temperature of 1,000°C, showing two distinct regions: mass-transfer limited (when the flow rate is low) and surface reaction limited.

velocity, distances from edges, load sizes, etc., while the surface reaction depends mainly on reactant concentration and temperature ($F_2 = k_s C_s$ with $k_s \propto e^{-E_a/kT}$). Thus, it is difficult with APCVD to control the uniformity and reproducibility of film deposition, and LPCVD is more commonly used.

Low-Pressure CVD

In LPCVD, as shown in Fig. 1.11, carrier gas is minimized or eliminated to make total pressure low, though reactant partial pressure is typically higher than that used in APCVD. Typical reactant gas pressure is 0.1–1.0 Torr with gas flow rates of 10–500 sccm (standard cc per minute). With the Grove model, $\overline{F_1} \approx D_G \frac{C_G}{\overline{\delta}}$. For gases, $D_G \propto 1/P$, where P is pressure. Thus, D_G increases by ≈ 1,000 times when P is reduced from 760 Torr to 1 Torr. Now, $\overline{\delta} \propto \sqrt{\frac{\mu}{\rho U L}}$, where $\rho \propto P$ from the ideal gas law, and U = free stream velocity ≈ 10–100 times higher for LPCVD (where forced convection dominates) than for APCVD. Thus, $\overline{\delta}$ increases by $\sqrt{1{,}000/100} \approx 3$, and $\frac{D_G}{\overline{\delta}}$ increases by ≈ 300 times. Since $\frac{D_G}{\overline{\delta}} \gg k_s$ at LPCVD, the deposition rate is limited by surface reaction, and less attention can be paid to the mass-transfer variables. Thus, the design of a LPCVD reactor is largely focused on obtaining good temperature uniformity and high wafer throughput (e.g., stand-up wafer loading).

1.6.4 LPCVD Low-Stress Silicon Nitride

Silicon nitride is an outstanding material for sturdy membrane or cantilever for MEMS. However, stoichiometric Si_3N_4 has tremendous tensile residual stress, so large that one cannot deposit thicker than 0.2 μm without crack in the film. This is one reason why the LOCOS process of complementary metal–oxide semiconductor (CMOS) uses only 0.1-μm-thick Si_3N_4. If 0.2-μm-thick Si_3N_4 is deposited on a bare Si wafer, the Si_3N_4 film will likely have cracks, at some time later after the deposition (if not immediately after the deposition). Consequently, the maximum thickness of Si_3N_4 on Si wafer without crack would be somewhere between 0.1 and 0.2 μm.

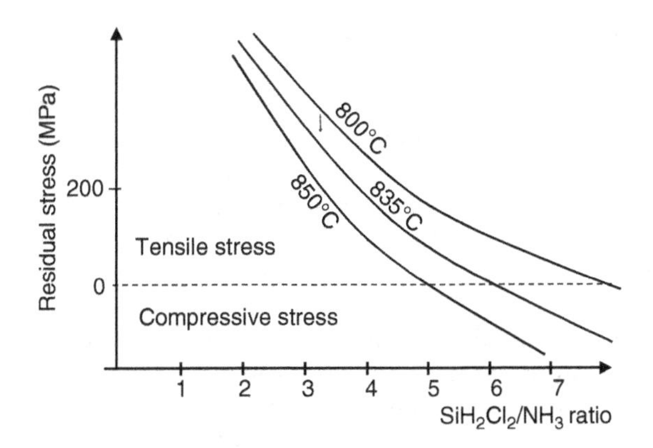

FIGURE 1.14 Residual stress of non-stoichiometric Si_xN_y as a function of deposition temperature and DCS/NH_3 gas ratio. (Adapted from [3].)

Consequently, MEMS typically uses low-stress Si_xN_y films that are deposited at 835°C and 300 mTorr with SiH_2Cl_2/NH_3 ratio of 5/1, according to Fig. 1.14a [3]. Such non-stoichiometric Si_xN_y has x and y equal to about 0.9 and 1, respectively, and is silicon rich, compared to Si_3N_4. By varying the gas ratio and deposition temperature, one can obtain various levels of residual stress (from tensile to compressive) with varying degree of silicon richness. One thing to note, though, is that SiH_2Cl_2 (dichlorosilane [DCS]) is corrosive, and the by-products (containing chlorine) decompose the pump oil, requiring routine oil change after 3–10 μm thick nitride deposition. Instead of DCS, SiH_4 can be used for the deposition, but at the cost of poor uniformity over a wafer. At room temperature, DCS (unlike SiH_4, which is explosive when exposed to oxygen) is in liquid form with relatively high vapor pressure (≥ 10 psi), and DCS vapor from liquid DCS (contained in a cylinder) is used for LPCVD Si_xN_y.

Question: Rank three deposition techniques for step coverage and deposition temperature (the lower, the better) from 1 (best) to 3 (worst): evaporation, sputtering, and CVD.

	Evaporation	Sputtering	CVD
Step coverage			
Deposition temperature			

Answer:

	Evaporation	Sputtering	CVD
Step coverage	3	2	1
Deposition temperature	1	2	3

Question: Low-stress or silicon-rich silicon nitride (Si_xN_y where x/y is 0.9–1) is deposited with SiH_2Cl_2/NH_3 ratio of 5 at 835°C with LPCVD. Which of the following three conditions is the one used to deposit stoichiometric silicon nitride (Si_3N_4): (1) SiH_2Cl_2/NH_3 ratio of 0.3 at 700°C, (2) SiH_2Cl_2/NH_3 ratio of 1 at 750°C, and (3) SiH_2Cl_2/NH_3 ratio of 5 at 850°C?

Answer: (1), Si_3N_4 is deposited with higher NH_3 gas flow than SiH_2Cl_2 flow at a lower deposition temperature.

Stress Gradient in Thickness Direction

Silicon-rich, non-stoichiometric silicon nitride has substantial amount of residual stress variation along the thickness direction due to the different amounts of time that each point along the thickness is at the deposition temperature (e.g., 835°C) in the LPCVD system. The silicon nitride at the bottom is at the high temperature, longer than the one at the top during the deposition, and has residual stress more compressive than the one at the top, as can be expected from Fig. 1.15a. The measured residual stress at different locations in the thickness direction of a 2-μm-thick silicon-rich silicon nitride (deposited with LPCVD at 835°C with SiH_2Cl_2/NH_3 ratio of 4/1) is sketched in Fig. 1.15a [4]. Since the deposition rate was about 0.4 μm/h, the bottom portion of the silicon nitride was at the high temperature about 5 hours longer than the top portion. Effect of the stress gradient is noticeable when a cantilever is formed, and its effect makes the cantilever warped; the more for a longer cantilever (Fig. 1.15). One way to avoid such warping is to micromachine a diaphragm with one layer of silicon nitride (first SiN) and then deposit a second SiN on both sides of the diaphragm, thus producing a stress gradient that is symmetric with respect to the mid-plane (Fig. 1.16) [4].

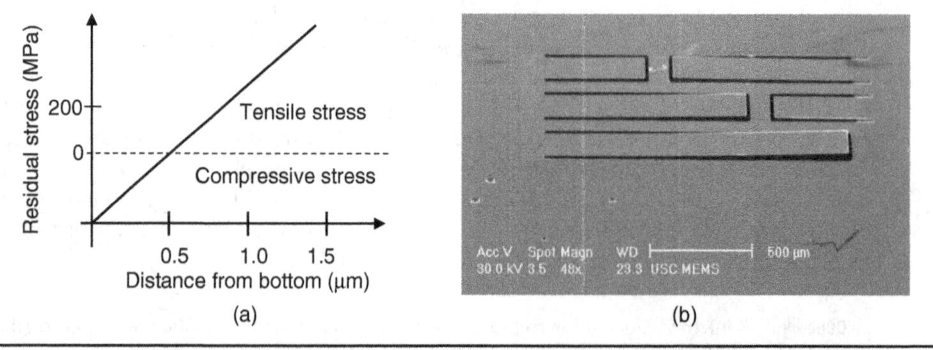

FIGURE 1.15 (a) Measured residual stress at different locations in the thickness direction of a silicon-rich silicon nitride deposited with LPCVD at 835°C with SiH_2Cl_2/NH_3 ratio of 4/1. (Adapted from [4]). (b) Effect of the residual stress gradient in (a) on a cantilever made of the silicon nitride.

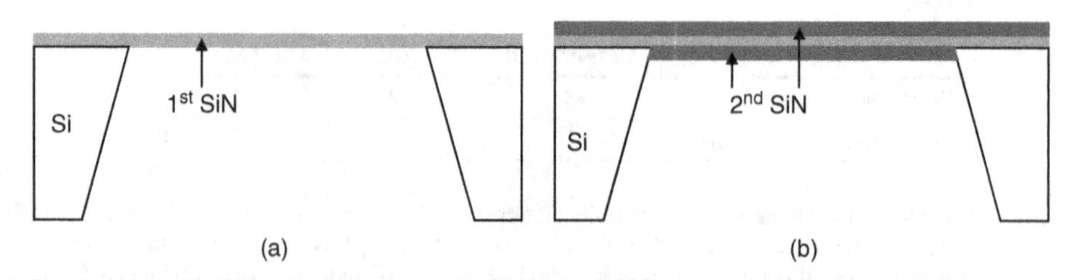

FIGURE 1.16 (a) A silicon-rich silicon nitride diaphragm with the stress gradient (along the thickness direction) that produces a bending moment and (b) a diaphragm made of one layer of first silicon-rich silicon nitride and two layers of second silicon-rich silicon nitride that are deposited after micromachining a diaphragm made of the first silicon-rich silicon nitride.

1.6.5 Amorphous Silicon, Polysilicon, and Epitaxial Silicon Depositions

Silicon thin films with various crystallinities can be deposited with silicon-containing gas such as SiH_4, $SiCl_4$, and SiH_2Cl_2 through CVD, largely depending on the deposition temperature. At a relatively low deposition temperature such as 400°C, deposited silicon film is amorphous with no crystallinity. At 600–650°C deposition temperature, though, the deposited film is polycrystalline, composed of many grains having crystalline structure in each grain. The grains are typically oriented in a certain crystalline direction with some randomness. Polysilicon (shortened name for polycrystalline silicon) deposited on thermally grown oxide with LPCVD at 620–650°C is preferentially oriented toward either <110> crystal direction (for undoped polysilicon with 160–320 Å grain size) or <311> direction (for polysilicon in situ doped with phosphorous with 240–400 Å grain size). With the relatively large grain size, the polysilicon surface is rough (>50 Å). As-deposited polysilicon has a compressive residual strain of ≈ −0.2%, but a high-temperature annealing can uniformize the residual stress gradient (along the thickness direction) and also reduce the residual strain (though it remains compressive).

On the other hand, an undoped polysilicon deposited on thermally grown oxide with LPCVD at 570–580°C has fine grains (that result in smooth surface with roughness <15 Å), and shows no texture, though <111> texture with 100 Å grain size shows up after 900–1000°C anneal. (Note: the grain size is 700–900 Å after anneal for polysilicon deposited at 620–650°C). The residual strain of the as-deposited fine-grain polysilicon is compressive (≈ −0.7%), but annealing can yield polysilicon film with tensile strain [5].

Monocrystalline silicon thin film with various dopants and doping level (e.g., n-type) can be deposited over a crystalline silicon substrate (e.g., lightly doped p-type) in CVD with silicon-containing gas such as $SiCl_4$, SiH_2Cl_2, etc. Since the deposited silicon film is crystalline with its lattice constant matched to that of the silicon substrate, there is no residual stress in the film. However, a very high deposition temperature of 1,150–1,250°C and ultra-clean silicon surface are required.

1.6.6 Atomic Layer Deposition and Atomic Layer Etching

High-quality film can be deposited conformally over a substrate, atomic layer by layer, through atomic layer deposition (ALD), similar to CVD. The difference between ALD and CVD is mainly in the self-limiting deposition in ALD after an atomic layer is deposited. This self-limiting characteristic slows down the deposition rate greatly, but allows extremely conformal deposition of a very high-quality film. For example, ZnO film can be deposited through ALD with a precursor, $(C_2H_5)_2Zn$ (diethyl zinc [DEZ]), and gaseous water by repeating deposition cycles illustrated in Fig. 1.17 through the following chemical reaction:

$$Zn(C_2H_5)_2^* + H_2O(g) \rightarrow ZnO^* + 2C_2H_6(g). \tag{1.3}$$

In the first cycle, DEZ is pulsed over a substrate having a monolayer of OH, and Zn in DEZ replaces H in OH, eventually resulting in a monolayer of ZnO covered by C_2H_5 (Fig. 1.17b). Once the monolayer is formed, the deposition stops, at which point excess DEZ and reaction by-products (C_2H_5) are purged through flowing inert gas or vacuuming. Then in the next cycle, gaseous H_2O is pulsed, and O in H_2O binds to Zn on the substrate, eventually forming another monolayer of ZnO, accompanied with another self-limiting deposition. When the excess H_2O and C_2H_5 are purged, two monolayers of

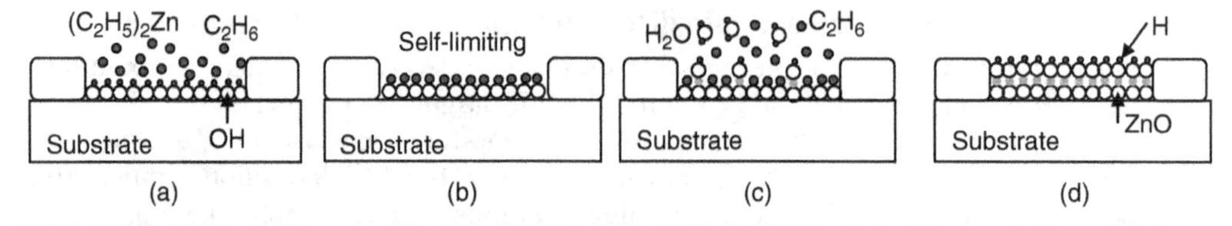

FIGURE 1.17 Process steps in ALD of ZnO: (a) pulse a precursor, diethyl zinc, $(C_2H_5)_2Zn$ (also known as DEZ), and let DEZ react with the substrate surface covered with OH, (b) purge or evacuate excess precursor DEZ and reaction by-products (C_2H_6), (c) pulse $H_2O(g)$, and let it react with the surface, and (d) purge excess $H_2O(g)$ and by-products to obtain a monolayer of ZnO. The steps (a)–(d) may be repeated to obtain a multilayer ZnO film. (Adapted from [6].)

ZnO has been deposited (Fig. 1.17d). Repeating the cycles increases the thickness of the deposited film.

Typical timings of the pulsing and purging along with the film thickness growth in ALD are illustrated in Fig. 1.18a. Heat (e.g., an elevated temperature of 100–200°C in case of ZnO ALD) or plasma can be used in ALD, similar to CVD (with LPCVD and plasma-enhanced CVD [PECVD] being based on heat and plasma, respectively). Plasma-enhanced ALD (PEALD) results in better stoichiometry at a lower temperature due to more reactivity of the precursors caused by plasma. The system for PEALD (e.g., Fig. 1.18b) is similar to a PECVD system but would require control electronics to cycle the pulsing and purging steps.

Using a similar idea of self-limiting characteristics present in ALD, one can develop atomic layer etching (ALE), as illustrated in Fig. 1.19. Reactants that are pulsed over a substrate produce monolayer of reaction product over the substrate surface. When the reaction product covers over the whole surface, the reaction stops and is self-limiting. After excess reactants and reaction by-products are purged, the reaction products are detached through bombardment of ions (or neural gas molecules) or any other means

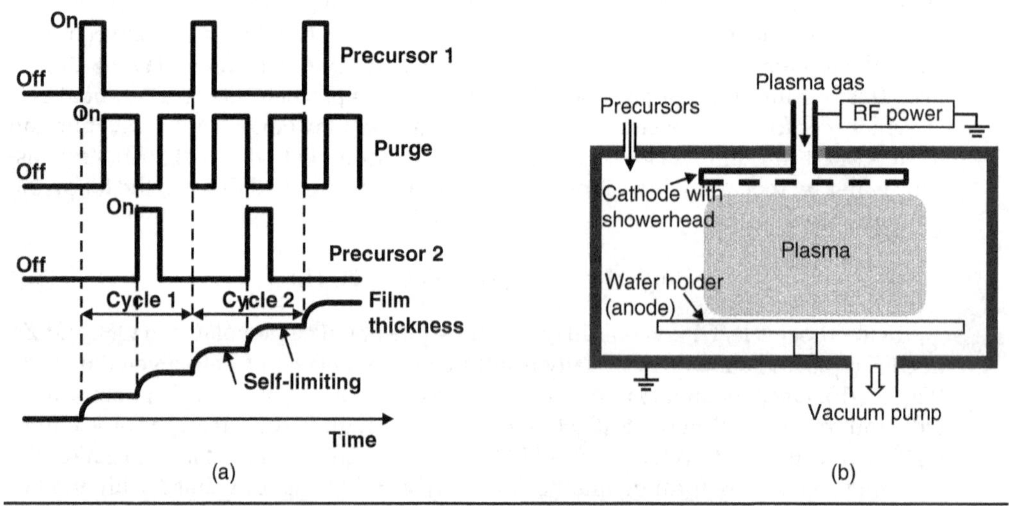

FIGURE 1.18 (a) Timing diagram of the ALD steps with two precursors over two cycles and (b) schematic of plasma-enhanced ALD (PEALD).

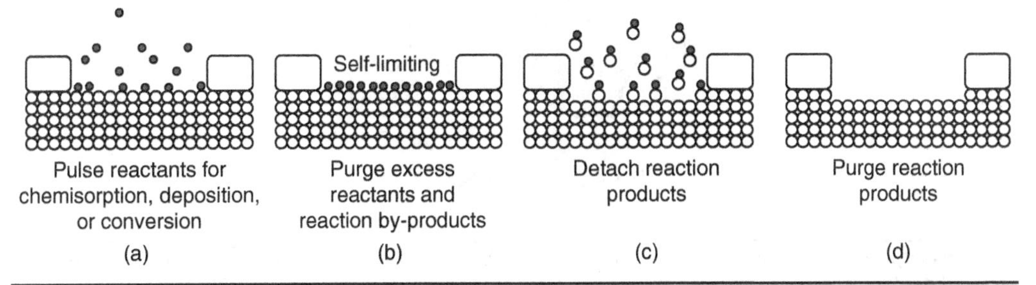

FIGURE 1.19 Generic process steps in atomic layer etching (ALE): (a) pulse reactants, and let them react with substrate surface, (b) purge or evacuate excess reactants and reaction by-products, noting the self-limiting characteristic at the surface when the surface is covered by reaction products, (c) detach reaction products with bombardment of Ar+ or any other means that allow selective detachment of the reaction products without etching or damaging the atoms beneath, and (d) purge reaction products.

that detach the reaction products without damaging the underlying atoms. One cycle of the steps in Fig. 1.19 that produces monolayer etching may be repeated to obtain etching of multiple atomic layers, one layer per cycle.

1.7 Mass Flow Sensing and Control

In CVD and ALD, mass flow controllers (MFCs) are used to control flows of reactant gases, and in this section, we will study how mass flow of gas can be sensed and controlled. Fluid means gas and liquid that deform under stress, and two most important parameters that characterize fluid are density ρ and viscosity μ. Viscosity is a measure of resistance of a fluid to flow (analogous to friction), and its unit is poise (= g/cm·s). In general, viscosity decreases rapidly with increasing temperature. In dealing with gases, the ideal gas law is important to remember, $PV = nRT$, where P = pressure in Pa (or N/m^2), V = volume in m^3, n = number of moles (mol), R = gas constant = 8.31451 N·m/(mol·K), and T = absolute temperature in K.

To sense mass flow rate, the following sensors have been developed: thermal flow sensor, rotary flow meter, flow sensing based on pressure sensing, and flow sensing based on acoustic Doppler effect. Mass flow rate can be expressed (1) in units of throughput Q (e.g., Torr·L/s) or (2) in terms of conservable quantities, mol/s or molecules/min (1 sccm = 2.69×10^{19} molecules/min). Note that $Q = kT \cdot dN/dt$, and depends on temperature. For example, 1 Torr·L/s = 2.13×10^{21} molecules/min at 273 K ≈ 79 sccm, while 1 Torr·L/s = 1.93×10^{21} molecules/min at 300 K ≈ 72 sccm.

Thermal Mass Flow Sensing

The amount of heat per unit time (required to raise the temperature of a gas stream by a known amount) depends on mass flow rate. Mass flow sensors can be made to be sensitive to heat capacity or thermal conductivity, as mass flow rate m′ (kg/s) = $m \cdot dN/dt = H/[C_p(T_2 - T_1)]$, where H and C_p are the amount of heat required to warm the gas stream and the specific heat, respectively. Note $Q = kT \cdot dN/dt \approx 1/\rho C_p$, where ρ is the gas density. In the mass flow sensor shown in Fig. 1.20, two thermocouples (or temperature sensors) measure the changes in temperature profile due to mass flow. An electronic control circuit can be used to keep the temperature profile constant by adjusting the heat generated by the heater. In that case, mass flow is proportional to the amount of the electrical power required to maintain a constant temperature profile.

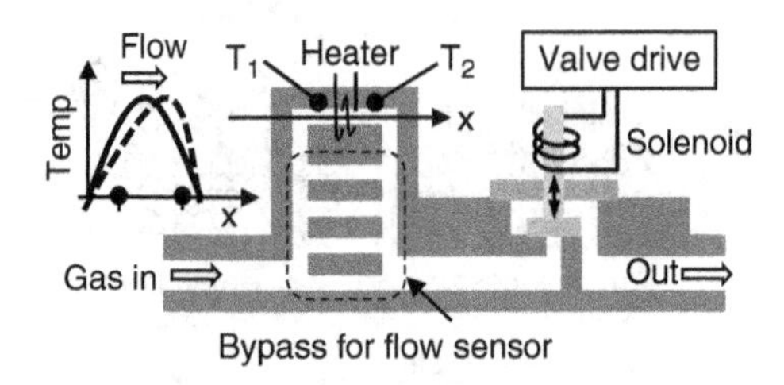

Figure 1.20 Mass flow controller (MFC) with a mass flow sensor (based on heat transfer by mass flow with a heater to generate a temperature profile and two temperature sensors to measure the temperature profile changed by mass flow) and a flow control valve.

If a flow sensor in an MFC shown in Fig. 1.20 is made with a tube of 0.2–0.8 mm internal diameter (ID), full-scale flow in the main line can be 2–150 Pa·L/s [7]. To increase the range of flow measurement, one can use a laminar flow bypass with a particular divider ratio (Fig. 1.20), as a flow divider allows a known fraction of the flow to pass through the flow-measuring tube. Accuracy of a thermal mass flow meter is typically ≈ 2% of the full scale. For example, a flow meter with 100 sccm of full-scale flow can measure flows down to 2 sccm.

Since $Q \approx 1/\rho C_p$, the thermal mass flow meter will have to be readjusted for different gases. Nitrogen (N_2) is commonly used for calibration, based on $Q_x = Q_{N2}(\rho_{N2}C_{N2})/(\rho_x C_x) = f_x Q_{N2}$, where f_x is the conversion factor relative to N_2. For example, the f_x for SiH_4 is 0.6, and 100 sccm of SiH_4 corresponds to $100 \times 0.6^{-1} = 166$ sccm of N_2. Thus, when calibrating an SiH_4 mass flow meter with N_2, make the full scale to be 166 sccm with N_2 so that the full scale with SiH_4 may be 100 sccm. For a mixture of 15%PH_3/SiH_4 gases, 100 sccm of the mixture corresponds to 15 sccm of PH_3 and 85 sccm of SiH_4, and consequently, N_2 flow $= 85 \times 0.6^{-1} + 15 \times 0.76^{-1} = 161.3$ sccm.

A mass flow controller (Fig. 1.20) is typically composed of a mass flow sensor, electrically actuated flow control valve (with the actuator being typically a solenoid) and electronics. The electronics determines how much the control valve needs to be opened, based on the set value and the real-time flow rate sensed by the sensor, and drives the valve.

1.7.1 Flow Control Valve

There are two types of valves: (1) passive valve that is opened and closed by the energy from fluid and (2) active valve that is actuated thermo-pneumatically, pneumatically, through thermal bimorph, electrostatically, electromagnetically (as used in a solenoid), and piezoelectrically. Various actuation methods can be compared with respect to actuation energy density defined to be actuation energy (actuation force times stroke distance) divided by actuator volume. Electrostatic actuators have low actuation energy density, while thermal actuators offer very high densities [8]. Shape memory alloy (SMA) actuators provide very large forces, but their linear deformation strain is limited to about 8%.

Thermopneumatically Actuated Microvalve

One of the early commercial MEMSs was thermopneumatically actuated microvalve [9] with a bulk-micromachined silicon sandwiched by two Pyrex glass plates through anodic bonding. The micromachined cavity is filled with fluid through a hole in the top Pyrex glass which contains also an embedded heater. When the fluid expands in volume due to the heat generated by the heater, the silicon membrane deflects, and the top

Pyrex and the silicon substrate pivot around the silicon pedestal, opening up the valve (that used to be closed). The opening size depends on the fluid's volume expansion, which depends on the heat.

1.8 Electroplating of Metals

Metals can be electroplated over a metal seed layer on a substrate, as electrical current carries metal ions (typically positive) from a metal anode to a cathode (where the substrate is in contact) in an electrolyte. Typically, photoresist, polyimide, or dielectric layer is patterned for selective electroplating for patterned metal deposition, as illustrated in Fig. 1.21. The thickness of electroplated metal is usually not uniform due to current crowding near the edges of patterned dielectric trenches, as the electrical field density is higher there [10]. To overcome this non-uniform thickness of the electroplated metal, one can use a process called Damascene process, as explained in the next paragraph.

In early 90s, IBM developed an electroplating process for copper interconnect, and named it Damascene process [11]. The process starts with a patterned polyimide (or dielectric), on which seed layer is deposited, followed by electroplating (Fig. 1.22). After electroplating metal thicker than the height of patterned trenches, the electroplated metal is selectively removed with chemical mechanical polishing (CMP) to thin it down

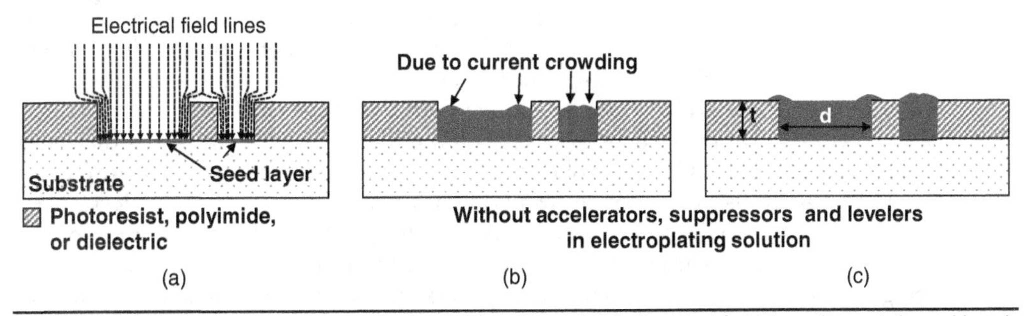

FIGURE 1.21 Basic process steps in electroplating of metal: (a) electrical field lines (to metal seed layer) that make current crowding near the edges of patterned insulating layer, (b) non-uniform electroplated metal thickness due to current crowding in a typical electroplating solution without any additives to make the thickness uniform, and (c) dependence of electroplated metal's topology on aspect ratio (d/t) between lateral size of plating area and dielectric thickness.

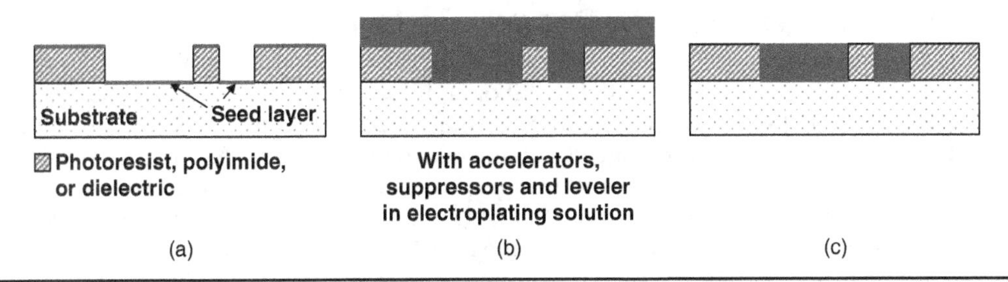

FIGURE 1.22 "Damascene process:" (a) seed layer over patterned dielectric trenches, (b) electroplating of metal with additives for accelerated plating near trench bottoms, suppressed plating on trench sidewalls, and level plating on trench tops, and (c) uniform thickness determined by trench depth, after chemical mechanical polishing (CMP).

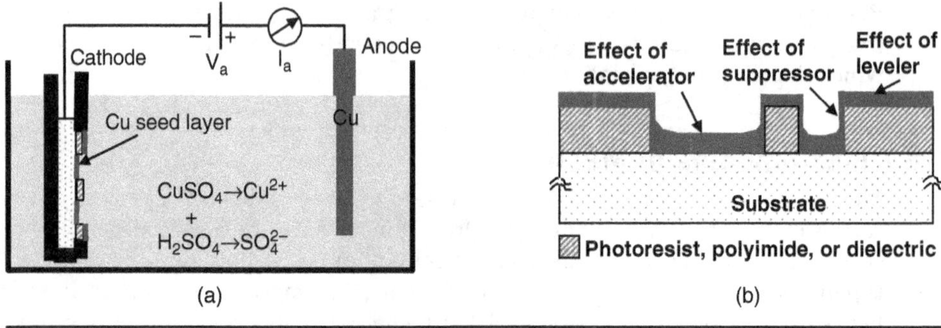

FIGURE 1.23 (a) Typical set-up for electroplating of copper in $CuSO_4 + H_2SO_4$ solution and (b) cross-sectional view of the electroplating in a solution that includes additives for accelerator, suppressor, and leveler (e.g., Cl- and organic suppressors to slow plating at selective areas). How the additives work depends on trench geometry and plating time, as the additives affect wetting and polarizing.

to the height of the trenches. As CMP works similar to mechanical lapping except that it removes different materials at markedly different rates, since CMP uses chemical etching effects, in addition to mechanical lapping, which are substantially different for different materials. Thus, CMP can be designed such that the removal is minimal for the patterned dielectric, and the thickness of the electroplated metal after CMP is determined by the height of the dielectric trenches (Fig. 1.22c). Based on this, a process called dual Damascene process has been developed with two dielectric layers and two etch-stop layers (one between the dielectric layers and the other between the lower dielectric layer and the substrate). Copper is one of the most common metals that is electroplated, and a basic apparatus for copper electroplating is illustrated in Fig. 1.23. In electroplating copper with Damascene process, seamless filling of the trenches without void is critical. Consequently, various additives are included in the electrolyte to increase the plating rate at the trench bottoms and to suppress the plating at the trench sidewalls, so that electroplating may continue from the bottom upward till the trenches are completely filled with electroplated metal. Otherwise, void may be enclosed in the trenches, as the sidewalls of the upper portions of the trenches get electroplated faster than those of the lower portions.

1.9 Soft Lithography and Its Derivative Technology

Soft lithography is a technology that uses a soft elastomer such as polydimethylsiloxane (PDMS) to form a replica of photolithographically defined patterns and then use it to stamp the patterns on another substrate using the soft elastomer replica. For example, Fig. 1.24 shows a contact printing of self-assembled monolayers (SAMs) of octadecyltrichlorosilane (OTS) [12] through soft lithography. In the process, a desired pattern is first made on a silicon substrate with photolithography or e-beam lithography, followed by spin-coating or casting of PDMS. When the PDMS is cured, it is peeled off from the silicon, and then its bottom face (having the replica of the photolithographic pattern on the silicon) is inked with OTS by dipping the PDMS in $CH_3(CH_2)_{16}CH_2\text{-}SiCl_3 + H_2O \rightarrow$ OTS-SAM + HCl. The OTS is then transferred to another substrate by making the PDMS contact that substrate. This way a few nanometer-thick SAMs can be patterned on any

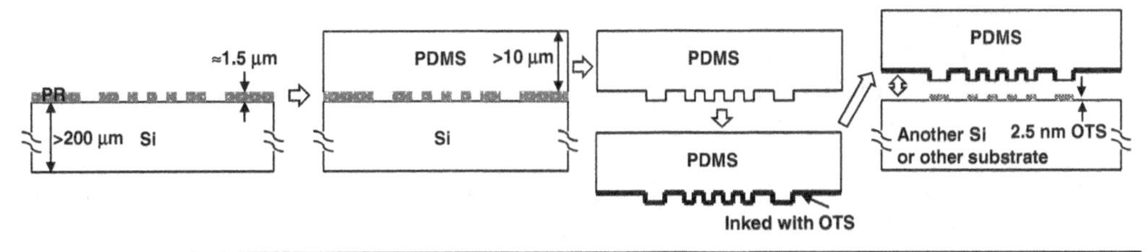

FIGURE 1.24 Micro-contact printing of OTS-SAMs through soft lithography.

substrate with a lateral dimension as small as what photolithography or e-beam lithography can produce.

Quite a few MEMS transducers have been fabricated with a fabrication process based on soft lithography. For example, Fig. 1.25 shows a focused ultrasound transducer made through steps involving soft lithography for PDMS membrane (Fig. 1.25a and b), patterning electrodes on PZT sheet followed by Parylene deposition (Fig. 1.25c and d), and bonding the PDMS membrane onto the PZT sheet with UV-curable epoxy (Fig. 1.25e and f) [13]. To prevent PDMS from sticking to the silicon substrate after casting, the silicon surface may be treated with SIT8174 (Gelest Inc.), a kind of silane consisting of Teflon-like carbon–fluorine groups that can be chemically bonded to the silicon surface to make it hydrophobic. To ensure good adhesion between Parylene and UV-curable adhesive as well as between PDMS and UV-curable adhesive, both PDMS and Parylene surfaces may be treated with another kind of silane (SIA0196, Gelest Inc.) to make them more hydrophilic. The UV-curable adhesive (NOA 60, Norland Products)

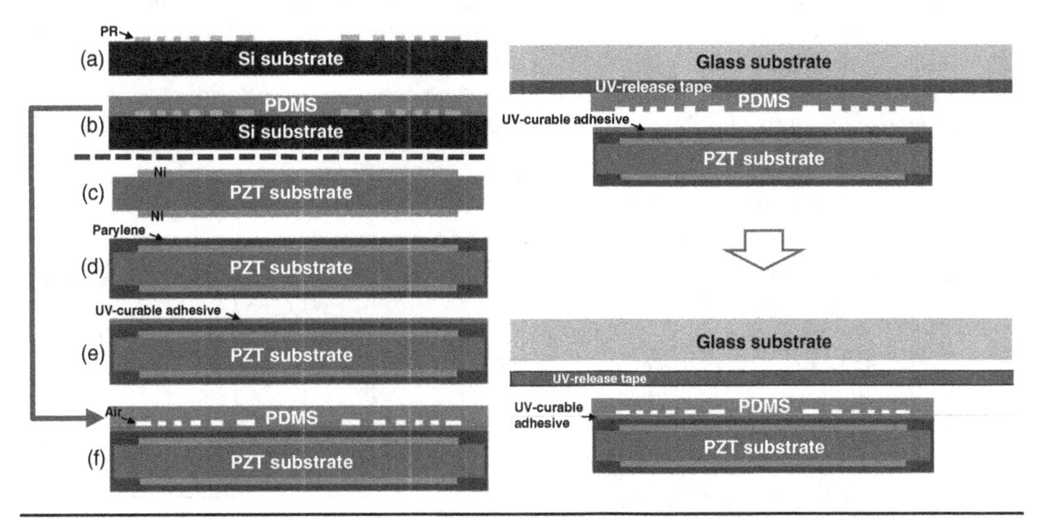

FIGURE 1.25 Fabrication process (not to scale): on silicon wafer, (a) create silicon mold through photolithography and (b) replicate PDMS membrane from the silicon mold. On PZT sheet, (c) pattern top/bottom electrodes, (d) deposit Parylene, (e) spin-coat UV-curable adhesive, (f) partially cure the adhesive, and bond the PDMS (peeled off from the silicon mold) onto PZT after alignment, then fully cure the adhesive. The step (f) is detailed at right: transfer PDMS membrane onto a glass plate covered with a UV-release tape (sticky side facing the glass), align with alignment markers on PZT and PDMS, bond with UV-curable epoxy with UV light, and release after UV exposure.

is spin-coated at 4,000 rpm for 2 minutes, which results in a cured thickness of 3 μm (Fig. 1.25e). The PDMS membrane is cut and peeled off from the silicon substrate and aligned under a stereo microscope for bonding to the PZT substrate. No pressure is needed because the PDMS membrane is automatically sucked onto the pre-cured UV-adhesive upon initial contact. In contact areas, the UV-adhesive wicks into the tiny gaps between PZT and PDMS. Then the UV-adhesive is fully cured with 5 minutes of 10 mW/cm² UV exposure (Fig. 1.25f, of which the detailed steps are shown at the right of the figure).

1.10 Wafer Bonding

In MEMS fabrication and/or packaging, two or more wafers or substrates are sometimes bonded for tall/thick or 3D structures, integration of multiple functions, low-cost packaging, etc. In this section, we will study various wafer bonding techniques that have been used in MEMS fabrication.

1.10.1 Direct Bonding between Silicon Wafers

A bare silicon wafer has 1–2-nm-thick native SiO_2, after a typical wafer cleaning procedure with oxidizing solution, which makes the surface hydrophilic, though crystalline silicon is hydrophobic. When two polished Si wafers are brought together with their mirror-like surfaces in contact with each other, the two wafers bond even at room temperature, whether there is a native SiO_2 or not, with a relatively low bonding energy (0.03–0.15 J/m²) or fracture surface energy. A subsequent high-temperature treatment (>800°C) increases the bonding energy to larger than 2 J/m², and a strong covalent bonding can be obtained between two silicon wafers. The bonding mechanisms are a little different for hydrophilic silicon wafers and hydrophobic silicon wafers, as illustrated in Fig. 1.26 for hydrophilic silicon wafers (with 1–2-nm-thick native oxide) and Fig. 1.27 for hydrophobic silicon wafers without any native oxide. A hydrophobic silicon surface is typically obtained by dipping a silicon wafer in hydrofluoric (HF) acid (to etch away native oxide) and has hydrogen bonded to silicon atoms on the surface.

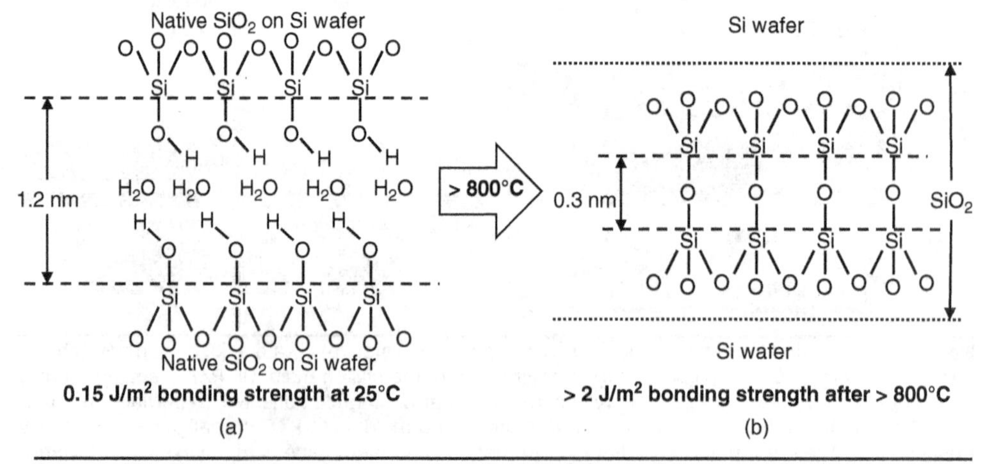

FIGURE 1.26 Direct bonding of two silicon wafers with 1–2-nm-thick native oxide (presenting a hydrophilic surface): (a) at room temperature and (b) after high-temperature (>800°C) treatment.

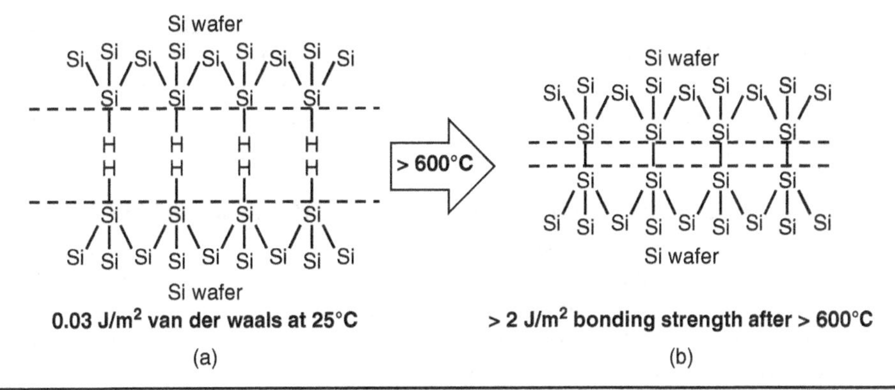

FIGURE 1.27 Direct bonding of two silicon wafers without any oxide (and thus, having hydrophobic surfaces): (a) at room temperature and (b) after high-temperature (>600°C) treatment.

On the other hand, a hydrophilic silicon surface with a native oxide on its surface has OH bonded to silicon atom of the native oxide, after a typical wafer cleaning procedure.

As illustrated in Fig. 1.26a, when two hydrophilic silicon wafers are brought together to be in contact with each other, the Si-OH groups interact with each other, producing Si-O-Si and H_2O, according to the following reaction [14].

$$-\mathrm{Si}-\mathrm{OH} + \mathrm{HO}-\mathrm{Si}- \Longrightarrow -\mathrm{Si}-\mathrm{O}-\mathrm{Si}- + H_2O$$

At room temperature, the bonding strength is about 150 mJ/m^2. The bonding strength increases more than 10 times when the wafers go through a high-temperature treatment for about tens of minutes, as H_2O molecules between the two wafers diffuse through the native oxides and thermally oxidize the silicon atoms in the wafers, forming a very thin SiO_2 between two Si wafers (Fig. 1.26b). In this thermal oxidation process, though, hydrogen gas is generated, through $Si + 2H_2O \rightarrow SiO_2 + 2H_2$, and may produce nanometer-sized voids in the SiO_2.

In case of bonding of silicon wafers with hydrophobic surface, as illustrated in Fig. 1.27a, the hydrogen-terminated silicon surfaces are weakly bonded through van der Waals force at room temperature. When the wafers go through a high-temperature treatment (Fig. 1.27b), hydrogens leave the interface as hydrogen molecules (H_2), diffusing along the interface or into the silicon wafers, leaving behind Si-Si bonds at the interface. Unlike the bonding of hydrophilic silicon wafers, no interlayer is formed between the two bonded wafers. However, bonding with hydrophobic surfaces may result in more nanometer-sized voids at the interface than that with hydrophilic surfaces.

The high-temperature treatment after the room temperature bonding increases the bonding strength more than 10 times and is usually carried out for a strong bonding. However, even without the high-temperature treatment, the bonding strength can be increased after storage at 25–200°C for many hours or days, surface treatment with O_2 plasma followed by heating at 200°C for many hours (resulting in bonding strength up to 0.8 J/m^2), or bonding of hydrophilic silicon wafers under vacuum at 150°C for > 60 hours (up to 3 J/m^2) [14].

Direct bonding between two silicon wafers offers a high bonding strength and hermetic sealing at the bonded interface, but requires very smooth surfaces (<6 nm surface roughness). The direct bonding does not happen if there is a nitride layer between the two wafers as the layer blocks any of the bonding mechanisms explained above.

1.10.2 Anodic Bonding between Silicon Wafer and Glass Wafer

A glass wafer (e.g., Pyrex 7740) containing mobile sodium and/or potassium ions can be bonded to a silicon wafer (covered with native oxide) when a 200–400 V is applied between the two wafers at 200–300°C, as shown in Fig. 1.28. This so-called anodic bonding produces Si-O-Si covalent bonds between silicon and glass wafers and can be used to hermetically seal MEMS devices made on a glass (Fig. 1.28a) or a silicon wafer (Fig. 1.28b).

The underlying mechanism of anodic bonding is generally thought to be due to SiO_2 growth between a silicon and a glass wafer [15] (similar to silicon–silicon direct bonding with native oxide illustrated in Fig. 1.26). The SiO_2 growth is greatly aided by a high electrical field across the interface (between the silicon and the glass) that pulls the two wafers together. The high electrical field is due to the depletion of mobile ions (Na^+ and/or K^+) in the glass near the interface, as illustrated in Fig. 1.29. Thus, anodic bonding needs mobile ions, and positive (or higher) voltage should be applied to silicon with respect to glass (Figs. 1.28 and 1.29). The mobile ions, though, may get into field-effect transistors or CMOS integrated circuits and affect their performance. The SiO_2 growth (and consequent bonding) was observed to propagate along the air gap from

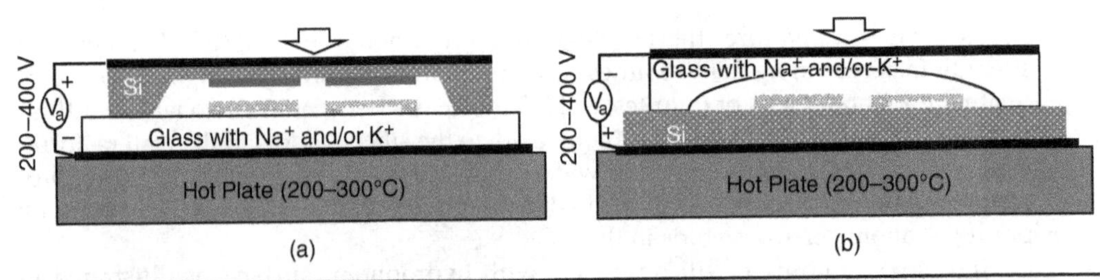

FIGURE 1.28 Anodic bonding between glass and silicon through heat, applied voltage, and pressure, with Na^+ playing a critical role in the formation of Si-O-Si covalent bonds: (a) MEMS structures on glass wafer and (b) MEMS structures on silicon wafer.

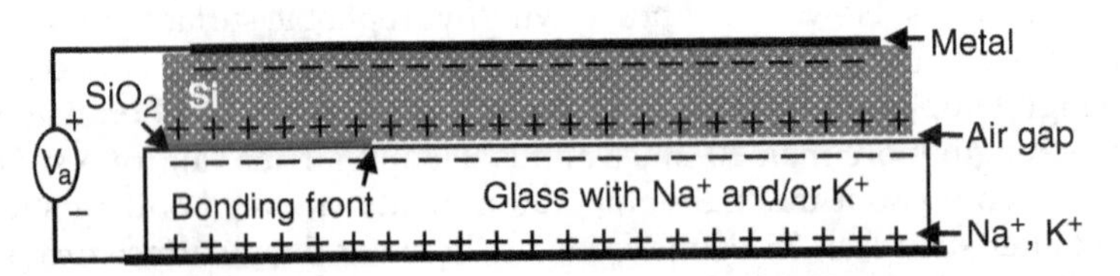

FIGURE 1.29 Bonding mechanism of anodic bonding that shows accumulation of Na^+ and/or K^+ ions (on one side of a glass wafer due to the applied voltage) that results in a depletion region near the nanometer air gap between a glass wafer and a silicon wafer.

one end to the other [16]. Anodic bonding offers a high bonding strength (~40 MPa) and hermetic seal with only about tens of minutes at a relative low temperature of 200–300°C. However, it requires very smooth surfaces (<10 nm surface roughness), similar to silicon–silicon direct bonding.

1.10.3 Bonding with Metallic Interlayer

With heat and applied pressure, bonding between wafers can be obtained with metallic interlayer, and various techniques (Fig. 1.30) have been developed. For example, using a low eutectic temperature (i.e., the lowest melting tempeature for alloy) of 363°C for Au-Si (eutectic composition being Au/Si = 97.15/2.85 wt%), one can hermetically bond two silicon wafers through Au-Si eutectic bonding, as illustrated in Fig. 1.30a, at 400°C with 0.8 MPa load in 30 minutes. Other bonding techniques with metallic interlayers include Cu-Cu thermocompression bonding at 250°C with >0.25 MPa (Fig. 1.30b), Cu-Sn solder bonding at 280°C with 0.4 MPa load (Fig. 1.30c), and advanced Cu-Sn solder bonding with 10-nm-thick Au layers (Fig. 1.30d). In all cases, metal adhesion layer such as Ti or Cr is needed to ensure good adhesion of Au or Cu to silicon substrate.

Wafer bonding with metallic interlayer offers hermetic seal with high bonding strength (up to ~60 MPa), and can also be done with rougher surfaces than the direct Si-Si bonding (except the Au-Si eutetic bonding). However, specific metal depositions are needed, and there may be reflow of solder (Sn) at a relatively low temperature in case of Fig. 1.30c and 1.30d.

1.10.4 Bonding with Insulating Interlayer

Two silicon wafers can be hermetically bonded through glass frit bonding with "glass solder" that has a relatively low melting temperature, as illustrated in Fig. 1.31. A glass solder is lead-glass powder (having a melting temperature <450°C) mixed with binders, fillers, and solvents, and can be spin-coated or screen-printed for 5–30-μm-thick glass frit. Glass frit bonding can tolerate some surface roughness, and its bonding strength can be as high as that obtainable with other techniques. In the process shown in Fig. 1.31, glass frit is screen-printed on a cap wafer, which is then bonded to a MEMS wafer through melting the glass frit at 350–450°C under a uniform pressure

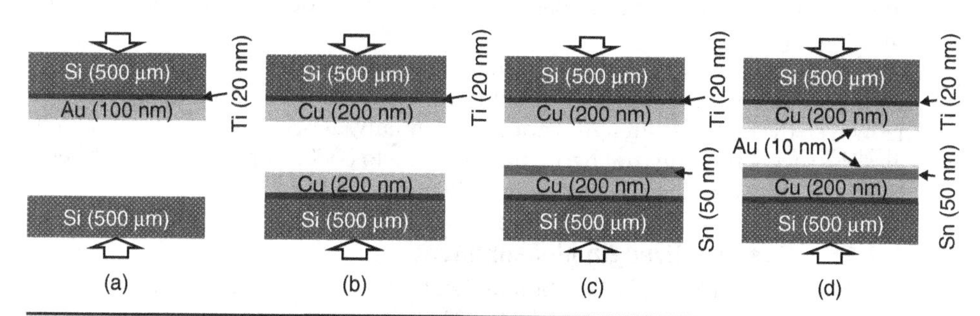

FIGURE 1.30 Bonding through metallic intermediate layer(s): (a) eutectic bonding between Au and Si, (b) thermocompression bonding between Cu and Cu, (c) solder bonding with Sn, and (d) advanced solder bonding with Sn.

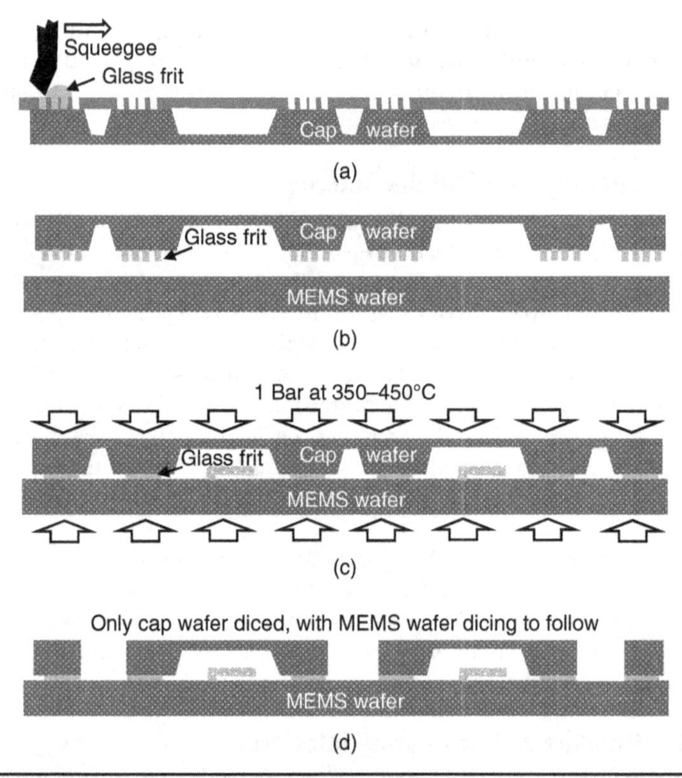

FIGURE 1.31 An exemplary set of process steps for glass frit bonding: (a) screen-print glass frit (or paste) over a capping wafer, (b) align the capping wafer to a MEMS wafer, (c) press the two wafers at 350–450°C to melt the glass frit and bond them, and (d) dice the capping wafer at room temperature, followed by dicing the MEMS wafer to separate the wafer into chips.

applied between the two wafers. The bonded wafers are diced twice in this case for a reason (i.e., the cap wafer is diced first, followed by dicing of the MEMS wafer), though the bonded wafers may be diced together. The cavity formed by the bonding can be under vacuum, if the bonding is done in vacuum; and since the leak through the glass frit seal is very low, glass frit bonding may be good for vacuum packaging MEMS [17].

Wafers can easily be bonded through adhesives (such as epoxy resin) at room temperature or a low temperature without much restriction on surface roughness and type. However, the bonding strength varies substantially, depending on the type of adhesive, and the seal is typically not hermetic. One way to coat a cap wafer with adhesive and bond the cap wafer to a MEMS wafer is illustrated in Fig. 1.32.

1.10.5 Bonding Strength Measurement

Bonding strength between two bonded wafers can be measured with any of the four destructive ways illustrated in Fig. 1.33. With a razor blade, a bonded interface between two wafers can be cracked, and as the blade is pushed through the interface, the cleavage at the interface causes the two rigid wafers to bend like cantilevers

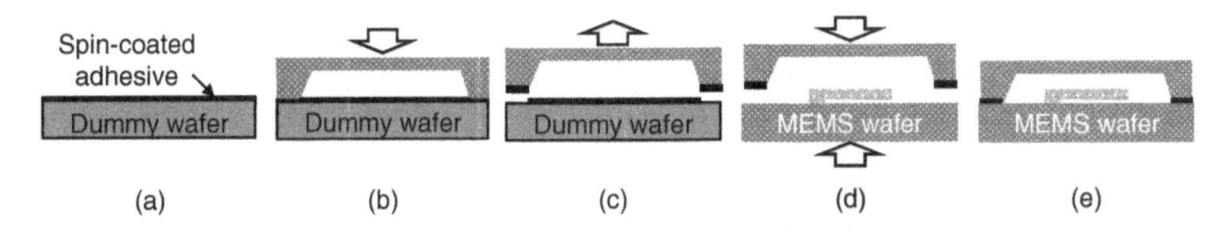

Figure 1.32 Process steps for adhesive bonding: (a) spin-coat adhesive over a dummy wafer, (b) press down a capping wafer over the dummy wafer, (c) pull up the capping wafer which is now inked with the adhesive at selected locations, and (d) press down the capping wafer over a MEMS wafer to obtain the two wafers bonded via adhesive (e).

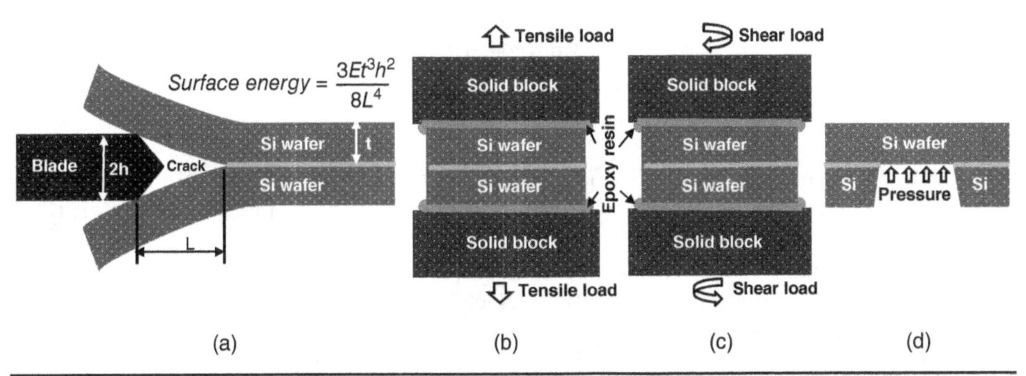

Figure 1.33 Measurement techniques for bonding strength between two wafers: (a) razor-blade test with controlled cleavage producing crack length L due to the blade with thickness 2h, allowing measurement of surface fracture energy in J/m^2, (b) tensile test (giving bonding strength in Pa, up to 80 MPa), (c) shear or torsion test (in Pa), and (d) pressure burst test (in Pa).

(Fig. 1.33a). The bending curvature is related to the adhesion energy (or surface fracture energy) γ, which is equal to $3Et^3h^2/8L^4$, where E, t, h, and L are the wafer's Young's modulus, wafer thickness, one-half the blade thickness, and crack length, respectively [18]. Bonding strength can also be measured with tensile (or shear) load applied, as shown in Fig. 1.33b (or Fig. 1.33c), up to 80 MPa which is the maximum bonding strength of the epoxy glue. Under a special case such as the one shown in Fig. 1.33d, uniform pressure can be used to burst-break the bonding to measure the bonding strength. The tensile/shear load and uniform pressure burst tests give bonding strength in Pa (or MPa), but the razor-blade test measures the adhesion energy and gives the bonding strength in J/m^2.

1.11 Flip-Chip Bonding for Electrical Interconnect

When the number of electrical interconnects between MEMS chips or between a MEMS chip and a printed circuit board (PCB) is large, flip-chip bonding with an array of solder bumps on a silicon chip can be used to avoid many wires and large wire-routing area on MEMS device. A solder composed of Pb (95 wt%) and Sn (5 wt%) with

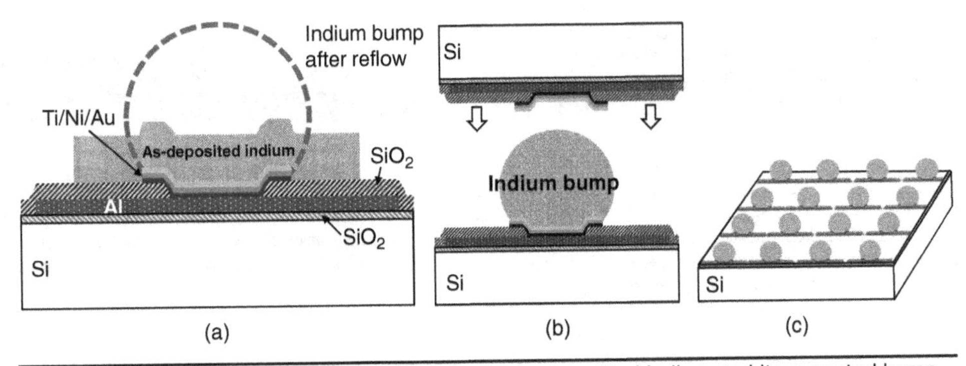

FIGURE 1.34 Parts of flip-chip bonding process: (a) as-deposited indium and its expected bump after a reflow with Ti/Ni/Au under bump metallization (UBM) with the Ni providing wettability during the reflow, (b) bonding of two silicon chips with indium solder bump, and (c) an array of solder bumps on a chip.

a melting temperature of 310°C is an option for solder bumps [19], but here we will study how to form solder bumps with indium which has a melting temperature of about 156°C.

In forming a solder bump, as illustrated in Fig. 1.34a (where the patterned SiO_2 layer is over the Al electrode, mainly for electrical insulation, not for forming the solder bump), a patterned under bump metallization (UBM) composed of three very thin metal layers (Ti/Ni/Au) is needed for adhesion and diffusion barrier (attainable with 10-nm-thick Ti), wettability of indium during a reflow step (50-nm-thick Ni), and protectability of Ni from oxidation (50-nm-thick Au) [20]. The gold layer needs to be deposited right after the Ni deposition in the same deposition system (either sputtering or evaporation system with multiple sources) without breaking the vacuum, in order to avoid oxidation of the Ni. After indium is deposited over the UBM and patterned to have a larger lateral dimension than that of the UBM, the indium solder is reflown at its melting temperature in N_2 environment with the UBM's Au/Ni layer providing a wetting surface for the indium to form a bump. The surface tension of the liquefied indium makes the bump look like a truncated sphere, while the lateral dimension of the UBM determines the lateral dimension of the bump after the reflow. Consequently, the bump height or diameter depends on the ratio of lateral areas between the patterned as-deposited indium and the UBM as well as the surface tension of the liquefied indium. A silicon chip with a solder bump (or many solder bumps) can be aligned and brought in contact with another silicon chip with UBM (which is flipped to be upside down), as illustrated in Fig. 1.34b.

The two silicon chips when brought together would have a finite amount of misalignment between the two UBMs, as shown Fig. 1.35a. However, during the solder reflow, the surface tension of the liquefied indium produces a lateral force to align the two UBMs (Fig. 1.35b), and the completed solder bump between the two chips would be located between the two well-aligned UBMs (Fig. 1.35c).

In depositing and patterning UBM and indium solder, a lift-off process with a patterned photoresist (PR) can be used to avoid the etching step of metal layers in a typical photolithography-based patterning process. For such a PR lift-off process, it is critical to have discontinuity in the deposited metal over the steps of a patterned

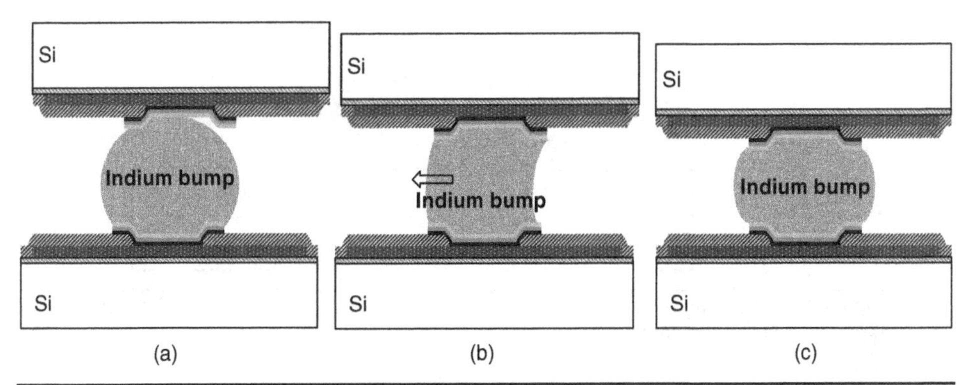

FIGURE 1.35 Two silicon chips being bonded through indium solder bump: (a) after alignment and contact, showing a finite misalignment between two UBMs on the two chips, (b) during reflow with surface tension (of the liquefied bump) aligning the chips in the lateral direction, and (c) completed bump with the two chips well aligned.

PR so that the wet PR remover may remove the PR completely. To ensure such discontinuity, one can use lift-off resist (LOR), which is a kind of photoresist, but is made to be insensitive to light and be soluble in photoresist developers and removers. A lift-off process with LOR is illustrated in Fig. 1.36; (a) after spin-coating negative PR over a spin-coated LOR, light is illuminated through a photomask over the soft-baked PR, and (b) the PR developer removes the unexposed area, and then removes LOR, isotropically, including the LOR under the light-exposed PR. The amount of the undercut region depends on the amount of the over-developing time and the LOR's pre-bake temperature and time. For example, one can subject a LOR-coated wafer to 175°C for 5 minutes to dry (or evaporate solvent out of) LOR film and also fix the undercut rate in a PR developer. When metal (or dielectric material) is sputter-deposited or evaporated over the silicon wafer, the deposited film has discontinuity over the etched steps (Fig. 1.36c), through which wet PR remover goes in to remove the PR and LOR (and thus lift off all the unwanted metal), leaving only a desirable patterned metal (Fig. 1.36d).

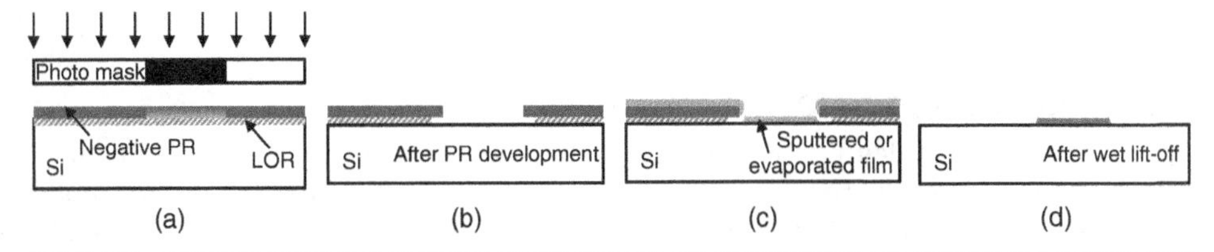

FIGURE 1.36 Lift-off process with lift-off resist (LOR): (a) after exposure to light through a photomask with the negative photoresist (PR) cross-linking in light-exposed areas, while the underlying LOR not being impacted by light exposure because LOR is made to be insensitive to light, (b) PR developer removes the exposed area of the negative PR and LOR underneath with lateral undercut depending on the time of over-development and the undercut rate, (c) discontinuous sputter-deposited (or evaporated) film, and (d) a patterned sputter-deposited film after the PR and LOR are removed in wet PR remover.

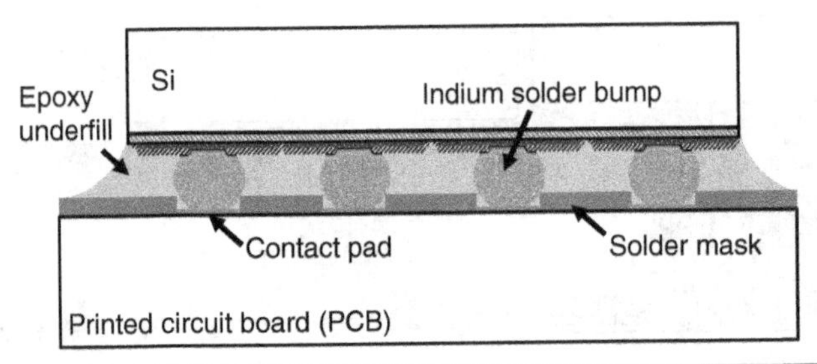

FIGURE 1.37 A silicon substrate flip-chip-bonded on a printed circuit board (PCB) through solder bumps with epoxy underfill.

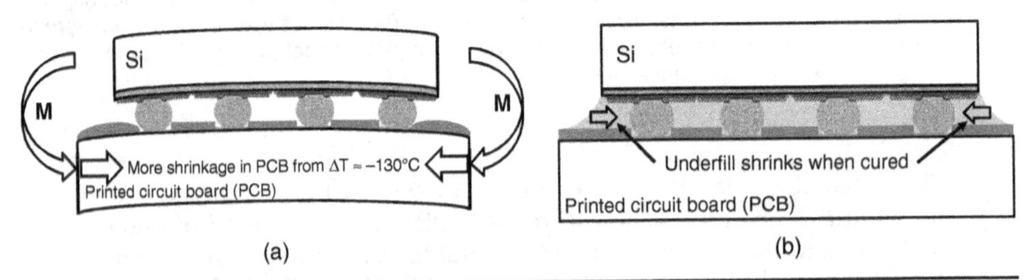

FIGURE 1.38 Bending of flip-chip-bonded substrates at room temperature: (a) after solder reflow at 156°C due to the mismatch of thermal expansion coefficients between silicon and PCB and (b) after curing of underfill which makes the underfill shrink.

Reliability of flip bonding with solder bumps can be enhanced with underfill between two substrates (Fig. 1.37). The gap between the two flip-chip-bonded substrates can be filled through capillary flow of epoxy resin (which is heavily filled with SiO_2) or any other advanced techniques [21], followed by curing of the epoxy. The hardened underfill has high Young's modulus due to the high content of SiO_2 and protects the solder joints from environment including moisture. With underfill, thermal stress on solder joints due to coefficient of thermal expansion (CTE) mismatch between silicon and PCB (Fig. 1.38a) is redistributed (and reduced up to 10 times) during the curing of the underfill epoxy resin (Fig. 1.38b) [22, 23].

1.12 Engineered Silicon Substrates

A silicon on insulator (SOI) wafer has a thin monocrystalline silicon layer over a buried silicon oxide on a silicon substrate, and it has been used to fabricate various MEMS devices. Standard SOI wafers with various thicknesses and sizes are commonly available through many vendors. However, at least one company (i.e., IceMOS Technology Ltd of Belfast Northern Ireland) offers unprocessed SOI-like starting wafers that have already gone through some processes such as micromachining and wafer bonding, so that MEMS researchers and developers may minimize the fabrication steps at their sites. For example, the wafer called cavity-bonded SOI (CSOI) contains buried air cavities under thin monocrystalline silicon diaphragms.

References

1. W.R. Runyan and K.E. Bean, "Semiconductor Integrated Circuit Processing Technology," Addison-Wesley, 1990.
2. E.J. Ng, V.A. Hong, Y. Yang, C.H. Ahn, C.L. M. Everhart, and T.W. Kenny, "Temperature dependence of the elastic constants of doped silicon," Journal of Microelectromechanical Systems, 24(3), June 2015, pp. 730–741.
3. M. Sekimoto, H. Yoshihara, and T. Ohkubo, "Silicon nitride single-layer x-ray mask," Journal of Vacuum Science and Technology, 21(4), Nov./Dec. 1982, pp. 1017–1021.
4. S.S. Lee, R.P. Ried, and R.M. White, "Piezoelectric cantilever microphone and microspeaker," Journal of Microelectromechanical Systems, 5(4), Dec. 1996, pp. 238–242.
5. H. Guckel, D.W. Burns, C.C.G. Visser, H.A.C. Tilmans, and D. Deroo, "Fine-grained polysilicon films with built-in tensile strain," IEEE Transactions on Electron Devices, 35(6), June 1988, pp. 800–801.
6. P.O. Oviroha, R. Akbarzadeh, D. Pan, R.A.M. Coetzee, and T.C. Jen, "New development of atomic layer deposition: processes, methods and applications," Science and Technology of Advanced Materials, 20(1), 2019, pp. 465–496.
7. http://fcon-inc.jp/en/en_MFC/Principle/Principle.html.
8. E.T. Carlen and C.H. Mastrangelo, "Electrothermally activated paraffin microactuators," Journal of Microelectromechanical Systems, 11(3), June 2002, pp. 165–174.
9. M.J. Zdeblick, R. Anderson, J. Jankowski, B. Klein-Schoder, L. Christel, R. Miles, and W. Weber, "Thermopneumatically actuated microvalves and integrated electrofluidic circuits," Solid-State Sensor and Actuator Workshop, Hilton Head Island, SC, June 1994, pp. 251–255.
10. L.T. Romankiw, "A path: from electroplating through lithographic masks in electronics to LIGA in MEMS," Electrochimica Acta, 42(20–22), 1997, pp. 2985–3005.
11. P.C. Andricacos, C. Uzoh, J.O. Dukovic, J. Horkans, and H. Deligianni, "Damascene copper electroplating for chip interconnections," IBM Journal of Research and Development, 42(5), Sept. 1998, pp. 567–574.
12. G.M. Whitesides and A.D. Stroock, "Flexible methods for microfluidics," Physics Today, 54(6), 2001, pp. 42–48.
13. Y. Tang, S. Liu, and E.S. Kim, "MEMS focused ultrasonic transducer with air-cavity lens based on polydimethylsiloxane (PDMS) membrane," The 33rd IEEE International Conference on Micro Electromechanical Systems (MEMS 2020), Vancouver, Canada, Jan. 2020, pp. 58–61.
14. Plößl, G. Kräuter, "Wafer direct bonding: tailoring adhesion between brittle materials," Materials Science and Engineering, R25, 1999, pp. 1–88.
15. Schmidt, P. Nitzsche, K. Lange, S. Grigull, U. Kreissig, B. Thomas, and K. Herzog, "In situ investigation of ion drift processes in glass during anodic bonding," Sensors and Actuators A: Physical, 67, 1998, pp. 191–198.
16. Thomas M.H. Lee, Debbie H.Y. Lee, Connie Y.N. Liaw, Alex I.K. Lao, and I. Ming Hsing, "Detailed characterization of anodic bonding process between glass and thin-film coated silicon substrates," Sensors and Actuators, 86, 2000, pp. 103–107.
17. G. Wu, D. Xu, X. Sun, B. Xiong, and Y. Wang, "Wafer-level vacuum packaging for microsystems using glass frit bonding," IEEE Transactions on Components, Packaging and Manufacturing Technology, 3(10), Oct. 2013, pp. 1640–1646.
18. W.P. Maszara, G. Goetz, A. Caviglia, and J.B. McKitterick, "Bonding of silicon wafers for silicon-on-insulator," Journal of Applied Physics, 64, 1988, pp. 4943–4950.

19. D.J. Pedder, "Flip chip solder bonding for microelectronic applications," Microelectronics International, 5(1), 1998, pp. 4–7.
20. C. Broennimann, et al., "Development of an Indium bump bond process for silicon pixel detectors at PSI," Nuclear Instruments and Methods in Physics Research Section A: Accelerators, Spectrometers, Detectors and Associated Equipment, 565(1), Sept. 2006, pp. 303–308.
21. Z. Zhang and C.P. Wong, "Recent advances in flip-chip underfill: materials, process, and reliability," IEEE Transactions on Advanced Packaging, 27(3), Aug. 2004, pp. 515–524.
22. Y. Tsukada, "Surface laminar circuit and flip-chip attach packaging," Proceedings 42nd Electronic Components and Technology Conference, San Diego, CA, 1992, pp. 22–27.
23. B. Han and Y. Guo, "Thermal deformation analysis of various electronic packaging products by moire and microscope moire interferometry," Journal of Electronic Packaging, 117, 1995, pp. 185–191.

Questions and Problems

Question 1.1 Dichlorosilane (SiH_2Cl_2) is liquid at room temperature, and contained in a cylinder as liquid. However, we can still deliver a gaseous SiH_2Cl_2 to low-pressure CVD (LPCVD) system for silicon nitride deposition. Explain how this is possible.

Question 1.2 What is the maximum flow of SiH_4 that can be read on a thermal flow meter that is calibrated for a full-scale nitrogen (N_2) flow of 500 sccm (844 Pa·L/s), assuming that the conversion factor for SiH_4 with respect to N_2 is 0.6? What is the minimum SiH_4 flow that the thermal flow meter can accurately measure?

Question 1.3 Dichlorosilane (SiH_2Cl_2) is a liquid with a vapor pressure of 167 kPa when stored in a cylinder at 20°C. (a) What is the maximum flow that can be read on a thermal flow meter that is calibrated for a full-scale nitrogen (N_2) flow of 500 sccm (844 Pa·L/s)? The conversion factor with respect to N_2 is 0.4. (b) A mass flow controller (MFC) along with the gas line is heated at 60°C and regulates the flow into a deposition chamber held at 40 Pa. If the dichlorosilane flow in the MFC is $Q(\text{Pa·L/s}) = 1 \times 10^{-7} P_{ave} \Delta P$ at 60°C, can this MFC provide a dichlorosilane flow equal to the full-scale value to the flow meter?

Question 1.4 Though a sputter target of TiNi (50%/50%) is used to deposit a TiNi (50%/50%) film with sputtering, why may a sputter-deposited film contain more Ni than Ti?

Question 1.5 At steady state during sputtering a TiNi (50%/50%) sputter target, (a) what will be the flux ratio between the sputtered Ti and Ni and (b) what will be surface composition at the target surface, if the sputter yields ($S \equiv$ number of target atoms sputtered ÷ incident ion) for Ti and Ni are: $S_{Ti} = 1.0$ and $S_{Ni} = 0.8$, respectively.

Question 1.6 Why does entrapment of sputtering gas (e.g., Ar) in a sputter-deposited film during RF sputtering increase as the magnitude of the negative substrate bias voltage ($|V_B|$) increases up to a certain value (e.g., 50 V) and decrease as the magnitude is increased higher?

Question 1.7 Why does viscosity of a gas increase with increasing particle temperature?

Question 1.8 How will the residual stress shown in Fig. 1.15a vary, if the silicon-rich silicon nitride is kept at 835°C for another 5 hours without any reactant gases but only with N_2 flow?

Question 1.9 If inlet gas flowing at 100 liters/minute at 27°C is input into a CVD tube at 1,100°C, what is the gas flow rate in the tube?

Question 1.10 The figure below shows a typical deposition rate as a function of temperature for an atmospheric pressure chemical vapor deposition with inert carrier gas. Qualitatively show how the curve will change when the deposition pressure is reduced to about 300 mTorr.

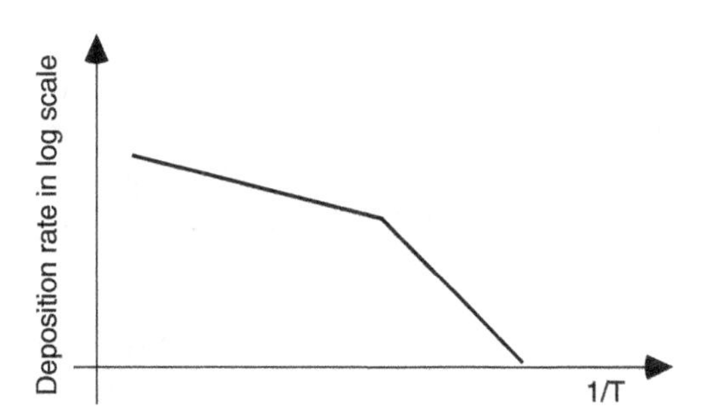

Question 1.11 The figure below shows a typical deposition rate as a function of gas flow rate for an atmospheric pressure chemical vapor deposition at 1,000°C. Qualitatively show how the curve will change when the deposition temperature increases to 1,200°C.

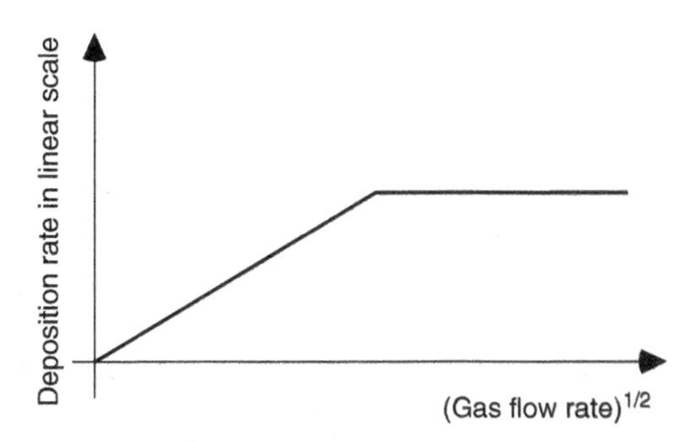

Question 1.12 Using symbols ↑ (increase), ↓ (decrease), and ↔ (negligible effect) to indicate how each of the parameters (in chemical vapor deposition of polysilicon using SiH_4 as a reactant gas) in the top row changes in response to the changes shown in the first column.

	Deposition Rate	**Grain Size**
At low pressure (≈ 0.3 T) and 700°C, gas flow rate ↑ (100–200 sccm, no gas depletion)		
At atmospheric pressure (with N_2 as a carrier gas) and 700°C, gas flow rate ↑ (100–200 sccm, no gas depletion)		
At atmospheric pressure with N_2 as a carrier gas, substrate temperature ↑ (700–900°C)		
At low pressure (≈ 0.3 T), substrate temperature ↑ (700–900°C)		

Question 1.13 In electroplating metal, special additives in the plating solution affect plating rates over different types of trench surfaces, and the following three different types of electroplating can happen over a trench made with polyimide. If the electroplating is carried out further, just enough to complete the filling of the trench, how will the final plating will look like for the three different cases, paying attention to whether there will be void or seam in the trench?

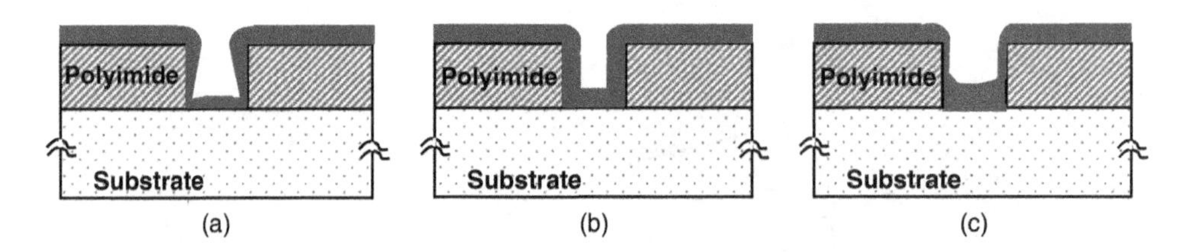

Question 1.14 Draw the missing steps in five figures or so between (a) and (b) in a dual Damascene process, assuming that the copper electroplating is done only once.

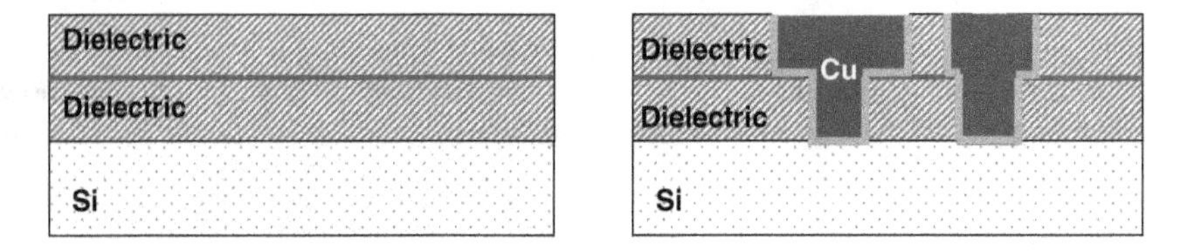

Question 1.15 In anodic bonding between silicon wafer and glass wafer (Section 1.10.2), why would a p-type silicon (such as the one doped with Boron) bond better to glass than an n-type silicon, as the electrostatic force across the air gap between the two wafers plays a major role in the bonding?

Problem 1.1 From the relative velocity distribution by Maxwell-Boltzmann, $f_v = \frac{1}{N}\frac{dn}{dv} = \frac{4}{\sqrt{\pi}}\left(\frac{m}{2kT}\right)^{3/2} v^2 e^{-mv^2/2kT}$, derive the average molecular velocity $\bar{v} = \sqrt{8kT/\pi m}$ through $\bar{v} = \left(\int_0^\infty v f_v \, dv\right) / \left(\int_0^\infty f_v \, dv\right)$.

Problem 1.2 If aluminum is sputter-deposited with 10 mTorr of Ar at a deposition rate of 1,000 Å/min at 300 K, what will be the Ar content in the deposited aluminum film in atomic percent, assuming that a sticking coefficient for Ar on aluminum is 0.01%? Note that atomic density of Al is $6.02 \times 10^{22}/\text{cm}^3$.

Problem 1.3 Answer the following for a polysilicon low pressure CVD (LPCVD) system with tube inner diameter = 6", tube length = 60", SiH_4 flow rate = 200 sccm, tube pressure = 300 mTorr, and tube temperature = 605°C, assuming SiH_4 decomposition inside the tube is negligible. (a) What is the silane's viscosity in the tube? Use 5 Å for silane's molecular diameter. (b) Calculate the "average" boundary layer thickness over the tube length. (c) Calculate the average boundary layer thickness over the tube length, if the pump is taken out and silane is forced to flow at 20 sccm by a nitrogen (N_2) carrier gas flowing at 100 liters per minute at atmospheric pressure (760 Torr), for atmospheric pressure CVD (APCVD). (d) How much lower is the average diffusion flux ($\bar{F} \approx D_G C_G/\bar{\delta}$) for APCVD, compared to that for LPCVD?

Problem 1.4 In LPCVD with SiH_4, if x_s is defined to be the fractional area that has adsorbed SiH_4, it is reasonable to take that the adsorption rate is equal to $An_o(1 - x_s)$ while the sum of the desorption rate and decomposition rate is equal to Bx_s, where A and B are constants, and n_o = silane input concentration. Based on this simple model, derive an equation for deposition rate, $R_D \approx \frac{n_o}{An_o + B}$, with C being another constant.

Problem 1.5 For polysilicon deposition by CVD at 1,000°C with a concentration of Si atoms in the gas stream being $4 \times 10^{16}/\text{cm}^3$, calculate the percentage change in growth rate if the deposition temperature changes by 1%, when the polysilicon is deposited at a surface-reaction-limited growth of 1.5 μm/min. Assume that the surface-reaction rate coefficient k_s is given by $k_s = 10^7 \exp(-E_a/kT)$ cm/sec.

CHAPTER 2

Micromachining

Fabrication of MEMS involves bulk micromachining and/or surface micromachining with wet chemistry and/or plasma-based etching. In this chapter, we will first study bulk micromachining of silicon with anisotropic and isotropic wet chemical etchants, and understand the etch-rate dependence on crystal planes as well as the importance of convex corners in anisotropic etching. Anisotropic etching of silicon (which has been used for commercial pressure sensors for decades) will be compared with isotropic etching. Then the ideas, techniques, and issues behind surface micromachining will be presented, and we will see how/why many of commercially successful MEMS (e.g., digital mirror arrays, inertial sensors, resonators) have benefited from surface micromachining. Combination of bulk and surface micromachining will be noted to be very useful. The third portion of this chapter will be on plasma-based dry etching (or micromachining), and we will learn the basics on plasma generation and how plasma can be used in etching. One of the most commonly used bulking micromachining techniques, deep reactive ion etching (DRIE) of silicon, will be studied. Other dry etching techniques based on vapor hydrofluoric acid and xenon difluoride also will be presented.

2.1 Bulk Micromachining

Bulk micromachining of silicon is typically done through etching silicon with wet etchants. Wet chemical etching of a material involves two major processes, mass transport and surface reaction, since chemical reactants need to be transported from bulk solution to the material's surface, followed by surface reaction and then mass transport of the reaction products from the surface to bulk solution. The mass transport is mostly through diffusion due to concentration gradient and is slow. Stirring that produces drifts affects the mass transport and can advantageously be used to enhance the etching rate and areal etching uniformity. The surface reaction depends on temperature exponentially and is proportional to exp $(-E_a/kT)$, where E_a, k, and T are activation energy, Boltzmann's constant, and temperature, respectively. The surface reaction also is typically slow, since wet etching in microelectromechanical systems (MEMS) processing is typically done below 100°C to avoid too much water evaporation during etching. Between mass transport and surface reaction, the slower of the two (if any one of the two is much slower than the other) limits the etching rate. In most cases, though, both of them are comparably slow and affect the etching rate.

With photolithography, wet chemical etching is often performed to produce an etch pattern in a certain layer of material. Thus, it would be desirable for a wet etchant to etch the desired material much faster than any adjacent materials by a factor of 5 or more.

2.1.1 Wet Etchants for Silicon Oxide, Silicon Nitride, Aluminum, and Polysilicon

In MEMS fabrication, the following thin films are commonly used: thermally grown silicon dioxide, chemical vapor–deposited (CVD) silicon dioxide, and CVD silicon nitride. Commonly used wet etchants for these materials are shown in Table 2.1, where the thermal SiO_2 is silicon dioxide grown on silicon through wet or dry oxidation at an elevated temperature, while the CVD SiO_2 is the one deposited at 450°C through chemical reaction of silane and oxygen gases with or without in situ phosphorus doping. The CVD Si_3N_4 is a stoichiometric silicon nitride deposited at 900°C with ammonia and dichlorosilane and is a little different from nonstoichiometric Si_xN_y (where x/y is about 1) that is often used in MEMS.

Hydrofluoric (HF) acid is a good etchant for silicon oxide and has been used for cleaning glassware. It has high enough vapor pressure at room temperature to be a gas but can be dissolved in water. Liquid HF solution with 49% HF concentration is commonly shipped to users in a well-sealed plastic container. [Note: Users must remember that HF at room temperature has high vapor pressure and should handle HF solution under a fume hood. Also, users must be extremely careful when handling HF, since it is not only toxic but also penetrates through human skin and attacks bones. Users must check materials safety data sheets (MSDS) before using any chemical.]

Various dilutions of HF acid are used in etching silicon oxide. Also, ammonia fluoride is used to produce a buffered HF (BHF) to maintain constant PH level during the etching. Various HF solutions (including P Etch that is a combination of nitric acid and HF acid) are effective in etching silicon dioxide. The etch rate of phosphosilicate glass (PSG) in HF or P Etch depends on P_2O_5 content in PSG and can be used to estimate its mol%. [Note: P Etch is composed of 3 HF (49%), 2 HNO_3 (70%), 60 DI H_2O.] Aluminum can be etched at 0.2 μm/min with a combination of 16 H_3PO_4, 2 DI H_2O, 1 HAc (98%), and 1 HNO_3 (70%).

Though there are usually good wet etchants for most dielectrics and metals, there is not a very good wet etchant for silicon nitride, which is usually etched with dry etching (based on plasma, as described later in the chapter). Hot phosphoric acid (at 150–180°C) as well as concentrated HF (49% HF) etch silicon nitride, but with those etchants photoresist cannot be used as an etch mask, since photoresist is damaged, dissolved, and/or

	HF (49%) 25°C	$HF:H_2O$ (1:10) 25°C	$HF:H_2O$ (1:100) 25°C	P Etch 25°C	NH4F:HF (BHF) 6:1 (40%:49%) 25°C	H_3PO_4 180°C
Thermal SiO_2	300 Å/s	300 Å/min	—	108 Å/min	1200 Å/min	Nil
Thermal SiO_2 with phosphorus predep.	—	575 Å/min	—	575 Å/s		
CVD SiO_2	1000 Å/s	1800 Å/min	180 Å/min	1080 Å/min		Nil
CVD SiO_2 with 10% Phosphorus (PSG)	—	7200 Å/min	360 Å/min	3240 Å/min	—	Nil
CVD Si_3N_4 (900°C)	150 Å/min	Nil	Nil	Nil	10 Å/min	65 Å/min

TABLE 2.1 Wet Etchants for Common Dielectric Materials Used in MEMS

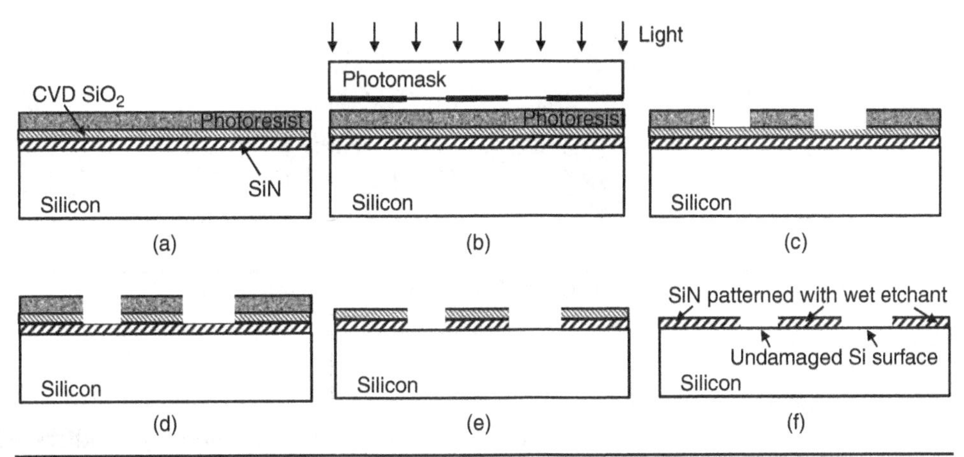

FIGURE 2.1 Patterning of silicon nitride (SiN) film with wet etchant: (a) deposit CVD SiO_2 and then spin-coat photoresist (PR) over a wafer with SiN film, (b) transfer pattern from a photomask to the wafer, (c) develop PR, (d) etch SiO_2, (e) etch SiN in hot phosphoric acid with the patterned SiO_2 as an etch mask, and (f) remove SiO_2. The exposed silicon surface is not damaged by hot phosphoric acid.

peeled off at a temperature higher than 150°C and in a strong acidic solution such as 49% HF (or in a basic solution). If silicon nitride (SiN) needs to be etched with a wet etchant (for some reason, such as for not damaging the underlying layer or substrate), one can use an extra layer of dielectric such as CVD SiO_2 as an etch mask to pattern SiN with hot phosphoric acid (without damaging silicon surface) through the steps illustrated in Fig. 2.1.

2.1.2 Crystallographic Notations

In wet etching of crystalline silicon with anisotropic etchants, silicon's crystal planes play a very important role. So, let us first learn how to denote crystal planes. A direction is described with [*abc*] where *a*, *b*, and *c* are smallest integers along the direction in *x*-, *y*-, and *z*-coordinates, respectively, as illustrated in Fig. 2.2. Negative components are denoted by putting a bar over the number (e.g., [$1\bar{1}0$]). Crystallographically equivalent sets of directions are written as <*abc*>.

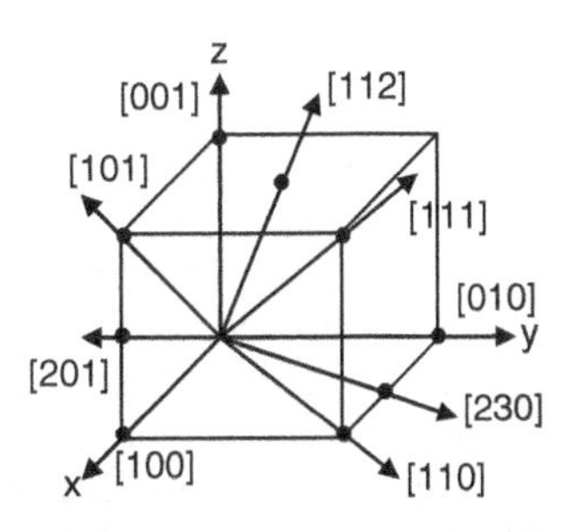

FIGURE 2.2 Typical denotation of a direction in Cartesian coordinates based on the smallest integers along the direction in *x*-, *y*-, and *z*-coordinates.

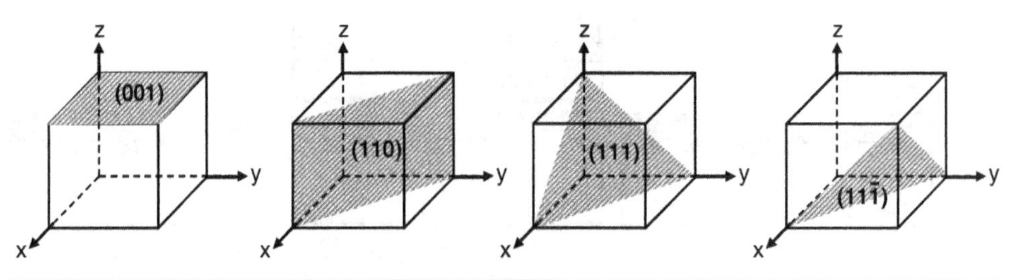

Figure 2.3 Some examples of crystal planes denoted with Miller indices.

Miller Indices for Planes

A plane is typically described with (*abc*) where *a*, *b*, and *c* are Miller indices. Miller indices for planes are obtained by (1) finding the intercepts of the plane on the crystal axes (*x*, *y*, and *z*), (2) taking the reciprocals of the intercepts, and then (3) clearing fractions, if there is any. Some examples of planes with Miller indices are shown in Fig. 2.3. A family of crystallographically equivalent planes is represented by {hkl}. One major reason for using Miller indices to denote a plane is that the three Miller indices for a plane are also the indices for the direction that is perpendicular to the plane. For instance, the direction perpendicular to a (100) plane is [100]. Thus, one can find out whether two planes meet at 90° by doing a dot product of their *normal* (perpendicular) directional vectors and seeing whether the dot vector product results in zero. For example, one can quickly know that (111) and (110) do not meet at 90°, but (111) and (1$\bar{1}$0) meet at 90° by noting [111]·[110] = 2 (nonzero) and [111]·[1$\bar{1}$0] = 0. In fact, one can calculate at what angle two planes meet through the dot product of their normal directional vectors. For example, (100) and (111) planes meet at an angle equal to $\cos^{-1}\left(\dfrac{1x1+1x0+1x0}{1x\sqrt{1^2+1^2+1^2}}\right)=\cos^{-1}\left(\dfrac{1}{\sqrt{3}}\right)=54.74°$.

For silicon, since it has cubic symmetry, all the {100} planes such as (100), (010), (001), etc. are crystallographically equivalent. Note that some of {100} planes meet with some of {110} planes at 90° while some others at 45°. For example, (001) plane meets with (110) plane at 90° while (010) plane meets with (110) plane at 45°.

2.1.3 Bulk Micromachining of Silicon

As listed in Table 2.2, crystalline silicon can be etched by hydrofluoric, nitric, and acetic acids (HNA), ethylene diamine pyrocatechol with water (EPW), potassium hydroxide (KOH), and tetramethylammonium hydroxide (TMAH); HNA is an isotropic etchant that etches silicon isotropically, independent of the crystal plane, while the others etch crystalline silicon anisotropically.

The isotropic etching by HNA is obtained through the following two functions performed by the two acids in HNA: the nitric acid oxidizes the silicon, producing silicon oxides, and the HF etches the oxidized silicon (i.e., silicon oxides). As HF/HNO_3 ratio increases, isotropicity of etching improves, but the roughness of concave bottom gets worsened. Thus, there is a trade-off between the isotropicity and mirror-likeness of the etched surface. At 50°C, HNA composed of 2:3:3 (HF: HNO_3: CH_3COOH) is reported to

Etchant	Composition	Etching Temp.	Etch Rate (mm/min)	(100)/(111) Etch Rate Ratio	Etch Mask
HF: HNO_3: water (or CH_3COOH) (HNA)	10 mL: 30 mL: 80 mL	22°C	0.7–3.0	1/1	Si_3N_4, SiO_2 (300 Å/min)
Ethylene diamine: pyrocatechol: water (EPW)	750 mL: 120 g: 240 mL	115°C	1.25	35/1	SiO_2, Si_3N_4, Ag, Ag, Cr, Cu, Ta
KOH: water	44 g: 100 mL	85°C	1.4	400/1	Si_3N_4, SiO_2 (14 Å/min)
TMAH	25 wt% TMAH	95°C	0.6	30/1	SiO_2, Si_3N_4, Ag, Ag, Cr, Cu, Ta

Source: Adapted from [2].

TABLE 2.2 Wet Etchants for Silicon

give the best isotropicity and mirror-like surface [1]. One difficulty with HNA etching is that HNA etches most dielectric materials (as well as photoresist) relatively fast. Silicon nitride can be an etch mask for HNA etching but is still etched at a nonnegligible rate and has to be quite thick for a long silicon etching (this issue will be covered in depth later, e.g., with Fig. 2.18).

Anisotropic etching of silicon is obtained with OH-based basic solution, with the (111) crystal planes being etched the slowest among all the crystal planes of silicon. One of the commonly used anisotropic etchants used to be EPW which can be masked by many different materials including silicon oxide and many metals, but its fume has been known to be carcinogenic. On the other hand, KOH is benign but etches most materials at a relatively high etch rate. As KOH etches SiO_2 at 14 Å/min, SiO_2 has to be at least 0.7 μm thick for it to be a good etch mask during a long KOH etching of 500-μm-thick silicon (in 500 minutes at 1 μm/min etch rate). Only silicon nitride (Si_xN_y) can withstand a long (several hours) KOH etching with a thickness less than 0.2 μm. Also, potassium ion in KOH can potentially affect transistor performance in complementary metal–oxide semiconductor (CMOS). [Note: any OH-based basic solution etches photoresist very fast, and photoresist cannot be a good etch mask for anisotropic etching of silicon.] Another popular anisotropic Si etchant is TMAH, which is known to produce uneven etch front. With TMAH, however, even Al can potentially be used as an etch mask. It is considered to be benign to health and free of CMOS-contamination concerns.

Fast Etching of (100) Silicon with Smooth Surface in High-Temperature KOH

The etching rates of {100} and {110} silicon can be increased 5–9 times and 4–20 times, respectively, by etching the silicon in KOH near its boiling point (rather than the usual temperature of 80°C), when the KOH concentration is more than 32 wt% [3]. The silicon surface after etching was reported to be smooth for KOH with its concentration greater than 30 wt% when the etching was done near KOH's boiling temperature.

Etching Mechanisms of Anisotropic Si Etching

Anisotropic etching of Si involves hydroxyl ions (OH^-) reacting with Si atoms to produce $Si(OH)_2$, which is then taken out as $Si(OH)_2(O^-)_2$ along with hydrogen gas as an etch by-product, as the following chemical reaction takes place: $Si + 2H_2O + 2OH^- \rightarrow Si(OH)_2(O^-)_2 + 2H_2$. Note that the etching produces dissolved Si-by-products (solid) and H_2 (gas). The H_2 gas may affect the mass transport of the etchant and etch by-products if the gas lingers around the etching surface or channel.

In anisotropic etching of silicon, the etch rate of {111} silicon crystal plane is 10–400 times slower than that of {100} or {110} plane. The reason why {111} Si planes are etched the slowest among the {100}, {110}, and {111} is not very well understood but can plausibly be seen in the diagrams depicting the atomic arrangements for the planes, as each of the planes are removed layer by layer. Silicon atoms on {111} planes appear to be arranged such that they present more "stereographical effects" than the other two planes. Yet, the difference seems to be not much.

The hydrogen gas by-product of an anisotropic etching of Si can limit the mass transport of the etchant and etch by-products, especially when the etch channel is narrow. The gas bubbles are more likely to stay in the channels when the channel is narrower. Ultrasonic waves can agitate the channel surface and were observed to make the gas to leave the surface faster [4]. However, single-frequency ultrasonic waves were not as effective as ultrasonic waves with multiple frequencies varying in time such as the one composed of 28–48–100 kHz (with each frequency lasting for 1 ms and the set of the frequencies being repeated in time). With such ultrasonic waves, the etch front was observed to be smooth, and 58-μm-deep grooves of 3 μm width were consistently obtained.

Etch-limiting Role of (111) Planes in Anisotropic Si Etching

In silicon, having cubic diamond structure, its (111) and (001) planes meet at an angle of 54.74° on a (100) wafer, as illustrated in Fig. 2.4. Thus, if there is an etch-window opening of a relatively narrow width w on (001) plane, then two crystallographically

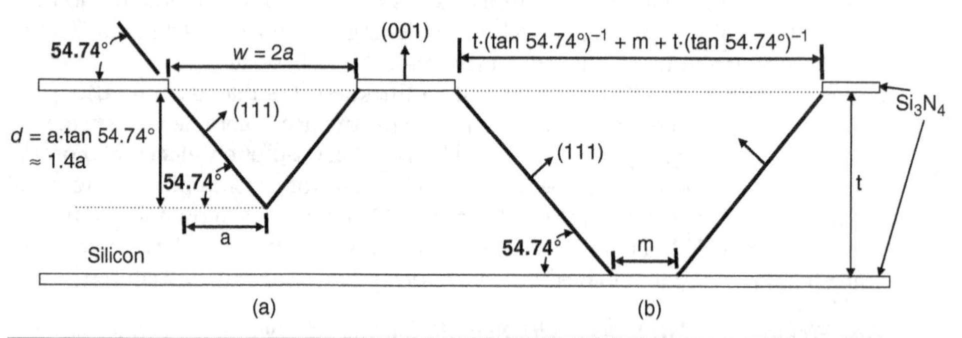

Figure 2.4 Anisotropic etchings of crystalline (100) silicon: (a) When the etch window w is relatively narrow, two {111} planes meet in a concave V-shape, and the etch depth d is virtually limited to $\frac{w}{2}\tan 54.74° \approx 0.7w$, assuming the etch rate for {111} planes is much slower than that for {100} planes. (b) When the etch window is wide enough, the whole silicon is etched from one side to the other through the thickness t. For this, the etch window has to be wider than $\frac{2t}{\tan 54.74°} \approx 1.4t$.

equivalent {111} planes meet at a depth d equal to $\frac{w}{2}\tan 54.74° \approx 0.7w$. Once two {111} planes meet in concave V-shape, the etchant sees two etching limiting {111} planes along those two directions, and consequently, the etch front progresses very slowly at the rate of the {111} etching rate, as illustrated in Fig. 2.4a. The etchings in the directions out of (or into) the paper are another matter and are governed by the size of the etch window along the direction out of (or into) the paper. If the etch-window opening is larger than $\frac{2t}{\tan 54.74°}$, as shown in Fig. 2.4b, the whole silicon wafer is etched from one side to the other through the wafer thickness t.

Some Common Anisotropically Etched Hole Geometries

With a square etch window on {100} Si wafer, anisotropic etching will produce {111} sidewalls as the etching removes Si atoms from {100} plane, as illustrated in Fig. 2.5. For a small square or a rectangular etch window (with the narrower side having a width equal to w), the {111} sidewalls form a V-shape at an etch depth (equal to $0.7w$), as illustrated in Fig. 2.5. The larger square etch window also will form a V-shape, if the etching proceeds long enough and if the wafer thickness is greater than 70% of the etch-window opening. For a {110} Si wafer, some of {111} planes meet with some of {110} planes at 90° angle, rather than 54.74°, and the etch shapes look like as shown in Fig. 2.5b. Thus, vertically oriented sidewalls (of {111} planes) can be obtained on {110} wafer, but two more {111} planes show up at a slanted angle. One can obtain a groove that is bound by four vertically oriented {111} planes that run parallel to each other, but those {111} planes do not meet at 90° angle when viewed from the top.

Anisotropic etching of silicon can easily produce square or rectangular cavity, nozzle, and diaphragm (or membrane made of Si or any other material). If a thin film is used as an etch mask, that thin film can also be a structural layer for bridge or cantilever over a V-shaped trench. One good usage of anisotropic etching has been on fabricating silicon pressure sensors, which have successfully been commercialized since mid-70s, largely for pressure sensing in automotive engine manifolds. Hundreds or thousands of 20–30-μm-thick silicon diaphragms (of 1×1 mm^2 in size) are formed by a single etching

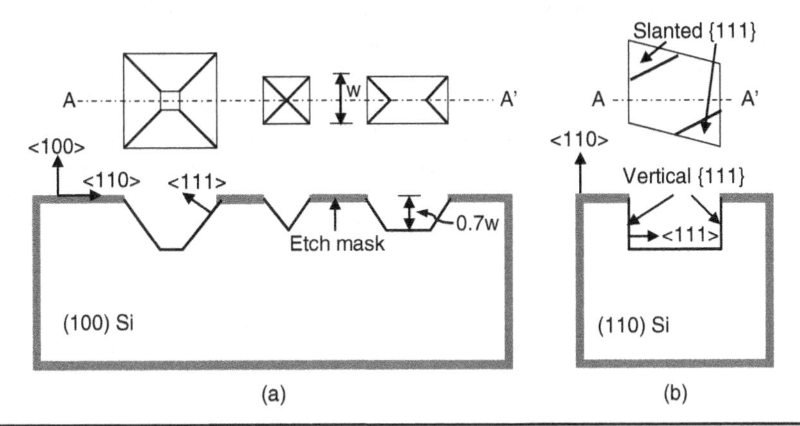

FIGURE 2.5 Top and cross-sectional (across A–A') views of anisotropic etching of silicon: (a) over (100) silicon wafer with various etch windows (Note: the edges of the square or rectangular etch windows are aligned to <110> directions.) and (b) over (110) silicon wafer.

step on a silicon wafer. Piezoresistors formed on the diaphragms by ion implantation of dopants act as the sensing elements since their resistances vary in response to the mechanical strain (in the diaphragm) caused by the diaphragm deflection due to an applied pressure. A circuit can be integrated with the pressure sensor on a single chip, but such monolithic integration does not usually offer economic advantage if the sensor area is large (e.g., 1×1 mm^2) and eats up the area that could have been used for circuits, since the fabrication process for integrated circuits is very complex, and an area on a circuits chip is expensive.

Effect of Boron Concentration on (100) Etch Rate

Heavy boron doping (of about 10^{19} cm^{-3}) reduces the etch rates for {100} or {110} silicon planes by a factor of about 100 and has been used to produce an etch stop layer and/or a diaphragm (or cantilever or bridge). The reduction can be more than 100 times when the doping concentration reaches 10^{20} cm^{-3} and can advantageously be used for various microstructures. However, one must be aware of the tensile residual stress in the boron-doped layer because the boron atoms occupying the substitutional sites of Si are smaller than Si atoms. The tensile stress in a boron-doped layer can cause a warping of a wafer or can cause a breakage of a diaphragm if a diaphragm is formed out of such boron-doped layer. One way to compensate the tensile residual stress in boron-doped silicon is to co-dope the silicon with germanium that is larger than Si. A thin diaphragm made of heavily boron-doped silicon is easily obtainable with anisotropic etching as the etching will stop at doped silicon. However, the maximum thickness of such a diaphragm is typically limited to several microns as it is difficult to obtain a deep depth of heavily doped region through diffusion.

2.1.4 Micromachining in (100) and (110) Silicon Wafers

Square or rectangular etch windows on (100) Si wafer would produce etch shapes as shown in Fig. 2.5a: (1) an inverse pyramidal pit for a small square due to four {111} planes showing up and limiting further etching, (2) a truncated inverse pyramidal pit for a larger square because the etching has not been performed long enough for the all the four {111} planes (that are forming the sidewalls) to meet with each other (or the wafer thickness is not thick enough), and (3) a long V-shaped pit because the two {111} planes along the length direction meet and limit the etching.

On (110) Si wafer one can obtain vertical sidewalls with four {111} planes that are perpendicular to (110) plane. However, the top-view shape will be a non-Manhattan geometry parallelogram (i.e., the intersecting lines meeting at an angle other than 90°), as illustrated in Fig. 2.5b. Also, two slanted {111} planes appear along with the four vertical {111} planes. Etched structures are bounded by four vertical {111} planes and two slanted {111} planes. Two vertical {111} planes intersect at 109.5° with each other.

Question: If (110) silicon is etched with KOH as shown in Fig. 2.5b, which of the following planes can be the vertical and slanted {111} planes: (111), $(\bar{1}11)$, $(1\bar{1}1)$, $(11\bar{1})$, $(\bar{1}\bar{1}1)$, $(1\bar{1}\bar{1})$, $(\bar{1}1\bar{1})$, $(\bar{1}\bar{1}\bar{1})$?

Answer: $(\bar{1}11)$, $(1\bar{1}1)$, $(1\bar{1}\bar{1})$, and $(\bar{1}1\bar{1})$ can be the vertical {111}.
(111), $(11\bar{1})$, $(\bar{1}\bar{1}1)$, and $(\bar{1}\bar{1}\bar{1})$ can be the slanted {111}.

2.1.5 Convex Corner and Beam Undercutting

One very important characteristic to note is that anisotropic etchant etches convex corners very fast even though the corners are formed by two {111} planes as illustrated in Fig. 2.6, since convex corner is prone to expose fast etching planes as a little of the protruding corner is etched. Convex corner behaves quite differently than concave corner though both types of the corners are formed by {111} planes. The inverse pyramidal pit shown in Fig. 2.5 has four concave corners and is very stable in maintaining its shape in the midst of further etching after the corners are formed. However, in Fig. 2.6 the two convex corners are not stable in maintaining their shapes in further anisotropic etching though they are all formed by {111} planes. Since the convex corners are etched very fast (Fig. 2.6), the silicon underneath the "sticking-out" beam is easily etched out, and one can form a suspended beam out of the etch mask.

For any etch window shape, anisotropic etching of (100) Si wafer will eventually produce an etch cavity bound by sidewalls of {111} planes if the etching is done long enough, since {111} planes are etched the slowest by far, while all the other planes (and convex corners) are etched fast. Top-view shape of the etched cavity will approach a rectangle that covers the outermost boundaries of the etch window. These points are illustrated in Fig. 2.7, which shows (a) a typical pyramidal etch pit bounded by (111) planes when (100) silicon is etched through a square etch window with its four edges aligned to <110> directions, (b) a larger and deeper pyramidal etch pit bounded by (111) planes when (100) silicon is etched through a square etch

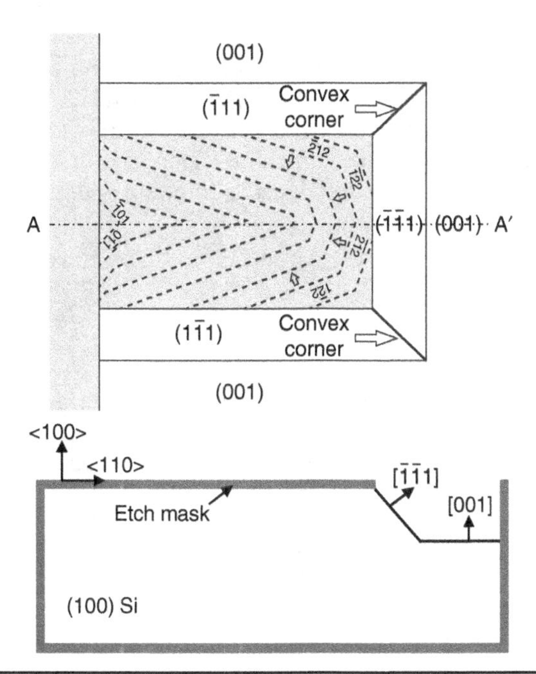

Figure 2.6 Top and cross-sectional (across A–A') views of an etch mask that produces convex corners on a (100) Si, showing how the convex corner is etched fast in Si anisotropic etching, though the convex corner is formed by two {111} planes.

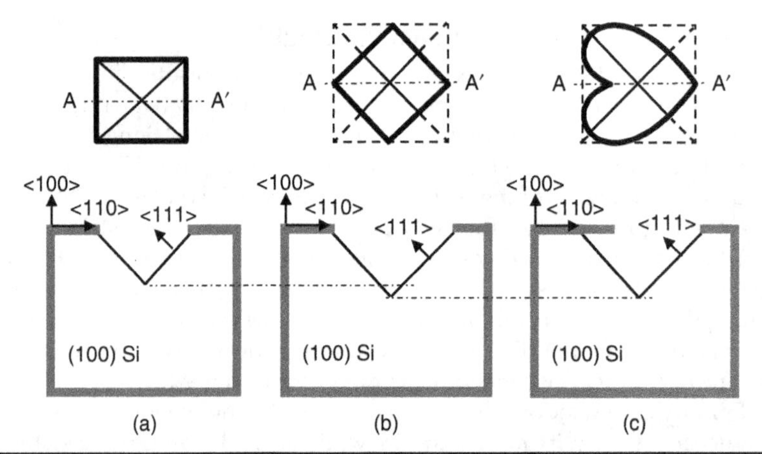

FIGURE 2.7 Etch shapes after sufficiently long anisotropic etching on a (100) silicon wafer with (a) a square etch window with its four edges aligned to <110> directions, (b) a square etch window with its four edges aligned to <100> directions, and (c) a heart-shaped etch window. (Adapted from [2].)

window with its four edges aligned to <100> directions, and (c) an etch pit bounded by (111) planes when (100) silicon is etched through a randomly shaped etch window after a "sufficiently long" etching time to allow all the convex corners to be etched away.

It is instructive to note that the essentially same etch shape appears different depending on how we draw the top views and which A–A′ lines to take for their cross-sectional views, as illustrated in Fig. 2.8. The figure shows the top and cross-sectional views of the etch shapes after sufficiently long anisotropic etching on a (100) silicon wafer with a square etch window with its four edges aligned to <100> directions.

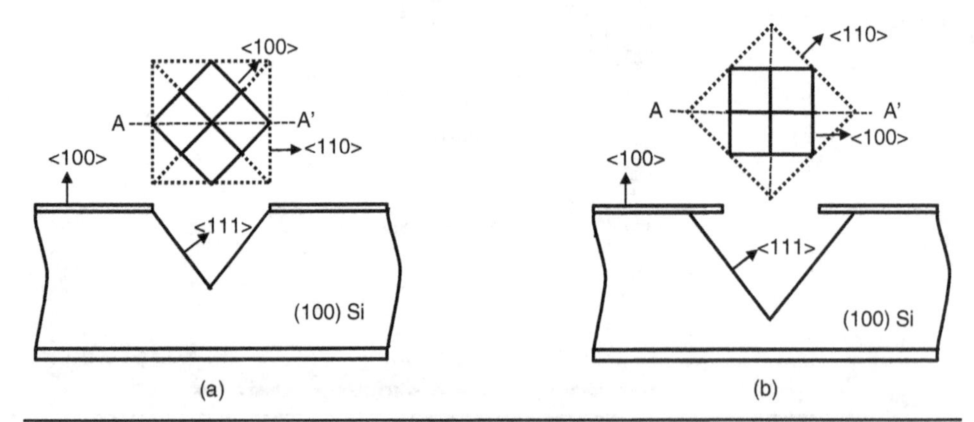

FIGURE 2.8 Top and cross-sectional views of the etch shapes after sufficiently long anisotropic etching on a (100) silicon wafer with a square etch window with its four edges aligned to <100> directions: (a) and (b) show the essentially same etch shapes except that those look different because how the top view is rotated and since the cross-sectional views are across the different A–A' lines.

Exercise Problem On a (100) silicon wafer (400 μm thick) a 1000-Å-thick Si_3N_4 has been deposited and patterned for a $40 \times 20\ \mu m^2$ *open* rectangle and an *open* rectangle plus triangle with 7 μm spacing between the two patterns as shown below. On the figure below sketch the cross-sectional views (across A–A′) of the resulting etch fronts if we etch the wafer in KOH for *2 hours*. Also, calculate the etch depth, undercut length, etc. and indicate those on your sketch. Assume that the KOH etches (100) silicon plane at 1 μm/min while (111) silicon plane is etched at 0.02 μm/min.

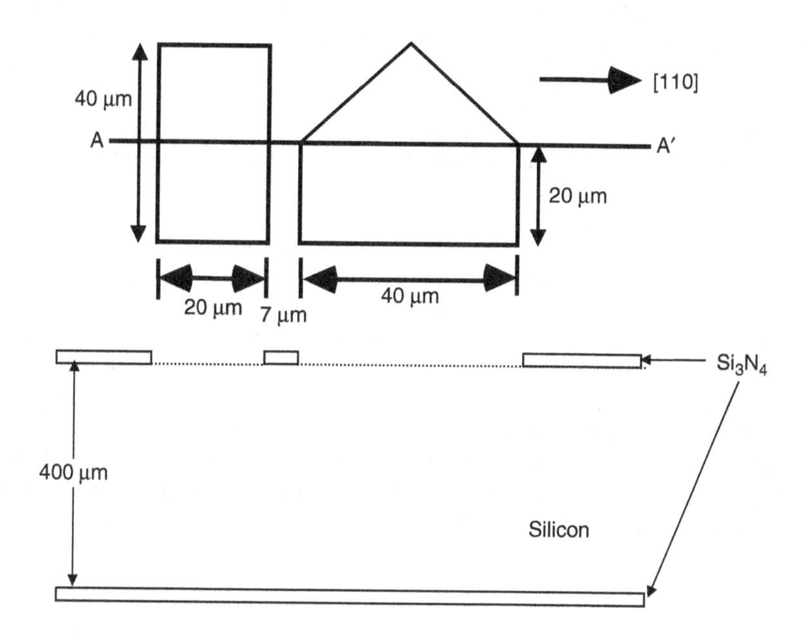

Answer:

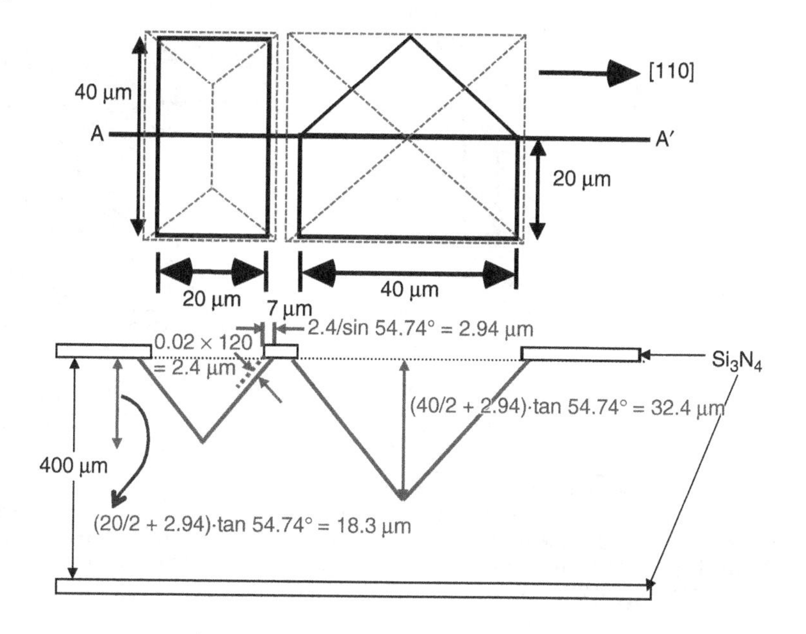

Convex Corner Compensation

A few techniques have been developed to delay convex corner etching. One technique uses legs added to the four corners to delay the appearance of convex corners in anisotropic etching of (100) silicon [5]. This technique needs a precise timing on when the anisotropic etching is stopped in order to avoid any of the convex corner etching. Another technique, called "street" corner compensation, also is based on time delay but can yield a mesa structure (a truncated pyramid) with sharp convex corners if the etching is stopped just at a right time (Fig. 2.9). This method produced the desire shape with KOH, but not with EPW, because EPW etches {110} planes twice slower than {100} and produces inverse V-shaped silicon at the four corners, which will become convex corners before the compensation "streets" are completely removed.

Convex Corner Protection

Anisotropic etching of Si can be used to produce long and narrow microfluidic channels as well as large cavities for liquid reservoirs. In forming and routing microchannels on a chip two or more microchannels often cross each other resulting in convex corners, as illustrated in Fig. 2.10. The convex corners are etched fast by anisotropic etchant and need to be protected, if one wants to preserve the crossing points sharp and not enlarged [7].

As shown in Fig. 2.11, a convex corner can be protected by first etching a V-shaped channel and then performing the following steps in sequence: deposit an etch-masking film (e.g., silicon nitride) all over the surface conformally, pattern the etch mask for the crossing channel, and then etch a crossing channel. The idea is that the etch-masking

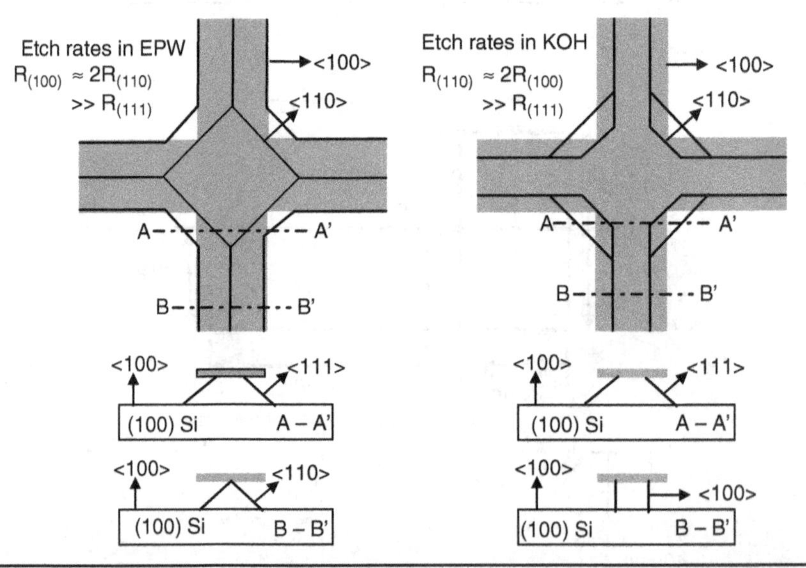

FIGURE 2.9 "Street" corner compensation by adding four "streets" at the corners to delay the appearance of convex corners; with the same mask pattern, the etching results are different for EPW and KOH because the ratio of the etching rates for {100} and {110} planes is different for EPW and KOH, though both etchants etch {111} planes much slower than any other crystal planes of silicon. (Adapted from [6].)

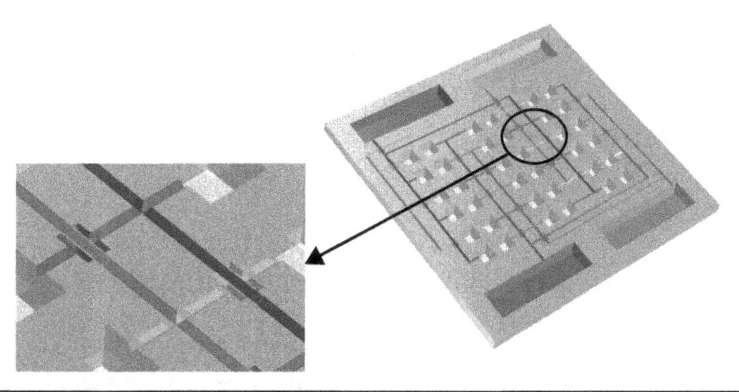

Figure 2.10 Microfluidic channel routing: 6 × 6 transducers or reservoirs with embedded channel routing system. (Adapted from [7].)

Figure 2.11 Channels with protected convex corners: (a) a second mask pattern on the {111} planes after the first anisotropic etching on a {100} wafer, (b) photo of protected convex corners, and (c) photo of protected convex corners with variable channel depth. (Adapted from [7].)

film protects one {111} plane of the two {111} planes forming the convex corner and basically presents no convex corner (formed by two {111} planes) to the etchant. This method, though, requires a spray coating of photoresist to cover over the sloped {111} plane (i.e., a sidewall of the first etched V-groove). The patterning of the etch-masking film over the sidewall allows one to obtain various etch depths on the sidewall, as shown by the two photos in Fig. 2.11.

Question: In forming an array of bulk-micromachined diaphragms shown in Fig. 2.10, what would be the minimum chip size if we want to create **four** 100 × 100 μm² square diaphragms on one side of the wafer (i.e., a 2 × 2 array)? Assume that the silicon wafer is 500 μm thick and is etched with KOH that etches (100) silicon plane at 1 μm/min, while (111) silicon plane is etched at a negligible rate.

Answer: $w = 2 \times (2d/\tan 54.74° + 100\ \mu m) = 2 \times (2 \times 500\ \mu m/1.4 + 100\ \mu m) = 1{,}629\ \mu m$. Thus, the minimum chip size will be 1.6 × 1.6 mm².

Usage of Convex Corners for Obtaining Different Etch Depths with One Mask and One Etching

Fast etching of convex corners formed by {111} planes can be used to produce different etch depths with one photomask and one etching step. For example, a photomask with one large square opening (size S_{1b}) and one mesh pattern (size S_{2b}) can be used to produce etch-depth profiles shown in Fig. 2.12 through one photolithography and one anisotropic etching. After sufficiently long enough etching, the mesh pattern produces pyramidal etch pits bound by {111} planes due to the small square openings (size S_m), while the large square opening (size S_{1b}) still has (100) plane to be etched, as illustrated in Fig. 2.12a and Fig. 2.13. Then as the {111} planes are etched very slowly, the (100) plane in the large square opening is etched at a much faster rate.

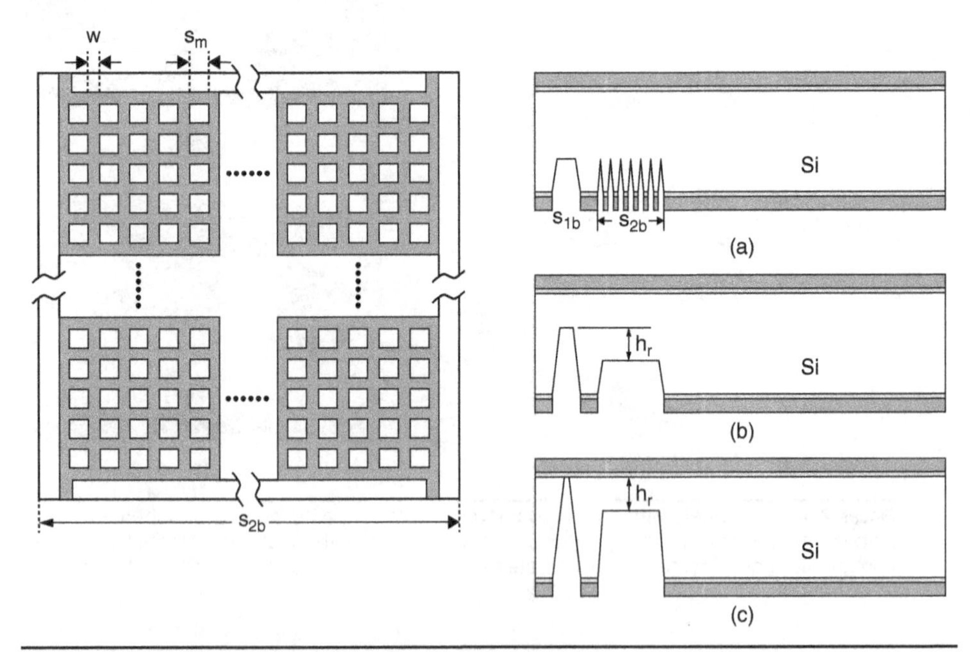

FIGURE 2.12 *Left:* Dimensions of the mesh pattern used to delay the etch rate. The pattern consists of squares having a side s_m in a mesh separated by a web of width *w*. *Right:* Cross-sections showing the etch-retardation scheme: (a) when {111} pyramidal pits make up all profiles in the mask area, (b) when all {111} planes have been removed after the {111} planes meet under the web to form convex corners, and (c) when a small front-surface diaphragm has appeared. (Adapted from [8].)

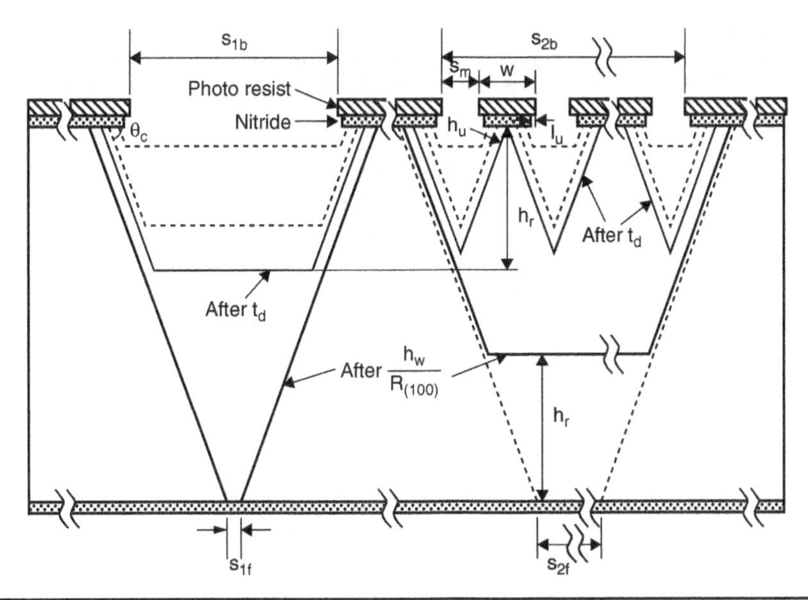

FIGURE 2.13 Cross-section showing etch fronts at varying times as anisotropic etching of silicon proceeds; the dimensional labels match those in Fig. 2.12. (Adapted from [8].)

At some point (the time of which is determined by the web width w), the {111} planes meet under the web width w and form convex corners (Fig. 2.13), which are etched fast. When the convex corners are etched, the large square area also has (100) plane to be etched (Fig. 2.12b), and the two different etch depths are achieved by one photolithography and one anisotropic etching. A further etching produces a tiny (e.g., 30 μm square and 1 μm thick) diaphragm (which can serve as a front-to-back alignment pattern) and, at the same time, a much thicker and larger diaphragm (made of the silicon that is partially etched with the "mesh-masking" pattern), as shown Fig. 2.12c and Fig. 2.13. The mesh-masking technique exploits the etch-rate difference between {100} and {111} planes to control the depths reached by etch pits in selected areas.

Questions: Answer the following for a 500-μm-thick silicon wafer that is etched with KOH that etches (100) silicon plane at 1 μm/min, while (111) silicon plane is etched at a negligible rate.

(a) What will be the size of the diaphragm on the front side of the silicon, if the size of the square pattern opened on the backside is 715 (±5) μm by 715 (±5) μm?

(b) If we use SiO_2 as an etch mask for the KOH etching (as well for the front-side diaphragm), what is the absolute minimum thickness of SiO_2 that is needed, assuming that KOH etches SiO_2 at 14 Å/min.

(c) If we over-etch the silicon 15% more than the minimum needed time for the small square diaphragm on the front side of the silicon in (a), how larger will the size of the square diaphragm turn out to be? For this question, assume that the etch rate for (111) plane is 100 times slower than that for (100) plane. Make sure that you convert the distance progressed in the [111] direction into the distance along the diaphragm dimension.

Answers:

(a) From the figure below, the opening on the front side is (715 ± 5 μm) – 2 × 353.5 μm = 8 ± 5 μm. → The diaphragm will be a square diaphragm of 8 ± 5 μm by 8 ± 5 μm, as illustrated below.

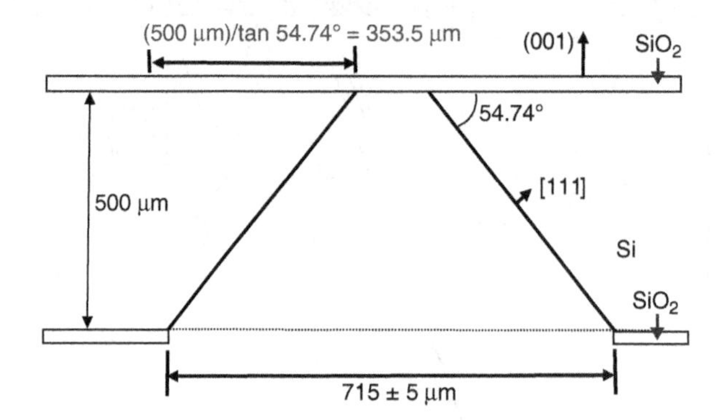

(b) Since KOH etches SiO_2 at 14 Å/min, the SiO_2 has to be at least 14 Å × 500 = 7,000 Å thick to withstand 500 minutes in KOH during the etching of 500-μm-thick silicon (at 1 μm/min etching rate).

(c) 15% more → 0.15 × 500 minutes = 75 minutes more → (75 min)(0.01 μm/min) = 0.75 μm more into [111] direction → (0.75 μm)/sin(54.74°) = 0.92 μm → 2 × 0.92 μm = 1.84 μm larger on one side of the square diaphragm, as illustrated below.

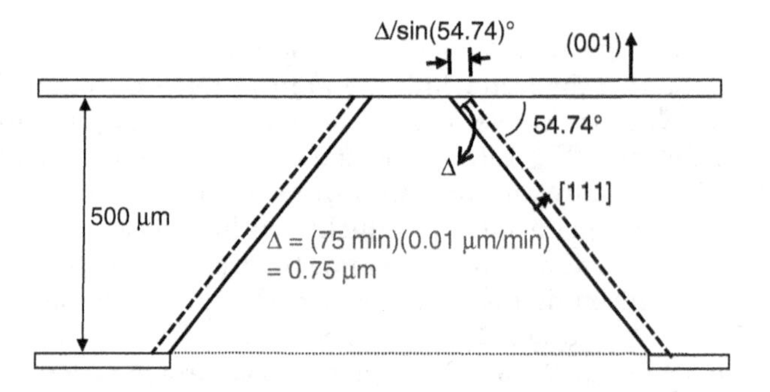

Etching from Two Sides

If anisotropic etching with KOH is done simultaneously from the front side and the backside on a (100) silicon wafer with an etch mask with openings (on both the front side and backside) indicated by solid lines in the top-view schematics in Fig. 2.14a [9], the cross-sectional views (across A–A′) of the resulting etch fronts will look like as shown in Fig. 2.14, depending on how long is the etching time. The shape shown in Fig. 2.14a (obtained by etching for just right amount of time) will become the shape shown in Fig. 2.14b if the etching continues, since the convex corners are quickly etched away.

2.1.6 Front-to-Backside Alignment

The wafer shown in Fig. 2.12c has a thin diaphragm (30 × 30 μm²) on the front side, which can be used for aligning front-side patterns to the patterns on the backside.

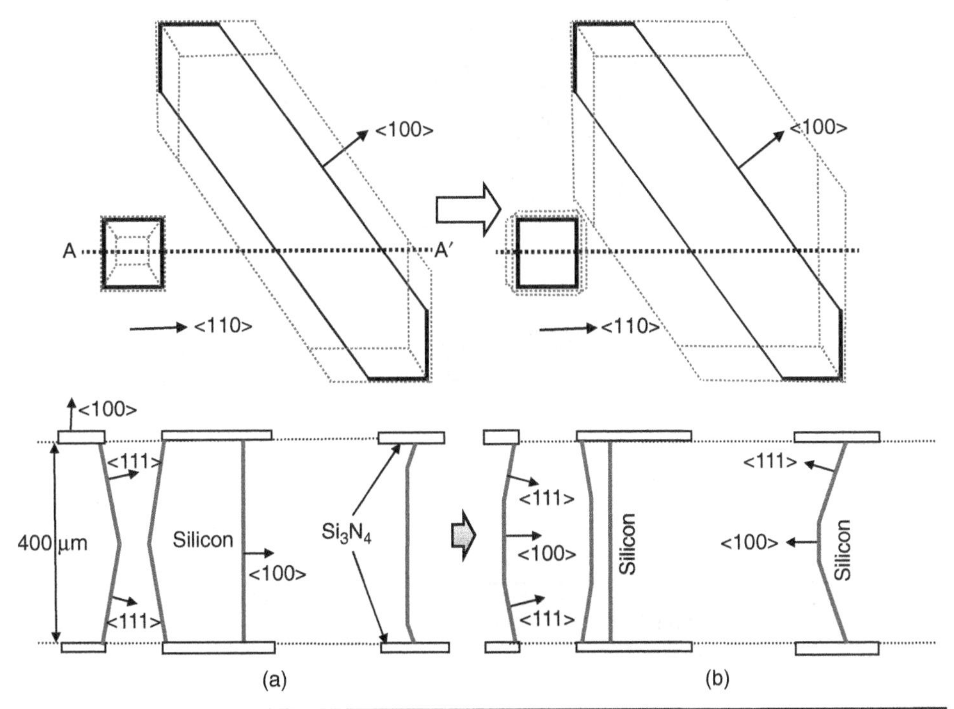

Figure 2.14 Top and cross-sectional views of double-side KOH etching on a (100) wafer with the openings (indicated by solid lines) aligned to the crystal planes as shown on the top-view schematics: (a) when the etching time is just little over the time needed to etch the half of the wafer thickness, that is, 200 μm/$R_{(100)}$ with $R_{(100)}$ being the etch rate of (100) plane and (b) when the etching is continued further after getting to the point shown in (a), as the convex corners are quickly etched away.

The wafer contains also a large 100-μm-thick diaphragm (2 × 2 mm^2) or many such diaphragms on which a pressure sensor or microphone can be built with CMOS circuits monolithically integrated on a single chip, as illustrated in Fig. 2.15 [8]. The 100-μm-thick silicon on the large diaphragm can be removed after completing the CMOS process on the front side for a thin diaphragm for a piezoelectric pressure sensor or microphone.

A tiny and thin diaphragm formed on the front side of a wafer can easily be seen from the front side along with the photomask patterns to be transferred to the front-side surface of the wafer. However, the diaphragm will appear very blurry from the backside in a mask aligner when the mask aligner is focused on the backside surface (for pattern transfer from a photomask to the backside). Thus, it may be necessary to form an extra alignment mark on the backside at the same time when the alignment diaphragm is formed. The extra alignment mark can easily be obtained by opening a very small square window (along with the window for the alignment diaphragm) that will produce an inverse pyramidal pit (bound by four {111} planes) after tens of minutes of anisotropic etching of Si. This concept for front-to-backside alignment is illustrated in Fig. 2.16.

Front-to-backside alignment can also be obtained with (1) an infrared aligner that allows a pattern on one side of a wafer to be transmitted through a wafer so that the

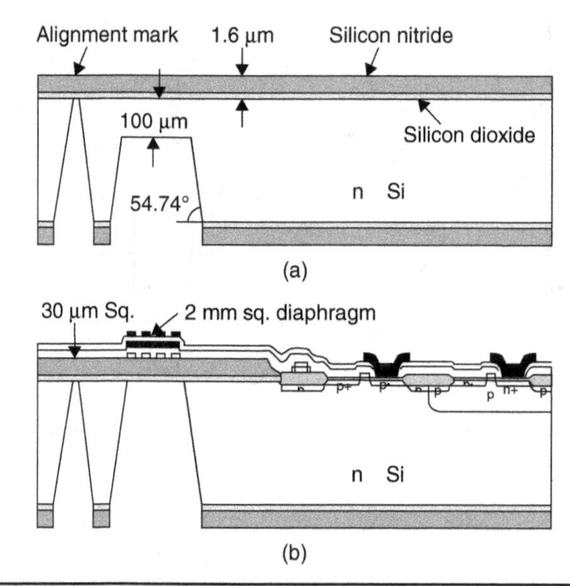

FIGURE 2.15 (a) Cross-sectional sketch showing anisotropically etched diaphragms in a (100) wafer, one for front-to-backside alignment and the other for a microphone, during CMOS processing. (b) Cross-sectional sketch of the completed wafer. (Adapted from [8].)

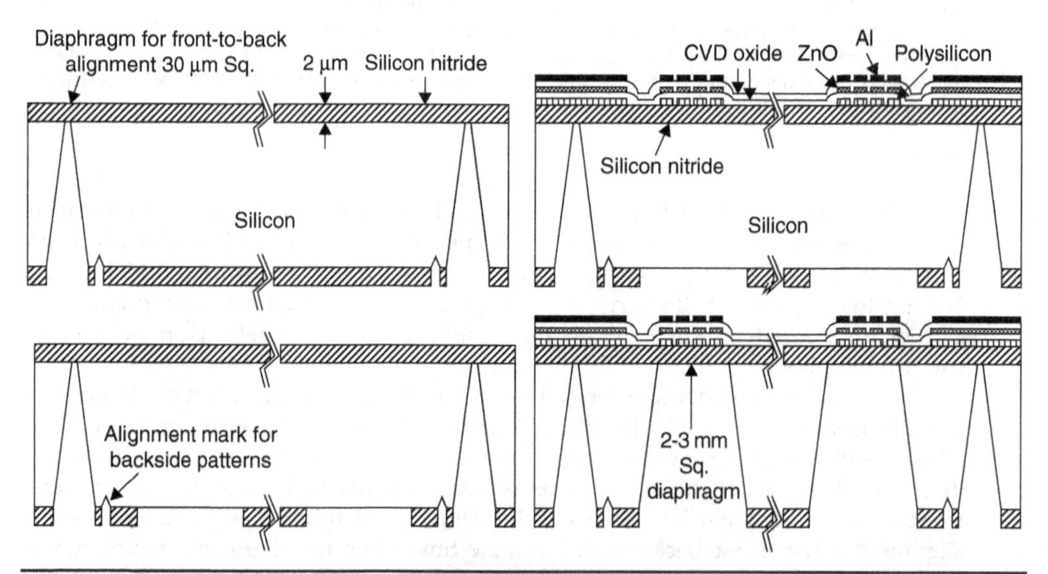

FIGURE 2.16 Cross-sectional sketches showing anisotropically etched diaphragms (on the front side) and pyramidal etch pits (on the backside) in a (100) wafer for front-to-backside alignment.

pattern may be viewed from the other side of the wafer for alignment (but infrared transmission through silicon wafer thicker than 300 μm may be poor), (2) an advanced mask aligner capable of showing both sides of a wafer so that a mask for one side (e.g., front side) may be aligned accurately with respect to pattern(s) on the other side (e.g., backside), or (3) a double-sided mask aligner that aligns two masks with respect to each other accurately before a wafer is inserted between the two masks and then transfers patterns on both the masks onto the front and backside of a wafer that is inserted between the two masks. If such an aligner is not available, one can use a mechanical jig to hold two masks and a wafer between the two masks, with the two masks and wafer aligned to three cylindrical posts forming a 90° angle [10], which may not give an alignment accuracy better than 25 μm. Or, one can form an alignment diaphragm at the beginning of the process to align features on the backside to those on the front side, as explained in the previous paragraphs.

2.1.7 Alignment of Pattern to Crystallographic Axes

Silicon wafer manufacturers indicate the crystal planes of a silicon wafer by the location and number of the flat cuts around the wafer's circular edge. However, the cut location may not be accurate; for instance, for a (100) wafer with a major cut made perpendicular to <110> direction, the alignment error can be 0.5°. Additionally, there is some misalignment to the wafer flat in a mask aligner. Thus, if an accurate alignment to crystal planes is needed, exact crystal planes can experimentally be determined by anisotropically etching grooves with patterns shown in Fig. 2.17.

2.1.8 Isotropic Etching of Silicon for Large Spherical Etch Cavity

For some MEMS (e.g., loudspeakers, rate-integrating gyroscopes, etc.), large spherical etch cavities may be needed and can be micromachined through isotropic etching of silicon with HNA (hydrofluoric, nitric, and acetic acids). One key challenge in producing a large etch cavity is in finding a good etch mask for HNA etching. Some potential etch-masking layers for HNA isotropic etching of Si are listed in Fig. 2.18a. The etch mask's adhesion and durability during the etching determines the maximum attainable etch-cavity size. Polyethylene-backing acrylic adhesive tape, though a bit too thick (75 μm), provides the best adhesion and durability and was used to produce the large etch cavity shown in Fig. 2.18b.

To produce a spherical etch front with a radius curvature of 2,640 μm on a 1,600-μm-thick Si substrate, two-step etching shown in Fig. 2.19a was used [12]. In the first step of using a 75-μm-thick tape during isotropic etching of silicon with HNA, the

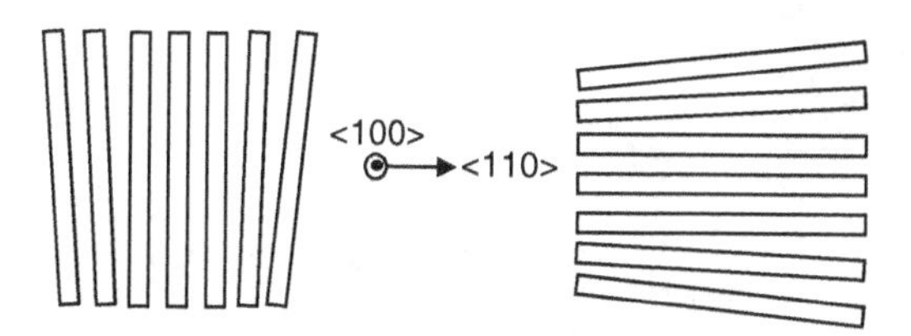

Figure 2.17 Mask patterns of rectangular windows on a (100) Si wafer for identifying <110> directions. (Adapted from [11].)

Etch mask materials	Durability	Adhesion	Attainable cavity radius	Remarks
SiN (2 μm thick)	Etched at 0.045 μm/min	No peeling off	0.5 mm	IC compatible, Pin hole problem
A174/ parylene or Cr/Au	Hardly etched	Peeling off after 20–30 min	0.5 mm	IC compatible
Tape (polyethylene backing acrylic adhesive)	Hardly etched	Rarely peeling off	> 2.5 mm	Too thick (75 μm), 5 hr RIE etching for patterning

(a)

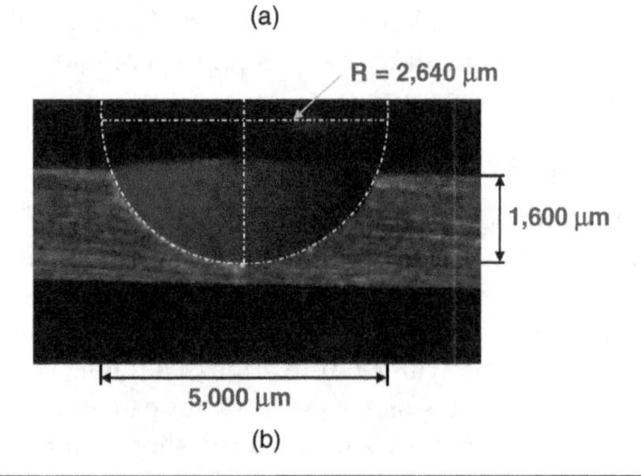

(b)

Figure 2.18 (a) Etch mask materials for HNA isotropic etching of silicon for a large spherical etch cavity. (b) Cross-sectional scanning electron microscopy (SEM) photo of a silicon with a large isotropically etched spherical etch cavity. (Adapted from [12].)

Isotropic etching	Mask layer	Etchant ($HF:HNO_3:CH_3COOH$)	Temp.	Etching time	Agitation
1st	75-μm-thick tape	2:3:3	50°C	self-etching stop	No
2nd	1-μm-thick SiN	1:4:3	20°C	30 min	No

(a)

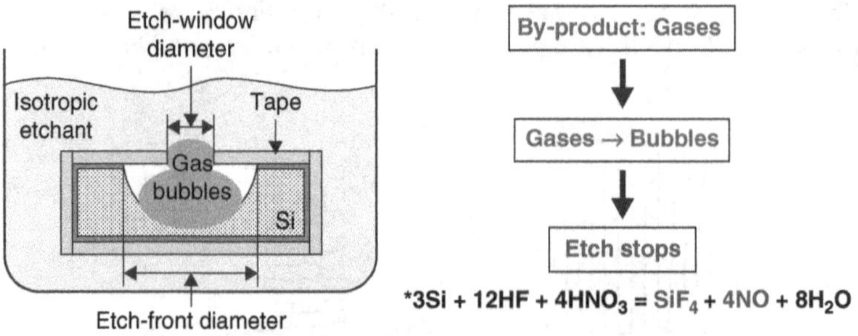

Etch-window diameter

Isotropic etchant

Tape

Gas bubbles

Si

Etch-front diameter

By-product: Gases

Gases → Bubbles

Etch stops

$$^{*}3Si + 12HF + 4HNO_3 = SiF_4 + 4NO + 8H_2O$$

(b)

Figure 2.19 (a) Two-step etching process with a 75-μm-thick tape as an etch mask for HNA etching of silicon. (b) Self-etching stop due to hydrogen gas (an etch by-product) filling up the etch cavity (under the etch-mask tape) and limiting mass transport of HNA etchant and etch by-products. (Adapted from [12].)

etching stops when gas by-product fills up the etched cavity and blocks HNA from reaching the silicon surface, as illustrated in Fig. 2.19b.

The measured etching rate versus the etching time as a function of the etch-window diameter is not uniform in time and becomes zero after a certain time, as shown in Fig. 2.20a. Since the etching stop happens at a later etching time for a larger etch-window diameter, the etched cavity size is determined by the etch-window diameter, as can be seen in Fig. 2.20b. The etch-front diameter is about 1.8 times the etch window size plus 0.4 mm, according to Fig. 2.20b, and is independent of the etching time as long as it is longer than the etch-stop time.

Using isotropic etching of silicon, one can fabricate dome-shaped diaphragm transducers such as the one shown in Fig. 2.21 for microspeaker, hemispherical gyroscope, etc. Its fabrication steps are briefly illustrated in Fig. 2.22.

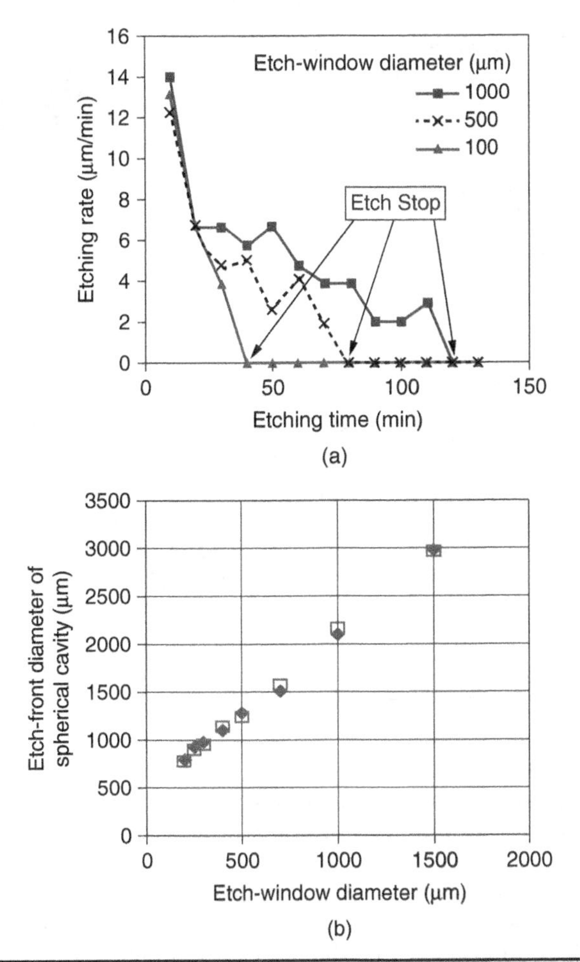

FIGURE 2.20 (a) Measured etching rate versus etching time as a function of etch-window diameter during HNA with a tape as an etch mask. (b) Measured etch-front diameter of spherical etch cavity versus etch-window diameter. The self-etching stop illustrated in Fig. 2.19b makes the etch-cavity size be mainly determined by the etch-window size. (Adapted from [12].)

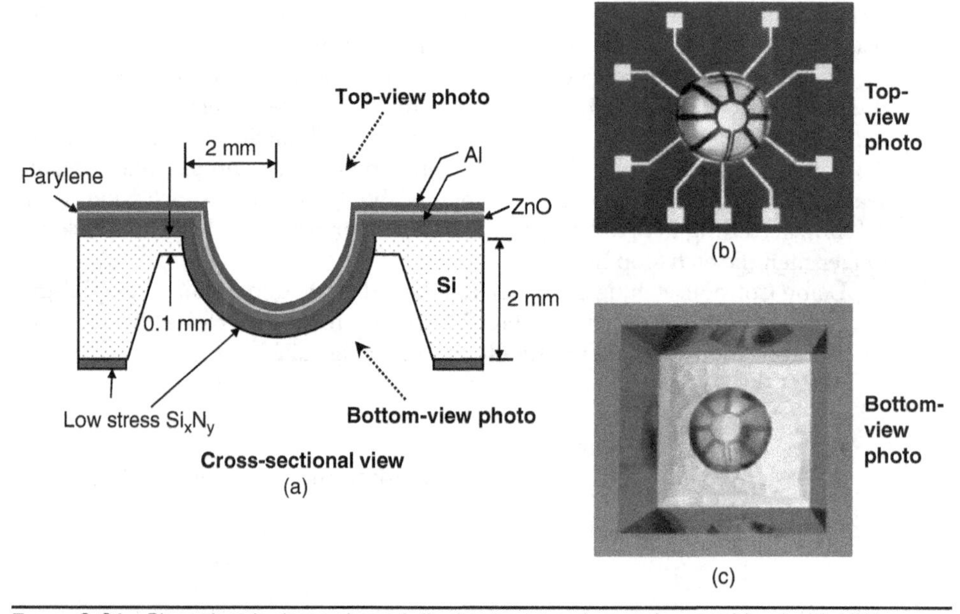

FIGURE 2.21 Piezoelectric dome-shaped diaphragm transducer micromachined with HNA etching via a 75-μm-thick tape as an etch mask: (a) cross-sectional view, (b) top-view photo, and (c) bottom-view photo. (Adapted from [13].)

2.1.9 Etching Apparatus

Since Si bulk micromachining usually requires a long etching time (often several hours to etch through the whole wafer thickness) at an elevated temperature (e.g., 80–120°C), the etching apparatus needs to have a good seal to minimize loss of the etchant and its water through evaporation. Thus, a reflux jar with a cooled sidewall and a top cap such as the one shown in Fig. 2.23 is desirable. (Note: as of this writing, there appears to be no commercial-off-the-self water-jacketed quartz beaker with no air pocket at the bottom, and one will have to custom-make the beaker in a quartz shop.) Stirring of the etching solution with a magnetic stirrer would enhance the mass transport of the etchant and etch by-products. In some elaborate setup, a motor may be used to produce up-and-down motions of wafers in etchant.

Wafer Holder for One-Side Etching

Protecting one side of a wafer while the other side is micromachined is often challenging due to a long etching time in an elevated temperature in case of silicon bulk micromachining. Parylene film is hardly etched by most of Si etchant, but tends to peel off within minutes in a Si etchant. Low-pressure chemical vapor deposited silicon nitride would provide a good protection, but is hard to be removed cleanly (after bulk micromachining) without damaging the underlying layers that it has protected during the micromachining. Thus, a mechanical jig such as the one shown in Fig. 2.24 may be very helpful in protecting one side while the other side is etched in a wet etchant, especially if the side one wants to protect has many sensitive layers (e.g., Al, polysilicon, etc.) [14].

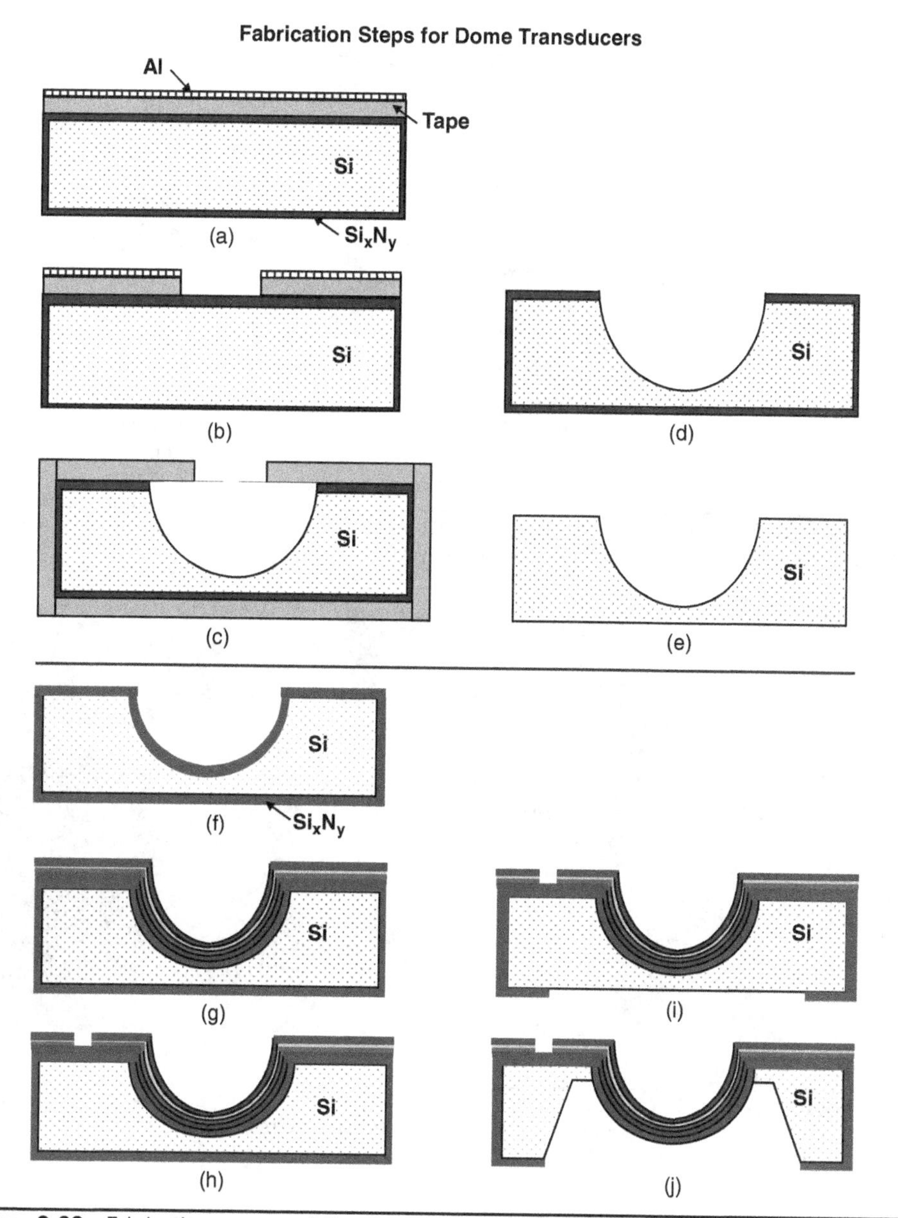

FIGURE 2.22 Fabrication process of the piezoelectric transducer shown in Fig. 2.21: (a)–(e) HNA etching of silicon with a tape as an etch mask, (f)–(j) thin-film depositions and delineations followed by releasing of the piezoelectric transducer. (Adapted from [13].)

Bulk micromachining is often performed near the end of a fabrication process, mainly because a micromachining typically produces movable or deflectable structures, and the wafer becomes difficult to be processed with such movable or deflectable structures. However, if it is possible to do bulk micromachining at an early stage of a fabrication process, then protecting one side while the other side is being micromachined would not be needed.

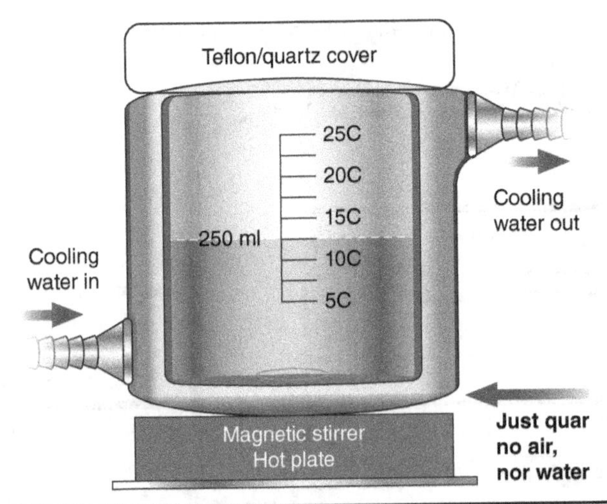

FIGURE 2.23 Schematic of an etching apparatus for bulk micromachining with cooling water jacketed around the sidewall of a cylindrical quartz beaker.

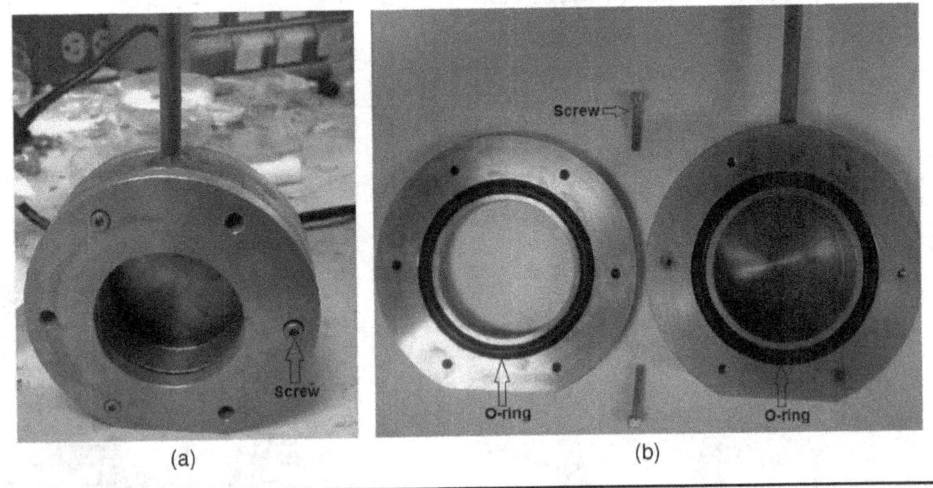

FIGURE 2.24 Photos of a mechanical jig for holding a wafer during one-side etching: (a) the whole jig (before any wafer is inserted) and (b) the two major pieces of the jig showing the O-rings that will be in contact with a wafer to be etched.

2.2 Surface Micromachining

Surface micromachining usually involves depositing and patterning a spacer (or sacrificial) layer, depositing and patterning a structure layer, and then removing the spacer layer to release a movable structure, as shown in Fig. 2.25. Even with the relatively simple steps with two photomasks, the released structure (the cross-sectional view of which is shown in Fig. 2.25d) is a movable bridge (with its two edges anchored) and two movable cantilevers with air gaps between the structures and silicon substrate.

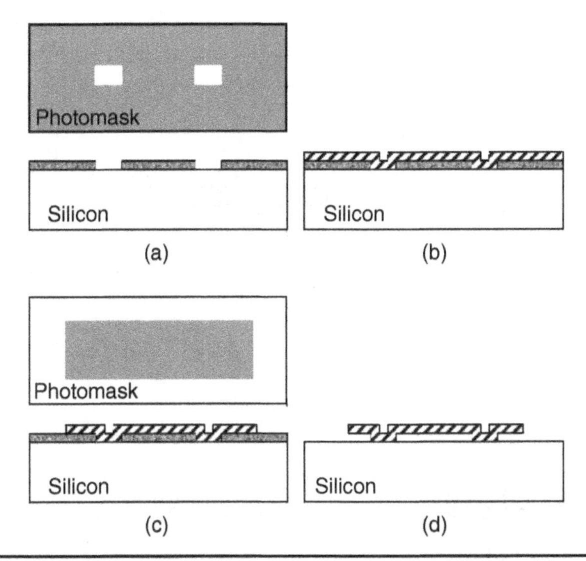

FIGURE 2.25 Basic surface micromachining process sequence: (a) deposition of a sacrificial layer followed by patterning of the layer after photolithography with a photomask, (b) deposition of a structural layer, (c) patterning of the structural layer with another photomask, and (d) removal of the sacrificial layer to release moveable structures.

In surface micromachining, while patterning a sacrificial layer (Fig. 2.25a), the Si substrate may get damaged by the etchant for the sacrificial layer, and a buffer or isolation layer may be needed. Also, a sacrificial layer is usually a few microns thick, and its patterned steps may not be easily covered by a thin structure layer. Evaporated film does not cover a step very well, since evaporation is typically done at microTorr and produces directional flux of evaporated molecules (see Fig. 1.9). Sputtered film covers a step better than evaporated film since sputtering is performed at milliTorr range, and sputtered molecules arrive at the deposition sites after many collisions with gas molecules in the sputtering system (thus, there is less directional energy in the sputtered molecules than in the evaporated molecules by the time they reach the deposition sites or the wafer). Chemical vapor deposition (CVD) (or low-pressure CVD [LPCVD]), on the other hand, offers a very conformal deposition of thin film and was a key technique in obtaining the first micromotor, as we will see in the next section. CVD not only deposits film over a step very conformally but also deposits film underneath an overhanging structure. Dielectrics (e.g., silicon dioxide, silicon nitride, etc.) and polysilicon or amorphous silicon are typically deposited with CVD. Metals, though, are usually deposited by sputtering or evaporation.

2.2.1 Double-Polysilicon Micromechanical Pin-joint Structures

Following a fabrication process shown in Fig. 2.26 with three simple photomasks, one can produce a movable pin-joint structure after removing the sacrificial layers (first and second PSG) [15]. After removing the sacrificial layer, the first polysilicon is completely released and free from the substrate. What is preventing the released polysilicon from

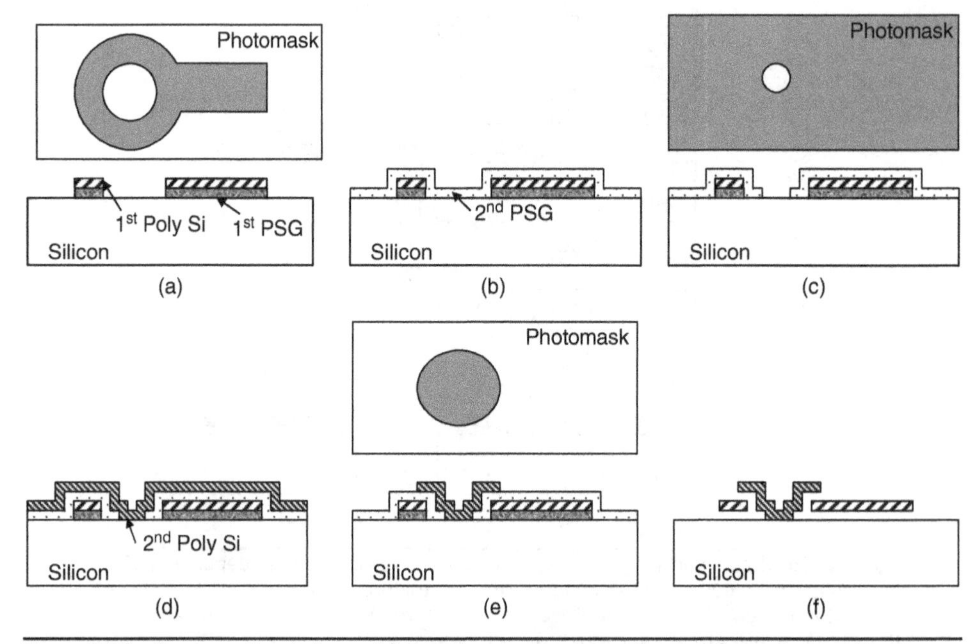

FIGURE 2.26 Cross-sectional views of fabrication process steps for a surface-micromachined movable pin around a circular hub: (a) after depositing and patterning first PSG and first polysilicon, (b) after depositing a second PSG, (c) after patterning the second PSG, (d) after depositing the second polysilicon, (e) after patterning the second polysilicon, and (f) after removing first and second PSG. Also shown are the three photomasks used in the patterning steps.

leaving the substrate is the second polysilicon that is anchored to the substrate and acts as a hub.

A little different pin-joint structure [16] can be fabricated with the process steps shown in Fig. 2.27. In this case, the first polysilicon is anchored to the substrate while the second polysilicon is completely released from the substrate. What is preventing the released polysilicon from leaving the substrate is the overhanging first polysilicon that provides interlocking mechanism with the protruding piece of the second polysilicon. The protruding piece is fabricated by using the fact that LPCVD provides extremely conformal deposition of PSG and polysilicon. The conformal deposition allows the second PSG to be deposited all over the steps as shown in the bottom drawing in Fig. 2.27c. Moreover, the conformal deposition is the key to obtaining the laterally protruding second polysilicon structure, which requires conformal deposition of polysilicon on the surfaces facing both upward and downward.

Based on the fabrication steps described in the previous paragraphs, an electrostatically actuated micromotor was demonstrated in late 80s [17]. As shown in Fig. 2.28, a circular hub is anchored to the substrate and restrains the movement of a completely released cross-bar rotor. An electrostatic voltage applied between the rotor and a nearby stator attracts the rotor to the stator, but due to the hub's restraint, the rotor can only rotate around the hub when actuated by the electrostatic force. The fabrication requires conformal deposition of the second PSG and the third polysilicon (by LPCVD) after etching the first PSG after Fig. 2.28b.

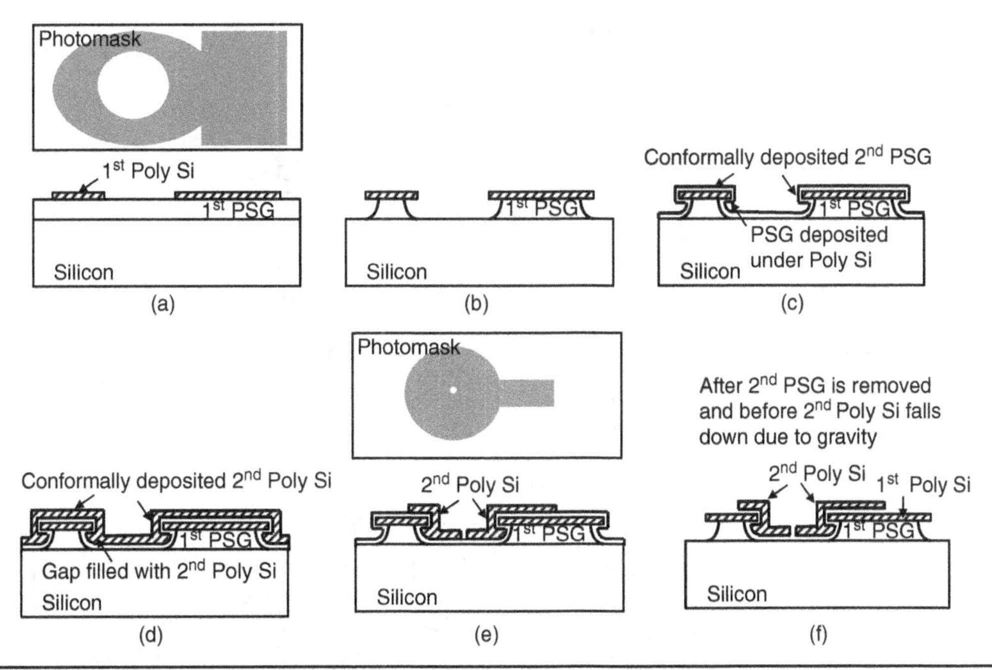

FIGURE 2.27 Cross-sectional views of the fabrication process steps for a surface-micromachined movable pin constrained by a laterally protruding hub: (a) after depositing the first PSG and first polysilicon and then patterning first polysilicon, (b) after etching the first PSG, (c) after depositing the second PSG, (d) after depositing the second polysilicon, (e) after patterning the second polysilicon, and (f) after removing the second PSG. Also shown are the two photomasks used in the patterning steps.

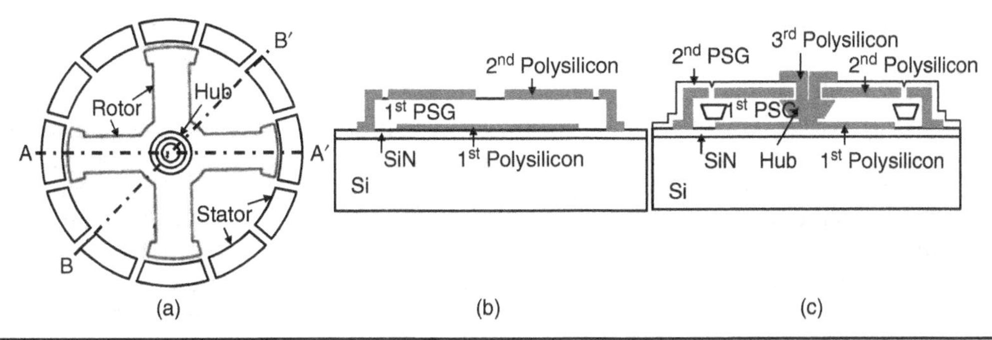

FIGURE 2.28 Top-view schematic (a) of a surface-micromachined electrostatic motor (b) and (c) cross-sectional views (across A–A') of the motor during the fabrication with (b) being earlier than (c). (Adapted from [17].)

Exercise: Draw the B–B′ cross-sectional view of the micromotor in Fig. 2.28a.

Answer:

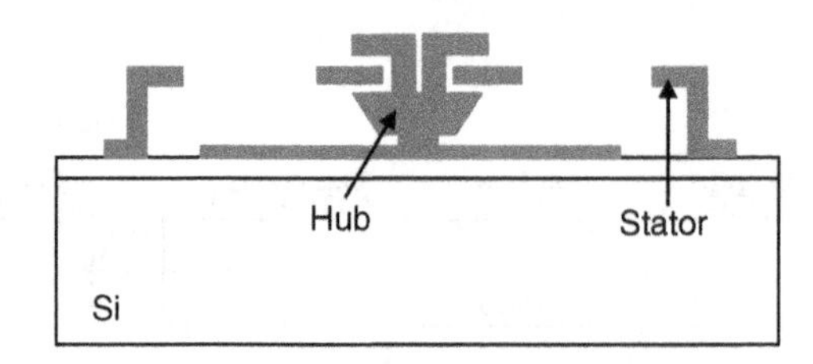

Exercise: Draw the missing steps between the two steps [(b) and (c)] shown Fig. 2.28. You should have four figures: one after etching the first PSG, another after depositing thin PSG, another after patterning the thin PSG for the hub to be electrically connected to the ground plane, and then one after deposition of the third polysilicon.

Answer:

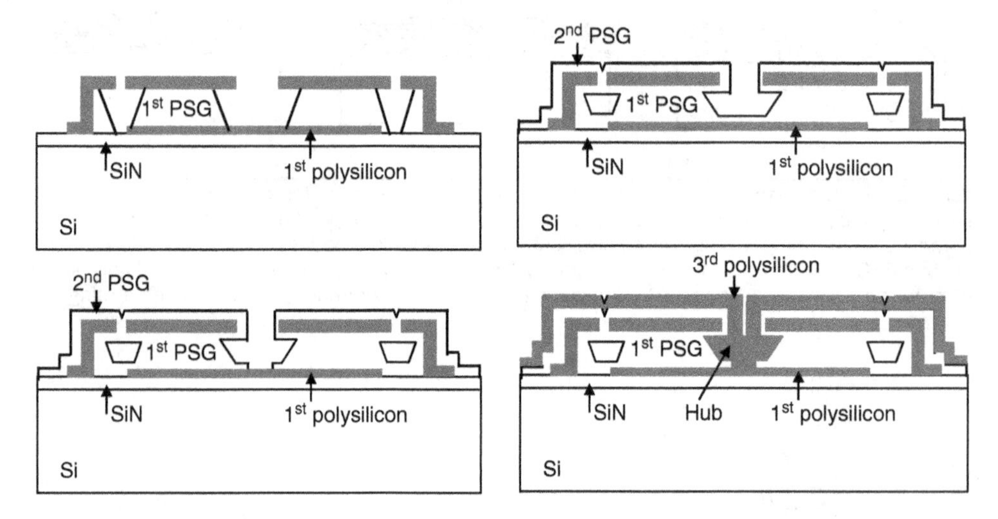

2.2.2 Hinged Plates

A hinged structure can be fabricated with the steps illustrated in Fig. 2.29, which involves depositing a sacrificial layer (e.g., PSG 1) and structure layer (e.g., Poly Si 1), patterning the structure (Poly Si 1), depositing a second sacrificial layer (e.g., PSG 2), patterning the two sacrificial layers for the anchoring of a hinge layer (e.g., Poly Si 2), depositing and patterning of the hinge layer (Poly Si 2), and then finally removing all the sacrificial layers (PSG layers) to release the structure layer (Poly Si 1). The released structure layer (Poly Si 1) can be lifted up manually with a probe tip and is restrained by the hinge structure formed by Poly Si 2.

Hinged plates shown here are all thin plates that are lifted up from a substrate, and can be folded with some kind of locking mechanisms, quite similar to origami. Since the plates are all thin, the structural stiffness can be a concern. Thus, a hollow triangular beam was fabricated by folding two of three hinged plates and shown to be able to withstand relatively large load with a spring constant, k = 1,368 N/m, five orders of magnitude higher than that of a flat beam [18].

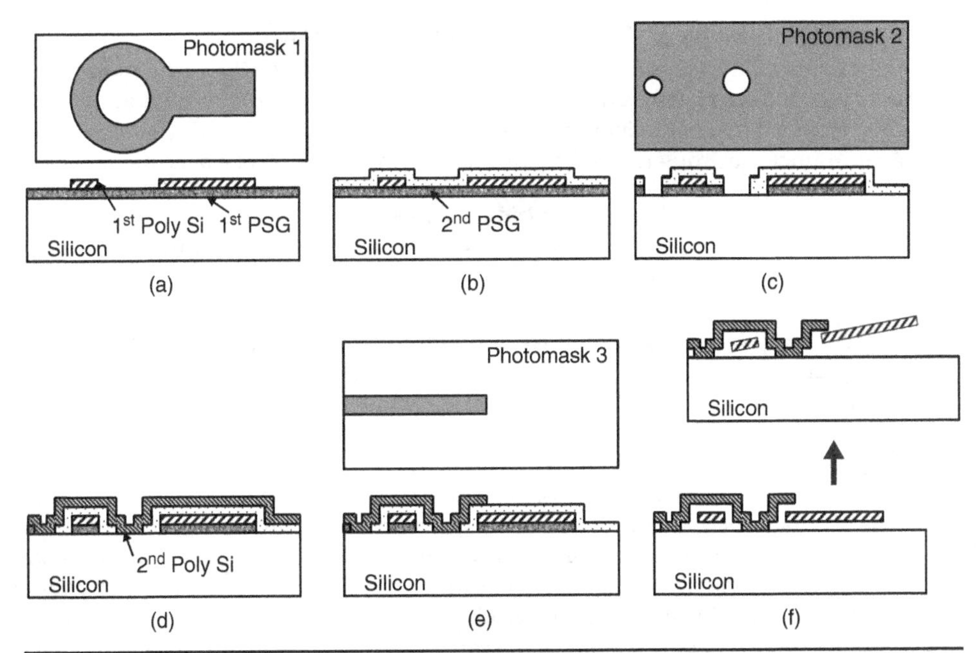

FIGURE 2.29 Fabrication steps along with three photomasks for a hinged structure: (a) depositing a sacrificial PSG and Poly Si 1 followed by patterning Poly Si 1, (b) depositing a second sacrificial PSG, (c) patterning the two PSG layers for the anchoring of Poly Si 2 to be deposited, (d) depositing Poly Si 2, (e) patterning Poly Si 2, and (f) then finally removing all the PSG layers to release Poly Si 1.

Some advanced hinge structures include a hinge that allows the plate to be lifted upward and a hinge that allows the plate to be pushed downward. Combination of these two hinge types can produce a fold-up structure with a flat surface at the top [19].

Automatic Release of Hinged Structures

Most of the hinged structures are manually assembled, but manual assembly of the micron-sized plates is very time demanding and labor intensive, and there have been efforts to automatically assemble the released plates, as described below.

Thermo-Kinetic Actuation for Release of Hinged Plates

When the ambient pressure is sufficiently low (e.g., tens of mTorr), gas molecules move in molecular flow regime (rather than viscous flow) with a velocity proportional to the square root of ambient temperature [20]. Thus, a substrate with a higher temperature than ambient would make the gas molecules from the substrate hit a hinged plate with higher kinetic energy than the gas molecules from the ambient. This differential kinetic energy (stemming from the difference in the molecular flow velocities) can be used to lift up a hinged plate as static friction is reduced with ultrasonic vibration energy.

This idea works in the molecular gas flow regime where Knudsen number ($K_n \equiv \lambda/d$, with λ and d being mean free path, which is inversely proportional to ambient pressure, and characteristic dimension, respectively) is greater than 1. Transition from the viscous to molecular flow regime depends on the size of a hinged plate.

Sequential Assembly of Hinged Plates by Magnetic Field

If magnetic film (such as permalloy) is allowed in a desired structure, one can deposit it over a hinged plate (usually by electroplating) and use magnetic field to lift up a plate [21]. External magnetic field of about 50 kA/m was shown to lift up hinged plates of 4.5-μm-thick electroplated permalloy and 0.2-μm-thick nickel elastic hinge.

Assembly of Hinged Plates by Centrifugal Force

Another approach to lift up hinged plates is to use centrifugal force by placing a released plate on a rotating cylinder [22].

2.2.3 Step Coverage, Selective Etching of Spacer Layer, and Sealing

In this section, we will study some issues that are commonly encountered in surface micromachining.

Step Coverage over Base Window

Since surface micromachining often requires a relatively thick sacrificial layer, a step over the patterned sacrificial layer tends to be tall and may not be covered well with a structural layer, as illustrated in Fig. 2.30a. If PSG is used as a spacer layer, its step can be tapered by varying the spacer etch rate from top to bottom and/or by reflowing etched spacer, as illustrated in Fig. 2.30b, in order to minimize step coverage problems.

If a different deposition technique can be used for the structural layer, keep in mind that evaporation offers a poor step coverage unless a wafer is rotated around to make the evaporated atoms to impinge on the wafer at multiple angles. Sputtering deposits film more conformally than evaporation but not like CVD that offers a very conformal deposition.

If PSG is implanted with phosphorous ions near its top layer, the lateral etch rate (or the undercut etch rate) is higher near the top than the bottom, and the step is tapered after PSG is etched in a wet etchant. Or, PSG (usually deposited with LPCVD) can be made to contain varying amount of phosphorous (from a few % to 10%) from the bottom to the top during the deposition, so that buffered HF would etch PSG faster (in all directions including the lateral direction) near the top than near the bottom.

Also, PSG, if its phosphorous content is larger than 5%, can easily be reflown at about 950°C (with nitrogen gas for about 30 min), and its etched step can be smoothened by a reflow step through viscous flow of the PSG at the high temperature. The higher the phosphorous content in the PSG, the lower is the needed temperature for reflow.

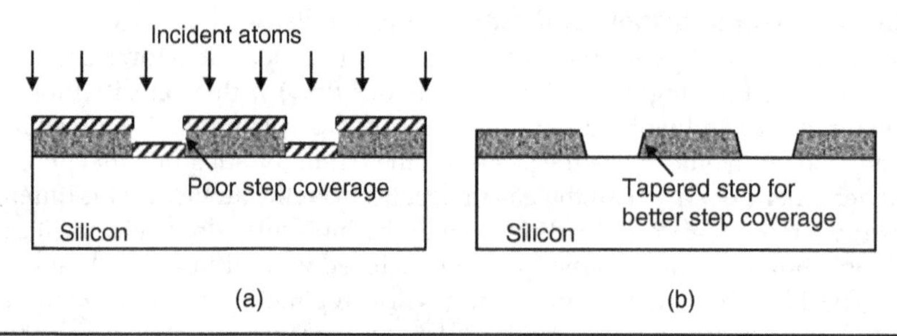

Figure 2.30 Step coverage issue and a potential solution: (a) when the step height is larger than the thickness of a film to be deposited over a step, there may be a poor step coverage, if the film deposition is not conformal. (b) Tapering the step can help on the step coverage.

Selective Etching of Spacer

Etching of a spacer (or sacrificial) layer usually takes a long time since the lateral dimension for the etchant to remove the spacer layer to release a structure is usually tens to hundreds of microns. During that long etching, the etchant used to remove a spacer layer etches any other layer that the etchant encounters. Thus, one has to know the etch rates (of the spacer-layer etchant) for the structure layer, isolation layer, etc., and estimate how much of those layers will be removed by the time the spacer-layer etching is completed. Ideally, one would want an etchant that etches the spacer layer much faster than any other layers exposed to the etchant and may have to use wet etchants (in spite of their inherent stiction issue), which typically offer more selective etching than plasma (except oxygen plasma that etches polymeric materials very well without etching dielectrics or metals).

To minimize the time for removing the sacrificial layer, one can make holes on the structural layer so that the etchant can access the spacer layer through the holes and etch out the spacer layer faster. However, this approach leaves the etch-access holes (on the structure layer) after the release of the structure, which may affect the performance of the released structure unless the holes are closed up after the release.

Question: Shown at *left* below is a cross-sectional view of layers on top of silicon for a surface micromachining. If the structural layer, sacrificial layer, and isolation layer are 2-μm-thick polysilicon, 1-μm-thick PSG, and 0.1-μm-thick silicon nitride, respectively, as shown at *right*, sketch a cross-section of the wafer that will result if the wafer shown at *right* below is etched in buffered HF for 30 minutes. Assume that the etch rates for polysilicon, PSG, and silicon nitride in buffered HF are 0.02 μm/min, 2 μm/min, and 0.003 μm/min, respectively. Make sure that you show the thickness variations along the horizontal directions as well as the thickness values.

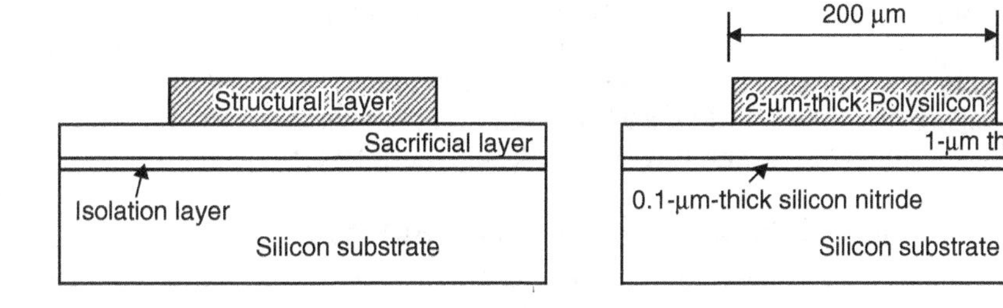

Answer:

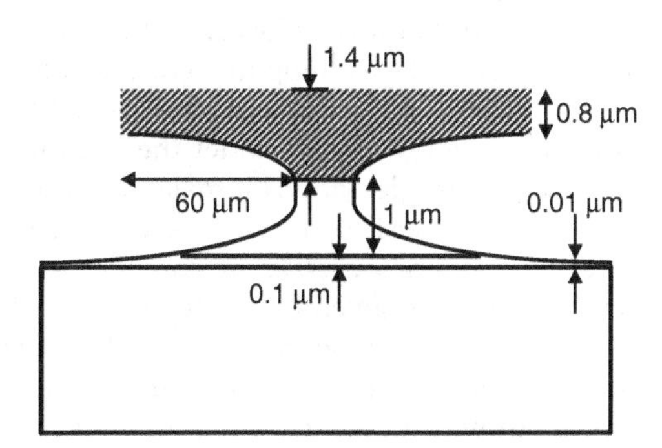

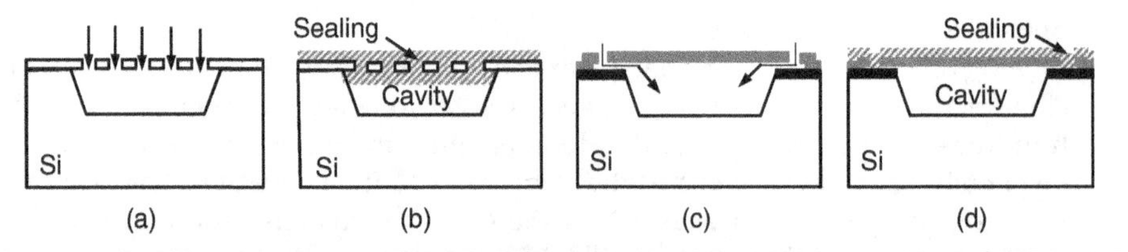

Figure 2.31 Vertical and horizontal etch-access channels for forming cavities: (a) cross-sectional view of anisotropic etching of silicon substrate through etch holes, (b) sealing of the etch holes, (c) cross-sectional view of horizontal etching of a sacrificial layer followed by anisotropic etching of silicon substrate, and (d) sealing of the horizontal etch channels.

Sealing Techniques

When holes are used to allow etchant to pass through (Fig. 2.31), one way to close up etch-access channels (or holes) is to oxidize polysilicon or silicon substrate, taking advantage of the fact that the volume of thermally grown silicon oxide is almost twice that of the silicon used for thermal oxidation (e.g., a 1-μm-thick silicon oxide is thermally grown by consuming 0.46-μm-thick silicon, as illustrated in Fig. 1.3). Note that thermal oxidation requires consumption of Si to grow silicon oxide (in oxygen environment), while CVD oxide deposition is done through chemical reaction of reactive gases such as silane and oxygen. This method of growing silicon oxide to seal up a gap works only with silicon (amorphous, polycrystalline, or monocrystalline) when the gap is relatively narrow.

Another way to close up etch-access channel is to use a sealant layer that is deposited by CVD, spin-coating, etc. Some sealant layers provide hermetic sealing. And if the sealant layer is deposited at a reduced pressure, the enclosed cavity is at that reduced pressure, when the sealant deposition is completed and the device is taken out of the deposition system. The differential pressure between the cavity and the atmosphere may collapse the cavity if the structure layer is not strong enough. Two approaches are shown in Fig. 2.31 for etch-access holes or channels, and one may be better than the other depending on application and device design.

2.2.4 Stiction of Surface-Micromachined Structures

After cantilevers are released by a wet etchant, the etchant is rinsed with deionization (DI) water, which is then dried in ambient with and without nitrogen blow. When the water drying processes under cantilevers having two different lengths are observed over time, the water under short cantilevers dries from the free end to the anchor, resulting in a completely released cantilever. However, the water under long cantilevers dries from the anchor to the free end, resulting in a stuck cantilever [23]. In case of the long cantilevers, as the water dries and gathers near the free ends of the cantilevers, the gathering (and forming of droplet) of water under the free end can bring all kinds of dirt and contaminants into the droplet, and can make the cantilever's free end be stuck to the substrate when all the water is dried. This is a very common stiction problem that happens when wet etchants are used to release surface-micromachined structures, especially cantilevers. There are several ways to avoid this kind of stiction. One way is to add a tiny area at the free end of the cantilever so that the gathering of the water during the drying process would occur over a very tiny area and the sticking force would be small due to the smallness of the area where the dirt and contaminants remain after the complete drying [23]. More general approaches to overcome the stiction problem are described below.

Supercritical CO_2 Drying of Surface-Micromachined Structures

A more generic approach to avoid the stiction during drying is to avoid the liquid–vapor interphase by taking the liquid to supercritical phase and then letting it to evaporate into vapor, as illustrated in Fig. 2.32a. For instance, one can take the following steps to release a structure without any stiction: (1) after wet etching to release a structure, rinse the wet etchant with water and then replace the water with isopropyl, keeping the processed wafer in liquid all the time; (2) place the wafer (along with isopropyl) in a chamber of a drying apparatus, close the chamber, input liquid CO_2, and replace the isopropyl with liquid CO_2 (at this point the chamber is at more than 1,100 psi at room temperature); (3) raise the chamber temperature to about 40°C to take the liquid CO_2 to supercritical CO_2; (4) then vent the chamber (while keeping the same temperature) so that gas CO_2 may come out of the chamber, and there is no more liquid in the wafer. The steps (2)–(4) make liquid CO_2 move into a supercritical phase and then a gas phase, as illustrated in Fig. 2.32a with 1, 2, and 3 denotations. This approach has been proven to be extremely effective in alleviating the stiction problem and has been used widely.

Freeze-Drying Method

Freeze-drying method also avoids the liquid–vapor interphase and can alleviate the stiction problem. This method first solidifies liquid (water, organic solvent, or t-butyl alcohol) by lowering the temperature and then sublimates the solid in vacuum. T-butyl alcohol is particularly attractive since its freezing temperature is 25.6°C, about the room temperature but damages pump oil [24].

The following steps are the typical steps taken: after wet etching of a sacrificial layer (with HF, for example) in a wafer and rinsing the wafer with water or organic solvent, freeze the wafer with the rinsing solution, and place the wafer inside a special vacuum system (with a cooler) so that the frozen rinsing solution is sublimated in the vacuum. More specifically, after HF etching and dehydrating with isopropyl alcohol, immerse the wafer in t-butyl alcohol (liquidized by warming with a hot plate), and place the wafer in a refrigerator to freeze the t-butyl alcohol. Then transfer the wafer to a vacuum system and evacuate with a rotary pump for sublimation of the frozen t-butyl alcohol [24].

Though this method is much simpler to implement than the supercritical CO_2 drying and requires much less capital investment, freezing liquid with a released structure may

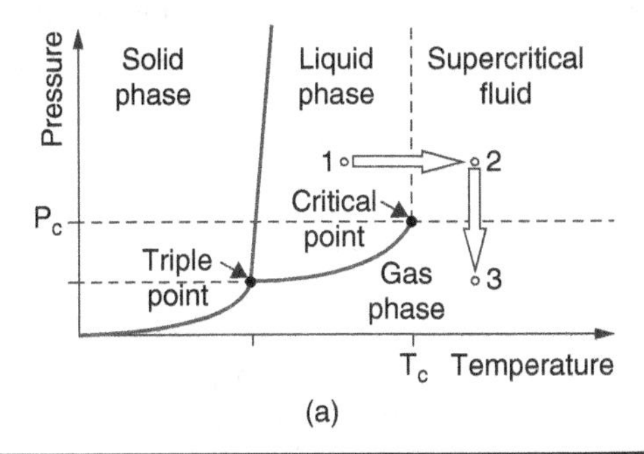

	Tc (°C)	Pc (atm)
CO_2	31.1	72.8
Methanol	240	79.9
Water	375	217.6

(b)

FIGURE 2.32 (a) Phase diagram showing the critical point (CP) at T_c and P_c. (b) Temperatures and pressures for critical points of CO_2, methanol, and water.

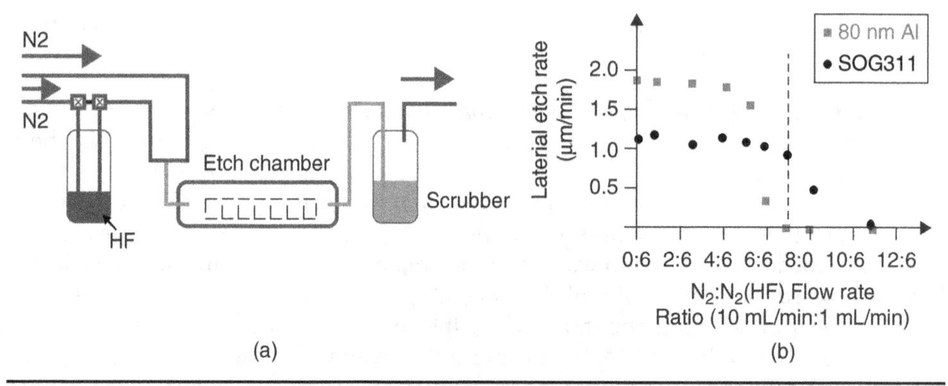

FIGURE 2.33 (a) Schematics of a vapor HF etching apparatus. (b) Selective etching of spin-on-glass with little etching of Al at a specific HF concentration level. (Adapted from [26].)

potentially damage the structure due to the relative large volume increase when liquid is solidified [25].

Vapor HF Etching of SiO_2

Using vapor HF can be a way to avoid a wet etchant to remove silicon dioxide spacer layer. If one cannot afford a commercial system for vapor HF etching, one can build an etching apparatus as shown in Fig. 2.33a. HF is a gas at room temperature but is made to be liquid by dissolving the gas into a water. Commonly available HF solution is 49% HF, though 75% HF is commercially available. With nitrogen flow into a container that holds HF solution, one can produce HF vapor, which is then mixed (for various concentration of vapor HF) with nitrogen coming from another gas line before being input into an etch chamber. A homemade apparatus without any electrically operated valve would work reasonably well.

An etch by-product of HF etching of silicon oxide is water, and there may be condensation of water by-product on the wafer surface. This water condensation can be alleviated by raising the temperature and/or increasing the flow rate of the nitrogen carrier gas. With vapor HF etching, many materials have their unique critical HF concentration below which there is no etching by vapor HF [26]. The critical HF concentration (below which the etching stops) depends on a type of a material. For example, Al is rarely etched at a sufficiently diluted vapor HF (i.e., 75 mL/min: 6 mL/min of N_2: HF), while spin on glass (SOG) is still highly etched, as can be seen Fig. 2.33b. Thus, with that HF concentration, mere 80-nm-thick Al can be used as an etch mask during etching SOG to release structures.

XeF_2 Dry Etching of Silicon and Polysilicon

If any silicon (amorphous, polycrystalline, or monocrystalline silicon) can be used as a sacrificial layer, then XeF_2 dry etching can be an excellent way to release a structure without any liquid involved. XeF_2 is a solid at room temperature but has a high vapor pressure. XeF_2 vapor is extremely effective in etching silicon (more than 1 μm/min) but hardly etches many other materials including Al, SiO_2, Si_3N_4, etc. With XeF_2 vapor etching, polysilicon can be a spacer layer, which is etched through a gas-permeable layer [27]. This particular method is explained further in the dry-etching section of this chapter.

O_2 Plasma Etching of Photoresist

If polymeric layer (e.g., photoresist) can be used as a sacrificial layer, O_2 plasma can be an incredibly good way to remove the sacrificial layer without affecting dielectrics or metals at all. Though most PR or polymers decompose at greater than 200°C, subsequent processing steps with PR must be done at low temperature. This greatly limits useable structure layer materials, if PR is used as a sacrificial layer. With PR as a sacrificial layer, Al (embedded with silicon up to a few percent for mechanical sturdiness) can be a structure layer, and Texas Instruments has taken full advantage of this in making their digital micromirror arrays, which are covered in Chap. 5.

2.2.5 Additional Issues of Surface Micromachining

Surface micromachining can be simple and economical, since the machining is done only on one side and also can be used to produce many micron-sized moveable structures in a batch of wafers. However, the reproducibility of mechanical properties (and long-term reliability) of thin films could potentially pose difficulties for some commercialization of surface micromachined devices. Also, it is known to be very difficult to engineer the residual stress of thin film by controlling the deposition parameters.

Another difficult issue with surface micromachining is surface effects (especially sticking of released structure) during drying step and/or operation. However, this issue has largely been resolved with various drying techniques and/or surface treatment.

Since the removal of the sacrificial layer (to release a structure) usually requires a long etching time, the etch selectivity of the sacrificial-layer etchant may present some difficulty. For example, the etch rates of buffered HF, that is, BHF 7:1 (NH_4F: 48 wt% HF) at room temperature for LPCVD silicon nitride, silicon oxide, and 7% PSG are 7–12, 700, and ≈10,000 Å/min, respectively. Consequently, while etching PSG in 7:1 BHF for 100 minutes, one may end up etching 700–1200-Å-thick silicon nitride as well.

Stress gradient along the thickness direction exists in as-deposited films such as polysilicon due to grain boundary imperfections that vary along the thickness direction owing to varying surface conditions that adatoms experience as the film is deposited. The stress gradient causes bending of a released structure and needs to be reduced. In case of polysilicon, post-deposition annealing at 1,050°C in nitrogen for about 30 minutes or rapid thermal annealing at 900°C for 7 minutes has proven to be very effective in reducing the stress gradient as the heat treatment anneals out the grain boundary imperfections.

2.2.6 Porous Silicon Micromachining

Bosch developed a fabrication process in forming an enclosed cavity through epitaxially depositing single-crystal silicon on top of porous silicon after anodizing silicon to produce porous silicon, which provides monocrystalline seeds for epitaxial formation of single crystal. During epitaxial layer formation, the porous silicon collapses and reflows, and a vacuum cavity is formed below the epitaxially deposited monocrystalline silicon membrane. This process makes a chip size smaller in lateral dimension and thinner than a conventional method of forming an enclosed cavity through silicon bulk micromachining followed by anodic bonding of silicon and glass.

2.3 Dry Micromachining

2.3.1 Plasma Etching

Plasma can be used to etch films and substrate without any liquid involved. There are basically three etching modes with plasma: sputtering, reactive ion etching (RIE), and plasma etching. As will be discussed a little later, plasma is generated by a high voltage between cathode and anode (positively biased with respect to the cathode) through ionization of gas molecules by high-energy electrons. Ionized gas molecules (that are positively charged) are drifted by high electrical field through cathode dark space region and can attain high velocity when they reach the cathode. The high-energy ions can hit the cathode with such an impact that the cathode material is sputtered off from the cathode.

Thus, if a wafer is placed on cathode, there always exists physical etching effect by positive ions. If the etching is done purely by physical etching effect, it is called sputtering or ion beam milling. For this, the ambient pressure is kept low so that there is minimum number of collisions between the ions and gas molecules as the ions are drifted to cathode through cathode dark space region. Also, the gases used to generate the plasma would have to be noble if no chemical etching effect is desired. However, if the plasma contains reactive radicals (through a proper choice of the gas), there exists chemical etching effect in addition to physical etching by the ions. Etching by both the physical impact and chemical reaction is called reactive ion etching (RIE).

If a wafer is placed on anode, only electrons (not ionized gas molecules) may potentially be drifted toward the wafer, which then experiences very little physical etching effect, since electron mass is very small, orders of magnitude smaller than ion mass. Thus, the etching is done exclusively through chemical reactions by reactive radicals in the plasma. This etching mode is called plasma etching. Table 2.3 summarizes the three different etching methods that are commonly used with plasma for dry etching.

Glow Discharge

With a large negative voltage at the cathode respect to the anode in a system shown in Fig. 2.34, electrons are emitted from the cathode and accelerated toward the anode. When the electrons gain enough energy, they can ionize gas molecules. The ionization produces more electrons, which are then accelerated by the electrical field and become energetic enough to ionize more gas molecules which produces more electrons. Once this avalanche process is ignited, the numbers of ionized molecules and electrons reach

	Plasma Etching (PE)	Reactive Ion Etching (RIE)	Physical Sputtering
Etching mode	Chemical Reaction	Chemical and physical (ion)	Physical bombardment of ions
Selectivity	Most selective	Less selective than PE	Not selective
Isotropicity	Isotropic	Directional	Most directional
Radiation damage	Least	More than PE	Most severe
Gas pressure	Higher than RIE	100 mT range	<100 mT
Wafer location	Anode	Cathode	Cathode

Table 2.3 Three Different Modes of Plasma-Based Etching

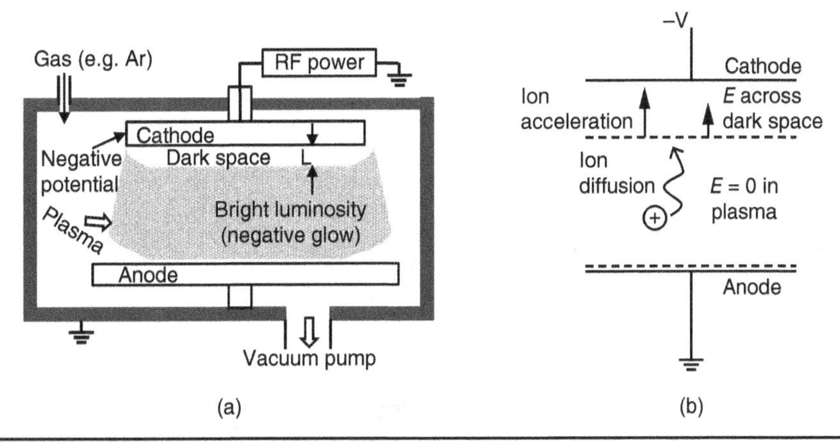

Figure 2.34 Plasma system: (a) details of glow discharge and (b) ions being accelerated by electrical field in the cathode dark space region.

very high quickly, and there exists so-called negative glow with bright luminosity between the cathode and anode. Negative glow contains equal numbers of electrons and ions (i.e., a plasma) along with many neutral molecules. The degree of ionization for a typical plasma is about 10^{-4} per molecule (i.e., 0.01%), but because of so many gas molecules, the numbers of the ions and electrons in plasma are very large.

At steady state, electric field inside plasma is zero, as plasma maintains charge neutrality. For if the electric field were not zero, electrons in plasma are accelerated faster than ions (due to their much smaller mass) and get out of plasma at a much faster velocity than ions, destroying the charge neutrality in plasma and resulting in no more plasma. The zero electric field makes electric potential be constant throughout the plasma. Thus, most of the large cathode voltage drops across the cathode dark space (i.e., optically dark sheath region between plasma and cathode).

In other words, the charge neutrality (i.e., net density of mobile charge $\rho = 0$) in plasma at steady state means that the electric potential V across the glow region is constant (i.e., at an equipotential) because $-\mathrm{d}^2V/\mathrm{d}x^2 = \mathrm{d}E/\mathrm{d}x = \rho/\varepsilon = 0$ as well as $E = 0$. Again, the electric field E inside plasma is zero to keep plasma neutral (i.e., if $E \neq 0$, the electrons get out of plasma at a much faster velocity than ions due to its much smaller mass).

Since most of the potential difference between the cathode and anode drops across the cathode dark space, there exists large electric field across the dark space. Due to the negative biasing at the cathode, the electric field in the cathode dark space accelerates positive ions toward the cathode but repels electrons from reaching the cathode. Thus, the ions near the interface between the dark space and the plasma are swept by the electric field across the dark space and accelerated toward the cathode. The accelerating ions may collide with neutral gas molecules and lose their kinetic energy through the collision process. Yet, there are sufficient numbers of ions that gain high energy through the acceleration by the electric field and hit the cathode with high physical impact.

The drifting of the ions from the interface (between the plasma and dark space) makes the ion concentration at the interface lower than that at the bulk of the plasma, and ions diffuse from the bulk to the interface (Fig. 2.34b). The electrons in the bulk of the plasma screen the ions from the potential between the cathode and anode. However,

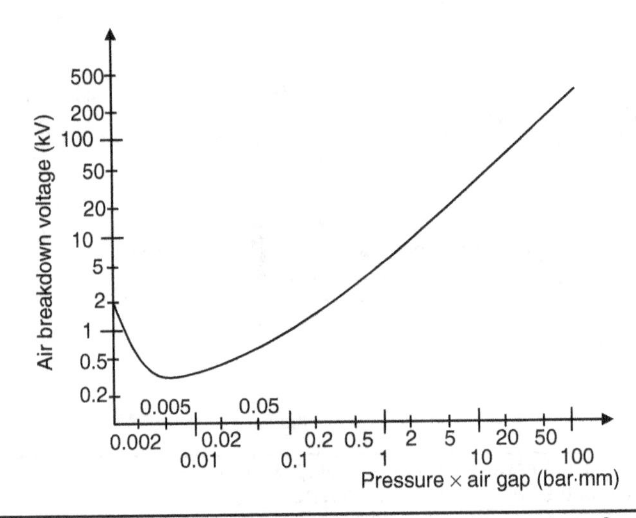

FIGURE 2.35 Paschen's curve for air breakdown voltage versus the product of pressure and air gap. (Adapted from [28].)

the ions (that have diffused to the interface between the dark region and the plasma) are swept by the *E*-field across the dark region toward the cathode.

The plasma is self-biased about 15 V above the highest positive potential in a given system (which typically grounds the anode and chamber wall and biases the cathode negatively) so that the electric fields in the cathode and anode dark spaces may be developed such as to repel electrons from reaching any of the cathode, anode, and chamber wall. Note that there also exists a dark space between the plasma and the anode (called anode dark space) where ions can be drifted but without gaining much energy due to such a low voltage difference (about 15 V) between the plasma and anode.

Breakdown *E*-field for gas atoms or molecules increases drastically, as shown in Paschen's relationship for air (Fig. 2.35), if the product of gas pressure and air gap (between two electrodes) is reduced below 2×10^{-3} bar·mm. The reason for such increase is that for a given ambient pressure (e.g., 1 bar), the traveling distance of electrons is now so much less (<2 μm) that it is harder for electrons to gain high enough energy to breakdown gas molecules. The breakdown *E*-field can potentially be increased by two orders of magnitude as the air gap is reduced submicron level.

Question: For a glow discharge system shown below, sketch the voltage as a function of distance from the cathode to the anode on the figure below, assuming that the anode (along with the chamber wall) is grounded. What is the electrical field in the cathode dark space region if the distance (L) of the cathode dark space region is about 2 cm?

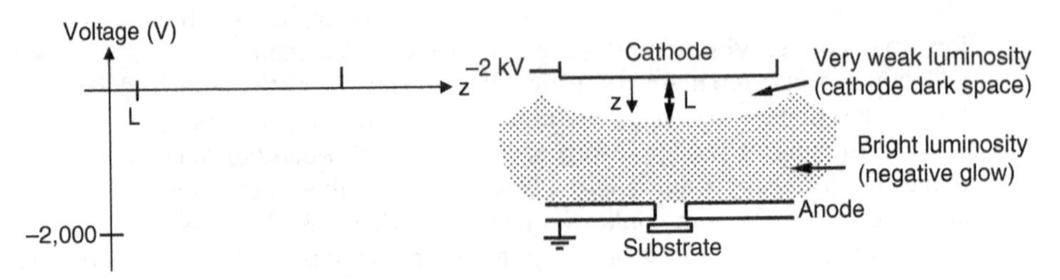

Answer:

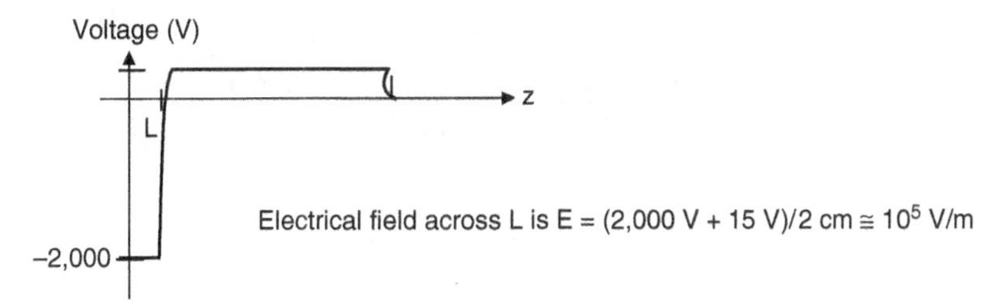

Question: In a glow discharge system shown above, how much will be the distance (L) of the cathode dark space region increase or decrease if the gas pressure in the system is increased from 10 mTorr (1.3×10^{-5} bar) to 100 mTorr (1.3×10^{-4} bar)? Note that almost all of the 2 kV (applied between the cathode and anode) drops across the cathode dark space region.

Answer: If the breakdown field remains constant over the pressure range, the L will remain constant. However, Paschen's curve shows that the breakdown voltage of 2 kV occurs at 10^{-3} or 0.5 bar·mm. Thus, the distance is equal to 7.7 cm (or 38 m) that is 0.001 (or 0.5) bar·mm divided by 1.3×10^{-5} bar. Since the distance between the cathode and anode is less than 1 m, the cathode dark space region is about 7.7 cm long at 10 mTorr and is reduced to 0.77 cm at 100 mTorr by a factor of 10.

2.3.2 Reactive Ion Etching

Anisotropy in RIE

In RIE, chemical etching effect by reactive radicals produces isotropic etching with lateral undercut, but at the same time, physical sputtering effect by high-energy positive ions produces anisotropic etching with vertical sidewall, as the impinging ions produce lattice damage extending several monolayers below the surface, as illustrated in Fig. 2.36. In RIE, both physical and chemical etching effects produce somewhat anisotropic etching with some amount of lateral undercut.

The anisotropicity can be enhanced by using carbon-containing gases for RIE. Carbon-containing gases (e.g., CHF_3, $CClF_3$, CF_4/H_2, etc.) decompose in plasma to form unsaturated species and polymer precursor radicals, which get adsorbed on the surface, becoming chemical-etching inhibitors. Though the etch-inhibiting polymers coat

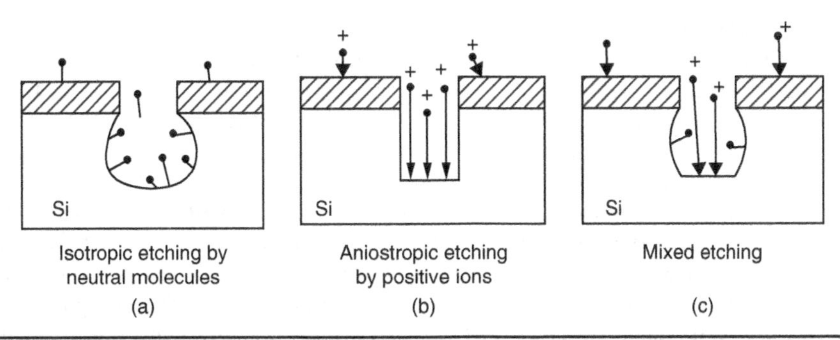

FIGURE 2.36 Cross-sectional views of what are happening in reactive ion etching (RIE) through (a) chemical reaction of reactive species, (b) physical impact of positive ions drifted by an electrical field, and (c) both chemical reaction and physical impact.

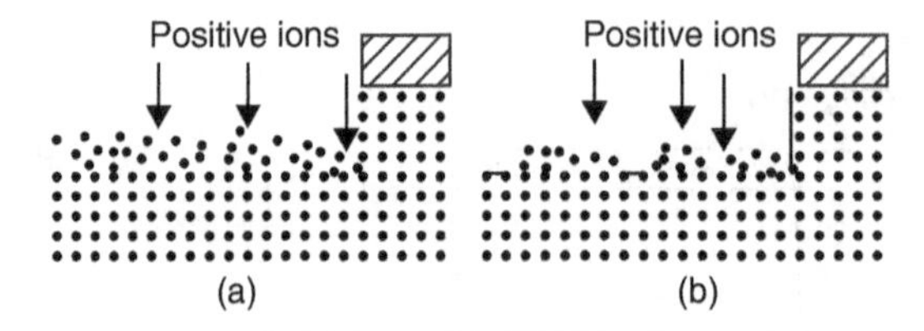

FIGURE 2.37 Physical etching effect by positive ions (a) without any carbon-containing gas and (b) with carbon-containing gas that produces polymers that get coated over the etch surface.

every surface equally, ion bombardment removes the inhibitors from the bottom surface (facing the ion flux head-to-head) much more effectively than from the sidewall surface (that runs in parallel with the ion flux), as illustrated in Fig. 2.37. Thus, the sidewall is well covered by the etch inhibitors and experiences very little chemical etching effect while the bottom surface experiences both physical and chemical etching.

2.3.3 Silicon Reactive Ion Etching and Deep Reactive Ion Etching

Silicon RIE with SF_6 and O_2

Silicon can effectively be etched in plasma with SF_6 that produces radical F^*, which reacts with Si to produce volatile SiF_4. If a wafer is placed on the anode of a plasma system, SF_6 plasma will etch Si almost isotropically with very little physical etching effect by positive ions [29]. If a wafer is placed on the cathode (rather than the anode) of a plasma system, the etching can be anisotropic, especially with an additional gas such as oxygen that oxidizes silicon, producing SiO_2 (on the sidewalls) that inhibits the sidewall etching. The oxidation indeed produces SiO_2 all over the silicon surface, but the ion bombardments remove the SiO_2 from the "bottom" surface more effectively than from the sidewalls. Addition of CHF_3 to SF_6 and O_2 in silicon RIE can produce all kinds of sidewall etch shapes as CHF_3 scavenges O_2 and reduces sidewall oxides [30].

With a wafer in RIE mode (i.e., a wafer placed on the cathode), a combination of SF_6 and O_2 gases can produce varying degree of anisotropicity, which can advantageously be used to release structures (often with reduced masking steps). For instance, a tall silicon cantilever can be fabricated using SF_6 (20 sccm) and O_2 (20 sccm) at 50 mTorr for anisotropic etching, followed by isotropic etching of silicon by turning off O_2 [31].

Deep Reactive Ion Etching

Using the anisotropicity obtainable with protective sidewalls, Bosch came up with an idea to use multiple number of etching and passivation cycles in order to etch silicon deep into the wafer thickness with little lateral undercut. The idea illustrated in Fig. 2.38 is to etch silicon with SF_6 a little and then coat the entire surface with polymeric layer by bringing carbon-containing gas (e.g., C_4F_8). The polymeric layer provides etch-inhibiting sidewall in the next etching cycle because of the ineffectiveness of the impinging ions in removing the polymeric layer from the sidewalls (unlike from the bottom surface). Each etching cycle is followed by a passivation cycle that coats the entire surface with polymeric layer. By repeating the cycles of etching and passivation, one can etch silicon deep with little lateral undercut, and can easily obtain an aspect ratio up to 20. This etching technique is commonly called deep reactive ion etching (DRIE) and can produce all kinds of high-aspect-ratio silicon structures because the

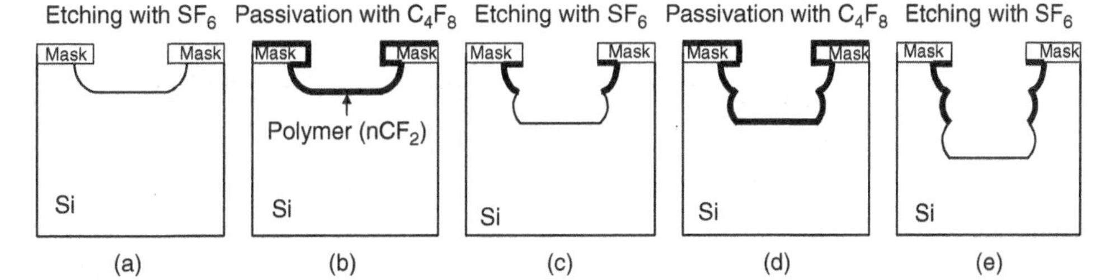

FIGURE 2.38 Deep reactive ion etching (DRIE) of silicon through repeated cycles of (a) RIE with SF_6, (b) passivation with C_4F_8, (c) RIE with SF_6, (d) passivation with C_4F_8, and (e) RIE with SF_6. (Adapted from [32].)

etching is independent of the crystal plane. Note, however, that the sidewall is not very smooth but jagged due to the multiple etching steps. What determines the roughness of the sidewall of a silicon structure made with silicon DRIE is the time of the isotropic etching step (i.e., Fig. 2.38a); the longer the time, the rougher the surface. By adding O_2 to a typical DRIE, an aspect ratio greater than 40 can be obtained since oxygen produces SiO_2 that adds protection of the sidewall (afforded by polymeric layer in a typical DRIE) [32].

Question: In silicon DRIE the etching rate depends on the silicon area to be etched (i.e., areas with more open silicon etch faster compared to areas with lower amounts of open areas), as indicated below. What are possible reasons for this so-called, "aspect ratio–dependent etching"?

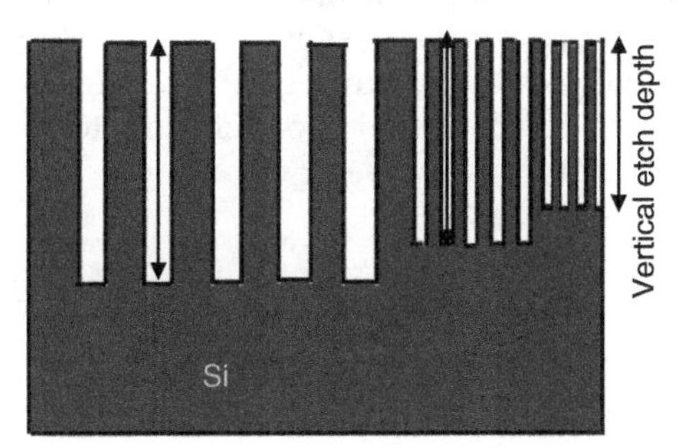

Answer: Mass transport of etchant and etch by-products depends on the cross-sectional area of the channel that those species need to pass through, similar to wet anisotropic etching of (110) wafer with vertical {111} side walls [4]. Consequently, the larger the open area, the faster the etching rate is. Also, the ion flux loss is greater at the bottom of a feature.

Question: In silicon DRIE with SF_6, C_4F_8, and O_2 indicate below which gas (or gases) (listed in the first row) provides the functions (listed in the first column) by putting a check mark or marks.

	SF_6	C_4F_8	O_2
Chemical etching			
Sputtering by positive ions			
Sidewall passivation			
Plasma generation			

Answer:

	SF_6	C_4F_8	O_2
Chemical etching	X		
Sputtering by positive ions	X	X	X
Sidewall passivation		X	X
Plasma generation	X	X	X

2.3.4 Dry Silicon Etching with XeF_2

Xenon difluoride (XeF_2) vapor etches silicon with excellent selectivity over many MEMS materials. The primary reaction between solid silicon and XeF_2 vapor is $2XeF_{2(g)} + Si_{(s)} \Rightarrow SiF_{4(g)} + 2Xe_{(g)}$, and there is no liquid involved or needed in the etching. The etching of silicon is isotropic and also more than a thousand times faster than those of many MEMS materials such as silicon oxide, silicon nitride, silicon carbide, Al, Au, Cu, Cr, TiN, photoresist, acrylic, etc.

Since XeF_2 is solid (a white crystal) with high-vapor pressure of about 4 Torr at room temperature, an expansion chamber is needed to collect XeF_2 vapor from the solid. The expansion chamber is then connected to an etching chamber where XeF_2 vapor etches silicon isotropically. Thus, valves are needed and operated to connect various chambers and pump so that the following steps are cycled through: filling up the expansion chamber with XeF_2 vapor from XeF_2 solid, using the XeF_2 vapor in the expansion chamber to etch silicon in the etching chamber, and then pumping out the etching chamber before the next cycle of these steps [33]. The etch rate depends on etching chamber pressure, volume of the etching chamber, total silicon area being etched, size of etch window, and proximity of etch windows. Etch pit roughness increases with the number of pulses, but variation of etch results is less with increasing number of pulses.

References

1. H. Hashimoto, et al., "Chemical isotropic etching of single-crystal silicon for acoustic lens of scanning acoustic microscope," Japanese Journal of Applied Physics, 32, May 1993, pp. 2543–2546.
2. K.E. Petersen, "Silicon as a mechanical material," Proceedings of the IEEE, 70, May 1982, pp. 420–457.
3. H. Tanaka, et al., "Fast wet anisotropic etching of Si {100} and {110} with a smooth surface in ultra-high temperature KOH solutions," Sensors and Actuators, A79, 2003, pp. 76–81. Transducers '03, Proceeding of the 12th International Conference on Solid-State Sensors and Actuators, pp. 1675–1678.
4. K. Ohwada, et al., "Groove depth uniformization in [110] Si anisotropic etching by ultrasonic wave and application to accelerometer fabrication," Proceedings of the 8th IEEE International Conference on Micro Electro Mechanical Systems (MEMS "95), Amsterdam, Netherlands, January 1995, pp. 100–103.
5. X.P. Wu and W. H. Ko, "A study on compensating corner undercutting in anisotropic etching of (100) silicon," The 4th International Conference on Solid-State Sensors and Actuators, Transducers '87, Tokyo, Japan, 1987, pp. 126–129.

6. S.C. Chang and D.B. Hicks, "Mesa structure formation using potassium hydroxide and ethylene diamine based etchants," IEEE Solid-State Sensor and Actuator Workshop, Hilton Head Island, SC, 1988, pp. 102–103.
7. J.W. Kwon and E.S. Kim, "Microfluidic channel routing with protected convex corners," Transducers '01, IEEE International Conference on Solid-State Sensors and Actuators, Munich, Germany, June 10-14, 2001, pp. 644–647.
8. E.S. Kim, R.S. Muller, and R.S. Hijab, "Front-to-backside alignment using resist-patterned etch control and one etching step," IEEE/ASME Journal of Microelectromechanical Systems, 1, June 1992, pp. 95–99.
9. S. Bütefisch, T. Weimann, A. Vierheller, V. Nesterov, and A. Dietzel, "Design and manufacture of a silicone pendulum for PTB's nanoforce standard facility," Transducers '19, The 20th International Conference on Solid-State Sensors, Actuators and Microsystems, Berlin, Germany, June 23-27, 2019, pp. 186–189.
10. R.M. White and S. W. Wenzel, "Inexpensive and accurate two-sided semiconductor wafer alignment," Sensors and Actuators, 13, 1988, pp. 391–395.
11. J. Frühauf, "Shape and Functional Elements of the Bulk Silicon Microtechnique," Springer-Verlag, Berlin, Heidelberg, 2005, p. 34.
12. C.H. Han and E.S. Kim, "Study of self-limiting etching behavior in wet isotropic etching of silicon," Japanese Journal of Applied Physics, 37, December 1998, pp. 6939–6941.
13. C.-H. Han and E.S. Kim, "Micromachined Piezoelectric Ultrasonic Transducers on Dome-Shaped-Diaphragm in Silicon Substrate," IEEE International Ultrasonics Symposium (Lake Tahoe, NV), October 17-21, 1999, pp. 1167-1172.
14. G. Kaminsky, "Micromachining of silicon mechanical structures," Journal of Vacuum Science and Technology, July/August 1985, pp. 1015–1024.
15. L.S. Fan, Y.C. Tai, and R.S. Muller, "Integrated movable micromechanical structures for sensors and actuators," IEEE Transactions on Electron Devices, 35(6), June 1988, pp. 724–730.
16. L.S. Fan, Y.C. Tai, and R.S. Muller, "Pin-joints, springs, cranks, gears and other novel micromechanical structures," 4th International Conference Solid-State Sensors and Actuators, Transducers "87, Tokyo, Japan, June 2-5, 1987, pp. 849–852 (U.S. Patent 4740410, April 26, 1988).
17. Y.C. Tai and R.S. Muller, "IC-processed electrostatic synchronous micromotors," Sensors and Actuators, 20, November 1989, pp. 49–55.
18. R. Yeh, E.J.J. Kruglick, and K.S.J. Pister., "Microelectromechanical components for articulated microrobots," Transducers '95, 2, pp. 346–349.
19. L. Fan, M. C. Wu, K. D. Choquette, and M. H. Crawford, "Self-assembled microactuated XYZ stages for optical scanning and alignment," Proceedings of International Solid State Sensors and Actuators Conference, Transducers '97, Chicago, IL, 1997, pp. 319–322.
20. V. Kaajakari and A. Lal, "Thermo-kinetic actuation for hinged structure batch microassembly," Technical Digest. MEMS 2002 IEEE International Conference. Fifteenth IEEE International Conference on Micro Electro Mechanical Systems, Las Vegas, NV, 2002, pp. 196–199.
21. E. Iwase, S. Takeuchi, and I. Shimoyama, "Sequential batch assembly of 3-D microstructures with elastic hinges by a magnetic field," Technical Digest. MEMS 2002 IEEE International Conference. Fifteenth IEEE International Conference on Micro Electro Mechanical Systems, Las Vegas, NV, 2002, pp. 188–191.

22. K. W. C. Lai, A. P. Hui and W. J. Li, "Non-contact batch micro-assembly by centrifugal force," Technical Digest. MEMS 2002 IEEE International Conference. Fifteenth IEEE International Conference on Micro Electro Mechanical Systems, Las Vegas, NV, 2002, pp. 184–187.
23. T. Abe, W.C. Messnel, and M.L. Reed, "Effective methods to prevent stiction during post-release-etch processing," Proceedings IEEE Micro Electro Mechanical Systems, Amsterdam, Netherlands, 1995, pp. 94–99.
24. N. Takeshima, K.J. Gabriel, M. Ozaki, J. Takahashi, H. Horiguchi, and H. Fujita, "Electrostatic parallelogram actuators," Transducers '91, 1991 International Conference on Solid-State Sensors and Actuators. Digest of Technical Papers, San Francisco, CA, 1991, pp. 63–66.
25. Y.L. Huang, H. Zhang, E.S. Kim, S.G. Kim, and Y.B. Jeon, "Piezoelectrically actuated microcantilever for actuated mirror array application," Solid-State Sensor and Actuator Workshop, Hilton Head Island, SC, June 2-6, 1996, pp. 191–195.
26. H. Guckel and J. J. Sniegowski, "Method to prevent adhesion of micromechanical mechanically resonant transducers," Sensors and Actuators, A21-A23, 1990, 346.
27. R. Toda, N. Takeda, T. Murakoshi, S. Nakamura, and M. Esashi, "Electrostatically levitated spherical 3-axis accelerometer," Technical Digest. MEMS 2002 IEEE International Conference. Fifteenth IEEE International Conference on Micro Electro Mechanical Systems, Las Vegas, NV, 2002, pp. 710–713.
28. E. Husain and R. S. Nema, "Analysis of Paschen curves for air, N2 and SF6 using the Townsend breakdown equation," IEEE Transactions on Electrical Insulation, EI-17(4), August 1982, pp. 350–353.
29. C. Linder, T. Tschan, and N. F. de Rooij, "Deep dry etching techniques as a new IC compatible tool for silicon micromachining," Transducers '91: 1991 International Conference on Solid-State Sensors and Actuators. Digest of Technical Papers, San Francisco, CA, 1991, pp. 524–527.
30. H. Jansen, M. de Boer, B. Otter, and M. Elwenspoek, "The black silicon method. IV. The fabrication of three-dimensional structures in silicon with high aspect ratios for scanning probe microscopy and other applications," Proceedings IEEE Micro Electro Mechanical Systems. Amsterdam, Netherlands, 1995, pp. 88–93.
31. Z. L. Zhang and N. C. MacDonald, "An RIE process for submicron, silicon electromechanical structures," Transducers '91: 1991 International Conference on Solid-State Sensors and Actuators. Digest of Technical Papers, San Francisco, CA, 1991, pp. 520–523.
32. F. Laerme, A. Schilp, K. Funk, and M. Offenberg, "Bosch deep silicon etching: improving uniformity and etch rate for advanced MEMS applications," Technical Digest. IEEE International MEMS 99 Conference. Twelfth IEEE International Conference on Micro Electro Mechanical Systems, Orlando, FL, 1999, pp. 211–216.
33. P. B. Chu et al., "Controlled pulse-etching with xenon difluoride," Proceedings of International Solid State Sensors and Actuators Conference, Transducers '97, Chicago, IL, 1997, pp. 665–668.

Questions and Problems

Question 2.1 Silicon nitride film is very difficult to be etched with a wet etchant and is traditionally etched by a reactive ion etching. However, according to Table 2.1 there are two etchants (namely, concentrated HF and hot phosphoric acid at 180°C) that can be used to etch silicon nitride. What are the problems with those etchants when you try to etch silicon nitride with a patterned photoresist on top of the silicon nitride?

Question 2.2 Does any of {111} planes meet with (110) at 90°? If so, name *one* {111} plane that meets with (110) plane at 90°?

Question 2.3 On a (100) silicon wafer (400 μm thick), a 1000-Å-thick Si_3N_4 has been deposited and patterned for *open* rectangles and *open* U-shape as shown below (with the drawings to scale). On the figure below *sketch* the cross-sectional views (across A–A') of the resulting etch fronts if the wafer is etched in KOH for *1 hour.* Assume that the KOH etches (100) silicon plane at 1 μm/min while (111) silicon plane is etched at 0.02 μm/min.

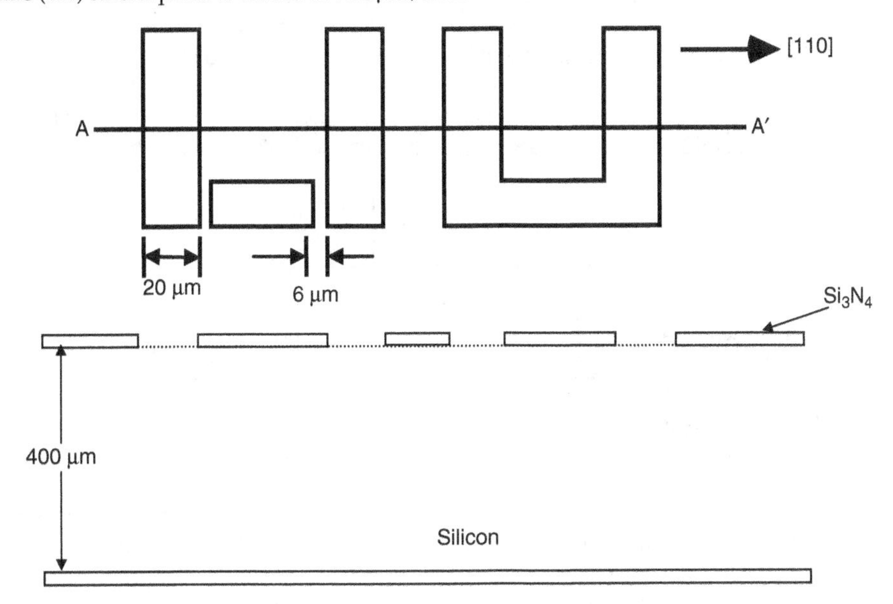

Question 2.4 In a glow discharge system shown in Fig. 2.34 with –2,000 V applied to the cathode, how much force does an ionized gas molecule (with a charge of 1.6×10^{-19} C) experience in the cathode dark space region? Assume that the distance (L) of the cathode dark space region is about 2 cm.

Question 2.5 To compare various dry-etching techniques, indicate below which effect(s) (listed in the first row) do the etching techniques (listed in the first column) use by putting a check mark or marks.

	Sputtering Effect by Positive Ions	Chemical Reaction Effect
O_2 plasma etching of photoresist		
Reactive ion etching (RIE)		
XeF_2 vapor etching of silicon		
Vapor HF etching of SiO_2		

Question 2.6 The following microstructure on a silicon wafer is to be deposited with 0.5-μm-thick polysilicon through low-pressure chemical vapor deposition (LPCVD). Sketch (on the figure below) the film deposited. Assume that the PSG and the silicon nitride in the figure are 2 μm and 1 μm thick, respectively.

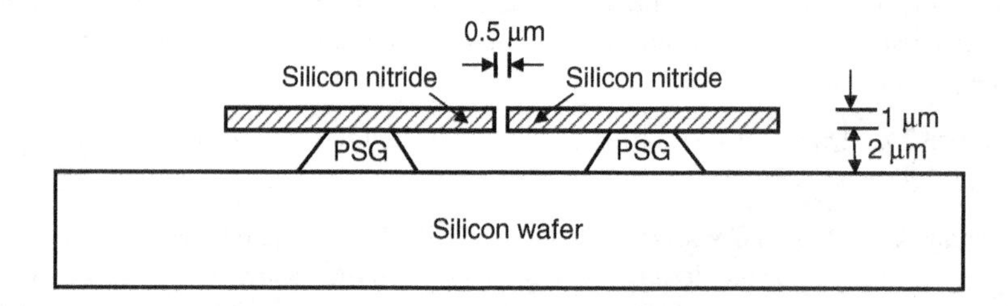

Question 2.7 A (111) Si wafer has gone through (a) low-pressure chemical vapor deposition (LPCVD) silicon nitride deposition and patterning (only on the front side), (b) RIE of silicon, (c) another silicon nitride deposition, and (d) RIE of silicon, as shown below. If the wafer is now etched in KOH for a very long time, what will be the resulting cross-section?

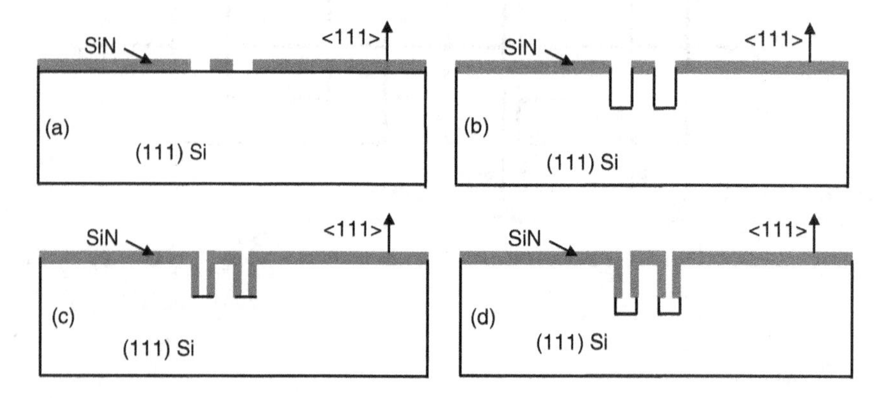

Question 2.8 For the following three questions assume that the silicon wafer is 500 μm thick and is deposited with 2-μm-thick silicon nitride on both sides of the wafer. And assume that KOH etches (100) silicon plane at 1 μm/min but (111) silicon plane at a negligible rate. Also, assume that HNA (hydrofluoric, nitric, and acetic acids) etches silicon *isotropically* at 1 μm/min.

(a) If we want to form a silicon nitride diaphragm of 10×10 μm² on the front side of the silicon wafer by etching the silicon wafer with KOH, how large will the size of an opening on the silicon nitride on the wafer backside have to be?

(b) If we want to form a silicon nitride diaphragm of 10×10 μm² on the front side of the silicon wafer by etching the silicon wafer with HNA, how large will the size of an opening on the silicon nitride on the wafer backside have to be?

(c) If we over-etch the silicon 15% more than the minimum needed time for the square diaphragms on the front side of the silicon for parts (a) and (b) above, how large will the sizes of the square diaphragms turn out to be?

Question 2.9 Answer the following for an anisotropic etching of (110) silicon wafer shown in Fig. 2.5**b**. (a) At what angle (in degrees) does the slanted {111} plane intersect the {110} plane? (b) If the two vertical {111} planes in Fig. 2.5**b** are $(\bar{1}11)$ and $(1\bar{1}1)$, what are the {110} planes? There are two possibilities.

Question 2.10 Your company, *an IC fabrication house*, has been producing surface-micromachined LPCVD silicon nitride cantilevers with PSG as a sacrificial layer. The PSG layer has been removed by BHF followed by rinsing in DI water and drying. However, because of relatively long length of the cantilevers, many of the released cantilevers are stuck after the fabrication. As a new fabrication engineer you have identified the following four alternatives (listed in the first column of the following table) and are about to compare them with respect to the issues listed in the first row of the following table. Fill in the blanks according to the instruction given in the first row. Assume that the company already has all the *typical* microfabrication equipment that an IC fabrication house would have.

	Need to change the sacrificial layer? Answer Yes or No. If the answer is "Yes," then what would be a good sacrificial layer?	Need to change the cantilever structural layer? Answer Yes or No. If the answer is "Yes," then what would be a good structural layer?	What is one noteworthy potential difficulty with the new approach? In terms of (1) the cantilever's structural sturdiness, (2) the sacrificial layer's surface roughness, or (3) the effect of the etch by-product. Pick only one of the above three to offer a note for each of the new approaches.
Vapor HF etching			
Using photoresist as a sacrificial layer			
Supercritical CO_2 dry release			
XeF_2 dry etching			

Question 2.11 If the top-view photomask pattern of Fig. 2.12 is slightly modified as shown below, how will the anisotropic Si etching be changed?

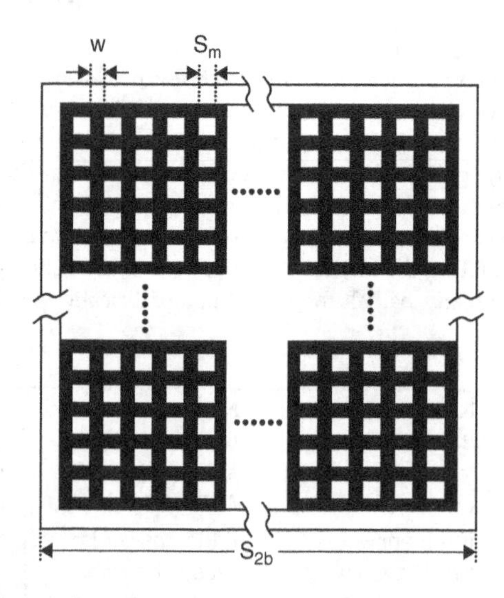

Question 2.12 A (100) silicon wafer has been etched with an anisotropic etchant to obtain a silicon shape as shown at right. What is the crystal plane marked as {???} at right?

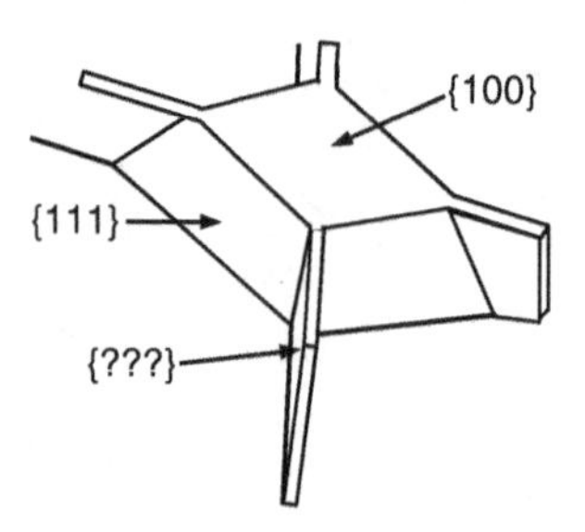

Question 2.13 On a (100) silicon wafer (400 μm thick), a 1000-Å-thick Si_3N_4 has been deposited and patterned for *open* rectangles and *open* U-shape as shown below (with the drawings to scale). On the figure below *sketch* the cross-sectional views (across A–A′) of the resulting etch fronts, if the wafer is etched in KOH for *1 hour*. Assume that the KOH etches (100) silicon plane at 1 μm/min, while (111) silicon plane is etched at 0.02 μm/min.

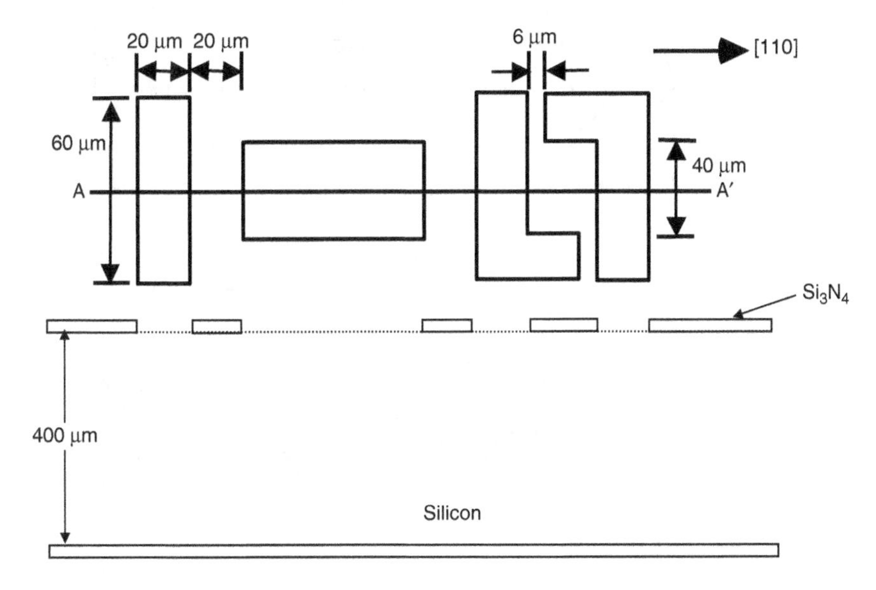

Question 2.14 On the following patterns made on a (100) silicon wafer, sketch the top-view shapes of the etched cavities after the silicon is etched in KOH long enough for the etch cavities to be bound by sidewalls of {111} planes.

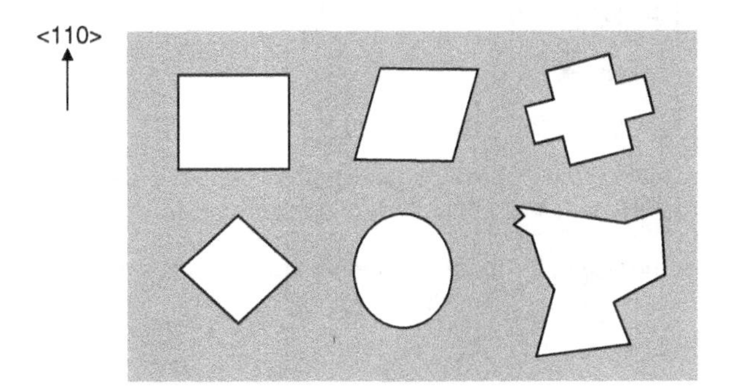

Question 2.15 If a square pattern is aligned to <100> directions, rather than to <110> directions, as shown below, sketch the top-view etch front that will result when the wafer is etched in KOH long enough that only the {111} planes are at the etch front.

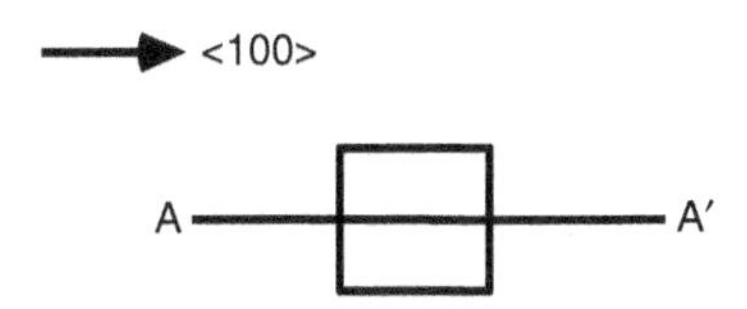

Question 2.16 Sketch two photo masks (one for the front side and the other for the backside of the silicon wafer) for one-time anisotropic etching of silicon that will produce the structures shown below.

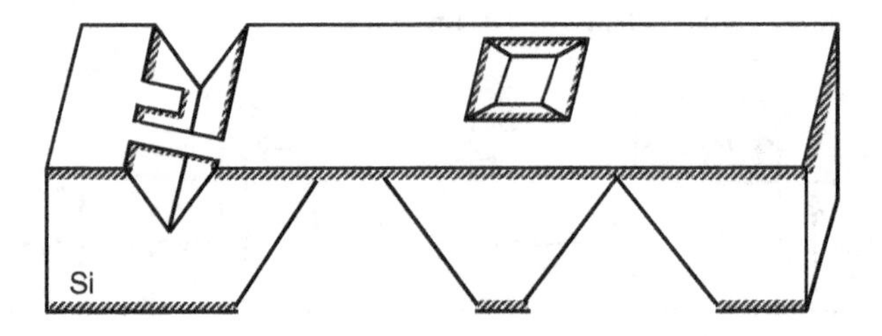

Question 2.17 Answer the following for KOH etching of a (100) Si wafer, shown below, assuming the silicon wafer is 500 μm thick and its {111} planes are etched at a negligible rate.

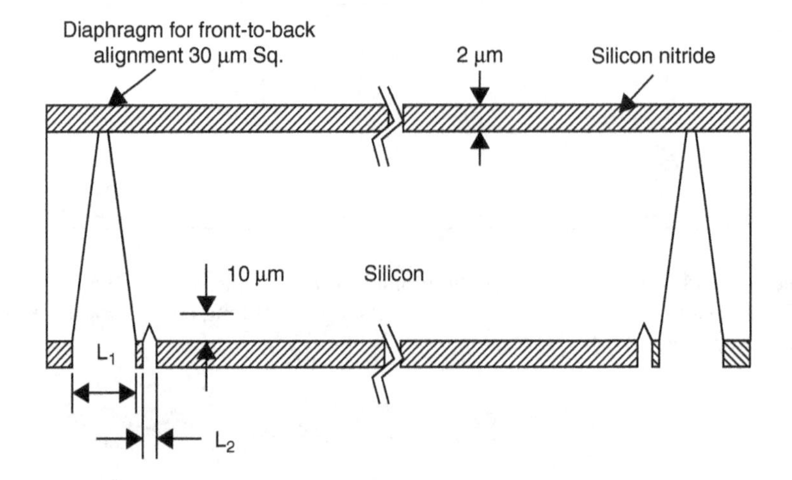

(a) Calculate the size (i.e., L_1) of the square pattern that should be opened on the silicon backside in order to create the $30 \times 30\ \mu m^2$ diaphragm on the front side.

(b) Calculate the size L_2 for the 10-μm-deep pyramidal pit as shown in the figure above.

Problem 2.1 On a (100) silicon (400 μm thick) wafer, 3000-Å-thick SiO_2 has been thermally grown and patterned for a 250-μm-diameter circle on both sides of the wafer (the circular area is where the SiO_2 is removed, and the circle on the front side is aligned with that on the backside). If the wafer is etched in EPW for (a) 1 hour, (b) 4 hours, and (c) 12 hours, what will be the resulting structures? Draw the top views and cross-sectional views for (a), (b), and (c). Assume that the EPW etches (100) silicon planes at 1 μm/min, while it etches the (111) planes 35 times slower (i.e., $\frac{1}{35}$ μm/min).

Problem 2.2 On a (100) silicon wafer (400 μm thick), 0.5-μm-thick polysilicon has been deposited and patterned to leave a polysilicon film over a $6 \times 20\ \mu m^2$ rectangle. On top of that, a 1000-Å-thick Si_3N_4 has been deposited and patterned for two $20 \times 20\ \mu m^2$ *open* squares with 6 μm spacing between the two squares as shown below. On the figure below, sketch the cross-sectional views (across A–A') of the resulting etch fronts if the wafer is etched in KOH for 1 hour. Assume that the KOH etches (100) silicon plane at 1 μm/min, while (111) silicon plane is etched at 0.02 μm/min. Calculate the etch depth, undercut length, etc. and indicate those on your sketch.

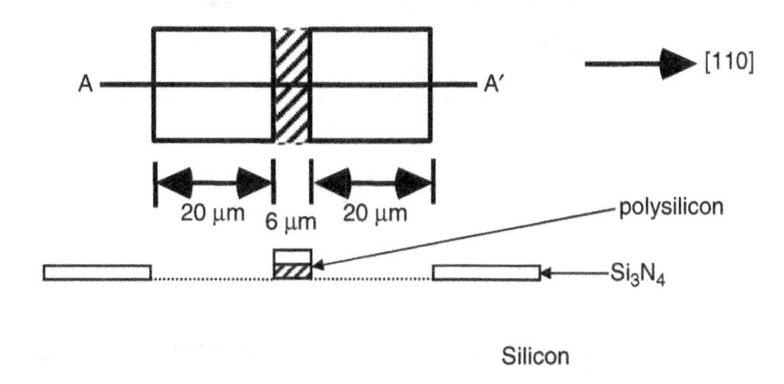

Problem 2.3 On a (100) silicon wafer (400 μm thick), 1000-Å-thick Si_3N_4 has been deposited and patterned for two $20 \times 20\ \mu m^2$ squares with 6-μm spacing between the two squares as shown below. Sketch the cross-sectional views (across A–A') of the resulting etch fronts if the wafer is etched in KOH for (a) 5 minutes, (b) 1 hour, and (c) 4 hours. Assume that the KOH etches (100) silicon plane at 1 μm/min while (111) silicon plane is etched at 0.02 μm/min. Calculate the etch depth, undercut length, etc. and indicate those on your sketches.

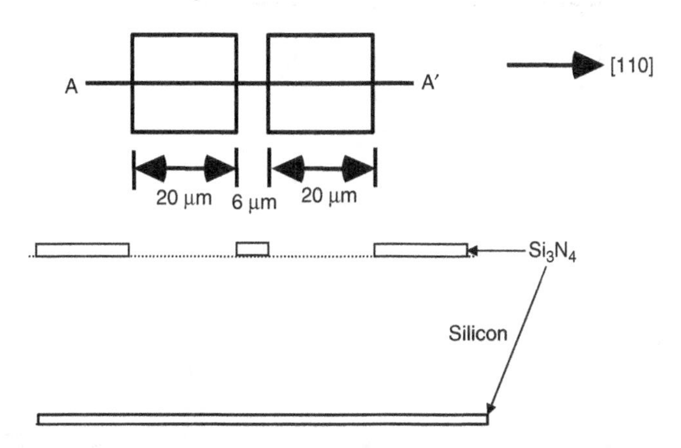

Problem 2.4 On a (100) silicon wafer (400 μm thick), a 1000-Å-thick Si_3N_4 has been deposited and patterned for a $40 \times 20\ \mu m^2$ *open* rectangle and an *open* square with 7-μm spacing between the two patterns as shown below. On the figure below sketch the cross-sectional view (across A–A′) of the resulting etch fronts if the wafer is etched in KOH for *2 hours*. Also, calculate the etch depth, undercut length, etc. and indicate those on your sketch. Assume that the KOH etches (100) and (110) silicon planes at 1 μm/min and 2 μm/min, respectively, while it etches (111) silicon plane at 0.02 μm/min.

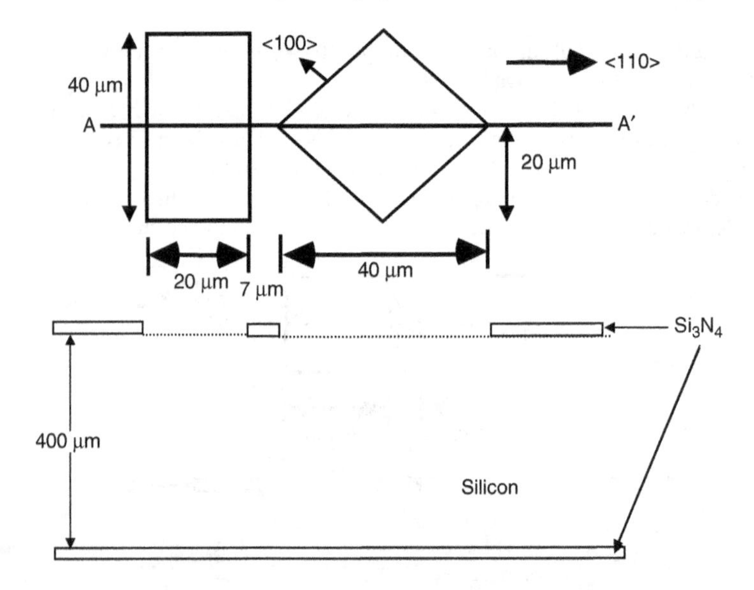

Problem 2.5 On a (100) silicon wafer (400 μm thick), a 1000-Å-thick Si_3N_4 has been deposited and patterned for a $40 \times 40\ \mu m^2$ *open* square and a $5 \times 5\ \mu m^2$ *open* square aligned to <100> direction as shown below. On the figure below sketch the cross-sectional views (across A–A′) of the resulting etch fronts if the wafer is etched in EPW for *10 minutes*. Also, calculate the etch depth, undercut length, etc. and indicate those on your sketch. Assume that the EPW etches (100) and (110) silicon planes at 1 μm/min and 0.5 μm/min, respectively, while (111) silicon plane is etched at a negligible rate.

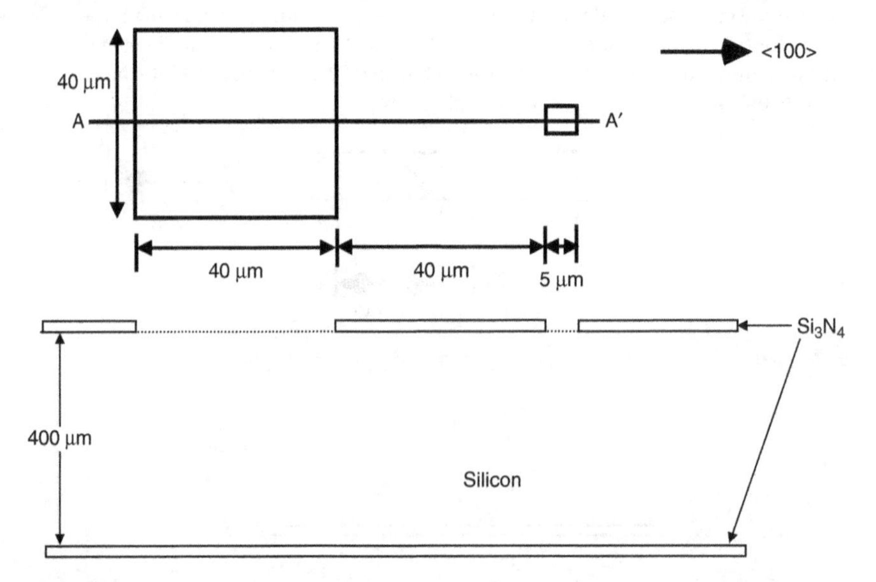

Problem 2.6 On a (100) silicon wafer (400 μm thick), 0.5-μm-thick silicon nitride is deposited and patterned as shown below. On the left, the silicon nitride is etched away over a 2-μm-wide rectangular track, i.e., only the 2-μm-wide "race" track area has the silicon nitride *removed*. On the right, four tiny areas of a 2-μm-wide rectangular race track have $1 \times 2\ \mu m^2$ silicon nitride still *remaining*. On the figure below sketch the cross-sectional views (across A–A′) of the resulting etch front, if the wafer is etched in KOH for *2 hours*. Also, calculate the etch depth and lateral undercut and indicate those on your sketch. Assume that the KOH etches (100) silicon plane at 1 μm/min, while (111) silicon plane is etched at 0.02 μm/min.

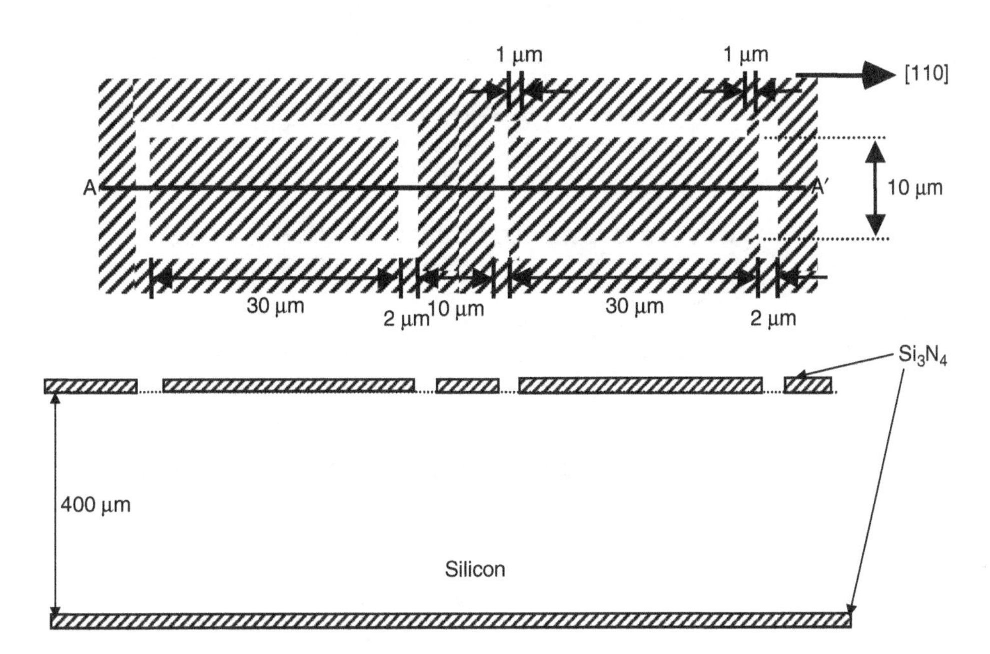

Problem 2.7 On a (100) silicon wafer (400 μm thick), 1000-Å-thick Si_3N_4 has been deposited and patterned as shown below. Sketch the cross-sectional views (across A–A′) of the resulting etch fronts, if the wafer is etched in KOH for (a) 2 minutes and (b) 200 minutes. Assume that the KOH etches (100) silicon plane at 1 μm/min while (111) silicon plane is etched at 0.01 μm/min. Calculate the etch depths and lateral undercut distance and indicate those on your sketches.

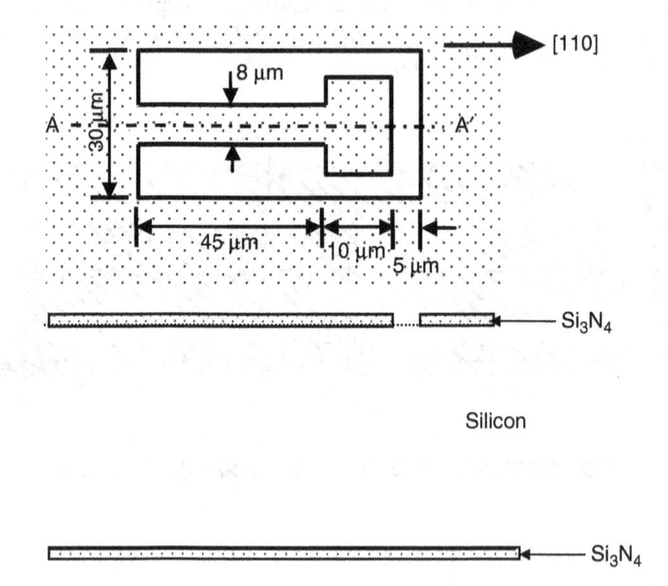

CHAPTER 3
Transduction Principles

There are many ways to covert electrical energy into mechanical or thermal energy (and vice versa), but we will study four most commonly used microactuator techniques in this chapter: electrostatic, electromagnetic, piezoelectric, and thermal actuation. Though the emphasis of this chapter is on learning the actuation principles, sensing principles also are covered as many microelectromechanical systems (MEMS) devices are sensors. In case of thermal transduction, for example, thermal actuation is inherently slow (with response to time typically slower than 1 ms) and/or too power hungry (especially for an array of many elements), but MEMS-based infrared (IR) sensing has offered unprecedented room-temperature night vision camera.

3.1 Electrostatic and Capacitive Transduction

Electrostatic actuation can be understood by considering attractive forces between two electrostatic charges of opposite polarities. If a stator and a movable piece are arranged as shown in Fig. 3.1a, there exist attractive forces between the opposite charges. If a movable electrode is placed between two stators (and the charges are produced through an applied voltage) as shown in the figure, the attractive forces are such that the vertical force components are all cancelled by each other, and there exists only a lateral force, which moves the movable piece to the left (if a voltage is applied to the left anchored electrode, as in Fig. 3.1a) or to the right (if the right anchored electrode is actuated, as in Fig. 3.1c). The charges shown in Fig. 3.1a can be produced by applying an electrostatic voltage between two electrodes, but an electrostatic voltage between two electrodes produces charges of opposite polarities on the two, thus resulting in only an attractive force. In other words, with an electrostatic voltage, it is not possible to produce a repulsive force unless there is an embedded charge nearby. When an applied voltage is removed, the movable electrode goes back to its original shape due to its mechanical spring constant.

3.1.1 Electrostatic Comb Drive

Arrangement of stators and movable structures in a comb configuration (Fig. 3.1b) can increase the attractive force by the factor corresponding to the number of the comb pairs (i.e., for n pairs, the force will be n times that of a single pair composed of two stators and one movable structure in between the two stators). If a large displacement is desired, a movable structure can be suspended by folded beams (with only two anchor

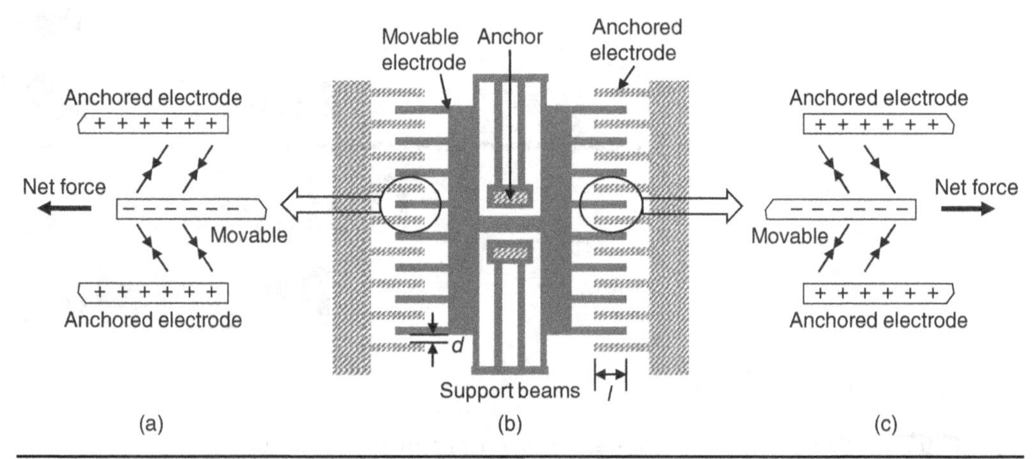

FIGURE 3.1 Electrostatically actuated comb drive: (a) and (c) expanded top views of portions of a comb drive showing two stators and one movable piece, (b) top-view diagram of a comb drive. (Adapted from [1].)

points as shown in Fig. 3.1b). The folded beam makes the structure very flexible by allowing a large displacement for a given electrostatic force, but cannot handle heavy weight nor deliver large force or torque. Comb fingers can also be arranged along a curved path to produce a rotational motion [1].

Since force is the first derivative of energy with respect to spatial distance, we can obtain a formula for the electrostatic force (on the movable comb fingers) exerted by an applied voltage as follows. The energy stored in a capacitor is $CV^2/2$, where C and V are the capacitance and applied voltage, respectively. For a comb finger surrounded by two stators (Fig. 3.1), the air-gap capacitance C is $2\varepsilon_o lt/d$, where ε_o = air permittivity, l = overlap distance between the comb finger and stators, t = thickness (or height) of the comb finger and stators (assuming that the comb finger and stators have same thickness), and d = air gap between the comb finger and stators (assuming that the two air gaps have same air gap). Differentiating the energy ($CV^2/2 = \varepsilon_o ltV^2/d$) with respect to l (the only variable that can change in response to an applied voltage), we obtain an equation ($F = \varepsilon_o tV^2/d$) for an electrostatic force exerted on the comb finger by an applied voltage.

If the number of the fingers (as well as the stator pairs) is n, the electrostatic force will be $n\varepsilon_o tV^2/d$. Thus, to maximize the force, one would want n and t (the thickness of the fingers and stators) to be as large as possible while keeping the air gap as narrow as possible.

Question: Calculate the electrostatic force (for an applied voltage of 10 V) between the fingers and stators in a comb drive actuator (Fig. 3.1) that has 2 μm for the height (or thickness) of the fingers and stators, 2 μm for the air gap between a finger and a stator, 10 for the number of the fingers, and 20 μm for the overlapping distance between the finger and the stator.

Answer: $F = n\varepsilon_o tV^2/d = 10 \times (8.85 \times 10^{-12}\ \mathrm{F/m}) \times (2 \times 10^{-6}\ \mathrm{m}) \times (10^2\ V^2)/(2 \times 10^{-6}\ \mathrm{m}) = 8.85 \times 10^{-9}\ FV^2/m = 8.85\ \mathrm{nN}$.

Question: In the comb drive shown in Fig. 3.1, does the electrostatic force in the x direction depend on the length (l) of the overlapping comb fingers in the x direction?

Answer: No.

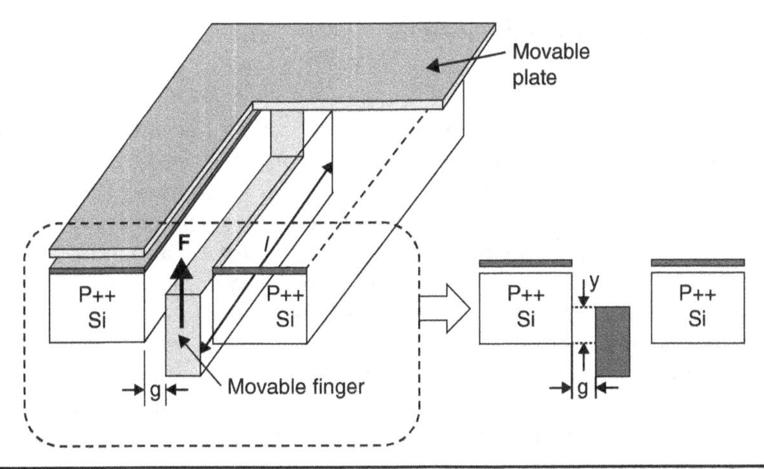

FIGURE 3.2 Conceptual diagrams of a vertical electrostatic comb drive. (Adapted from [2].)

Vertical Comb Drive

Vertically directed motion can be produced by comb fingers and stators that are arranged such that there is net electrostatic force that is directed vertically, as shown in Fig. 3.2 [2]. The comb fingers need to have a support mechanism that accommodates a vertical motion more readily than a lateral motion. In Fig. 3.2, the air-gap capacitance C is $2\varepsilon_o ly/g$, where ε_o = air permittivity, l = length of the movable finger that overlaps with the stators, y = overlapping vertical distance between the comb finger and stators (assuming that the adjacent stators have same height), and g = air gap between the comb finger and stators (assuming that the two air gaps have same gap). Differentiating the energy ($CV^2/2 = \varepsilon_o lyV^2/g$) with respect to y (the only variable that can change in response to an applied voltage), we obtain an equation ($F = \varepsilon_o lV^2/g$) for an electrostatic force exerted on the comb finger by an applied voltage. If the number of fingers (as well as the stator pairs) is n, the electrostatic force will be $n\varepsilon_o lV^2/g$. Thus, to maximize the force, we want n and l to be as large as possible, while keeping the air gap as narrow as possible. Electrostatic force of the vertical comb drive is equal to $n\varepsilon_o lV^2/g$, while that of the lateral comb drive is $n\varepsilon_o tV^2/d$, with g and d being the air gap between the movable and stationary fingers for each case. Thus, it is easier to make the force large with the vertical comb drive than with the lateral one, since l (the length in the vertical drive) can be made large more easily than t (the thickness in the lateral drive).

Question: For the vertical comb drive shown in Fig. 3.2, *sketch* applied voltage V as a function of the vertical distance y (for y between 0 and y_f). [Hint: The spring force of the movable polysilicon beam is equal to ky where k and y are the spring constant and the distance in vertical direction as indicated in Fig. 3.2, respectively. The electrostatic force (acting vertically on the movable beam) is equal to $\varepsilon_o V^2 l/2g$, where l and g are the beam length and the air gap, respectively, as indicated in Fig. 3.2.]

Answer: $ky = \varepsilon_o V^2 l/2g \rightarrow V = \sqrt{\dfrac{2gky}{\varepsilon_o l}}$

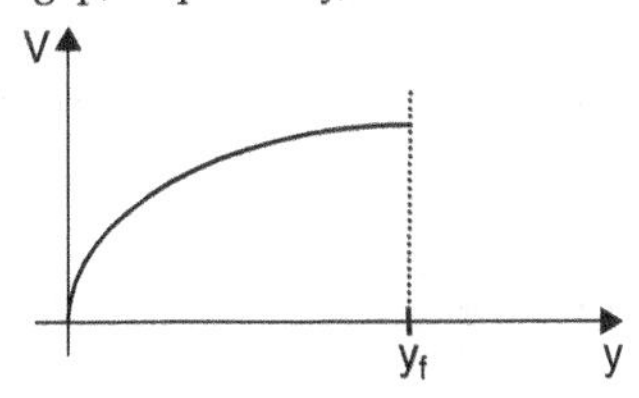

3.1.2 Electrostatic Micromotors

The electrostatic micromotor shown in Fig. 2.28 has 12 stators and four-pole rotor. The rotor, after being released, sits around a hub and on the substrate due to gravity. When an electrostatic voltage is applied between the rotor and any one of patterned stators, the rotor is attracted to the stator and rotates a little, being confined by the hub that allows the rotor to rotate but not to translate [3]. The applied electrostatic voltage produces an asymmetric fringing field (due to the rotor being at a lower position than the stator) that provides a little elevational force for the rotor and eases the friction between the rotor and the substrate. For continuous rotation of the rotor, the stators (12 of them in this case) can be grouped into three sets (#1, #2, and #3 as shown in Fig. 3.3), each consisting of four stators according to the four-pole rotor. Each of the three sets is applied with an electrostatic pulse only for one-third of a period, and the pulses are phased in time such that #1, #2, and #3 groups receive the pulses in sequence (in that order) without any overlap between them, so that the rotor is pulled close to #1, #2, and then #3 stators sequentially.

Another type of electrostatic micromotors that have been explored is a wobble motor with its rotating axis wobbling around a circular bearing as a rotor rotates around the bearing [4]. The rotor is one contiguous annular ring, which contacts the bearing as the rotor is pulled to stators (located outside and around the rotor with the air gap between the rotor and stator variable) by electrostatic voltages. Wobble motor has a larger initial air gap (between the rotor and stator) than the motor described in the previous paragraph, and the attractive electrostatic force between the rotor and stator (and thus the torque) can be higher because the air gap becomes very narrow when an electrostatic voltage is applied between them.

Application of Micromotors

Though micromotors can be boon for microrobots, their torque is very small, and there is difficulty of translating the rotor's motion out of the planar structure. One easy application, though, was considered to be on optical scanning by fabricating diffraction grating on top of a rotor [5]. The idea was to have an electrically rotatable diffraction grating on a rotor, so that diffracted light can be rotated around for an optical scanner. [Note: When a light with wavelength λ is illuminated on a diffraction grating with

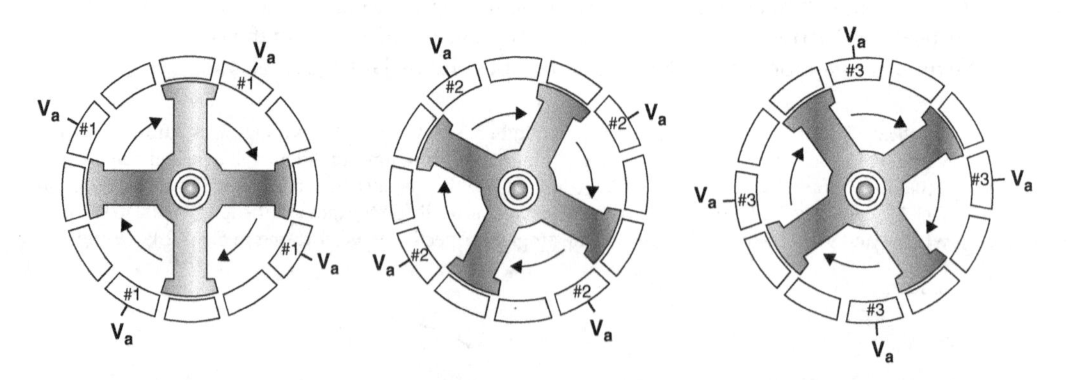

FIGURE 3.3 Schematics showing how the electrostatic micromotor (shown in Fig. 2.28) is actuated for continuous rotation of the rotor [3].

grating period Λ, a row of diffracted lights with diffraction orders q is spaced according to diffraction angle given by $\theta_q = \theta_i + q\lambda/\Lambda$, where $q = 0, \pm1, \pm2, \ldots$, θ_i = angle of incident beam relative to the normal.] If the rotor of a micromotor is made of polysilicon, the rotor surface is a bit too rough (for diffraction grating) due to the polysilicon's large grain size of 200–500 Å. Chemical mechanical polishing (CMP) on polysilicon was shown to reduce the surface roughness from R_a = 420 Å to 17 Å after removing 1500-Å-thick polysilicon [5].

3.1.3 "Pull-in Effect" in Electrostatic Actuation

In case of two parallel plates shown in Fig. 3.4a, the air-gap capacitance is $\varepsilon_o A/g$, where ε_o = air permittivity, A = overlapping area of the two parallel plates, and g = air gap between the two plates. Differentiating the energy ($CV_a^2/2 = \varepsilon_o AV_a^2/2g$) with respect to g (the only variable that changes in response to an applied voltage V_a), we obtain an equation ($F = \varepsilon_o AV_a^2/2g^2$) for an electrostatic attractive force acting on the two plates (separated by an air gap of g) for an applied voltage. This electrostatic force is balanced by spring force equal to $K(g_o-g)$, as the two plates are pulled closer by a distance equal to (g_o-g), where g_o and g are the initial and final gap, respectively. We, thus, have an equation $K(g_o-g) = \varepsilon_o V_a^2 A/2g^2$, from which we can obtain a curve for g/g_o versus V_a/V_c, up to V_c, as shown in Fig. 3.4b.

Interestingly, when the applied electrostatic voltage V_a is greater than V_c (commonly called the pull-in voltage), the equation shows that g decreases only with V_a reduction, which is not physically happening. What happens physically when the voltage reaches the pull-in voltage is that the increase of the electrostatic force (which is inversely proportional to the square of the gap) happens at a much higher rate than that of the spring force, as the gap is reduced, so much higher that the spring force is no longer strong enough to withstand the accelerating pace of the gap closure. The end result is that the movable plate is then "snapped" (or pulled) into the fixed plate, and we see the sharp decrease of the gap distance at the pull-in voltage. It turns out that the "pull-in effect" in electrostatic actuation happens when the two plates (separated by an air gap) are drawn together by *one-third* of the initial gap between the two plates. We can prove this with $K(g_o-g) = \varepsilon_o V_a^2 A/2g^2$ by calculating V_a as a function of g (starting from g_o) and then obtaining the critical value of g below which V_a becomes less as g gets less. The critical

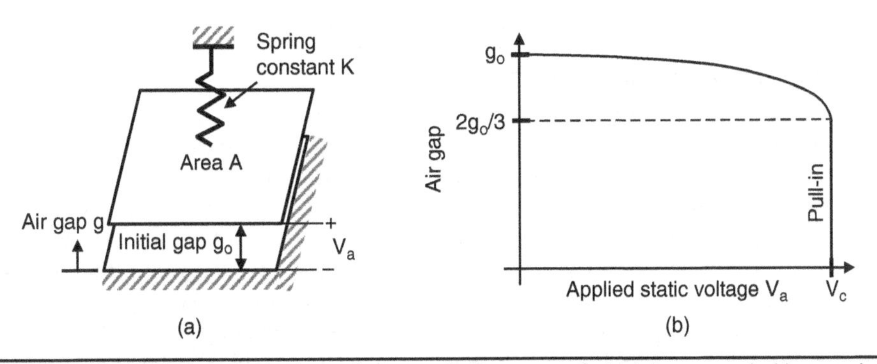

Figure 3.4 (a) Parallel plates with air gap between the two, one of which is suspended by a spring while the other is anchored to a substrate and (b) air gap g versus applied electrostatic voltage V_a, showing the pull-in effect when the gap is reduced more than one-third of the initial gap g_o.

g can be used to obtain the pull-in voltage, $V_c = [(8Kg_o^3)/(27\varepsilon_o A)]^{1/2}$. The actual derivations of the critical g and V_c are left as an exercise for Question 3.4 at the end of this chapter.

Question: The parallel plates in Fig. 3.4a have an initial air gap of 3 μm and has been measured to be electrostatically pulled down to the substrate by an applied voltage of 10 V (i.e., the on-set of the pull-in effect happens at 10 V applied). *Qualitatively* sketch the deflection of the cantilever as the applied electrostatic voltage is varied from 0 to 15 V.

Answer:

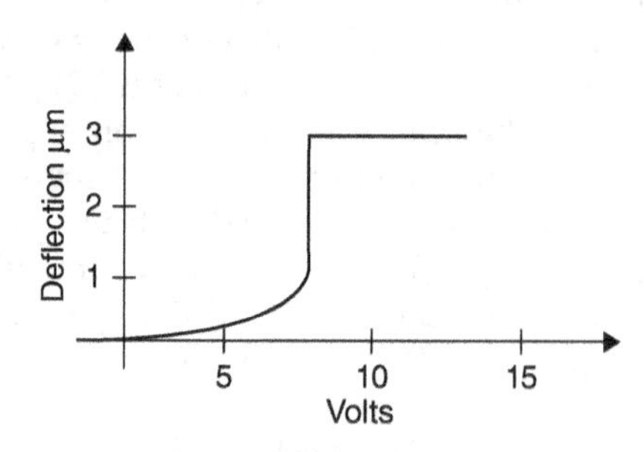

Question: For a parallel-plate, air-gap capacitor with one of the plates being suspended with a spring constant K equal to 10 N/m, while the other plate is fixed (unmovable), what will be the resulting air gap, if 75 V is applied? Assume that the top-view area and initial air gap of the plates are $100 \times 100\ \mu m^2$ and 1 μm, respectively, and $\varepsilon = 8.85 \times 10^{-12}$ F/m.

Answer: $V_c = [(8Kg_o^3)/(27\varepsilon_o A)]^{1/2} = \sqrt{\dfrac{8 \times 10 \times 10^{-18}}{27 \times 8.85 \times 10^{-12} \times 10^{-8}}} = 16.4\ \text{V}$

Since the applied voltage is greater than the critical voltage, the suspended plate is pulled-in, and the resulting air gap is zero.

Question: A bridge has an air gap of 1 μm between it and the underlying substrate, and has been measured to be electrostatically pulled down to the substrate by an applied voltage of 10 V (i.e., the on-set of the pull-in effect happens at 10 V applied). If the air gap is increased to 3 μm, at what voltage will the bridge be electrostatically pulled down?

Answer: Since $V_c \propto g_o^{3/2}$, it will be pulled down at 52 ($= 10 \times 3^{1.5}$) V.

Pull-in Effect in Cantilever

The pull-in effect can be picturized in a cantilever that is electrostatically attracted to a nearby fixed stator through in-plane bending, as illustrated in Fig. 3.5. In this case, a good physical insight can be obtained by considering charge distribution in the cantilever and stator as the cantilever is bent due to electrostatic force. The cantilever's free end portion that is bent most is closest to the stator and experiences largest attractive force (accompanied by largest induced charge density). As the applied voltage approaches the pull-in voltage (the same equation as the one shown in the previous section except that the spring constant is an equivalent spring constant for a cantilever going through the bending), the electrostatic force increases at a much higher pace than the spring force due to the gap reduction, and at a particular point, the spring force no longer withstands the electrostatic force, resulting in a snap-in (or collapse) of the cantilever into the substrate [6].

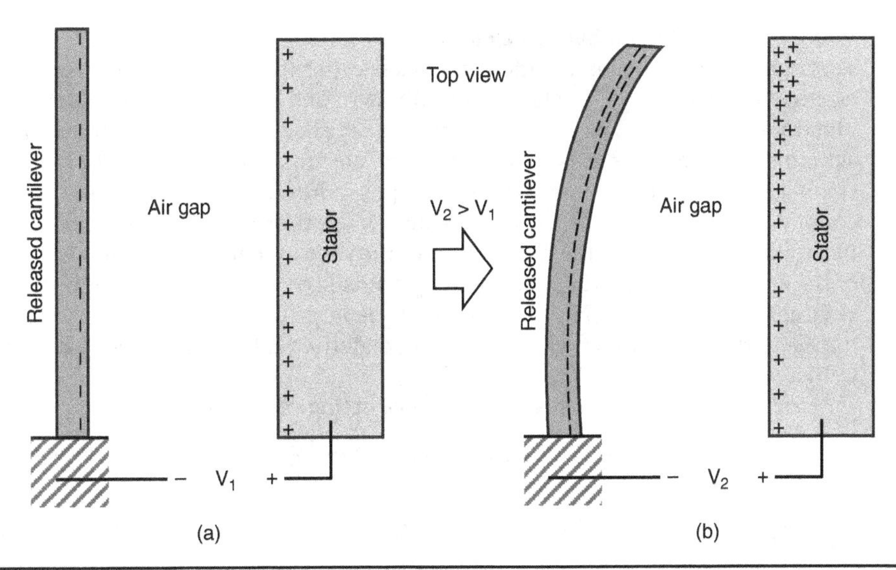

FIGURE 3.5 Top-view schematics of a released cantilever and a fixed stator with a voltage applied between the two: (a) under little cantilever deflection at a low applied voltage and (b) under a larger cantilever deflection with a larger applied voltage.

One can avoid the pull-in effect by making the gap nonuniform (i.e., making the gap wider where the cantilever displacement is larger) so that the electrostatic force does not increase so rapidly as the gap closes, as illustrated in Fig. 3.6. If a fixed stator is curved according to $s(x) = \delta_{max}(x/L)^n$ with x being the distance a long the length L direction, no pull-in effect is reported to be observed for $n \geq 3$ for the released cantilever [7]. There is a pull-in effect for $n = 0, 1, 2$, but not for $n \geq 3$.

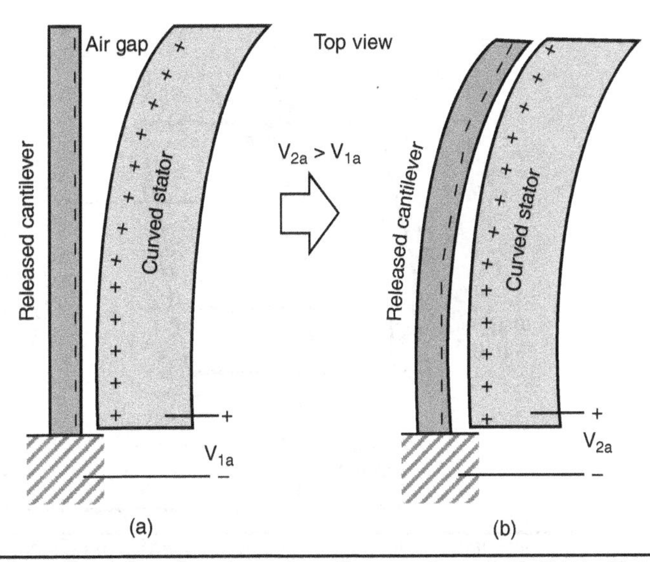

FIGURE 3.6 Top-view schematics of a released cantilever and a curved stator with a voltage applied between the two: (a) under little cantilever deflection at a low applied voltage and (b) under a larger cantilever deflection with a larger applied voltage.

No Pull-in Effect with a Series Fixed Capacitor

The pull-in effect can be avoided if a fixed capacitor is integrated in series with an electrostatic actuator (i.e., electrostatically actuated air-gap narrowing structure), as illustrated in Fig. 3.7. The idea is to divide an applied voltage between the electrostatic actuator and the series capacitor so that the electrostatic actuator gets proportionally smaller amount of the voltage when the gap is reduced. This can be seen easily with the equation for the voltage (V_o) appearing across the electrostatic actuator as a function of the applied voltage (V_a), the capacitance of the series capacitor (C_1), and the capacitance of the electrostatic actuator (C_o). When C_o is small compared to C_1 (i.e., when the air gap is large), V_o is about V_a. However, when C_o is large due to the air-gap reduction produced by the applied voltage, V_o is approximately equal to C_1V_a/C_o, which decreases as C_o increases.

In this case, electrostatic attractive force acting on the two plates (separated by an air gap of g) is $F = \frac{\varepsilon_o A V_o^2}{2g^2}$, and we have $K(g_o - g) = \frac{\varepsilon_o A V_o^2}{2g^2}$, as before. Since $V_o = \frac{C_1}{C_o + C_1} V_a = \frac{C_1}{\varepsilon_o A/g + C_1} V_a$, the equation becomes $(g_o - g)g^2 = \frac{\varepsilon_o A}{2K}\left(\frac{C_1 V_a}{\varepsilon_o A/g + C_1}\right)^2$, from which we get $\pm(\varepsilon_o A + C_1 g)\sqrt{g_o - g} = C_1 V_a \sqrt{\frac{\varepsilon_o A}{2K}}$. Of the two possible equations, only one is physically valid since the gap g has to be always positive for any given V_a. Now, from $(\varepsilon_o A + C_1 g)\sqrt{g_o - g} = C_1 V_a \sqrt{\frac{\varepsilon_o A}{2K}}$, we can obtain a curve for g/g_o versus V_a/V_c up to the pull-in voltage V_c similar to Fig. 3.4b. It can be shown that with a proper value of C_1 with respect to the initial capacitance of the air-gap capacitor before any voltage is applied, i.e., C_i $(= \varepsilon_o A/g_o)$, the pull-in effect can be avoided.

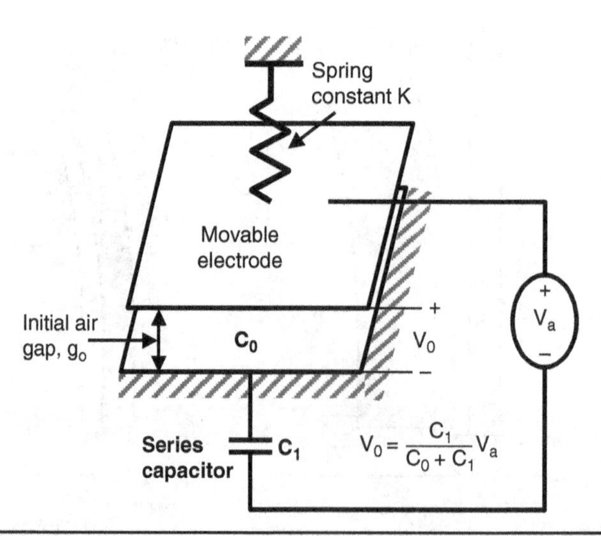

FIGURE 3.7 For a controllable deflection of g_o, the initial gap can be increased to $3g_o$ or a series capacitor can be added to have control over the g_o without pull-in effect, as shown here. (Adapted from [8].)

Question: For an air-gap varying capacitor integrated with a series capacitor (shown in Fig. 3.7), what is the maximum C_1 that will make the movable plate (of the air-gap varying capacitor or "original actuator") not experience any electrostatic pull-in effect as the applied voltage V_a is varied? Give your answer in terms of the initial capacitance of the "original actuator" (i.e., $C_i = \varepsilon_o A/g_o$, where ε_o, A, and g_o are air dielectric constant, and the top-view area and initial air gap of the "original actuator," respectively).

Answer: If we view Fig. 3.4b as a curve for V_a versus g/g_o (rather than g/g_o versus V_a), we can see that the pull-in happens when $dV_a/dg = 0$. So, differentiating the left side of $(\varepsilon_o A + C_1 g)\sqrt{g_o - g} = C_1 V_a \sqrt{\frac{\varepsilon_o A}{2K}}$ with respect to g and setting it to zero, we obtain $C_1\sqrt{g_o - g} - \frac{\varepsilon_o A + C_1 g}{2\sqrt{g_o - g}} = 0$, and $g = \frac{2g_o}{3} - \frac{\varepsilon_o A}{3C_1}$, from which we see that $g \le 0$ for $C_1 \le \frac{\varepsilon_o A}{2g_o}$. Consequently, if C_1 is less than or equal to $C_i/2$, there is no pull-in effect.

Example of Air-Gap Varying Capacitor Integrated with Series Capacitor

For air-gap varying capacitor (C_s) integrated with a series capacitor (C_1) formed with ZnO film, shown in Fig. 3.8, the following equation can be obtained for the electrostatic pull-in voltage for the bridge with its two edges clamped

$$V_{pi} = \sqrt{\frac{8K_{\text{eff}}\left(g_o + \frac{h_o \varepsilon_o}{\varepsilon_{\text{ZnO}}}\right)^3}{27A\varepsilon_o}}$$

where K_{eff}, g_o, h_o, and A are the spring constant, initial air gap between the top movable bridge and ZnO top surface, ZnO film thickness, and actuation area, respectively, while ε_o and ε_{ZnO} are the dielectric constants of air and ZnO, respectively. The instability between the electrostatic force and the restoring force due to the stiffness of the beam occurs when the air gap (g) is

$$g = \frac{2g_o - \frac{h_o \varepsilon_o}{\varepsilon_{\text{ZnO}}}}{3} \quad \left(\text{for } g_o \ge \frac{h_o \varepsilon_o}{2\varepsilon_{\text{ZnO}}}\right)$$

It is possible to avoid the pull-in effect (for a continuous capacitance variation) if the initial air gap (g_o) is smaller than the ZnO thickness (h_o) times $\varepsilon_o/2\varepsilon_{\text{ZnO}}$ (≈0.06). However, that requires the initial C_s to be larger than $2C_1$.

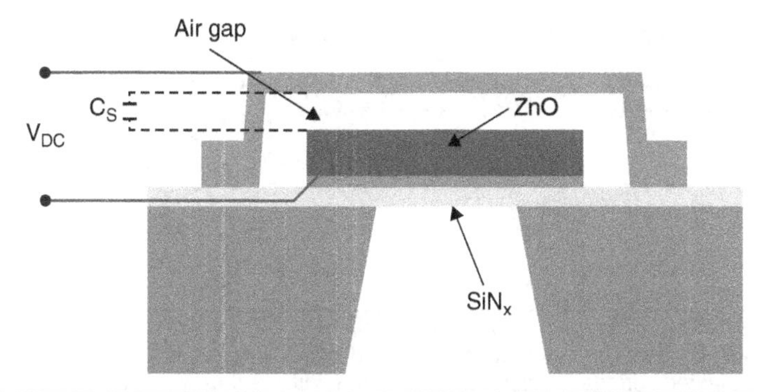

FIGURE 3.8 Cross-sectional view of an air-gap varying capacitor (C_s) integrated with a series capacitor based on a dielectric layer such as ZnO.

Question: For an air-gap varying capacitor integrated with series capacitor shown in Fig. 3.8, what is the maximum initial air gap, if we want to avoid the pull-in effect (for a continuous capacitance variation), for ZnO thickness and $\varepsilon_o/\varepsilon_{ZnO}$ being 2 μm and 0.12, respectively?

Answer: $g_{o,max} = h_o\varepsilon_o/2\varepsilon_{ZnO} = 2\ \mu m \cdot 0.12/2 = 0.12\ \mu m$

Electrostatic Scratch Drive Actuator Using Pull-in Effect

The working principle of electrostatic scratch drive actuator (SDA) which takes advantage of the pull-in effect is illustrated in Fig. 3.9 [9]. When a released structure shown in the figure is electrostatically pulled down (with a voltage V) and then goes back to its original form through mechanical restoring force (with the voltage reduced to zero), the structure has moved laterally by a certain distance Δx. If the cycle of the voltage application followed by zero voltage is repeated, the structure keeps moving laterally.

In Fig. 3.9 there is only one anchor for the released structure that consists of an SDA attached to a link frame and a buckling beam. The right end of the buckling beam is directly anchored on the silicon substrate through a conductive layer (sandwiched by insulating layers at its top and bottom), while the left end is connected to a link frame. For this connection, two narrow and short torsion bars are provided at the left end of the buckling beam. The driving voltage applied to the conductive layer reaches the SDA through the buckling beam and link frame. Since the conductive layer and the buckling beam are at the same potential, there is no electrostatic force on the buckling beam nor on the mechanical links, but there is electrostatic force between the SDA and the silicon substrate. As the applied voltage is cycled between 0 and the pull-in voltage, the SDA moves forward (or laterally rightward) and pulls the link frame, which in turn pulls the left part of the buckling beam, causing it to buckle.

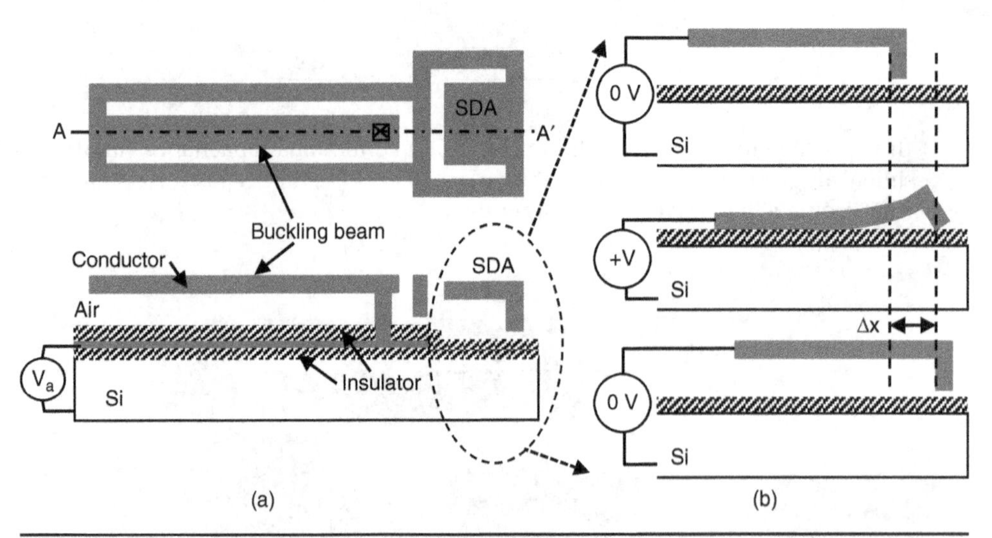

FIGURE 3.9 (a) Top view and cross-sectional (across A–A') views of an electrostatic scratch drive actuator (SDA) based on released fold-up conductive film and (b) cross-sectional views of SDA when the applied voltage is raised from 0 V to a voltage that will pull down the SDA beam to the silicon substrate and then back to 0 V. (Adapted from [9].) The cycling of the voltage from 0 V to a pull-in voltage and then back to 0 V makes the SDA to advance by Δx in one direction.

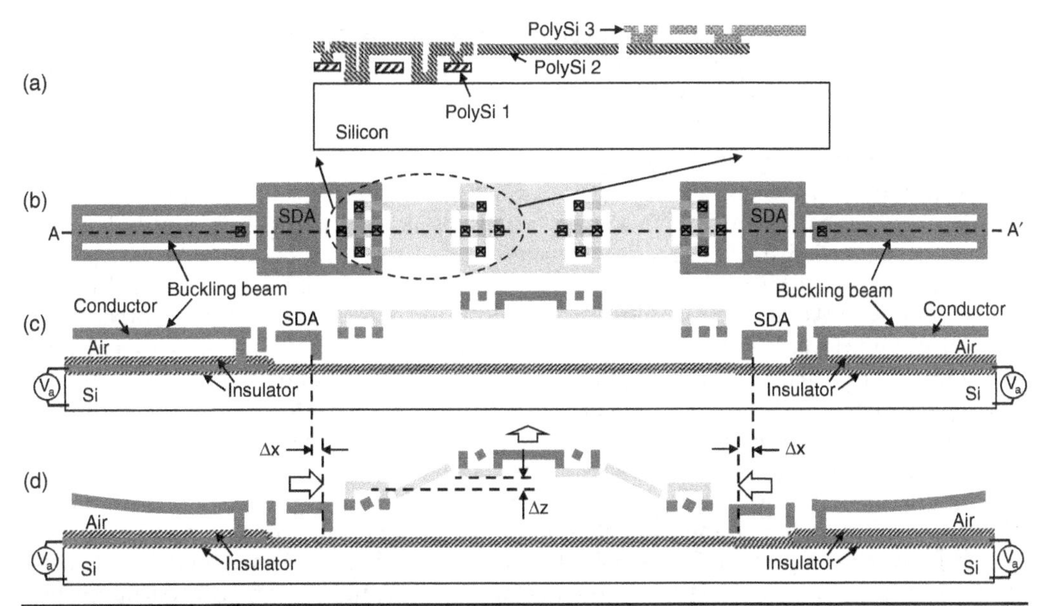

Figure 3.10 Vertically movable stage with a pair of SDAs and two types of hinges: (a) cross-sectional view of two types of hinges across A–A′, (b) top-view schematic showing the SDAs, two types of hinges, and the vertically movable stage (in the center), (c) cross-sectional view across A–A′ after fabrication, before driving SDAs, and (d) across A–A′ after electrically actuating SDAs to generate a linear translation of Δx, which causes the hinged plates to rotate upward, lifting up the vertically movable stage.

Question: In an SDA shown in Fig. 3.9, how will the "buckling beam" bend when the SDA is electrostatically pulled down?

Answer:

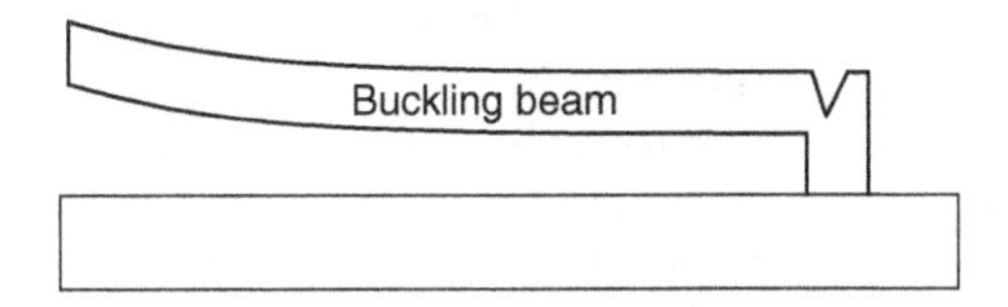

Such an actuator has been used to move a hinged structure upward or downward, as shown in Fig. 3.10. The plates are directly connected to each other with two types of hinges shown in Fig. 3.10. The stage cannot be moved in both x and y directions unless a sliding joint is incorporated to allow independent movement of the stage in both directions [10].

3.1.4 Electrostatic Repulsion Force through Nonvolatile Charge Injection

An electrostatic voltage applied between two objects produces only an attractive force between those two. However, if an object with embedded charge is nearby, a repulsive force can be obtained between it and one of the two objects being applied with an electrostatic voltage. In one example, a floating gate of an electrically erasable programmable read only memory (EEPROM) structure is charged with electrons, and a pair of electrodes is placed over it [11]. When an electrostatic voltage is applied between the two electrodes, one of them is negatively charged and experiences a repulsive force from the floating gate that has already been negatively charged. A repulsive force of 0.2 μN across 3-μm gap for 360-μm-long beams is reported after electrically floating beams are charged with 45 V for 10 seconds [11].

3.1.5 Capacitive Sensing

Variation of the air gap between two parallel plates results in capacitance change and has been used for many MEMS sensors, some of which will be covered in later chapters. Here, we will just overview some ways to covert such a capacitance change into electrical signal for capacitive sensors. Capacitance bridge such as Wien or Schering bridge with AC voltage would be a straightforward way of converting capacitance change into an electrical signal, but power consumption through the resistive element(s) in the bridge may not be negligible. An approach used in condenser microphone is to apply a DC voltage across a variable capacitor and convert a time-varying capacitance into an AC voltage, as described in Chap. 7 for MEMS microphone. Another easy approach is to use an op-amp-based charge amplifier (Chap. 10) with a variable capacitor between the input and output of the op amp, but how low frequency the amplifier can cover depends on the equivalent input resistance (of the charge amplifier) which cannot be too high for DC bias stability. A relaxation oscillator based on RC or LC with a variable capacitor C offers frequency change as a function of varying capacitance, but its temperature sensitivity and electronics tend to be higher and more complex, respectively, than other approaches. Or one can use a capacitance-to-digital converter (e.g., Texas Instruments' FDC1004) that charges a variable capacitor with step pulses (e.g., at 25 kHz) to positive level on a half cycle (followed by negative level on the other half), which is then transferred to a sample-and-hold circuit via a switched capacitor circuit before being converted to digital signal by a sigma-delta analog-to-digital converter (ADC). In all these approaches, though, parasitic capacitance plays a significant role and needs to be minimized or well characterized for the lowest minimum detectable capacitance.

A capacitive sensor with a variable air gap allows a force feedback to maintain the air gap constant as the gap varies in response to the external input (e.g., acceleration). Thus, by monitoring the needed voltage to maintain the gap constant, one can measure a wide range of sensing measurand, though the air gap may be very small (for the sake of good sensitivity). We will see a good example of this force-balance feedback when we study a force-balance capacitive MEMS accelerometer in Chap. 6.

3.2 Electromagnetic Transduction

3.2.1 Magnetic Actuation versus Electrostatic Actuation

The energy density in a gap for a conventional electromagnetic motor is $U_M = B^2/2\mu_o \approx 1\ \text{MJ/m}^3$, a remarkably large value. However, scaling such a motor to micron scale has been hampered by difficulty in fabricating large-turn coils on a chip. Also, as the dimension is scaled down, ohmic loss in the coil as well as leakage of magnetic flux in a planar structure reduce the energy conversion efficiency.

On the other hand, MEMS technology naturally renders structures that can be electrostatically actuated, and the first micromotor was an electrostatic one. However, the energy density, $U_E = \varepsilon_o E^2/2$, that can be stored in an air gap of an electrostatic motor is limited by the breakdown E-field of air, which is about 3 MV/m for macroscopic gaps. Thus, U_E is four orders of magnitude lower than U_M for macroscale motors, and thus electromagnetic actuation is far more efficient than electrostatic actuation for macromotors. However, as the air gap is reduced to narrower than 4 μm, the breakdown E-field increases drastically, as shown in Paschen's relationship (Fig. 2.35), at standard atmospheric pressure. The reason for such increase is that the traveling distance of electrons is now so much less that it is harder for electrons to gain high enough energy to ionize gas molecules. The breakdown E-field can potentially be increased by two orders of magnitude as the air gap is reduced submicron level, and U_E can be close to U_M for microscale motors. Smooth metal surfaces are experimentally observed to breakdown at over 150 MV/m. Note, however, the breakdown field cannot be higher than 1 GV/m since there will be cold emission of electrons by an electrical field around 1 GV/m.

Question: At an ambient pressure of 1 bar, what are the breakdown electrical fields for parallel plates separated by 2, 20, and 200 μm air gap.

Answer: From Fig. 2.35, the breakdown voltages are 500, 300, and 1,000 V for 2, 20, and 200 μm air gap, respectively. The breakdown electrical fields are then 250 MV/m (= 500 V ÷ 2 μm), 15 MV/m (300 V ÷ 20 μm), and 5 MV/m (= 1,000 V ÷ 200 μm).

Question: For two parallel plates separated by a 2-μm air gap, calculate the breakdown electrical fields in ambient pressures of 1, 0.1, and 1000 bar.

Answer: With $d = 2$ μm $= 2 \times 10^{-3}$ mm, we get the following from Paschen's curve:

for 1 bar, $V_{BD} = 500$ V $\rightarrow E_{BD} = 500$ V ÷ 2 μm = 250 MV/m
for 0.1 bar, $V_{BD} > 10$ kV $\rightarrow E_{BD} > 10$ kV ÷ 2 μm = 5 GV/m $\rightarrow$ 1 GV/m (emission field)
for 1,000 bar, $V_{BD} = 7$ kV $\rightarrow E_{BD} = 7$ kV ÷ 2 μm = 3.5 GV/m $\rightarrow$ 1 GV/m (emission field)

Question: For the comb drive with the comb fingers separated from the stationary electrode by 2 μm air gap (Fig. 3.1), what is the *maximum* DC bias voltage that can be applied between the comb fingers and the stationary electrode in ambient pressure (i.e., 1 bar) without breaking down the air?

Answer: The Paschen curve shows about 500 V breakdown voltage for 0.002 bar·mm.

3.2.2 Electromagnetic Actuators

There are three major approaches to move an object with electromagnetic actuation. The first approach is to use reluctance force stemming from energy gradient as a function of space, similar to electrostatic force. The energy stored in an inductor is $Li^2/2$, where L and i are the inductance and applied current, respectively. Differentiating the energy with respect to x (the variable that changes in response to an applied current), we obtain an equation $\left(F = \frac{1}{2} i^2 \frac{dL}{dx}\right)$ for a force.

The force can be large for large inductance gradient and current. Reported force and traveling distance by this approach have been orders of magnitude larger than those by electrostatic actuation. Note, however, the relatively large power consumption owing to the unavoidable series resistance (R) that consumes power (P) through i^2R heating. Also, for large inductance gradient, coils have to be wound and may present difficulties for miniaturization and mass production.

For a movable plunger (or a plunger spring), permalloy which is a combination of Ni and Fe (e.g., 78% Ni and 22% Fe) is often used since it can easily be electroplated. A 3-mm-long electromagnet formed by winding 25-μm-thick wire was shown to have $L \approx 2\text{–}50$ mH per 100 turns below 100 kHz and $R \approx 10\ \Omega\ /100$ turns, and to produce $F > 1$ mN and travel distance > 250 μm at a power consumption of < 1 mW [12].

Question: For the electromagnetic actuator reported in [12], the paper reported that the electromagnetic force was greater than 1 mN with power consumption < 1 mW. What was dL/dx if the electromagnet had 100 turns of 25-μm-thick wire (yielding 10 Ω)?

Answer: $\frac{dL}{dx} = \frac{2F}{i^2} = \frac{2F}{(P/R)} = \frac{2\times 1}{1/10} = 20\ \text{N/A}^2$

Example 1: In-Plane Electromagnetic Actuator for Flexible Metallic Microstructures
A mechanically tunable IR filter for wavelength cutoff was made with LIGA (based on x-ray lithography, electroplating, and plastic molding) which allowed the tall high-aspect-ratio microstructures of electroplated permalloy (78% Ni and 22% Fe), as shown in Fig. 3.11 [13]. The cutoff wavelength was designed to depend on the air-gap spacing (between the permalloy plates due to waveguide effect), which was varied a large amount as the plunger was moved laterally through electromagnetic actuation by an electromagnetic drive. The air gap between the plunger and the electromagnetic drive was 5 μm, and the plunger was moved a large distance without any pull-in effect as an electrical current was applied to the electromagnetic drive.

Question: For the electromagnetic actuator made of a plunger and an electromagnet as shown in Fig. 3.11, calculate the electromagnetic force that the plunger experiences, if a 10 mA DC current flows through the electromagnet. Also, indicate the direction of the force on the figure. Assume that the magnitude of dL/dx is equal to 10 H/m.

Answer: $F = \frac{i^2}{2}\frac{dL}{dx} = \frac{(10\times 10^{-3}\ \text{A})^2}{2} \times (10\ \text{H/m}) = 0.5$ mN. The force direction is indicated with hollow arrow in Fig. 3.11.

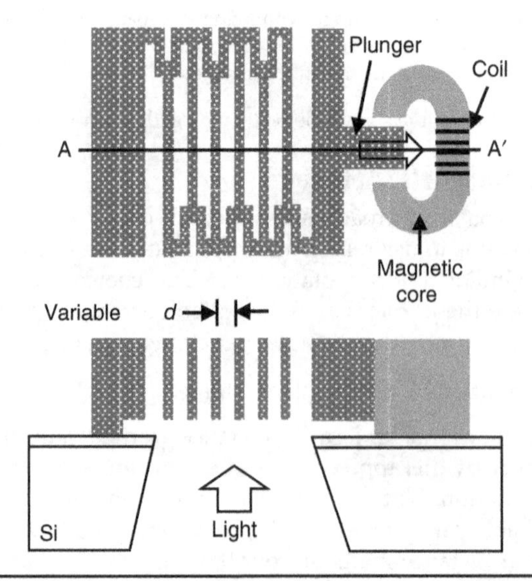

FIGURE 3.11 Top-view and cross-sectional (across A–A′) views of an electromagnetically driven tunable light filter with spring fixtures joining the ends of the filter plates. (Adapted from [13].)

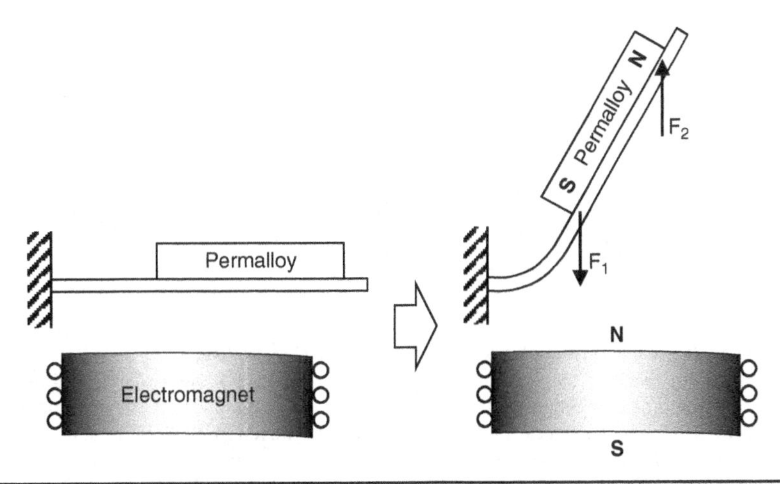

FIGURE 3.12 Out-of-plane permalloy magnetic actuator.

Example 2: Out-of-Plane Magnetic Actuator

Another approach for electromagnetic actuation is to use a soft magnetic film such as permalloy in a nonuniform magnetic field (that is produced by an electromagnet for electrical control), as shown in Fig. 3.12. One end of the soft magnetic film is magnetized into a pole (e.g., south in case of Fig. 3.12) opposite to the pole which exists near that end where the magnetic field is the highest. The other end of the magnetic film is then automatically magnetized into the other pole (north in Fig. 3.12) due to duality of magnetic poles. Once the two ends of the magnetic film are magnetized this way, the film experiences repulsive and attractive forces (in the nonuniform magnetic field), which bend the beam according to the bending moments that are proportional to the product between the force and the arm. With this approach, a 70° angular deflection was demonstrated with 5-μm-thick electroplated permalloy over a 1.8-μm-thick polysilicon [14].

Question: A long and narrow polysilicon cantilever with permalloy on top is aligned in two different positions with respect to an electromagnet in the length direction of the cantilever, as shown below, while in the width direction the cantilever is aligned in the center of the electromagnet in both cases. Sketch the bending curves for both cases when the electromagnet is actuated to produce magnetic field.

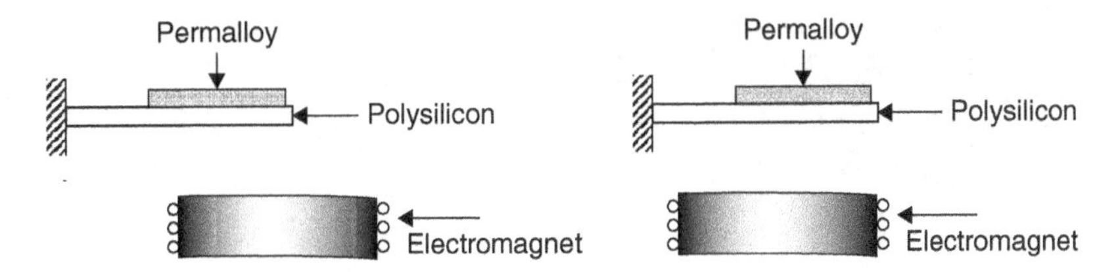

Answer: The left one will bend downward, while the right one will bend upward (similar to Fig. 3.12). The permalloy will be polarized with the opposite pole at a point that is closest to the pole in the electromagnet. Thus, the left one will have its opposite pole (say, south pole in response to north pole induced on the electromagnet) at the free end of the cantilever, as illustrated below, while the right one will have its opposite pole (say, south pole for the same north pole induced on the electromagnet) in the center of the cantilever in its length direction. The center of the left cantilever will have north pole induced on it due to the south pole in its free end. Since the opposite poles attract each other and because the force F_2 acting on the free end will be more effective than the force F_1 on the center due to leveraging arm effect, the left cantilever will be bent downward, as shown below.

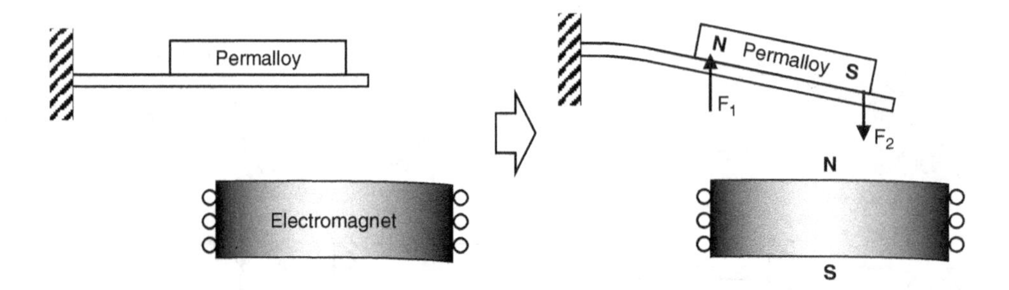

Example 3: Electromagnetic Actuation with Planar Coil on Movable Plate

A third approach to obtain electromagnetic actuation is to use a planar coil on a movable plate. Magnetic moment $\boldsymbol{m}$ generated by the coil works with a magnetic field $\boldsymbol{B}$ to produce a torque ($\boldsymbol{\tau} = \boldsymbol{m} \times \boldsymbol{B}$) and force [$\boldsymbol{F} = (\boldsymbol{m} \cdot \nabla)\boldsymbol{B}$]. For the force to be nonzero, the magnetic field must be nonuniform in space. For a rotatable mirror shown in Fig. 3.13a, a checkerboard array of permanent magnets can be placed under an array of the deflectable mirrors for high magnetic field gradient, as done in [15]. Spiral coils can be patterned on the nonreflective side of the square "mirror" plate that is suspended by a set of two torsional beams that are supported by a narrow frame suspended by another orthogonal set of torsional beams. The other side of the suspended mirror plate is deposited with Al for a reflective mirror. A similarly designed mirror was shown to rotate up to ±10° when a current was flown through the spiral coils [15].

Question: For the electromagnetic mirror described in Fig. 3.13, calculate the power dissipation in the coil, if 1 mA of DC current flows through the coil. Assume that the coil is made of patterned Al film that is 1 μm thick, 10 μm wide, and 10 μm long. Also, assume that the Al resistivity is 3×10^{-6} Ωcm.

Answer: Since $R = \rho\frac{L}{A} = (3 \times 10^{-6}\ \Omega\text{cm})\frac{1\ \text{cm}}{1 \times 10 \times 10^{-8}\ \text{cm}^2} = 30\ \Omega$, the power dissipation is $i^2R = (10^{-3})^2 \times 30 = 30\ \mu\text{W}$.

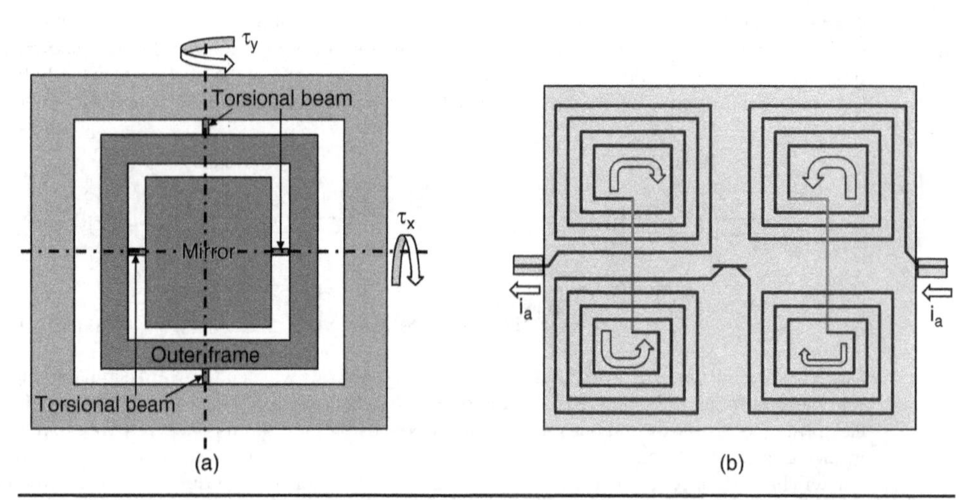

Figure 3.13 (a) Top view of electromagnetically actuated mirror supported by a set of two torsional beams that are supported by an outer frame which also is supported by an orthogonal set of two torsional beams and (b) a set of four spiral coils deposited on the nonreflective side of the square mirror plate. (Adapted from [15].)

Question: For the electromagnetic mirror described in Fig. 3.13, if we simplify the mirror to have only two spiral coils as shown below, in what direction (x, y, or z) does the magnetic field have to be applied if we want a torque to rotate the mirror around the torsional support beams (along the x direction) by a constant magnetic field only (i.e., a field with zero field gradient).

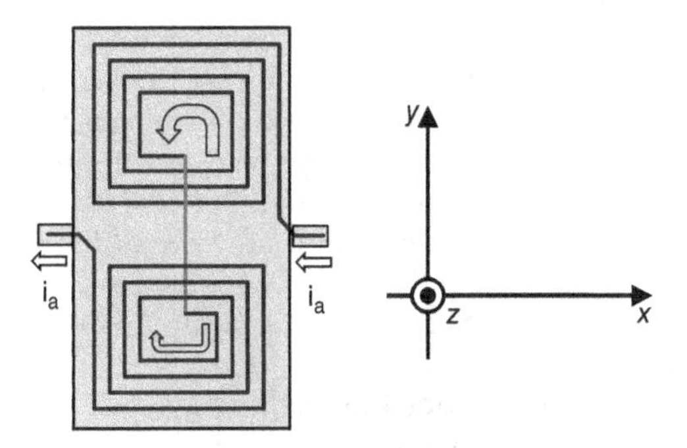

Answer: y direction, since each of the coils produces magnetic moment m in the z direction.

Question: For the electromagnetic mirror described in Fig. 3.13, would the torque τ_x rotate the outer frame by any significant amount?

Answer: No.

3.2.3 Magnetic Field Sensing

Magnetic field can be sensed through a variety of ways including Hall effect–based semiconductor, magnetoresistor, magnetodiode, magnetotransistor, flux gate, superconducting quantum interference device (SQUID), optically pumped magnetometer based on Zeeman effect, and induction coil [16]. Induction coil is the most versatile as it can detect wide range of field strengths but cannot detect *static* DC magnetic field. All the others listed above can detect static magnetic field but over different ranges of field strength and frequency with different resolutions. The most commonly used magnetic sensor is the one based on Hall effect, which is essentially a piece of a doped semiconductor with electrical current applied along one direction (e.g., x direction) and magnetically induced voltage measured along the orthogonal direction (e.g., y direction) to both the current direction and the magnetic field direction (e.g., z direction). The minimum detectable field by Hall effect magnetic sensor is about 100 nT, which is pretty good in light of the fact that the earth's magnetic field on the equator is 31 μT. A magnetoresistor can offer a better resolution at a cost of higher power consumption, and anisotropic magnetoresistors (AMRs) have been widely used. Flux gate can detect even 0.1 nT but requires a pair of orthogonal coils around a core [17].

Quantum-effect–based SQUID and optically pumped magnetometer can detect even 0.1 and 5 pT, respectively. However, SQUIP requires cooling to 4.2 K (–269°C) for current to flow between two superconductors separated by a thin insulation layer under 0 V, while optically pumped magnetometer requires elaborate vapor cell and electronics (similar to what are needed for atomic clock). Just as a reference, a magnetometer in a smartphone has about ±2 mT range. A typical fridge magnet has 5 mT of magnetic field strength, while magnetic resonance imaging (MRI) uses 1.5–3 *T* magnetic field.

3.3 Piezoelectric Transduction

3.3.1 Piezoelectric Effects

Piezoelectric phenomena can very well be understood by considering four ideal cases: two for actuation (with applied voltage) and two for sensing (with applied force). When a voltage is applied between two plates sandwiching a piezoelectric medium, stress (T) and strain (S) will be developed in the medium through a "converse" piezoelectric effect. If the medium is not bound by anything, it is free to expand or contract, and the stress (T) will be zero, while there is finite strain (that is a shape change), which is accompanied by an induced polarization, as shown in Fig. 3.14. In this case, $S = d_t E + s^E T$ is the best equation to use to calculate the strain (S) as a function of applied electrical field (E), with $T = 0$.

Now, if the medium is somehow held to a constant shape so that S is zero, as in Fig. 3.15, then the applied voltage produces stress (T) in the medium. In this case, $T = -e_t E + c^E S$ is the best equation to use to calculate the piezoelectrically induced stress (T) as a function of applied electrical field (E) with $S = 0$. Note that in this case there is no induced polarization in the piezoelectric medium, since there has been no strain.

In case of sensing with applied force, if the two electrodes are shorted with an external wire, as shown in Fig. 3.16, there is actual flow of electrons from one electrode to the other through the wire due to piezoelectric effect (i.e., mechanical strain being converted to electrical polarization). Actually, the induced polarization is the one that attracts electrons from one side to the other side through the wire. In this case, $D = e^T E + dT$ is the best equation to use to calculate electric displacement (or electric flux density) D as a function of T with $E = 0$.

Now, in case of sensing with applied force, if the two electrodes are open-circuited as shown in Fig. 3.17, the induced polarization has no external path to draw electrons from one electrode to the other. Consequently, there is no free charge on the electrodes, if the piezoelectric medium is infinitely resistive. The induced polarization produces an

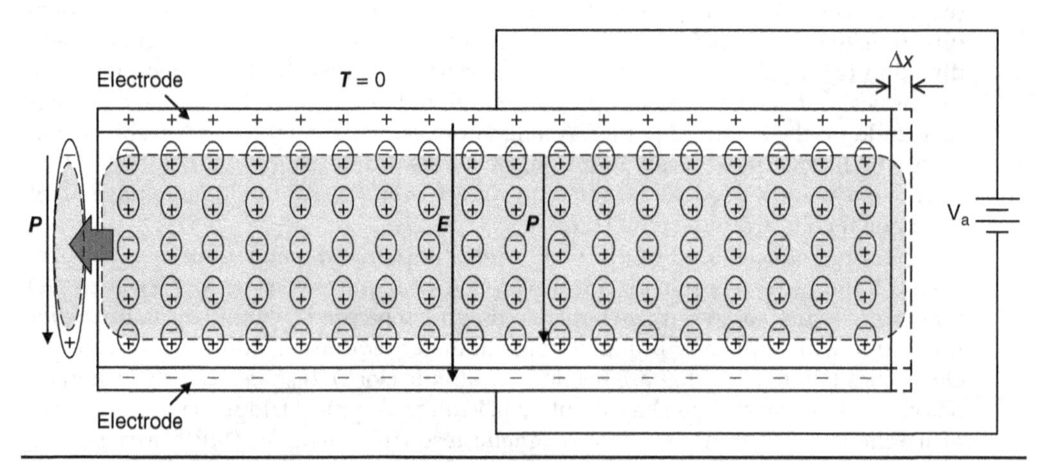

FIGURE 3.14 Piezoelectric actuation, when the medium is free to expand or contract (i.e., $T = 0$).

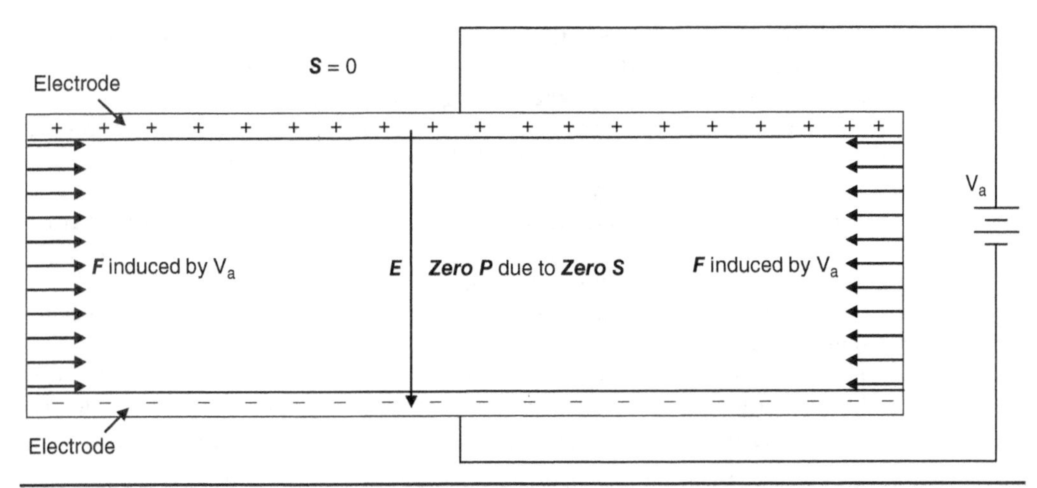

Figure 3.15 Piezoelectric actuation in a hypothetical case of the medium being bound completely so that it cannot expand nor contract (i.e., S = 0).

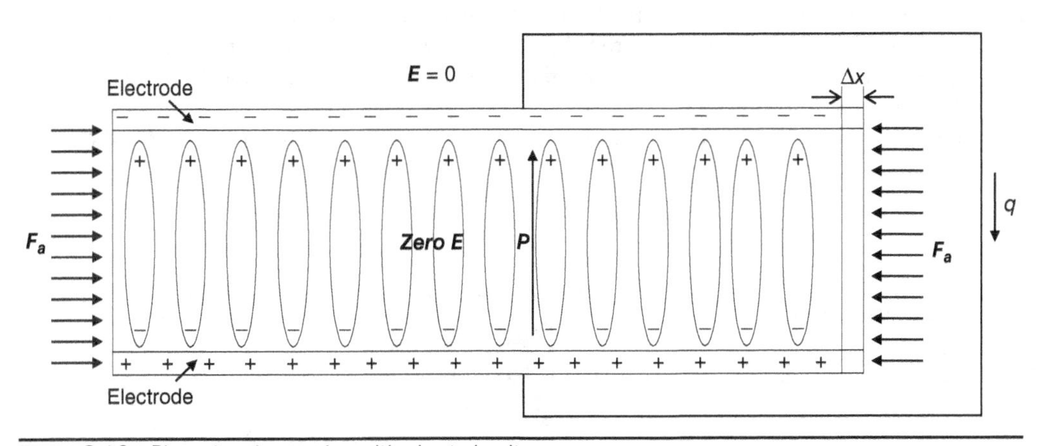

Figure 3.16 Piezoelectric sensing with short circuit.

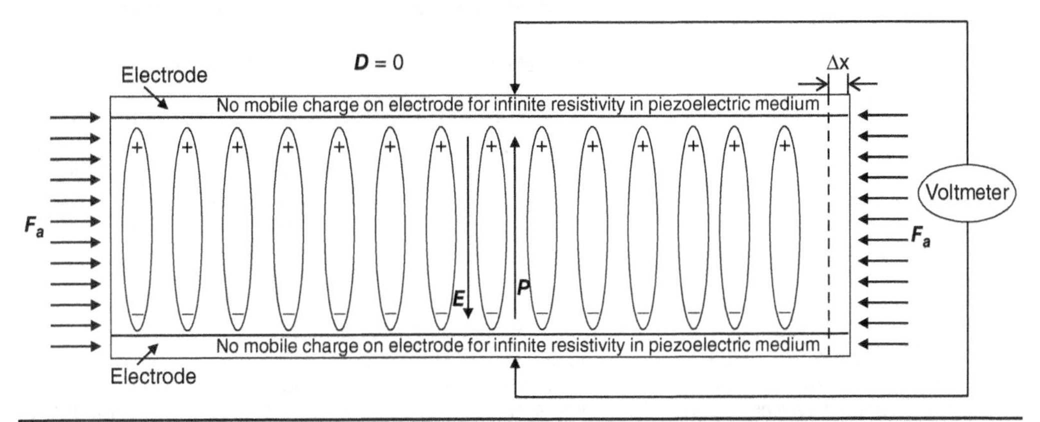

Figure 3.17 Piezoelectric sensing with open circuit for an infinitely resistive piezoelectric medium.

electrical field between the two electrodes, which can be measured with a voltmeter. However, no piezoelectric medium is infinitely resistive, and consequently, electrons move from the one electrode to the other through the piezoelectric medium, according to RC time constant. That time constant is $\tau = RC = (\rho L/A)(\varepsilon A/L) = \rho\varepsilon$, where ρ and ε are resistivity and permittivity, respectively, while L and A are the thickness and area of the piezoelectric medium, respectively.

There are basically four equations for the four cases shown in Figs. 3.14 to 3.17: $D = \varepsilon^T E + dT$, $S = d_t E + s^E T$, $D = \varepsilon^S E + eS$, and $T = -e_t E + c^E S$, where D and E are electric displacement and field, respectively, S is strain, T is stress, and s and c are elastic compliance and stiffness constants, respectively. For dielectric (nonpiezoelectric) materials, electric flux density or electric displacement, D, can be obtained by $D = \varepsilon_o E + P = \varepsilon E$ (C/m^2). Free charges (ρ) can be obtained by $\nabla \cdot D = \rho$ (C/m^3), which indicates that free charge is nonzero only where there is divergence of D vector. Electrical potential at b with respect to that at a is $V_{ba} = V_b - V_a = -\int_a^b E \cdot dl$ (V), from which we see that there is no potential difference where the electrical field is zero. These fundamental equations are useful in understanding the four cases illustrated in Figs. 3.14 to 3.17.

How an external electrical field induces polarization in a dielectric medium can be picturized with Fig. 3.18a. All the dipoles are aligned as shown in the figure for the given external field, and one can see that all the dipole charges cancel with each other except at the two surfaces. Thus, one can consider polarization (P) as a vector originating from a negative fixed charge (on one surface) and ending at a positive fixed charge (on the other surface).

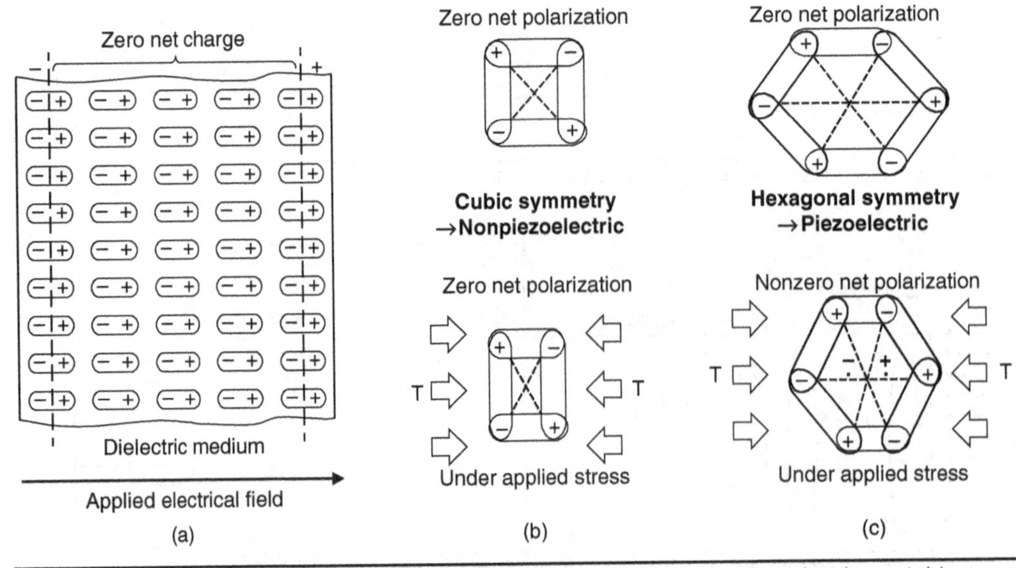

FIGURE 3.18 (a) Dielectric material under an applied electrical field, (b) nonpiezoelectric crystal (e.g., silicon with cubic symmetry) without and with a mechanical strain, and (c) piezoelectric crystal (e.g., ZnO or AlN with hexagonal symmetry) without and with a mechanical strain (showing piezoelectrically induced polarization charge).

Mechanical strain ($S = \Delta l/l$, i.e., change of length over original length for one-dimensional case) changes the shapes of the dipoles of dielectric materials. If the medium has a center symmetry, the shape change does not induce any electrical polarization, as can be seen in Fig. 3.18b. However, for media with certain crystal symmetries, the strain induces an electrical polarization, as illustrated in Fig. 3.18c [18]. Electrical polarization change due to mechanical strain is called piezoelectric effect, which has "converse" effect that produces mechanical strain when electrical field is applied.

One application of piezoelectric material can be to pick up vibration energy and convert it to electrical energy, and then use the electrical energy to produce mechanical strain. A so-called "smart ski" can potentially be designed to reduce the vibration of the ski by converting vibration into an electrical energy, which then is used to reduce the vibration through straining/stressing some key points on the ski.

Question: A PZT block with its initial polarization as indicated with *c*-axis arrow below is applied with a uniform load, as shown below. Assuming that the PZT has *infinite* resistivity (for this example), d_{33} = 300 pC/N, d_{31} = −100 pC/N, d_{15} = −140 pC/N, and $\varepsilon^T = 990\varepsilon_o$, calculate the voltage that is developed between the top and bottom electrodes if the load is 1 N (~0.1 kg).

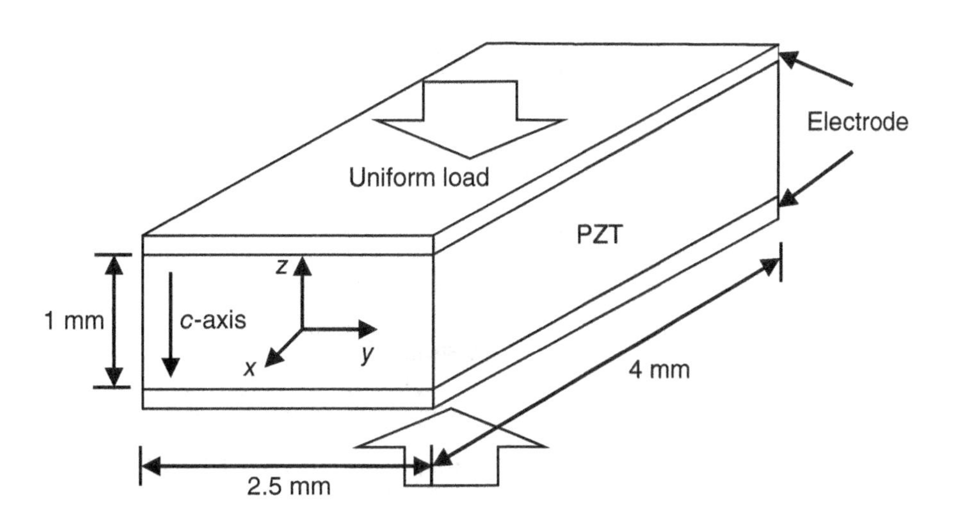

Answer: This example is similar to Fig. 3.17, and $D = \varepsilon^T E + dT$ is the best equation to use to calculate electrical field E as a function of T with $D = 0$. The stress T is equal to F divided by the area, while ε^T is permittivity at a constant T (usually zero). Once the electrical field is obtained, the induced voltage (between the two metal plates) can be calculated by integrating the electrical field along the z direction. $|E| = d_{33}T/\varepsilon = (300 \times 10^{-12})\cdot[1/(2.5 \times 4 \times 10^{-6})]/(990 \times 8.85 \times 10^{-12}) = 3{,}424$ V/m $\rightarrow V = Eh = 3{,}424$ V/m × 1 mm = 3.424 V.

Question: A PZT block with its initial polarization as indicated with *c*-axis arrow below is applied with a static electrical voltage, as shown below. Calculate the shape change (in *x*-, *y*-, and *z*-directions) of the PZT block due to the 10 V applied, assuming that the PZT has h = 1 mm, d_{33} = 100 pC/N, d_{31} = −50 pC/N, and d_{15} = −140 pC/N.

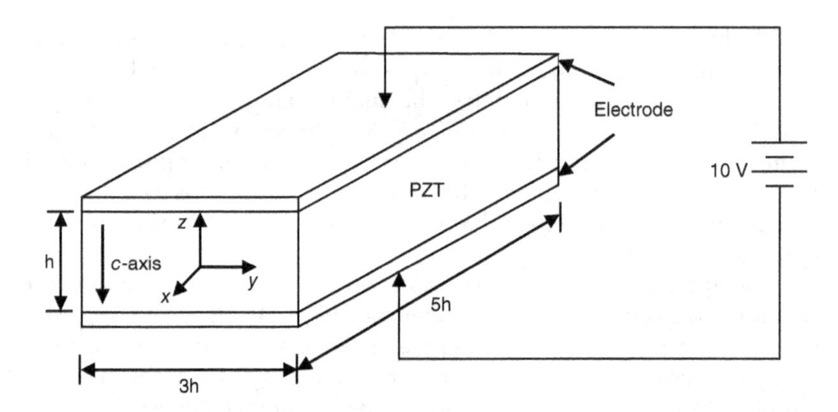

Answer: This case is similar to Fig. 3.14, and $S = d_tE + s^ET$ is the best equation to use to calculate the strain (S) as a function of applied electrical field (E) with $T = 0$. The length change along the z direction is equal to the strain times the original thickness. The applied voltage produces only z-directed electrical field (E_3) and $E_1 = E_2 = 0$, and $E_3 = 10\ \text{V} \div 1\ \text{mm} = 10^4\ \text{V/m}$.

$$S_1 = \Delta x/5\text{h} = d_{31}E_z = d_{31}\text{V/h}$$
$$S_2 = \Delta y/3\text{h} = d_{31}E_z = d_{31}\text{V/h}$$
$$S_3 = \Delta z/\text{h} = d_{33}E_z = d_{33}\text{V/h}$$

$$\begin{pmatrix} S_1 \\ S_2 \\ S_3 \\ S_4 \\ S_5 \\ S_6 \end{pmatrix} = \begin{pmatrix} 0 & 0 & d_{31} \\ 0 & 0 & d_{31} \\ 0 & 0 & d_{33} \\ 0 & d_{15} & 0 \\ d_{15} & 0 & 0 \\ 0 & 0 & 0 \end{pmatrix} \begin{pmatrix} 0 \\ 0 \\ E_3 \end{pmatrix}$$

Thus,

$$\Delta z = d_{33}\text{V}$$
$$\Delta x = d_{31}\text{V}(5\text{h/h})$$
$$\Delta y = d_{31}\text{V}/(3\text{h/h})$$

$\text{Q} = \text{CV}$

Energy $= \text{CV}^2/2 = \text{QV}/2 =$ Force times Distance $\rightarrow$ Unit-wise: (C/N) times V = meter

$$\Delta z = 100 \times 10^{-12} \times 10 = 1\ \text{nm}$$
$$\Delta x = d_{31}\text{V}(5\text{h/h}) = -50 \times 10^{-12} \times 10 \times 5 = -2.5\ \text{nm}$$
$$\Delta y = d_{31}\text{V}/(3\text{h/h}) = -50 \times 10^{-12} \times 10 \times 3 = -1.5\ \text{nm}$$

The medium size is reduced in the x and y directions but is increased in the z direction.

As can be seen in this numerical example, extensional or longitudinal strain produced by piezoelectric actuation is usually very little. Still, such actuation is commonly used for precision positioning of heavy objects. If a large strain is desired, one can consider using a bending displacement with a piezoelectric bimorph.

Question: A piezoelectric ZnO crystal (with $h = 1$ mm, $w = 5$ mm, and $l = 10$ mm) is covered with two electrodes as shown below and is applied with a time-varying force (F) that is uniformly distributed over the left and right faces. Assume $d_{33} = 10$ pC/N, $d_{31} = -5$ pC/N, $d_{15} = -14$ pC/N, $\rho = 10^7\ \Omega$cm, and $\varepsilon = 10^{-12}$ F/cm. For open-circuit type, sketch (on the figure below, *ignoring the relative sign issue*) the polarization field P within the ZnO, the electrical field E within the ZnO, the electrical charge Q

developed on the top electrode, and the voltage V developed between the two electrodes, as a function of time, due to the applied force. **Calculate** the *peak* electrical field and voltage, and indicate those on the E and V curves.

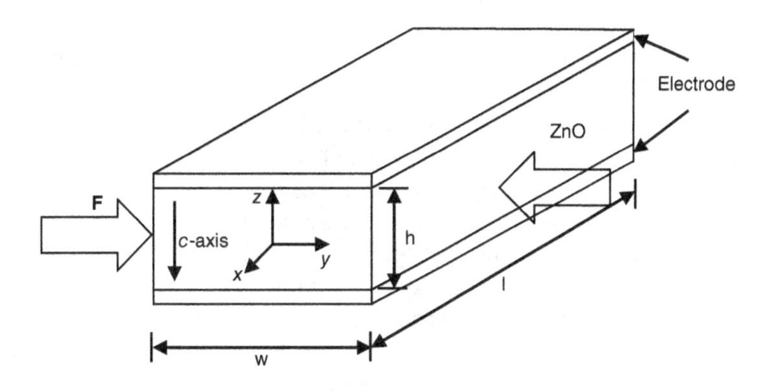

Answer: The answers are given below (in lightened darkness) for the time-varying applied force (F).

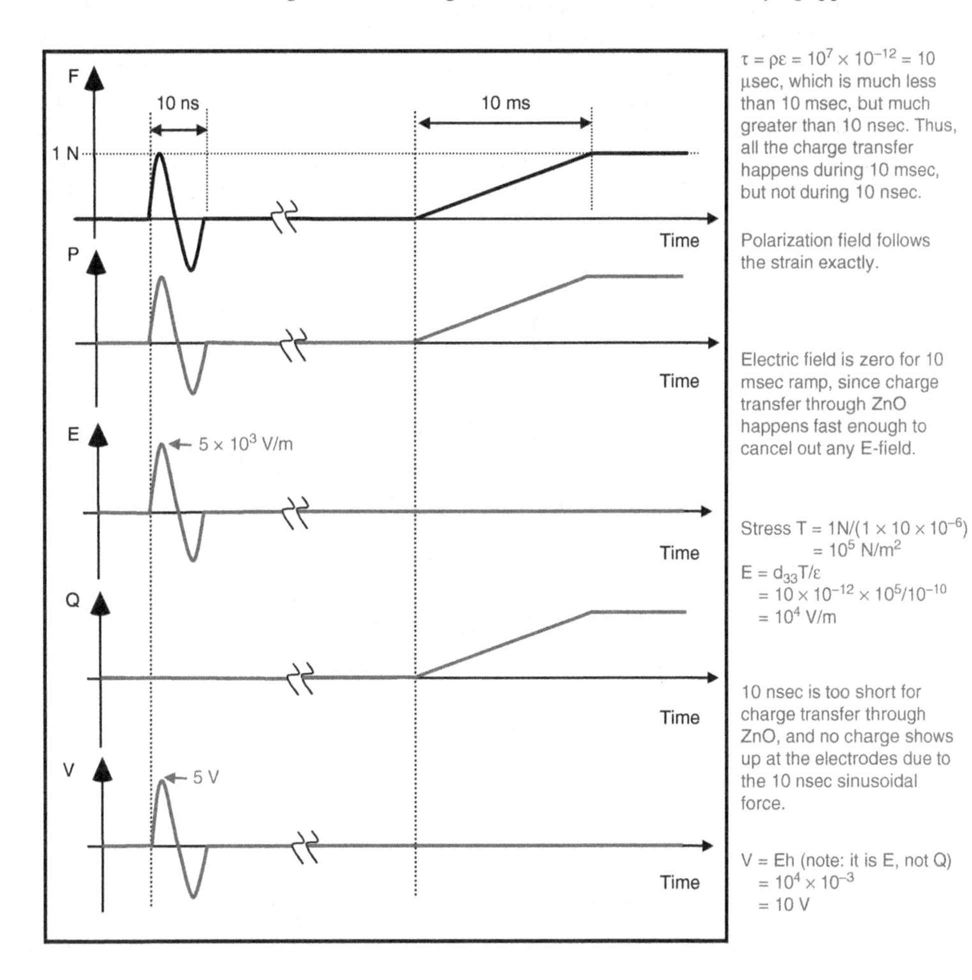

3.3.2 Stress and Strain in Piezoelectric Medium

Stress is a second-rank tensor that needs two indices to specify (1) the plane on which the force is acting and (2) direction of the force (Fig. 3.19). The unit for stress is same as pressure (i.e., force per unit area), Pa (= N/m^2) being the most commonly used unit. Since each index takes three variables (in x, y, and z), there are total of nine stress components: σ_{xx}, σ_{xy}, σ_{xz}, σ_{yx}, σ_{yy}, σ_{yz}, σ_{zx}, σ_{zy}, and σ_{zz}. (The first index indicates the plane, while the second index indicates the direction. For example, σ_{xy} is the force [per unit area] that acts on a plane cutting the X axis perpendicularly and also is directed along the y direction.) From Onsager theorem that is based on the principle of microscopic reversibility (according to which any microscopic process, and its reverse will take place at the same rate on the average under equilibrium conditions), $\sigma_{ij} = \sigma_{ji}$, provided a proper choice is made for the "fluxes" and the linearly connected "fields." Thus, there are only six independent variables for the stress; σ_{xx}, σ_{yy}, and σ_{zy} (three normal stresses) and σ_{xy} $(=\sigma_{yx})$, σ_{xz} $(=\sigma_{zx})$, and $\sigma_{yz}(=\sigma_{zy})$ (three shear stresses). The repeated indices of the normal stresses are usually reduced to one, while the shear stresses are sometimes denoted with τ, rather than σ.

For anisotropic, **nonpiezoelectric** solid, $S_i = s_{ij}\, T_j$ and $T_i = C_{ij}\, S_j$, where C_{ij} and s_{ij} are the elastic stiffness coefficients and the elastic compliance coefficients, respectively. These equations come from general Hook's law for elastic medium. For anisotropic, **piezoelectric** solid, $S = s^ET + d_tE$ and $T = c^ES - e_tE$, where S is strain; T is stress; s and c are elastic compliance and stiffness, respectively; d and e are piezoelectric coefficients; and E is electric field. The piezoelectric coefficients link electrical field to stress and strain. The equation $S = s^ET + d_tE$ can be expressed in a matrix form, as shown in Fig. 3.20.

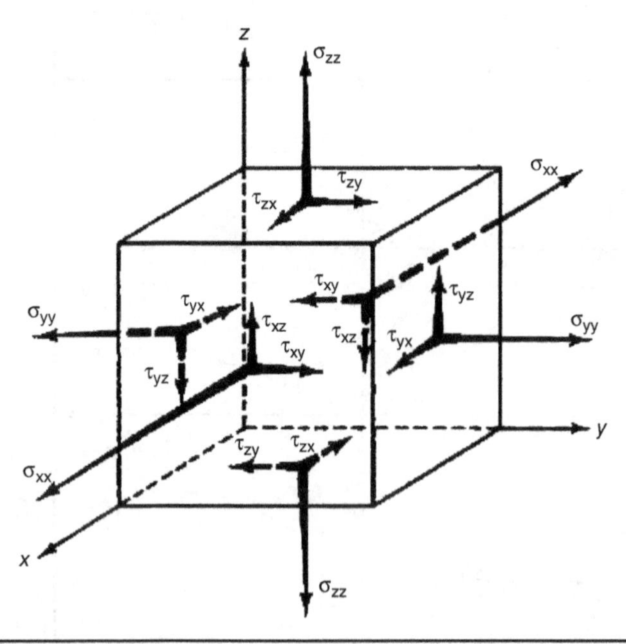

FIGURE 3.19 Diagram to illustrate the definition of stress as a second-rank tensor with two indices indicating the direction of the force (per unit area) as well as the plane on which the force is acting.

$$
\begin{pmatrix} S_1 \\ S_2 \\ S_3 \\ S_4 \\ S_5 \\ S_6 \end{pmatrix} =
\begin{pmatrix}
s_{11} & s_{12} & s_{13} & s_{14} & s_{15} & s_{16} \\
s_{21} & s_{22} & s_{23} & s_{24} & s_{25} & s_{26} \\
s_{31} & s_{32} & s_{33} & s_{34} & s_{35} & s_{36} \\
s_{41} & s_{42} & s_{43} & s_{44} & s_{45} & s_{46} \\
s_{51} & s_{52} & s_{53} & s_{54} & s_{55} & s_{56} \\
s_{61} & s_{62} & s_{63} & s_{64} & s_{65} & s_{66}
\end{pmatrix}
\begin{pmatrix} T_1 \\ T_2 \\ T_3 \\ T_4 \\ T_5 \\ T_6 \end{pmatrix} +
\begin{pmatrix}
d_{11} & d_{12} & d_{13} \\
d_{21} & d_{22} & d_{23} \\
d_{31} & d_{32} & d_{33} \\
d_{41} & d_{42} & d_{43} \\
d_{51} & d_{52} & d_{53} \\
d_{61} & d_{62} & d_{63}
\end{pmatrix}
\begin{pmatrix} E_1 \\ E_2 \\ E_3 \end{pmatrix}
$$

Figure 3.20 Matrix equation for $S = s^E T + d_t E$.

Elasto-piezo-dielectric matrix								
C_{11}	C_{12}	C_{13}	0	0	0	0	0	e_{31}
C_{12}	C_{11}	C_{13}	0	0	0	0	0	e_{31}
C_{13}	C_{13}	C_{33}	0	0	0	0	0	e_{33}
0	0	0	C_{44}	0	0	0	e_{15}	0
0	0	0	0	C_{44}	0	e_{15}	0	0
0	0	0	0	0	$\frac{C_{11}-C_{12}}{2}$	0	0	0
0	0	0	0	d_{15}	0	ε_{11}	0	0
0	0	0	d_{15}	0	0	0	ε_{11}	0
d_{31}	d_{31}	d_{33}	0	0	0	0	0	ε_{33}

Figure 3.21 Matrixes for elastic stiffness, piezoelectric strain and stress constants, and permittivity of ZnO.

Note that in Fig. 3.20, the stress and strain are written with one index (its values varying from one to six) rather than with two indices, since there are indeed only six independent stress (or strain) components due to Onsager theorem (as explained earlier). The first three components (1–3) are normal stresses (or strains), while the last three components (4–6) are shear stresses (or strains). Many of the 18 piezoelectric coefficients (as well as the 36 stiffness or compliance coefficients) are zero for many crystals.

For ZnO which has hexagonal close-packed wurtzite structure (C_{6v}), many of the stiffness coefficients (C_{ij}), piezoelectric coefficients (d_{ij} and e_{ij}), and permittivity coefficients (e_{ij}) are zero as shown in the elasto-piezo-dielectric matrix (Fig. 3.21). Two other common piezoelectric materials used in MEMS, AlN and PZT, also have a matrix for the piezoelectric coefficients (d_{ij} and e_{ij}) same as ZnO.

The symbols, units, and equations that are commonly encountered in analyzing piezoelectric sensors and actuators are summarized in Table 3.1.

Question: A 0.5-mm-thick PZT wafer with its initial polarization as indicated with arrows below is applied with a static electrical voltage, as shown below. Calculate the normal (ε_x, ε_y, ε_z) and shear (ε_{xy}, ε_{yz}, ε_{zx}) strains that will be produced by 10 V applied, assuming that the PZT has $d_{33} = 100$ pC/N, $d_{31} = -50$ pC/N, and $d_{15} = -140$ pC/N.

	Symbol	Unit	
Stress	T	N/m²	$T = -e_t E + c^E S$
			$T = -h_t D + c^D S$
Strain	S	Unitless	$\mathbf{S = d_t E + s^E T}$
			$S = g_t D + s^D T$
Electric field	E	V/m	$E = \beta^T D - gT$
			$E = \beta^S D - hS$
Electric displacement	D	C/m²	$\mathbf{D = \varepsilon^T E + dT}$
			$D = \varepsilon^S E + eS$
Elastic compliance	s	m²/N	$e_t \equiv$ transpose of e
Elastic stiffness	c	N/m²	$d_t \equiv$ transpose of d
Permittivity	ε	F/m	$g_t \equiv$ transpose of g
Dielectric impermeability	β	m/F	$h_t \equiv$ transpose of h
Piezoelectric strain constant	d	C/N or m/V	$s^E \equiv$ s with constant E
Piezoelectric stress constant	e	C/m² or N/Vm	$c^D \equiv$ c with constant D $\varepsilon^T \equiv \varepsilon$ with constant T
Piezoelectric strain constant	g	Vm/N or m²/C	$\beta^S \equiv \beta$ with constant S
Piezoelectric stress constant	h	V/m or N/C	

Table 3.1 Symbols, Units, and Equations for Stress, Strain, Electric Field, and Electric Displacement

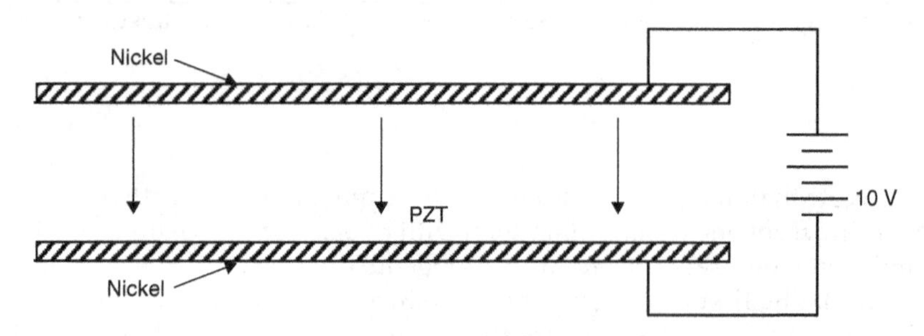

Answer:

In this case, $E_1 = E_2 = 0$

$$E_3 = 10\ \text{V}/0.5\ \text{mm} = 2 \times 10^4\ \text{V/m}$$

$$\varepsilon_x = \varepsilon_y = d_{31}E_3 = -50 \times 10^{-12} \times 2 \times 10^4 = -10^{-6}$$

$$\varepsilon_z = d_{33}E_3 = 100 \times 10^{-12} \times 2 \times 10^4 = 2 \times 10^{-6}$$

$$\varepsilon_{xy} = \varepsilon_{yz} = \varepsilon_{zx} = 0$$

$$\begin{pmatrix} \varepsilon_x \\ \varepsilon_y \\ \varepsilon_z \\ \varepsilon_{xy} \\ \varepsilon_{yz} \\ \varepsilon_{zx} \end{pmatrix} = \begin{pmatrix} 0 & 0 & d_{31} \\ 0 & 0 & d_{31} \\ 0 & 0 & d_{33} \\ 0 & d_{15} & 0 \\ d_{15} & 0 & 0 \\ 0 & 0 & 0 \end{pmatrix} \begin{pmatrix} E_1 \\ E_2 \\ E_3 \end{pmatrix}$$

In other words, the medium size is reduced in the x and y directions but is increased in the z direction.

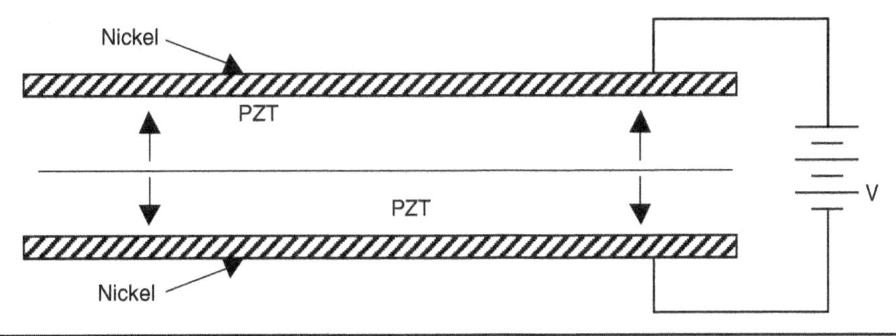

FIGURE 3.22 Cross-sectional view of PZT bimorph.

3.3.3 Piezoelectric Bimorph

Piezoelectric actuation in extensional mode produces very small strain, as seen in the numerical example in the previous section. If a large displacement (or strain) is desired, one can use a bending displacement, which can easily be produced by a bimorph (or unimorph). The large displacement by bending, though, is obtained at the cost of a reduced force (or torque) that the actuator can generate.

A piezoelectric bimorph can be formed by two PZTs as shown Fig. 3.22. The polarization directions for the two substrates are opposite to each other, so that the applied electrical field produces opposite strains in the two substrates. For example, if the top layer is strained in tension in the in-plane direction, the bottom layer is strained in compression in the in-plane direction. The opposite strains in the two layers force both of them to bend, producing much larger bending displacement (at the free end of the bimorph) than the strains themselves.

Question: For two PZT substrates (each of which is polarized as indicated with arrows) that are glued together and applied with a static electrical voltage as shown Fig. 3.22, sketch the deformation shape of the whole structure. Assume d_{31}, d_{33}, and d_{15} are –200, 500, and 600 pC/N for the PZT, respectively.

Answer:

Question: A PZT substrate (poled with an initial polarization, P_s, as indicated with an arrow below) is glued to a silicon substrate of about the same thickness. The Al electrode on top of the silicon is segmented as shown, and static electrical voltages of opposite sign are applied as shown below.

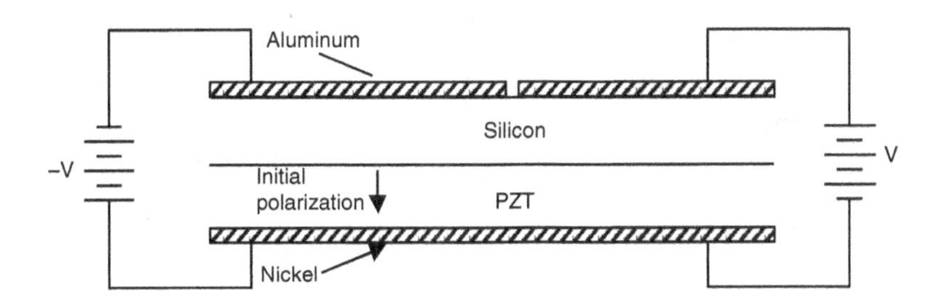

Sketch the deformation shape of the whole structure, assuming that d_{31}, d_{33}, and d_{15} are −200, 500, and 600 pC/N for the PZT, respectively, and noting that plus sign corresponds to a tensile stress or strain.

Answer:

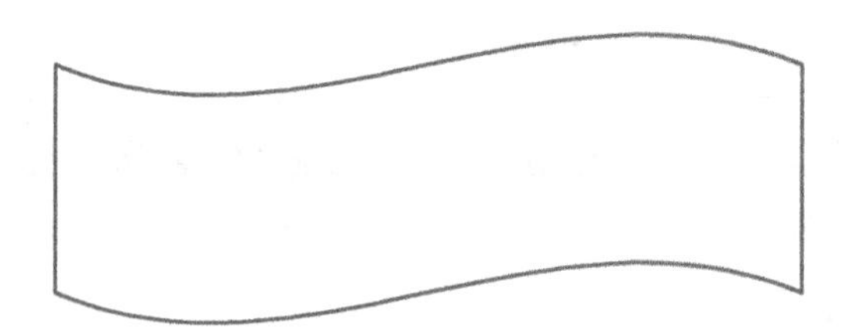

3.3.4 Progressive Flexural Wave

By connecting a number of bimorphs side by side as shown in Fig. 3.23, one can generate flexural wave (i.e., wave due to bending of a plate) [19]. The side-by-side connection of the bimorphs produces a bending with multiple cycles of sinusoidal wave. For example, the ABAB in Fig. 3.23 produces a bending with two sinusoidal cycles, when a voltage is applied between the electrode over the ABAB and the ground electrode under it. The BABA also produces a bending with two cycles; and if it is applied with a voltage that is 90° out of phase to that applied to ABAB with the gap between ABAB and BABA being a quarter wavelength (Fig. 3.23), then the bendings from the ABAB and the BABA become a propagating flexural wave.

Flexural wave can move an object either along the same direction as the wave's propagation direction or in the opposite direction, depending on the rigidity of the object. If the object is rigid, it moves in the direction opposite to the wave's propagation direction. If the object is compliant, it moves along with the wave.

Question: For the two sets of piezoelectric bimorphs (ABAB and BABA) arranged as shown in Fig. 3.23, what does the frequency ω (= $2\pi f$) have to be for a 100-μm-long ABAB–BABA to produce the traveling flexural wave with 3 m/s?

Answer: According to Fig. 3.23, 100 μm = 4.25λ. → λ = 23.5 μm → $\omega = 2\pi v/\lambda = (2\pi \times 3)/(23.5 \times 10^{-6})$ = 0.8 Mrad/s.

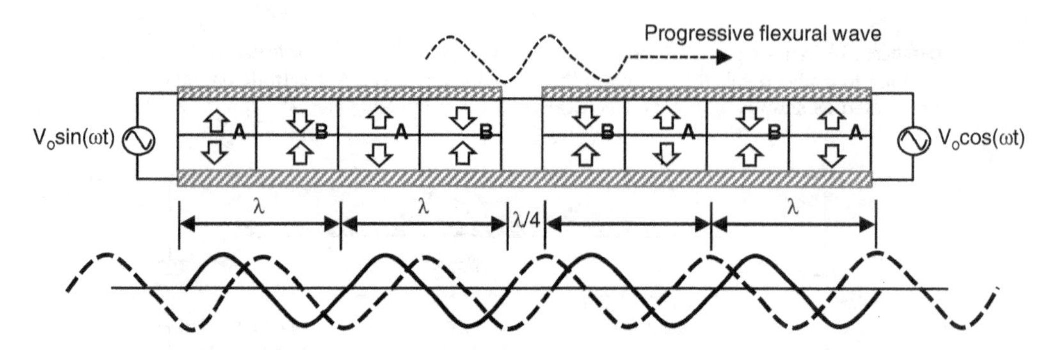

FIGURE 3.23 How progressive flexural wave is produced by a set of piezoelectric bimorphs.

3.3.5 Ultrasonic Motor

Ultrasonic motors (first proposed in 1973 by H.V. Barth of IBM) use ultrasonic elastic waves to move a rotor that is in contact with a stator carrying the waves, and are friction-driven. They offer the following advantages over conventional electromagnetic motor: high torque and high efficiency at low speed, simple structure and shape flexibility, nonvolatility, excellent control at start and stop, no electromagnetic-interference induction, silent operation, etc. Ultrasonic motors, however, need (1) relatively expensive high-frequency electrical power source and (2) a strong wear-resistant material due to the inevitable wear and tear at the contacts. Due to the advantages, ultrasonic motors have already been commercially used in many applications. One notable and well-known application has been in the autofocus mechanism of cameras. Additionally, ultrasonic motors have been used in the drive of a watch, plastic-card-forwarding device, paper-sheet-forwarding device, the drive to wind window blinds, etc.

A movable object can be rotated by a traveling ultrasonic wave. Since for a rotary motor, the traveling wave must continue around a ring in one direction without being reflected, the piezoelectric element that generates the traveling wave is divided into two major regions that are driven with $V_o \sin\omega t$ and $\pm V_o \cos\omega t$ ($+V_o$ for a wave propagating to the right while $-V_o$ for the opposite direction), respectively (Fig. 3.24). Each of those two regions of the piezoelectric element are segmented into the ↑ and ↓ sections in Fig. 3.24a (corresponding to the A and B sections in Fig. 3.24b) to increase the number of waves around the ring by either poling the piezoelectric ceramic after electrode patterning or properly applying $\pm V_o \sin\omega t$ and $\pm V_o \cos\omega t$ to segmented electrodes during motor operation.

In a rotary motor, a rotor ring is pressed onto a metal stator having comb teeth [20]. The points on the surface of the stator move elliptically with an amplitude of submicron to several microns depending on the amplitude (V_o) of the applied voltage and the frequency. In order to maximize the transfer of the elliptical surface motion to the rotor motion, an engineered polymer may be coated over the rotor for good frictional contact (as well as for wear resistance). Moreover, the stator surface is patterned to have comb teeth in order to produce L-leveraged tangential force by the surface motion in the

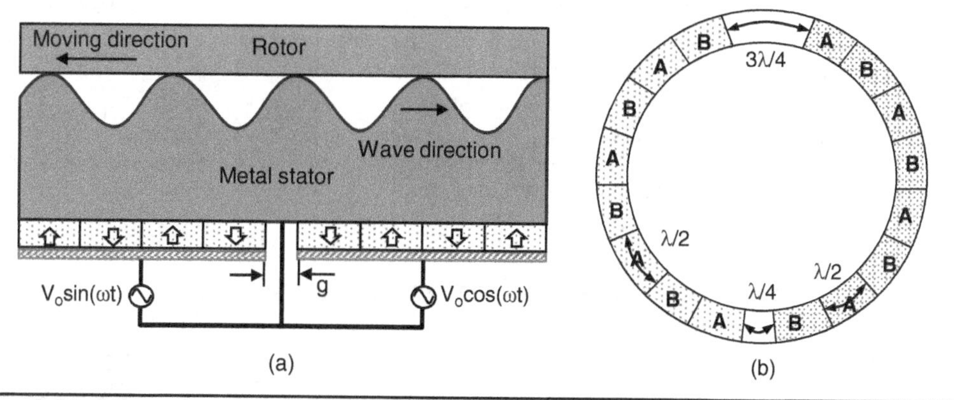

FIGURE 3.24 Ultrasonic motor: (a) cross-sectional view illustrating how a traveling wave is generated on a metal stator from a set of piezoelectric unimorphs and moves a rotor ring and (b) top view of the stator.

transverse direction. If there are nine cycles of sinusoidal wave around the ring, there are at least nine contact points between the rotor and stator at any given moment. The nine teeth that are in contact with the rotor are replaced by a next set of nine teeth while those are still in contact. Thus, the rotor's wobble is very small.

Question: For the ultrasonic motor (shown in Fig. 3.24), what is the gap distance (g) between two piezoelectric actuators if the motor's traveling wave propagates at 3 m/s and the frequency of the applied voltage is 300 Hz?

Answer: The gap has to be a quarter of the wavelength and is equal to $\frac{3\text{ m/s}}{4\times(300\text{ Hz})} = 2.5\text{ mm}$

Ultrasonic Motor Based on Bending Cylindrical PZT Thin-Film Transducer

If PZT film is deposited on a sidewall of a cylinder, the cylinder can be bent along the cylinder's axial direction due to the in-plane strain (in this case, the contraction or extension in the sidewall plane), which is produced through piezoelectric effect by an applied electrical field [21]. By segmenting the electrode into four sections, four different bending shapes can be produced. When the four bendings are sequentially generated, a disk sitting on one end of the cylinder rotates. The rotational torque can be high, as a higher contact force makes the energy transfer from the cylinder bending to the rotational motion better. Also, the rotation is bidirectional. To deposit the relatively thick PZT film, an autoclave was used for hydrothermal synthesis on solid Ti with $Pb(NO_3)_2$, $ZrOCl_4$, and KOH at 160°C and 6 atm [21]. Deposition of 7–9-μm-thick PZT film took 48 hours.

SAW Motor

As we will study in Chap. 4, surface acoustic wave (SAW) has an elliptical particle motion (on a surface of a solid substrate), amplitude of which drops rapidly to zero within a half of the wavelength in the depth direction. Though SAW has a very small vibrational amplitude of ≈ 8.1 nm at 10 MHz, it was still shown to be able to move ruby balls (1.56 mm in diameter) in contact pressure of ≈ 80 MPa [22]. With a set of four interdigital transducer (IDT) electrodes on one surface of piezoelectric $LiNbO_3$ substrate, a SAW was produced over an annual ring and used to generate a maximum ball-moving speed of ≈ 22 cm/s with 80 V drive. A SAW launched on one surface of a thin substrate becomes a lamb wave (antisymmetric mode of which is a flexural wave), if the thickness of the substrate is small enough with respect to the wavelength. Thus, an IDT electrode on one surface of a thin substrate (or membrane) with a piezoelectric thin film can be used to launch a flexural wave over the membrane, which can be used to move a small object.

Question: Two PZT substrates with three different sections (each of which is polarized as indicated with arrows below) are glued together and applied with a static electrical voltage, as shown below. Sketch the deformation shape of the whole structure.

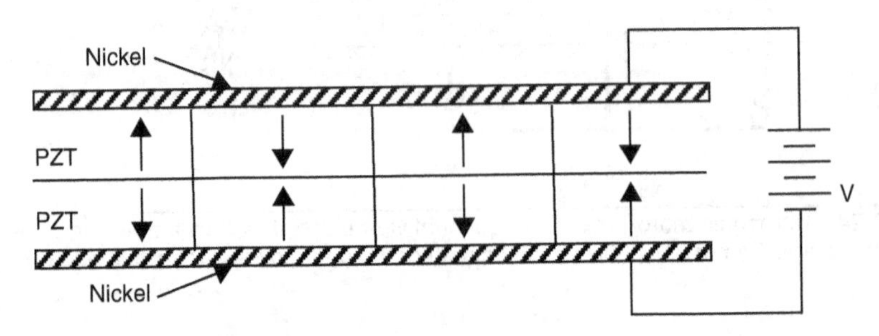

Answer:

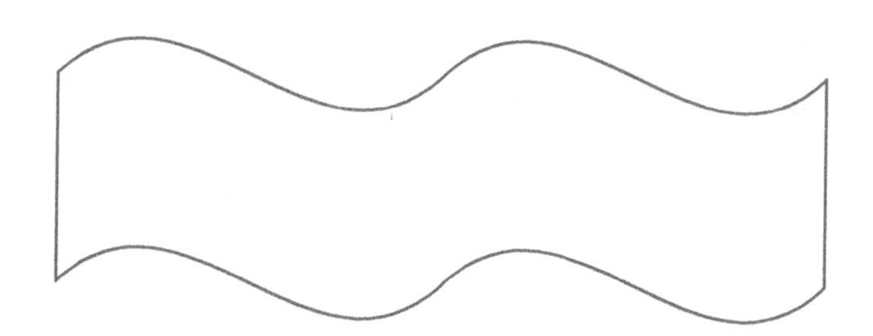

Question: One PZT substrate (which is divided into three regions that are polarized as indicated with arrows below) is applied with a static electrical voltage, as shown below. Sketch the deformation shape of the whole structure, assuming that $d_{33} = 300$ pC/N and $d_{31} = -100$ pC/N.

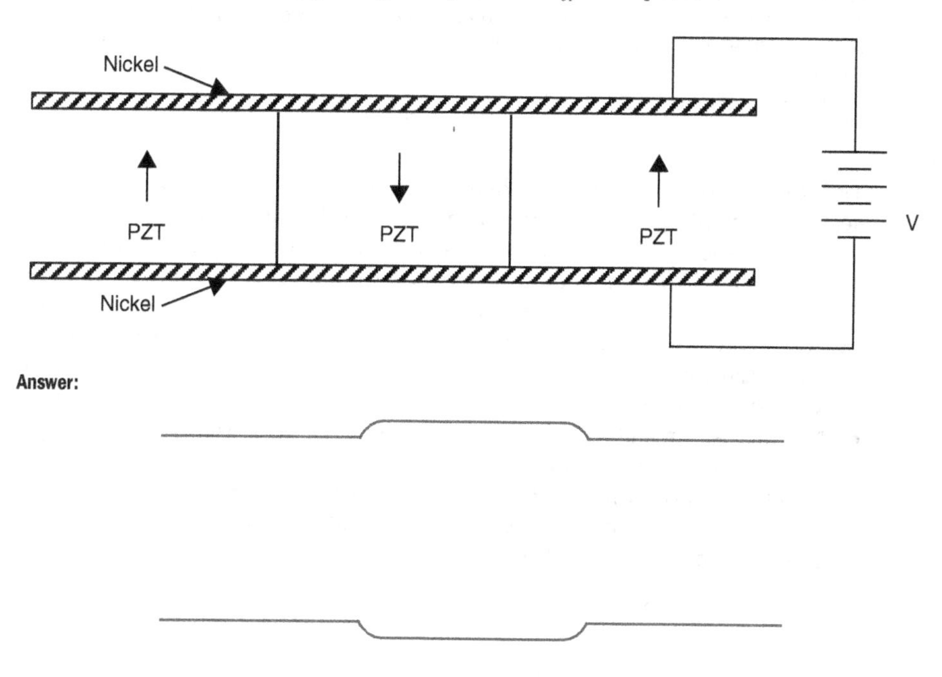

3.4 Thermal Transduction

Electrothermal actuation offers strong actuation force and has been used with and in micromachined structures. Sensing temperature or heat especially IR imaging for night vision camera has recently been revolutionized by microelectromechanical systems (MEMS) technology. In this section, we will first study MEMS-based electrothermal actuation followed by uncooled IR imaging based on MEMS-based thermopiles and bolometers.

3.4.1 Electrothermal Actuation

With surface micromachining, an electrically conductive cantilever-like structure with two arms and two anchors (Fig. 3.25a) can easily be fabricated. One arm, called *hot arm*, has narrow width (and high electrical resistance R), while the other arm has wide width

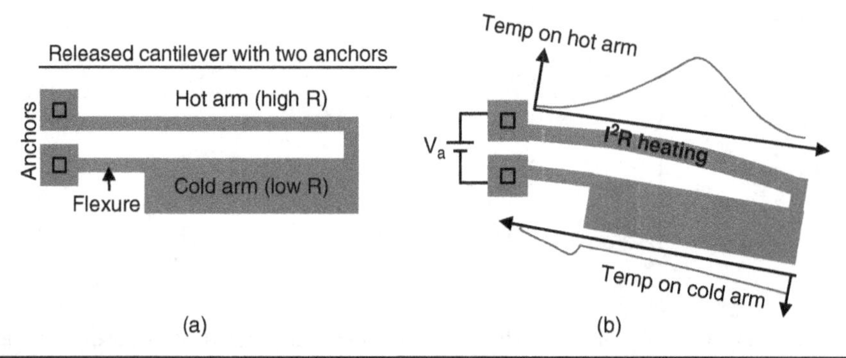

FIGURE 3.25 Top views of an electrothermal actuator based on a released conductive cantilever with two anchors: (a) one arm called "hot arm" is narrow and has a high electrical resistance *R*, while the other arm is composed of a so-called "cold arm" with wider width for a lower *R* and a "flexure" with narrow width for easy lateral bending. (b) The bending of the actuator when electrical voltage is applied between two anchors due to much higher I^2R heating on the hot arm. The higher the current *I*, the larger the bending.

(and low electrical resistance). When a voltage V_a is applied between the two anchors, electrical current *I* flows from one anchor to the other and heats up the arms due to I^2R heating, producing temperature profiles along the arm length, as illustrated in Fig. 3.25b. With two anchors fixed to the substrate, the temperature rise makes the hot arm to bend in plane, as shown, since the expansion of the hot arm due to the temperature rise is much more than that in the cold arm. The bending amount will increase as the applied voltage (or current) is increased but only in one direction (i.e., clockwise direction from the top view in the case of Fig. 3.25). When the current is reduced, the electrical I^2R heating becomes less, resulting in less thermal expansion as the temperature drops, and the bending displacement becomes less.

Another approach for electrothermal actuation with surface-micromachined structure is to use chevron-shape released structure between two anchors with a shuttle that connects the arms on the left with those on the right, as shown in Fig. 3.26. When a

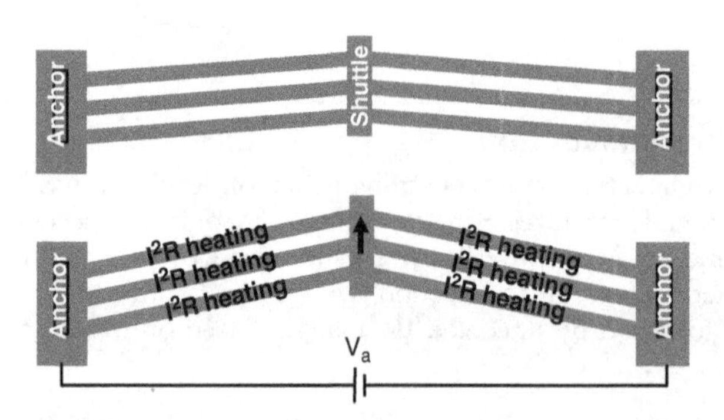

FIGURE 3.26 Top views of an electrothermally actuated chevron-shape released structure before (top) and after (bottom) an electrical voltage is applied between the anchors. The thermally induced displacement of the shuttle can be increased with higher current and/or more pairs of the chevron arms.

voltage is applied between the anchors, the arms are heated due to the current flow, and expand, pushing the shuttle in the direction indicated with an arrow in Fig. 3.26. The higher the applied voltage (and/or the larger the number of the arms' pairs), the larger the lateral displacement of the shuttle will be. When the applied voltage is removed, the I^2R heating stops, and the temperature drops to room temperature in 1 millisecond – 1 second, depending on the thermal mass, amount of temperature change, etc. The chevron-type and cantilever-like electrothermal actuators are typically used for in-plane actuation [23], though those can be used for out-of-plane actuation with some design variation. For out-of-plane actuation, a bimetal structure, described next, is typically used.

Bimetal composed of two conductive layers with different coefficients of thermal expansion (CTE) is yet another approach for electrothermal actuation. Since two layers (one with high CTE and the other with low CTE) are typically deposited at an elevated temperature, the bimetal structures will be warped at room temperature, as illustrated in Fig. 3.27b, upon release, since the high CTE layer goes through a larger amount of shrinkage over $\Delta T = T_{\text{deposition}} - T_{\text{room}}$. If an opposite bending curvature is desired, the low CTE metal can be deposited over the high CTE metal. When the temperature is increased electrothermally, the bending curvature will become less, as shown in Fig. 3.27b, since the expansion due to the temperature rise is larger on the high CTE metal than that on the low CTE metal.

Electrothermal actuation offers strong actuation force, large displacement, easy fabrication, and no need for nonstandard material. However, it is inherently slow in response since the temperature change in a structure is inherently slow with its thermal mass. The lower the thermal mass, the faster the response is, since the temperature change can happen faster as the actuation current is varied. Consequently, micromachined

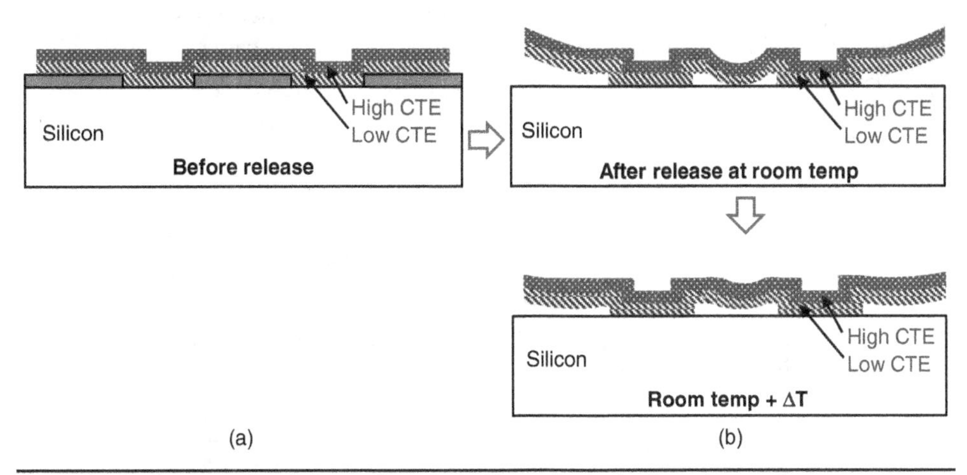

FIGURE 3.27 Cross-sectional views of an electrothermal actuator based on surface-micromachined bimetals: (a) before release after two different metals have been deposited at an elevated temperature, and (b) after release, showing the bending of the released structures due to the coefficient of thermal expansion (CTE) difference between two metals, as the high CTE metal contracts more than the low CTE metal when those metals are brought down to room temperature from their elevated deposition temperatures. Also shown is the bending shape when the temperature is raised through current flow I in the bimetal as the I^2R heating makes the high CTE metal expand more than the low CTE metal.

structures can be responsive up to 1–10 kHz in electrothermal actuation, which is quite good in comparison to human muscle's response speed of 5 Hz. The response speed is sufficient for some applications but is too slow for some other applications.

Also, the power consumption is inherently large in electrothermal actuation, since temperature has to be raised. As each micromachined electrothermal actuator consumes anywhere between 0.1 and 100 mW, the maximum number of such actuators is limited. Finally, a long-term reliability and reproducibility is a concern with electrothermal actuation, if the temperature excursion is large as the materials go through a large amount of thermal expansion (followed by contraction) in repeated cycles.

3.4.2 Uncooled Infrared Imaging

All affordable night vision cameras are currently made with MEMS technology for reasons that will be clear after studying this section. For a night vision camera, one will have to rely on IR detection, since the most useful wavelength range due to black body radiation (say, from −55 to +125°C) is in mid-wave IR (3–5 μm), as can be seen in Fig. 3.28.

Since light generates electron–hole pairs in semiconductors, there have been so-called photonic IR detectors based on photodiode or photoconductor (of which the resistivity changes due to electron–hole pair generation) such as mercury-cadmium-telluride (MCT) and quantum well infrared photodetectors (QWIP) [24]. Photonic detectors, however, need to be cooled to 77–100 K to get rid of noise currents that are present even without any incident IR for high performance suitable for night vision camera. Consequently, night vision cameras based on photonic IR detectors are expensive and bulky.

With IR absorber, IR can be turned into heat, which then can be sensed through thermal detectors. Two types of MEMS-based thermal detectors have been pursued: (1) thermocouple (or thermopile which is nothing but serially connected thermocouples) based on thermal electric effect and (2) bolometer based on resistivity change due to heat. As illustrated in Fig. 3.29b, a voltage V_{out1} develops between two dissimilar metals, measurable at one end, when two ends of the metals are at different temperatures

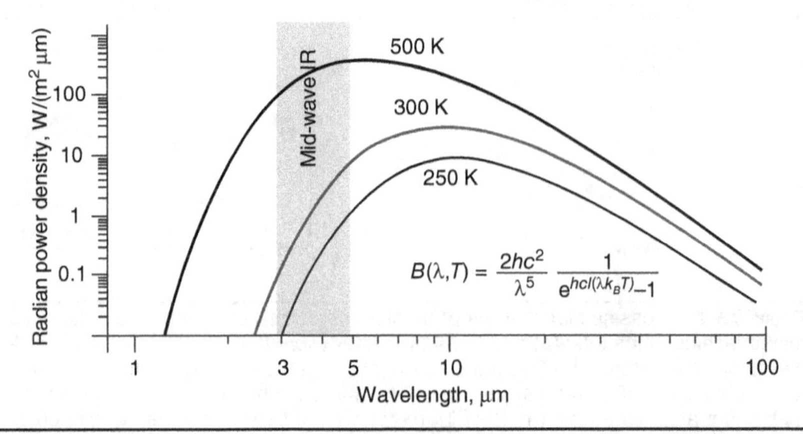

FIGURE 3.28 Radiation power density of black body radiation as a function of temperature. For thermal imaging (e.g., for night vision), mid-wave IR that has wavelength between 3 and 5 μm is the best for most thermal imaging applications.

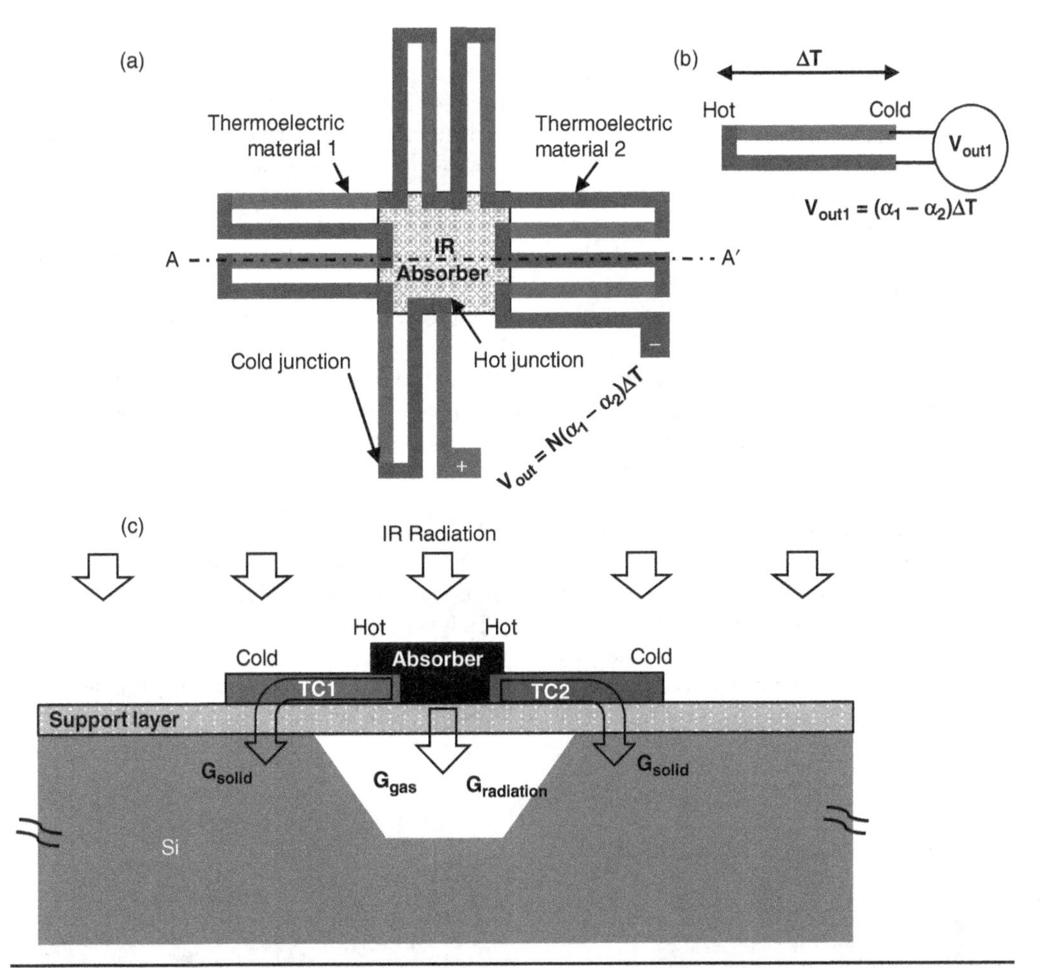

Figure 3.29 IR sensor based on thermopile (which is a serial connection of thermocouples): (a) top view with eight couples of two thermoelectric materials (such as Bi-Te and Bi-Sb-Te) having Seebeck coefficients α_1 and α_2, (b) thermocouple voltage $V_{out} = (\alpha_1 - \alpha_2)\Delta T$ due to temperature difference ΔT at hot and cold junctions, and (c) cross-sectional view of the thermopile across A–A′ showing the hot and cold junctions and heat transfer paths: heat conduction through solid into the substate (G_{solid}), which is dominant, heat conduction into gas (G_{gas}), and heat radiation ($G_{radiation}$).

with the temperature difference of ΔT, due to the Seebeck effect. This so-called thermocouple voltage can be obtained with $V_{out1} = (\alpha_1 - \alpha_2)\Delta T$, where α_1 and α_2 are Seebeck coefficients for thermoelectric metal 1 and 2, respectively [25]. Note that the two metals must be in contact and form a junction on one end (Fig. 3.29b) so that the voltage developed on one metal may be added up to the one developed on the other. By forming such junctions at the hot and cold spots and connecting them serially, as shown in Fig. 3.29a, one can increase the thermocouple voltage level by N times for N such pairs. The larger the ΔT, the higher the thermocouple voltage is, and a micromachined structure such as the one illustrated in Fig. 3.29c would be ideal since it can be designed to maximize ΔT. The heat produced by IR absorption by the absorber is dissipated into the silicon substrate by heat conduction through solid (G_{solid}), which is usually dominant, heat conduction through gas (G_{gas}), and heat radiation ($G_{radiation}$). Consequently, by minimizing G_{solid} through suspending the diaphragm (i.e., separating it from the substrate), the hot area can be kept as hot as the absorbed IR produces. Since thermocouple or thermopile produces a voltage, a night vision camera based on a 2D array of micromachined thermopiles will consume very low power. However, the size of such a thermopile cannot be made too small as there is a trade-off between ΔT and the size. Bolometer, on the other hand, consumes relatively large power, through its resistance R, as an applied current I flows through the resistor, and I^2R power is consumed. However, it can be made very small, and micromachined bolometers have dominated low-cost night vision camera market.

Micromachined bolometers are typically based on a micromachined plate that is suspended by support beams (oftentimes folded), as illustrated in Fig. 3.30a and b. Uncooled IR focal plane arrays (FPAs) based on micromachined bolometers currently use amorphous silicon (a-Si) or vanadium oxide (VO_x) as the thermally sensitive resistor. Amorphous silicon is sturdy enough to be freely suspending without a support layer and also is a good IR absorber. However, VO_x is fragile and does not absorb IR well, and needs a supporting and IR absorbing layer for which silicon-rich silicon nitride (SiN) has commonly been used (Fig. 3.30c). The sensitivity R_i of a "scaled pixel" in FPA based on micromachined bolometer is $R_i = (\text{FF} \cdot \varepsilon \cdot \text{TCR} \cdot V_{bias})/(G_{th} \cdot R)$, where FF = fill factor, ε = absorption coefficient, TCR = thermal coefficient of the resistance, V_{bias} = applied voltage, G_{th} = thermal conductance, and R = thermistor resistance [26]. Increasing V_{bias} leads to more power consumption and also I^2R heat generation. In case of VO_x, the TCR is

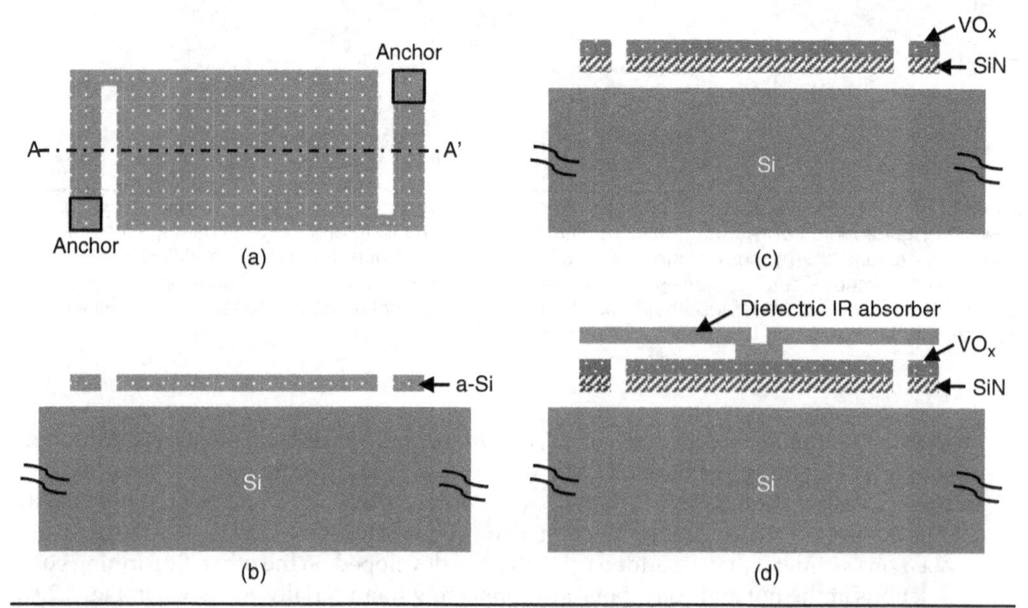

FIGURE 3.30 Basic structures of resistive bolometers for uncooled (room temperature) IR imaging: (a) and (b) top and cross-sectional views of a bolometer with freestanding amorphous silicon (a-Si) plate thermally isolated from substrate, (c) cross-sectional view of a bolometer-based VO_x resistor supported by silicon-rich silicon nitride (SiN) which also acts as an IR absorber, and (d) cross-sectional view of a bolometer with a dielectric IR absorber added on top of the structure shown in (c).

about −2%/°C at room temperature and can be even higher with VO_x ($x > 2$), which has higher resistivity. Commercial uncooled IR bolometer arrays are currently all fabricated with surface micromachining for structures similar to those shown in Fig. 3.30. Most of them use VO_x, while several others use a-Si, and are 320 × 240 – 1024 × 768 arrays with a pixel pitch of 17–28 μm [26].

References

1. W.C. Tang, T.H. Nguyen, and R.T. Howe, "Laterally driven polysilicon resonant microstructures," Sensors and Actuators, 20, 1989, pp. 25–32.
2. A. Selvakumar, K. Najafi, W.H. Juan, and S. Pan, "Vertical comb array micro-actuators," Proceedings IEEE Micro Electro Mechanical Systems, Amsterdam, Netherlands, January 29–February 2, 1995, pp. 43–48.
3. Y.C. Tai, L.S. Fan, and R.S. Muller, "IC-processed micromotors: Design, technology and testing," Proceedings IEEE 2nd Workshop on MEMS, Salt Lake City, UT, February 1989, pp. 1–7.
4. Mehregany and Y.C. Tai, "Surface micromachined mechanisms and micromotors," Journal of Micromechanics and Microengineering, 1, 1991, 73–85.
5. A.A. Yasseen, S.W. Smith, M. Mehregany, and F.L. Merat, "Diffraction grating scanners using polysilicon micromotors," Proceedings IEEE Micro Electro Mechanical Systems, Amsterdam, Netherlands, January 29–February 2, 1995, pp. 175–180.
6. J. R. Gilbert, R. Legtenberg, and S. D. Senturia, "3D coupled electro-mechanics for MEMS: Applications of CoSolve-EM," Proceedings IEEE Micro Electro Mechanical Systems, Amsterdam, Netherlands, January 29–February 2, 1995, pp. 122–127.
7. R. Legtenberg, E. Berenschot, M. Elwenspoek, and J. Fluitman, "Electrostatic curved electrode actuators," Proceedings IEEE Micro Electro Mechanical Systems, Amsterdam, Netherlands, January 29–February 2, 1995, pp. 37–42.
8. E.K. Chan and R.W. Dutton, "Electrostatic micromechanical actuator with extended range of travel," JMEMS, 9(3), September 2000, pp. 321–328.
9. T. Akiyama, D. Collard, and H. Fujita, "Scratch drive actuator with mechanical links for self-assembly of three-dimensional MEMS," Journal of Microelectromechanical Systems, 6(1), March 1997, pp. 10–17.
10. L. Fan, M.C. Wu, K.D. Choquette, and M.H. Crawford, "Self-assembled microactuated XYZ stages for optical scanning and alignment," Proceedings of International Solid State Sensors and Actuators Conference, Transducers '97, Chicago, IL, 1997, pp. 319–322.
11. Z. Liu, M. Kim, N. Shen, and E. C. Kan, "Novel electrostatic repulsion forces in MEMS applications by nonvolatile charge injection," Technical Digest. MEMS 2002 IEEE International Conference. Fifteenth IEEE International Conference on Micro Electro Mechanical Systems, Las Vegas, NV, January 20–24, 2002, pp. 598–601.
12. H. Guckel, T. Earles, J. Klein, D. Zook, and T. Ohnstein, "Electromagnetic linear actuators with inductive position sensing for micro relay, micro valve and precision positioning applications," Proceedings of the International Solid-State Sensors and Actuators Conference, Transducers '95, Stockholm, Sweden, 1995, pp. 324–327.
13. T.R. Ohnstein, et al., "Tunable IR filters using flexible metallic microstructures," Proceedings IEEE Micro Electro Mechanical Systems, Amsterdam, Netherlands, January 29–February 2, 1995, pp. 170–174.

14. C. Liu, T. Tsao, Y.C. Tai, T.S. Leu, C.M. Ho, W.L Tang, and D. Miu, "Out-of-plane permalloy magnetic actuators for delta-wing control," Proceedings IEEE Micro Electro Mechanical Systems, Amsterdam, Netherlands, January 29–February 2, 1995, pp. 7–12.
15. J. Bernstein, W.P. Taylor, J. Brazzle, G. Kirkkos, J. Odhner, A. Pareek, and M. Zai, "Two axis-of-rotation mirror array using electromagnetic MEMS," Technical Digest. MEMS 2003 IEEE International Conference. Sixteenth IEEE International Conference on Micro Electro Mechanical Systems, Koyto, Japan, January 19–23, 2003, pp. 275–278.
16. The Measurement, Instrumentation, and Sensors Handbook by J.G. Webster, CRC Press LLC, 1999, pp. 48.1–48.4.
17. F. Primdahl, "The fluxgate magnetometer," Journal of Physics E: Scientific Instruments, 1, 1979, pp. 242–253.
18. Sensors and Signal Conditioning by Pallas-Areny and Webster, Wiley & Sons, Inc., pp. 247–257.
19. S. Miyazaki, T. Kawai, and M. Araragi, "A piezo-electric pump driven by a flexural progressive wave," Proceedings IEEE Micro Electro Mechanical Systems, Nara, Japan, 1991, pp. 283–288.
20. An Introduction to Ultrasonic Motors by T. Sashida and T. Kenjo, 1993.
21. T. Morita, M. Kurosawa, and T. Higuichi, "An ultrasonic motor using bending cylindrical transducer based on PZT thin film," Proceedings IEEE Micro Electro Mechanical Systems, Amsterdam, Netherlands, January 29–February 2, 1995, pp. 49–54.
22. M. Takahashi, M. Kurosawa, and T. Higuchi, "Direct frictional driven surface acoustic wave motor," Proceedings of the International Solid-State Sensors and Actuators Conference, Transducers '95, Stockholm, Sweden, 1995, pp. 401–404.
23. A. Potekhina and C. Wang, "Review of electrothermal actuators and applications," Actuators, 8, 2019, 69; doi:10.3390/act8040069.
24. A. Karim and J.Y. Andersson, "Infrared detectors: Advances, challenges and new technologies," IOP Conference Series: Materials Science and Engineering, 51, 2013, 012001.
25. D. Xu, et al., "MEMS-based thermoelectric infrared sensors: A review," Frontiers of Mechanical Engineering, 12(4), 2017, 557–566.
26. A. RogalskI, "History of infrared detectors," Opto–Electronics Review, 20(3), 2012, pp. 279–308.

Questions and Problems

Question 3.1 In atmospheric pressure (i.e., 760 Torr ≈ 1 bar) the air breaks down when about 7,000 V is applied between two electrodes separated by about 2 mm. What will be the breakdown voltage, if the ambient pressure is reduced by two orders of magnitude (i.e., to 7.6 torr)?

Question 3.2 For the comb drive shown in Fig. 3.1 with the comb fingers separated from the stationary electrode by 1 μm air gap, what is the *maximum* DC bias voltage that can applied between the comb fingers and the stationary electrode in ambient pressure (i.e., 1 bar)?

Question 3.3 Starting from a formula for the energy stored in a capacitor ($CV^2/2$), derive the equation ($F = \varepsilon_0 V^2/2d^2$) for an electrostatic force (per unit area) acting on the two plates (separated by an air gap of d) for a given voltage. [Hint: Force is the first differentiation of energy with respect to spatial distance, and the capacitance per unit area in this case is ε_0/d.]

Question 3.4 The "pull-in effect" in electrostatic actuation happens when the two plates (separated by an air gap) are drawn together by one-third of the initial gap between the two plates. Prove it using the spring force between the two plates to be equal to $k(g_o-g)$, where k, g_o, and g are the spring constant, initial gap, and electrostatically varied gap, respectively. [Hint: Use the electrostatic force obtained in Question 3.3 and equate it with the spring force. Then calculate g as a function of V and g_o for the quadratic equation, and obtain the critical value for g above which V becomes an imaginary number. Finally, use the critical g to obtain V_c.]

Question 3.5 A certain diaphragm of 100 μm² area (with an effective spring constant k equal to 1 N/m) has an initial air gap of 3 μm between it and the underlying substrate. Sketch the center gap distance (between the diaphragm and the underlying substrate) as the applied electrostatic voltage is varied from 0 to 150 V. Calculate any critical values and indicate them on the graph.

Question 3.6 A bridge has an air gap of 1 μm between it and the underlying substrate, and has been measured to be electrostatically pulled down to the substrate by an applied voltage of 10 V (i.e., the on-set of the pull-in effect happens at 10 V applied). If the air gap is increased to 3 μm, at what voltage will the bridge be electrostatically pulled down?

Question 3.7 A bridge (2 μm wide and 5 μm long) has an initial air gap of 1 μm between it and the underlying substrate, and is being electrostatically deflected by an applied voltage V_a through a capacitor C_1 in series with the bridge, as shown below. What is the maximum C_1 that will allow the deflection of the bridge without any "pull-in" effect over the whole air gap range?

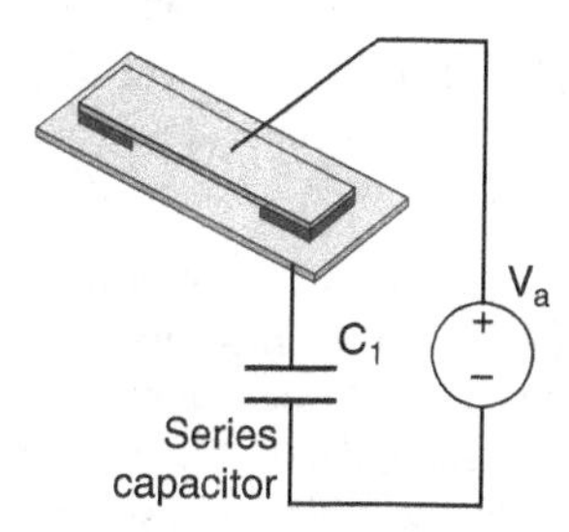

Question 3.8 A cantilever shown below has an initial air gap of 3 μm between it and the underlying substrate and has been measured to be electrostatically pulled down to the substrate by an applied voltage of 10 V (i.e., the on-set of the pull-in effect happens at 10 V applied). *Qualitatively* sketch the deflection of the cantilever's (a) free end and (b) center when the applied electrostatic voltage is varied from 0 to 15 V.

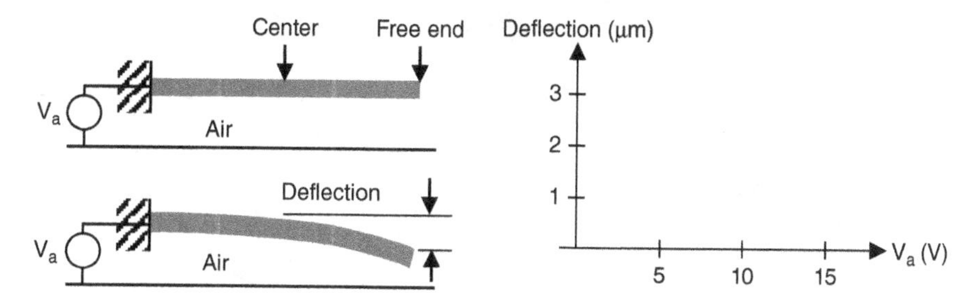

Question 3.9 *Sketch* applied voltage V versus air gap g (for g between 0 and g_o) for two parallel plates separated by an air gap as shown in Fig. 3.4a. [Hint: The spring force of the movable plate is equal to $k(g_o-g)$, where k, g_o, and g are the spring constant, initial gap, and electrostatically varied gap, respectively. The electrostatic force is equal to $\varepsilon_o V^2 A/2g^2$, where ε_o, V, and A are vacuum

permittivity, applied voltage, and plate area, respectively, and the spring force and the electrostatic force must be equal at static equilibrium.]

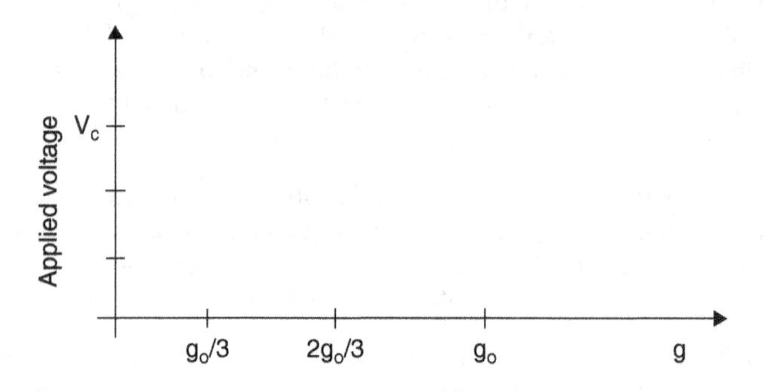

Question 3.10 From the equation $k(g_o - g) = \varepsilon_o V^2 A/2g^2$ (Fig. 3.4), we obtain $V = k\sqrt{(g_o - g)g^2}$ where $k = \sqrt{2K/\varepsilon_o A}$. If we are to sketch V versus g, we will see that V is zero at $g = 0$ and g_o. At what g is V the maximum?

Question 3.11 Across the air gap of an air-filled parallel capacitor (C), there exists the following electrostatic attractive force F due to voltage V owing to RF power passing through the capacitor: $F = \frac{\varepsilon_o A V^2}{2g^2}$, where A, g, and ε_o are the area of the capacitor, the air gap of the capacitor, and vacuum permittivity, respectively. Calculate the electrostatic force if a 1 GHz RF signal with a power level of 1 W passes through the air-gap capacitor with an air gap of 1 μm. [Hint: RF power is equal to V^2/Z, where Z is the impedance and is equal to $1/\omega C$ for a capacitor.]

Question 3.12 The *left* figure below shows an electromagnetic actuator made of a plunger and an electromagnet (see [13] for details). If the measured coil inductance as a function of the plunger position is as shown below at *right*, (a) calculate the linear reluctance force that the plunger will experience for a current of 40 mA applied to the coil in the electromagnet, (b) calculate the linear displacement (in μm) that the plunger will move due to a 40 mA current applied to the coil in the electromagnet, assuming the plunger's spring constant to be 100 N/m. In what direction would the plunger move when a drive current is applied to the electromagnet?

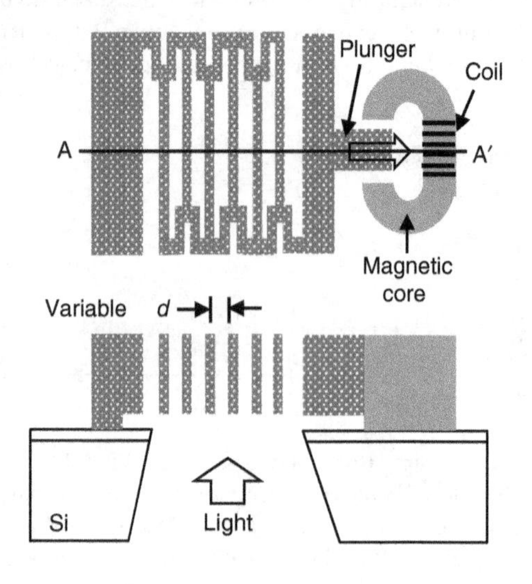

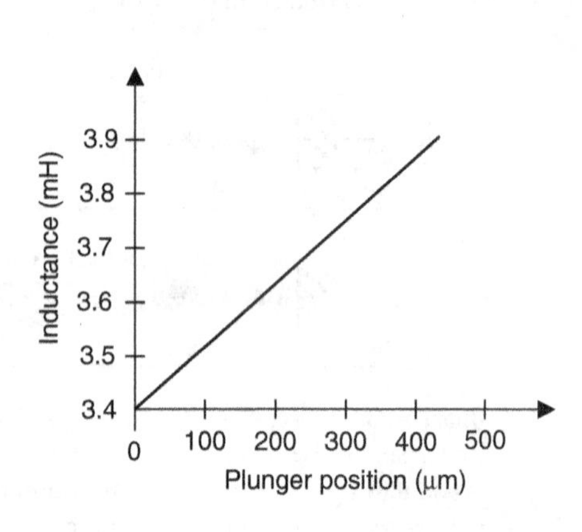

Question 3.13 Two PZT substrates with four different sections (each of which is polarized as indicated with arrows below) are glued together and applied with AC electrical voltages, as shown below. What will the applied frequency (ω) have to be in order to produce a traveling flexural wave at 1 m/s? What frequency will have to be used for the sinusoidal voltage sources if we want to produce a flexural wave traveling at 300 m/s?

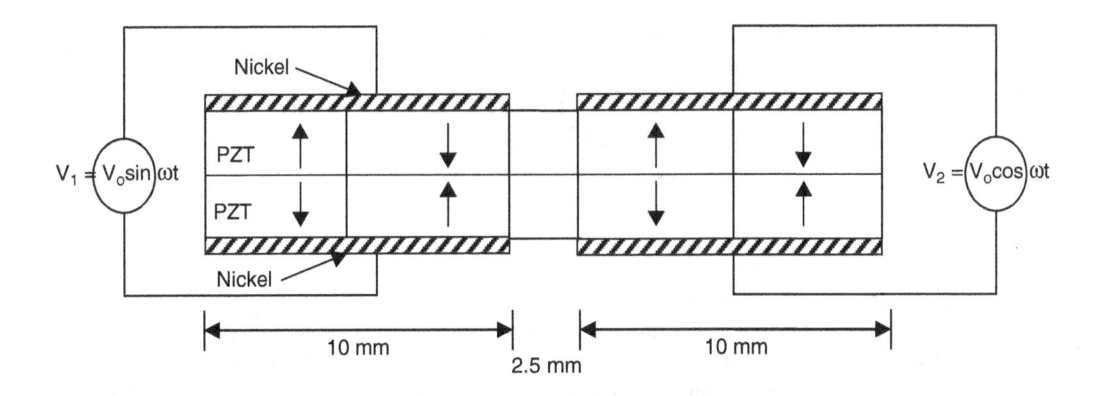

Question 3.14 Two PZT substrates with their initial polarizations as indicated with arrows below are applied with a static electrical voltage, as shown below. (a) *Sketch* how the PZT as a whole will *bend* due to the 10 V applied, assuming that the PZT has $d_{33} = 100$ pC/N, $d_{31} = -50$ pC/N, and $d_{15} = -140$ pC/N. (b) For the left PZT of the two PZTs below, calculate the displacement in the thickness direction (i.e., the PZT's polarization direction) due to the applied 10 V, assuming the thickness to be 1 mm.

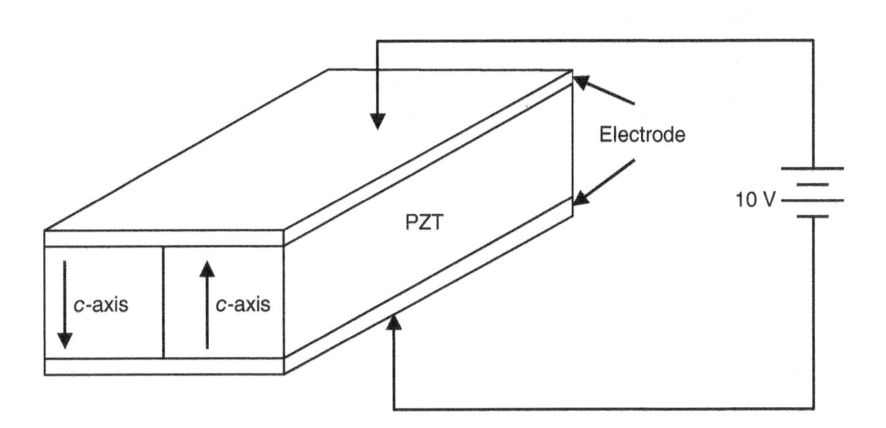

Question 3.15 A PZT block with its initial polarization as indicated with arrows below is applied with a static electrical voltage, as shown below. Calculate the shape change (in *x*-, *y*-, and *z*-directions) of the PZT block due to the 10 V applied, assuming that the PZT has $d_{33} = 100$ pC/N, $d_{31} = -50$ pC/N, and $d_{15} = -140$ pC/N.

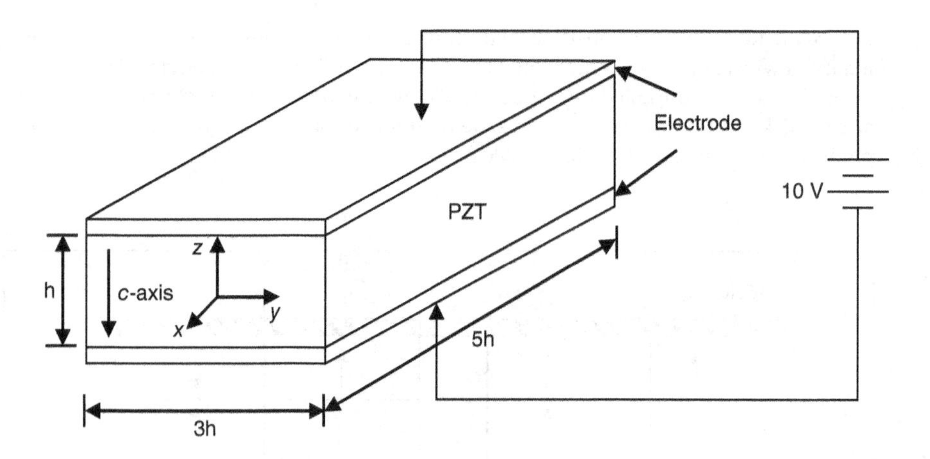

Question 3.16 A piezoelectric PZT block with its initial polarization as indicated with arrows below ($h = 1$ mm, $w = 5$ mm, and $l = 10$ mm) is mechanically unbound (so that no stress can be sustained) and applied with a static electrical voltage, as shown below. Calculate the length changes for h, w, and l due to the 10 V applied, assuming that the PZT has $d_{33} = 100$ pC/N, $d_{31} = -50$ pC/N, and $d_{15} = -140$ pC/N.

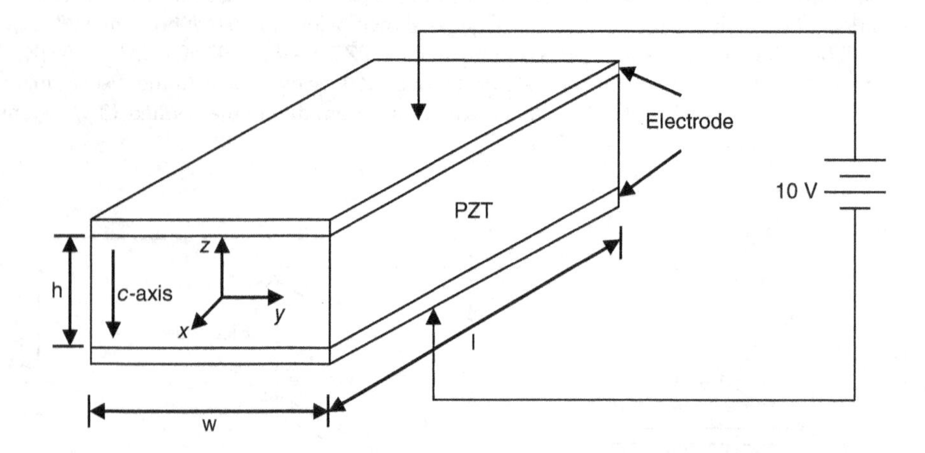

Question 3.17 If the voltage applied to the PZT block above is sinusoidal in time (i.e., $V = V_o \sin\omega t$) with amplitude $V_o = 10$ V and $\omega = 2\pi \cdot 100$ rad/s (which means that the frequency is 100 Hz), calculate the maximum instantaneous power consumption as the PZT vibrates in response to the applied voltage. Assume 10^{-10} F/cm for PZT's dielectric constant and infinitely large value for PZT's resistivity. [Hint: Instantaneous power consumption can be obtained by differentiating the capacitive energy with respect to time.]

Question 3.18 A zinc oxide crystal (with $h = 1$ mm, $w = 5$ mm, and $l = 10$ mm) is being pressed with a static force $F = 100$ N, as shown below. Calculate the open-circuit voltage that would develop between the two metal plates, if ZnO's resistivity were infinite. Assume that $d_{33} = 11$ pC/N, $d_{31} = -5$ pC/N, $d_{15} = -14$ pC/N, and $\varepsilon_{11} \cong \varepsilon_{33} \cong 10\varepsilon_o = 10 \times 8.854 \times 10^{-12}$ F/m.

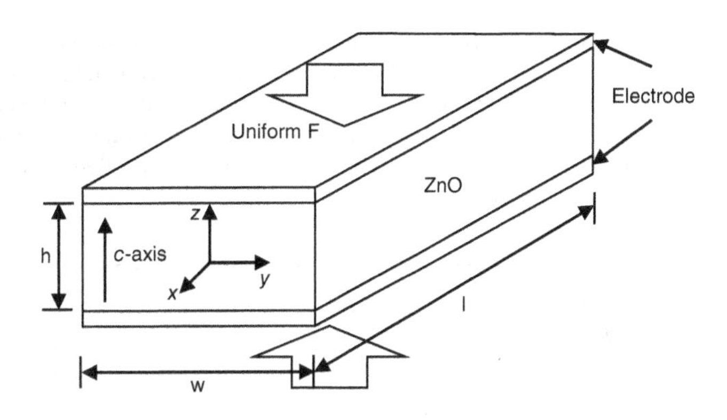

Question 3.19 For bending a cantilever, rank the three different actuation methods (listed on the first row) for the issues listed in the first column with 1 being the best and 3 being the worst.

	Electrostatic actuation	Electromagnetic actuation	Piezoelectric actuation
Power consumption			
Maximum obtainable bending displacement			
Actuation voltage for bending without pulll-in			
Easiness in microfabrication			

Question 3.20 For an array of vertically deflectable cantilevers (i.e., cantilevers that can be pulled down [or pushed up] to [from] the substrate), rate the following three actuation techniques between "good" and "no good" for the issues listed in the first row of the table below.

	Electrical power consumption	Bidirectional motion generation	Dynamic range (i.e., linear control range)	Required voltage level for a large deflection	IC process compatibility
Electrostatic					
Electromagnetic					
Piezoelectric					

Problem 3.1 A piezoelectric ZnO film ($10 \times 10 \times 1$ μm) is covered with two electrodes as shown below. Do the following for a sinusoidally varying *stress with peak-to-peak value of* 2×10^7 *Pa* applied as shown below.

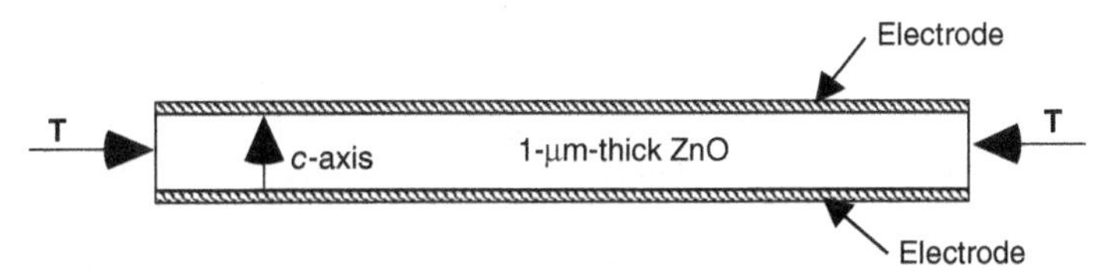

For the open-circuit type, ignoring mechanical resonance effect, sketch on the figure below (a) the electrical charge developed on the top electrode and (b) the voltage developed between the two electrodes as a function of time. (c) Calculate the maximum peak-to-peak voltage developed between the two electrodes and indicate it in the figure below. Assume that the resistivity and the dielectric constant of ZnO are 107 Ωcm and 8 × 10–13 F/cm, respectively. And $d_{33} = 11 \times 10^{-12}$ C/N and $d_{31} = -5 \times 10^{-12}$ C/N for ZnO.

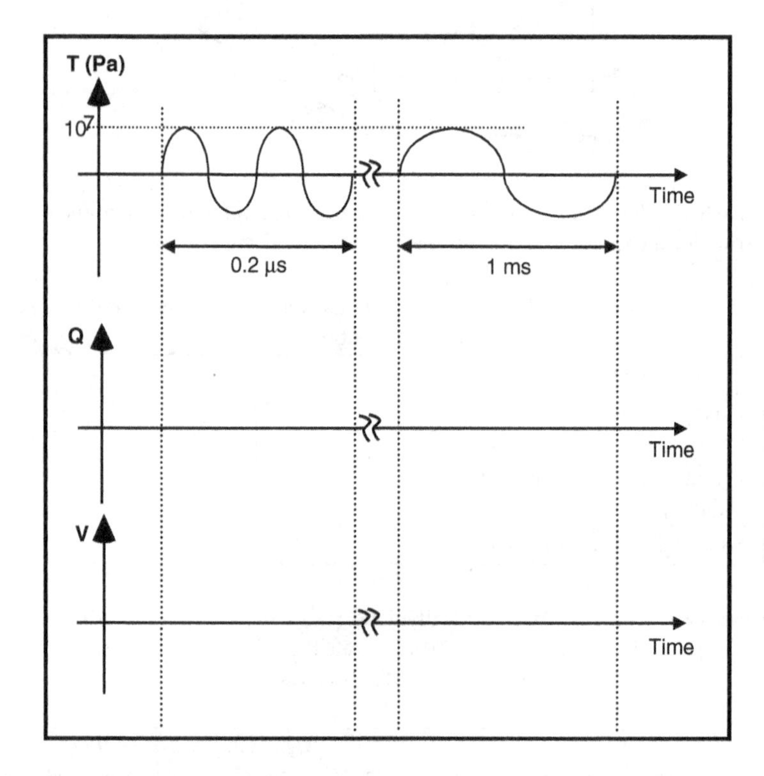

Problem 3.2 A piezoelectric ZnO crystal (10 × 10 cm × 1 mm) is covered with two electrodes and two layers of 0.2-mm-thick SiO_2 as shown below. Do the following for a sinusoidally varying *stress with peak-to-peak value of* 2×10^7 *Pa* applied uniformly on the sides as shown below.

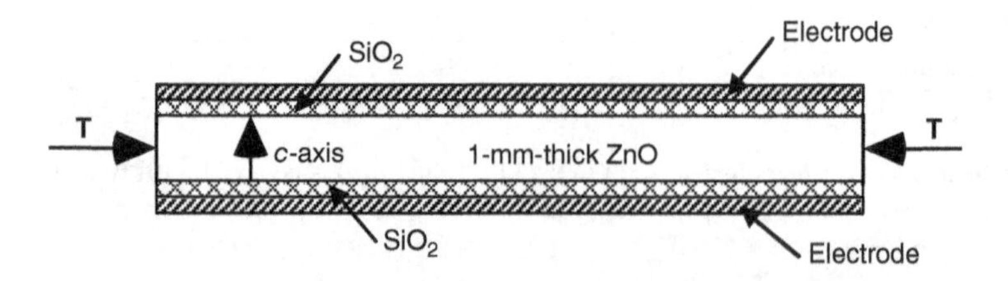

For the open-circuit type, ignoring mechanical resonance effect, sketch on the figure below (a) the electrical charge developed on the top electrode and (b) the voltage developed between the two electrodes as a function of time. (c) Calculate the maximum peak-to-peak voltage developed between the two electrodes. Assume that the resistivity and dielectric constant of ZnO are 10^7 Ωcm and 8×10^{-13} F/cm, respectively. And $d_{33} = 11 \times 10^{-12}$ C/N and $d_{31} = -5 \times 10^{-12}$ C/N for ZnO. Also, assume that the resistivity and dielectric constant of SiO_2 are infinite and 3.5×10^{-13} F/cm, respectively.

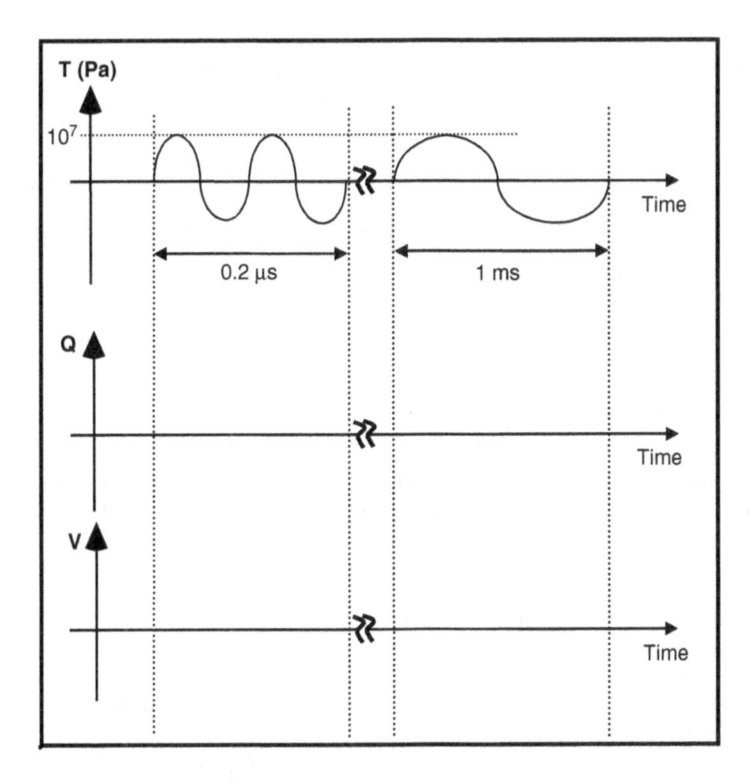

Problem 3.3 In the figure below, a piezoelectric ZnO film is covered with two electrodes, and is applied with a force that is uniformly distributed over the sidewalls and varying sinusoidally in time (i.e., $F = F_o \sin\omega t$) so that a sinusoidally varying voltage (between the two electrodes) $V = V_o \sin\omega t$ is developed between the two top and bottom electrodes. Assume that the resistivity, dielectric constant, and d_{31} of ZnO are 10^7 Ωcm, 10^{-12} F/cm, and 5 pC/N, respectively. Also, ignore any mechanical or acoustic resonance effect.

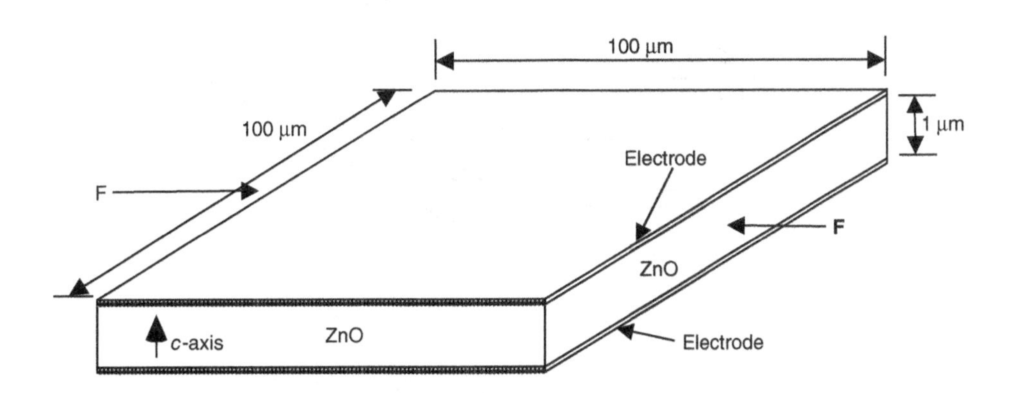

For the open-circuit type, ignoring mechanical resonance effect, calculate the magnitude of the voltage (V_o) developed between the two electrodes for forces varying at (a) 100 MHz, (b) 1 MHz, and (c) 0.01 Hz. Assume that the magnitude of the applied force is 1 N (i.e., $F_o = 1$ N) in all three cases.

Problem 3.4 A piezoelectric ZnO crystal is covered with two electrodes as shown below and is applied with a force varying sinusoidally in time (i.e., $F = F_o \sin\omega t$).

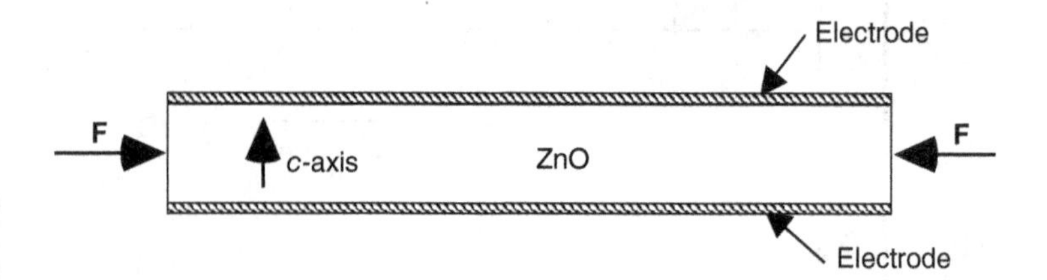

For the open-circuit type, *sketch* (on the log–log scale) the voltage developed between the two electrodes, as the frequency ($f = \omega/2\pi$) of the sinusoidal force is varied between 1 Hz and 1 GHz. Assume that the resistivity and the dielectric constant of ZnO are 10^7 Ωcm and 10^{-12} F/cm, respectively. And ignore any mechanical or acoustic resonance effect but consider only the charge transfer effect due to the finite resistivity of ZnO crystal.

Problem 3.5 A piezoelectric ZnO crystal (with $h = 1$ mm, $w = 5$ mm, and $l = 10$ mm) is covered with two electrodes as shown below and is applied with a sinusoidal force that is uniformly distributed over the top and bottom faces. Assume $d_{33} = 10$ pC/N, $d_{31} = -5$ pC/N, $d_{15} = -14$ pC/N, ρ (resistivity) = 10^7 Ωcm, and ε (dielectric constant) = 10^{-12} F/cm for ZnO. For the open-circuit type, ignoring mechanical resonance effect, (a) *sketch* (on the figure below) the polarization field P within the ZnO, the electrical field E within the ZnO, the electrical charge Q developed on the top electrode, and the voltage V developed between the two electrodes as a function of time. (b) Calculate the *peak* electrical field and voltage and indicate those on the E and V curves.

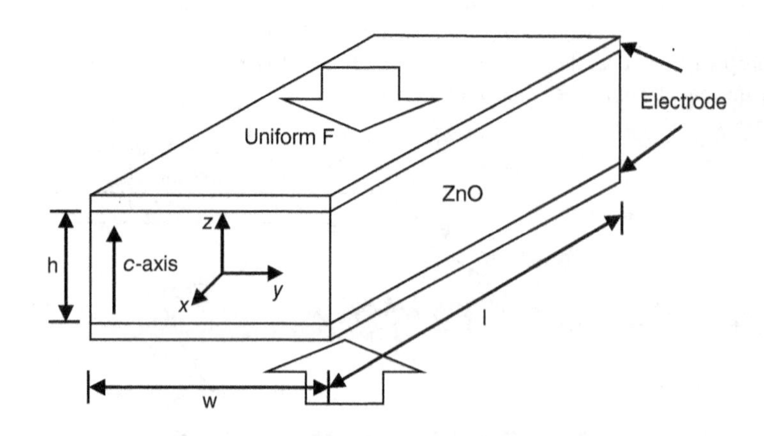

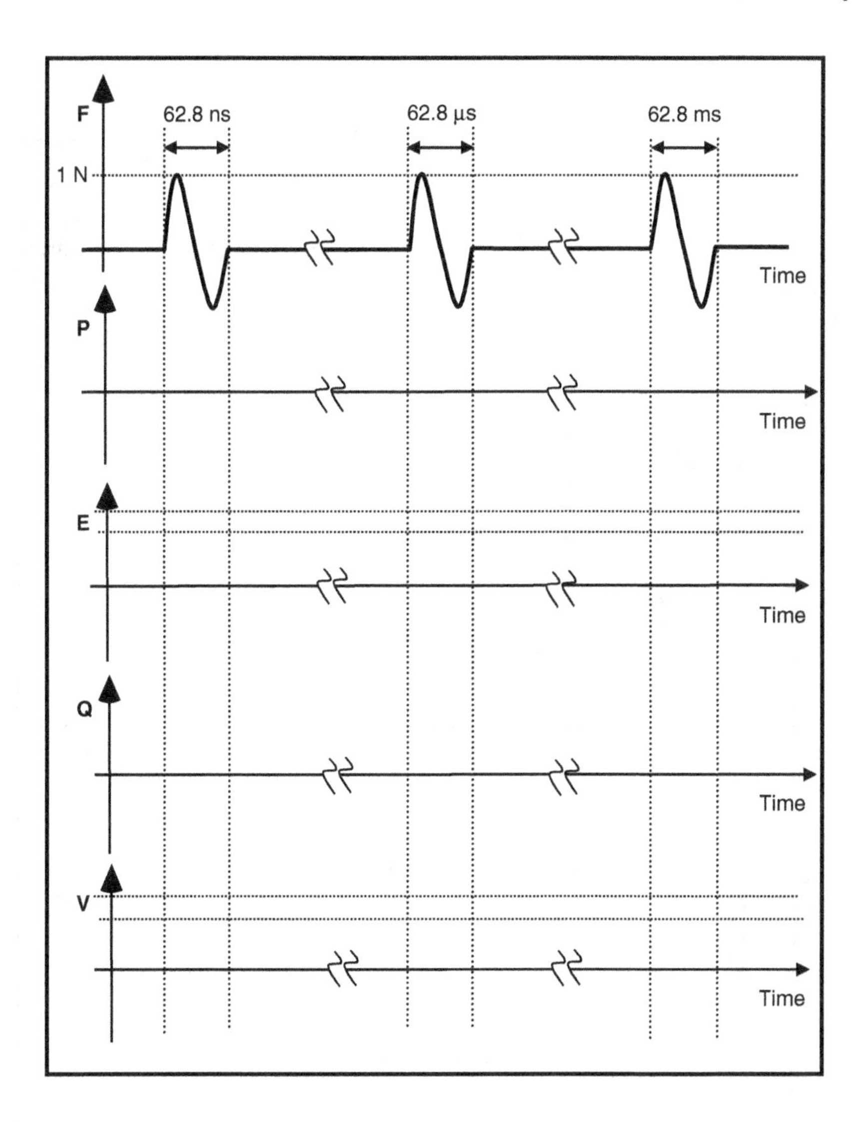

Problem 3.6 As shown below, a piezoelectric ZnO crystal ($h = 1$ mm, $w = 5$ mm, and $l = 10$ mm) is covered with two Al electrodes and is applied with a force that is uniformly distributed over the top and bottom faces. The electrodes are connected to a unity gain amp with 1 MΩ input resistance. Ignoring mechanical resonance effect, (a) *sketch* (on the figure below) the magnitudes of the polarization field P within the ZnO, the electrical field E within the ZnO, the electrical charge Q developed on the top electrode, and the voltage v_o after the unity gain amplifier as a function of time; (b) calculate the *peak* electrical field and voltage and indicate those on the E and V curves. Assume $d_{33} = 10$ pC/N, $d_{31} = -5$ pC/N, $d_{15} = -14$ pC/N, ρ (resistivity) = 10^7 Ωcm, and ε (dielectric constant) = 10^{-12} F/cm for ZnO.

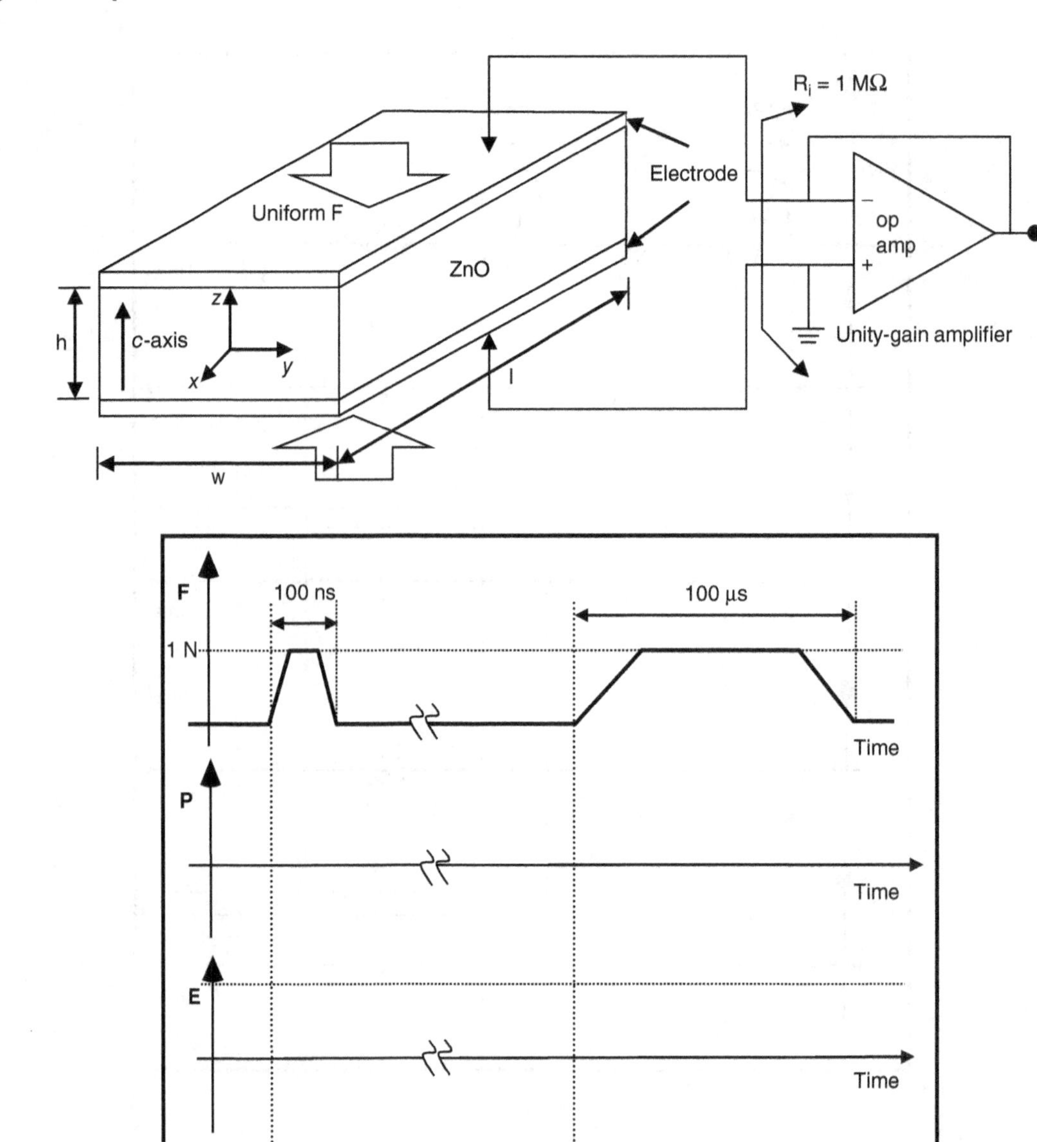

R_i = 1 MΩ
Electrode
Uniform F
ZnO
op amp
Unity-gain amplifier
h
c-axis
z
y
x
l
w
F
100 ns
100 μs
1 N
Time
P
Time
E
Time
Q
Time
V
Time

CHAPTER 4

RF MEMS

Since 2001, filters based on film bulk acoustic resonator (FBAR) and surface acoustic wave (SAW) have been used for the bandpass filtering at the RF front end for cell phones, completely replacing the dielectric filters that are much larger in volume. With advancement of long-term evolution (LTE) for smartphones, the number of bandpass filters needed at the RF front end (i.e., immediately adjacent to antenna) started skyrocketing around 2014 and is expected to grow to 100 per LTE smartphone by 2020 as the number of the frequency bands (Fig. 4.1) becomes large [1]. As we will see in this chapter, one of the FBAR filter types is based on suspended microelectromechanical systems (MEMS) diaphragm.

There are other opportunities for RF MEMS to contribute to smartphones or wireless transceivers. For example, RF MEMS switches perform far better than transistor-based switches, while tunable MEMS capacitors are far superior to varactors based on reverse-biased pn junction in terms of insertion loss, on/off ratio, tuning range, etc. Also, MEMS-based inductors have quality factor (Q) greater than several tens, many times higher than on-chip spiral coil, at GHz. However, commercial applications of these have not taken off as MEMS-based FBAR filters, and we will try to gain insight to the reasons for this lack of commercial success, as we study them in this chapter.

In this chapter, we will study silicon MEMS resonators, followed by acoustic-wave resonators and filters. Since FBAR filters have provided the critical RF front-end technology for smart phones, we will study fundamentals of acoustic waves in solid that are directly related to FBAR and FBAR-based filters. In the sections, many equations will be derived from basic equations and physical principles and should be relatively easy to follow through. Some of the equations and physical concepts (e.g., wave propagation, reflection, and dispersion) are applicable to other waves such as electromagnetic waves, and it would be good to learn them well. The remaining portion of the chapter is on tunable MEMS capacitors and MEMS switches.

4.1 Electromagnetic Wave Spectrum

The frequency (and thus the wavelength) of electromagnetic (EM) wave spans over a very wide range, as shown in Fig. 4.2, and the wave's propagation and absorption are markedly different for different frequencies of EM wave. The ranges of the frequency and wavelength that are used by cell phones are currently 1–2 GHz and 0.15–0.3 m, respectively. As the frequency goes beyond tens of GHz, EM wave behaves like light (i.e., optical wave) requiring "line-of-sight" for wave propagation.

Electromagnetic waves from 0.23 to 40 GHz are grouped into various bands with their names listed in Table 4.1.

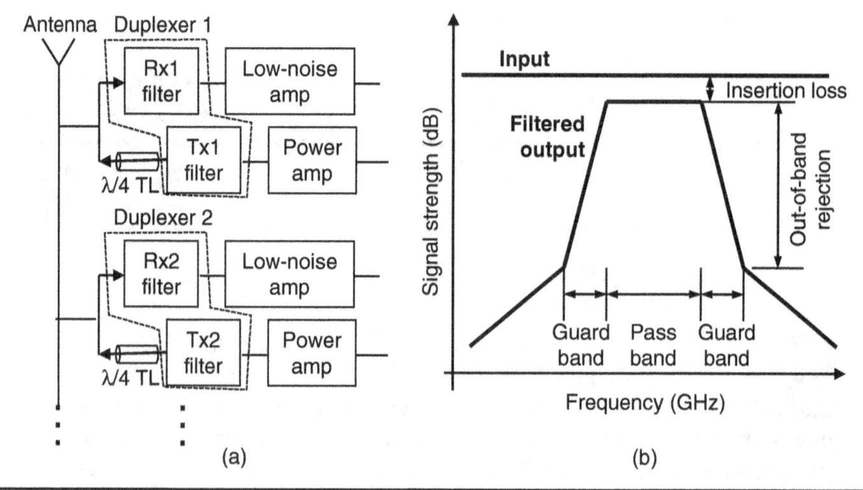

Figure 4.1 (a) RF front end of a smartphone with a duplexer composed of two bandpass filters, one for incoming signal (denoted as *Rx* filter before low-noise amplifier [LNA]) and the other for outgoing signal (denoted as *Tx* filter after power amplifier [PA]), for each band, and (b) characteristics of a bandpass filter. The number of bands in high-end LTE can be more than 15, resulting in more than 30 filters per LTE smartphone.

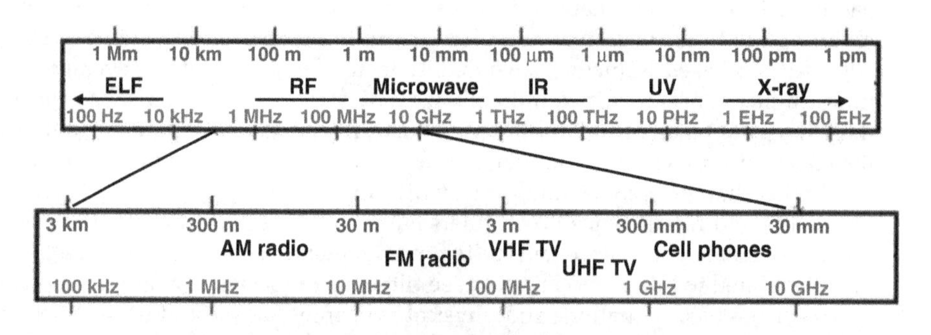

Figure 4.2 Ranges of the wavelength and frequency of electromagnetic wave.

Various cell phone systems use frequency from 0.8 to 2 GHz, as shown in Table 4.2, with advanced mobile phone system (AMPS), global system for mobile communication (GSM), Nippon total access communication system (NTACS), personal communication network (PCN), and personal communication system (PCS). GSM is popular in Europe, while AMPS and PCS are popular in the states.

P band	0.23–1 GHz	L band	1–2 GHz	S band	2–4 GHz	C band	4–8 GHz
X band	8–12.5 GHz	Ku band	12.5–18 GHz	K band	18–26.5 GHz	Ka band	26.5–40 GHz

Table 4.1 Frequency Ranges for Various Bands for Communication

System	Transmission Band [MHz]	Receive Band [MHz]	Bandwidth [MHz]
AMPS	824–849	869–894	25
GSM	890–915	935–960	25
NTACS	898–925	843–870	27
PCN	1710–1785	1805–1880	75
PCS	1850–1910	1930–1990	60

TABLE 4.2 Various Cell Phone Communication Systems

4.2 Silicon Micromechanical Resonator

4.2.1 Resonators at 1–10 MHz

Silicon micromechanical resonator can potentially compete with quartz resonator for clocking and timing at 10 MHz. A company called SiTime has been commercializing silicon resonators by using its Epi-Seal fabrication process illustrated in Fig. 4.3. Since the process requires a polysilicon deposition step at a very high temperature for cleaning, sealing, and encapsulation (Fig. 4.3d), there is no metal till near the end (Fig. 4.3f), and the electrodes of the resonator are doped silicon. One commercial resonator in built-in vacuum pressure resonates in a wineglass mode at a frequency of 5 MHz with Q of 75,000.

Silicon's temperature coefficient of frequency (TCF) is about –30 ppm/°C, unlike quartz which can be cut in a certain way to have its TCF near zero at room temperature. Consequently, the resonator needs a temperature compensation circuit, and SiTime has integrated complementary metal–oxide–semiconductor (CMOS) circuit chip with the resonator on a single package. SiTime sells oscillators (with the high Q resonator providing the precise control of the oscillation frequency). One example is a 5 MHz oscillator made (with a 0.2×0.2 mm^2 and 10-μm-thick silicon resonator) on a $2.0 \times 2.5 \times 0.85$ mm^3 package with temperature compensation achieved by sensing the temperature with a bandgap-reference circuit on the CMOS chip and adjusting the phase locked

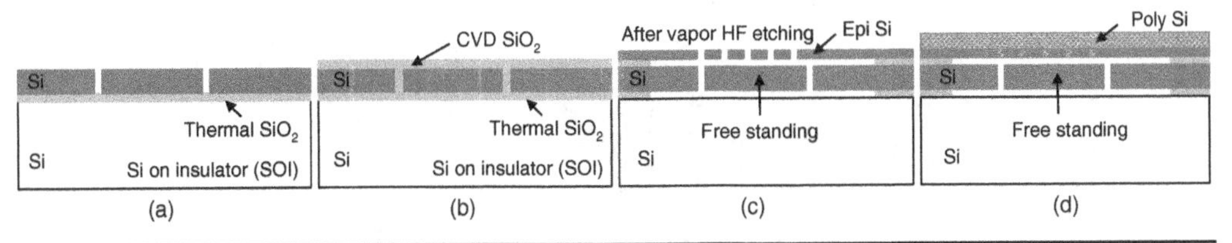

FIGURE 4.3 SiTime's Epi-Seal fabrication process for silicon micromechanical resonator that is under vacuum through on-chip encapsulation with polysilicon: (a) after silicon deep reactive ion etching (DRIE) of silicon on insulator (SOI), (b) after chemical vapor deposition (CVD) SiO_2 deposition and patterning, (c) after epitaxial silicon deposition and patterning of etching holes, followed by vapor HF etching of SiO_2, (d) after depositing polysilicon with gases containing hydrogen, chlorine, and silicon to clean the resonator, seal the etch holes, and encapsulate the resonator, after which the remaining process steps are to form vias for electrical connection to the resonator, followed by metallization. (Adapted from [2].)

loop (PLL) based on the oscillator. The oscillation frequency fluctuates by ±50–100 ppm over the operating range of –45 to +85°C. With a careful calibration over temperature, the frequency change is reported to be about ±5 ppm from –45 to +85°C, but each device needs to be calibrated after packaging.

4.2.2 Filters Based on Silicon Micromechanical Resonators

A comb drive has been studied for electrical filtering due to its potentially large tunability, low insertion loss, and high stopband-to-passband rejection ratio. In [3], two comb drives are mechanically coupled through a coupling spring. Each comb drive can be represented with an equivalent spring and mass, while the coupling can be conceptually viewed as a coupling spring between the two spring–mass systems [3]. The mechanical resonance can be modeled with an equivalent circuit composed of inductor, capacitor, and resistor. Two or more comb drives can be combined to produce a very impressive band pass filter, and an electrical filter made of the comb drives shows a very low insertion loss and high passband-to-stopband rejection ratio [4], but at about 340 kHz, too low for wireless communication. For a comb drive to operate at GHz, its lateral dimension must be reduced to nanometer range, and it is not clear whether nanometer-scale comb drive is feasible.

4.2.3 In Pursuit of GHz Silicon Resonators

A "flexural-mode beam" suspended by four quarter-wavelength torsional beams [5] is based on the idea that if torsional beams are in contact with a flexural-mode beam at their nodal points, the flexural-mode beam has "free" boundary conditions on its four support points. A fundamental resonant frequency of 71.5 MHz with such a beam was shown to have a Q of 8,250 in vacuum. A DC bias voltage V_P is applied to make the electrode-to-resonator gap spacing small enough for adequate electromechanical coupling.

Since the resonant frequency ω_o is equal to square root of spring constant k divided by mass m, i.e., $\omega_o = \sqrt{k/m}$, the resonant frequency can be increased by increasing the spring constant (or the structural stiffness). Thus, to push the resonant frequency to GHz, in Fig. 4.4, a circular disk is vibrated (in extensional mode) with an applied electrostatic voltage between the disk and surrounding electrodes that are separated

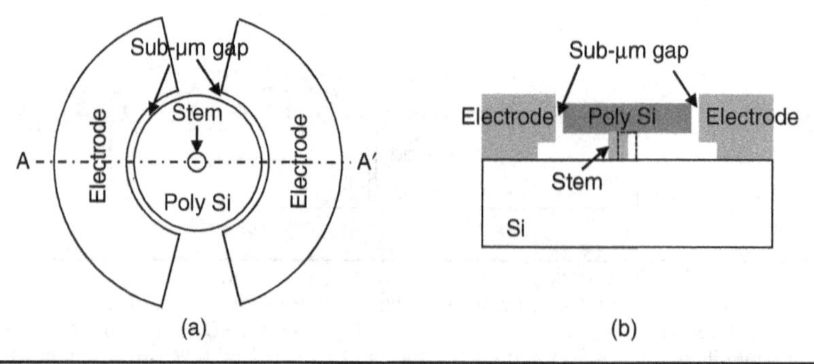

FIGURE 4.4 (a) Top-view schematic of a polysilicon disk vibrating in radial extensional mode and (b) cross-sectional view of the polysilicon disk (across A–A′) with misaligned stem. (Adapted from [9].)

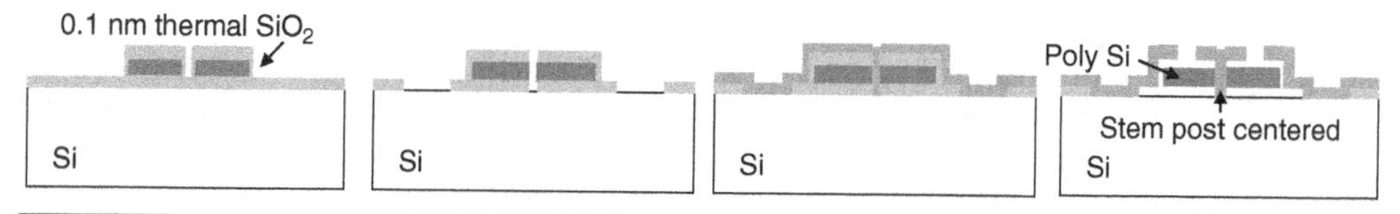

FIGURE 4.5 Brief fabrication process of a self-aligned disk resonator. (Adapted from [7].)

from the disk by submicron air gap. A polysilicon disk in extensional mode has a high stiffness constant and can resonate at a high frequency, especially if the mass is small. With electroplated metal electrodes, the submicron air gap is obtained by thermally oxidizing polysilicon and then later removing the oxide.

As the disk size is made smaller (for smaller mass) to push the resonant frequency into GHz range, the misalignment of the stem post (near the disk's center, as illustrated in Fig. 4.4b) for the disk has more pronounced effect on the disk's vibrational behavior. Thus, how the stem post is fabricated with respect to the disk is very critical. The fabrication sequence that achieves a self-aligned disk is shown in Fig. 4.5, where the disk layer is patterned with a hole in the center first before filling the hole with a deposited polysilicon. In this way, the accuracy of making a hole in the center of the disk layer depends on the accuracy of the patterns on the photomask (which is usually very good for most pattern generators). With this new fabrication process, the same group reported a polysilicon disk resonator with its observed resonant frequencies (including harmonics) beyond 1 GHz [7]. As the resonant frequency is made to be greater than 1 GHz, the quality factor Q becomes lower since a resonator is typically governed by a constant fQ product. At such a high resonant frequency (with mechanically stiff structure), viscous damping is negligible, and vacuum packaging does not increase the Q substantially.

Question: The transducer gap in the disk transducer (shown in Fig. 4.4) is submicron. How was the submicron gap achieved without any submicron lithography?

Answer: The polysilicon was thermally oxidized to produce submicron-thick silicon oxide, which was later etched away to produce the submicron gap.

Question: Sketch the first photomask pattern used for the process shown in Fig. 4.5.

Answer:

Effect of Material-Mismatched Isolation Support on Q

The reported Q's at 1–2 GHz with polysilicon disk resonators have been 1,500–15,000 [6–8]. The wide range of the Q's is due to various attempts to reign over the Q. The Q of a vibrating resonator depends on four major loss mechanisms, and the following equation can be used for the Q: $1/Q = 1/Q_{defect} + 1/Q_{TED} + 1/Q_{viscous} + 1/Q_{support}$, where Q_{defect}, Q_{TED}, $Q_{viscous}$, and $Q_{support}$ are the quality factors accounting energy loss due to material defects, thermoelastic damping (TED), viscous damping (which is negligible in vacuum), and loss to the substrate through anchor, respectively. The Q of the disk resonators was reported to be dominated by the anchor loss [8], as the losses due to material defects, TED, and viscous damping are negligible due to the facts that the Q was not high enough for material defects to play a role; TED characteristic frequency was 20.4 MHz, much lower than the resonant frequency; and the resonator was stiff enough for a viscous damping to be a minor factor, respectively.

Now that high Q resonators at GHz have been demonstrated with polysilicon microstructure through electrostatic actuation [8], the remaining challenges or issues are (1) relatively low passband-to-stopband ratio and electromechanical coupling coefficient that have practical implication for filter applications (though not for oscillator applications), (2) a very high motional resistance ($>> 50\ \Omega$) at the resonant frequency due to low electromechanical coupling coefficient that is inherent in electrostatically actuated resonators, and (3) relatively low power handling capability.

Reducing Motional Resistance

One major issue of an electrostatically actuated resonator is its high motional resistance that is typically in the range of tens of kΩ. It is too high for matching to 50 Ω characteristic impedance of coaxial or coplanar lines in most RF systems. Most of RF antennas also have low impedance, and signals will be reflected back, if a high impedance filter is connected to an antenna.

Thus, attempts have been made to reduce the resistance by reducing the air gap and increasing the coupling area and the DC bias voltage since the motional resistance R_x is $R_x \propto \frac{\text{Gap}}{\text{Coupling Area} \cdot V_P}$. An approach of combining several resonators in parallel (to increase the coupling area) reduced the motional resistance down to a few kΩ [10].

Electrostatic Attractive Force due to RF Power

When an RF power passes through an air-gap capacitor, the power produces an electrostatic attractive force. The narrower the gap is, the higher the attractive force is for a given power. The attractive force can be large enough to pull in the suspended structure and can seriously limit the power handling capability of a electrostatic resonator as well as RF MEMS switch (or tunable MEMS capacitor) made of an air-gap capacitor.

Across the air gap of an air-filled parallel capacitor (C), there exists the following electrostatic attractive force F due to voltage V owing to RF power (P) passing through the capacitor:

$$F = \frac{\varepsilon_o A V^2}{2g^2} = \frac{CV^2}{2g} = \frac{\omega C V^2}{2g\omega} = \frac{P}{2g\omega} \tag{4.1}$$

where A, g, ε_o, and ω are the capacitor area, the air gap of the capacitor, vacuum permittivity, and RF frequency, respectively.

Question: Estimate the electrostatic attractive force due to RF power passing through the disk resonator shown in Fig. 4.4, assuming that the air gap (between the PolySi and electrode), RF power, and RF frequency are 0.5 μm, 1 W, and 1 GHz, respectively.

Answer: $F = P/(2g\omega) = 1/(2 \times 0.5 \times 10^{-6} \times 2\pi \times 10^9) = 0.16$ mN

Question: In case of the disk micromechanical resonator shown in Fig. 4.4, would you increase or decrease the parameters listed in the first column to meet the desired properties listed on the first row below? Please indicate (in the blanks below) ↑ for increase, ↓ for decrease, and ↔ for no effect to the first order.

	Decrease impedance to 50 Ω	Increase power handling capability	Increase resonant frequency
Coupling area			
Transducer gap			
DC bias voltage			
Disk stiffness			

Answer:

	Decrease impedance to 50 Ω	Increase power handling capability	Increase resonant frequency
Coupling area	↑	↔	↔
Transducer gap	↓	↑	↔
DC bias voltage	↑	↓	↔
Disk stiffness	↔	↑	↑

4.3 Acoustic Wave Resonators and Filters

Unlike electrostatically actuated resonators, acoustic wave resonators have low motional resistance as the acoustic waves are generated and coupled through a piezoelectric film which offers much larger electromechanical coupling coefficient (than air through which an electrostatic resonator operates). Also, a resonance at 1–2 GHz is easily obtainable with SAW resonators or FBARs. A SAW resonator needs just an electrode patterned into interdigital transducer (IDT) on top of a piezoelectric substrate, while FBAR is made with air-backing (at the top and bottom) or with solid mounting (with an acoustic reflector at the bottom). The resonant frequencies of SAW resonator and FBAR are determined by IDT's comb finger period and piezoelectric film's thickness, respectively. There are two types of FBAR, air-backed FBARs fabricated with MEMS technology and solid mounted resonators (SMRs) with an acoustic reflector formed by multiple layers of high and low impedances.

Currently, SAW-based and FBAR-based filters are the only two types of filters being used for RF front-end filtering for cell phones, and there is no other competing technology that challenges those acoustic wave filters. In this section we will learn some key fundamentals on acoustic wave resonators and filters.

4.3.1 Acoustic Wave Resonator Concept

A resonator based on piezoelectric quartz crystal (which is nothing but crystalline SiO_2) has a very high Q and also, with a proper cut of the crystal, has zero TCF near room temperature. Thus, quartz has been the key material for time-keeping oscillators for many electronic systems. However, its fabrication processes are not compatible with IC fabrication process, while a resonance at greater than 500 MHz requires the quartz to be too thin for typical machining process, though nonconventional thinning processes such as chemical polishing, plasma etching, or ion beam machining can be used to produce several μm-thick quartz. One can certainly use harmonic frequencies for high-frequency applications, but electromechanical-coupling coefficient decreases as the harmonic order increases. For oscillator applications electromechanical-coupling coefficient does not play a major role, but for filters it is directly related to the achievable bandwidth of the filter composed of the resonators.

The reflection coefficient R for a wave normally incident at the interface between two different media with impedances Z_1 and Z_2 is

Z_1 | Z_2
Incident wave

$$R = \frac{Z_1 - Z_2}{Z_1 + Z_2}. \tag{4.2}$$

The magnitude of R can be between 0 and 1; 0 for $Z_1 = Z_2$ and 1 for Z_1 and Z_2 being so much different from each other that either Z_1 or Z_2 can be ignored with respect to the other. An acoustic wave resonator is based on R being close to 1 so that the wave may be reflected (back into the resonator) from the interface, as explained below.

An applied RF voltage between two electrodes sandwiching a piezoelectric layer (such as the one shown in Fig. 4.6) produces an acoustic wave that travels between the two electrodes. Most of the waves are reflected at the solid–air interface because of large impedance mismatch between a solid and air. And if the frequency is right, there could be a resonance of the waves, and a standing wave is formed between the two electrodes. The fundamental resonance is when the standing wave has its wavelength equal to two times the thickness of the resonator, as illustrated in Fig. 4.6 (assuming that the electrode is negligibly thin). Thus, the fundamental resonant frequency is equal to the wave velocity divided by the wavelength (that is two times the thickness of the resonator).

The speed of a longitudinal acoustic wave in piezoelectric ZnO is ~3,000 m/s (~10^5 times lower than the speed of the light), and the resonant frequency can be GHz if the thickness is around 1.5 μm. Consequently, it is relatively easy to form a GHz resonator with a piezoelectric thin film, and FBAR was introduced in early 80s.

4.3.2 Quartz Resonator

Quartz crystal (crystalline SiO_2) is piezoelectric (with relatively low piezoelectric coefficients) and has some amazing properties such as existence of zero-temperature-coefficient cuts, Q of 1,000,000 at around 1 MHz without any vacuum package, and similarities to silicon in terms of the material sturdiness, abundance in nature, and

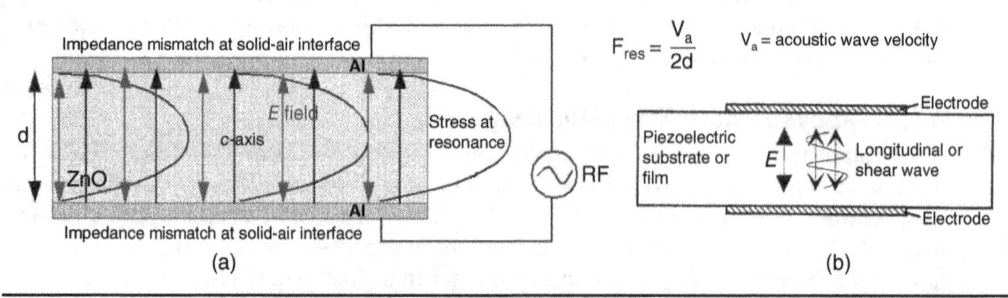

Figure 4.6 Cross-sectional views of bulk acoustic wave resonators (BARs) showing (a) stress distribution along the thickness at the fundamental thickness-mode resonance and (b) wave propagation directions.

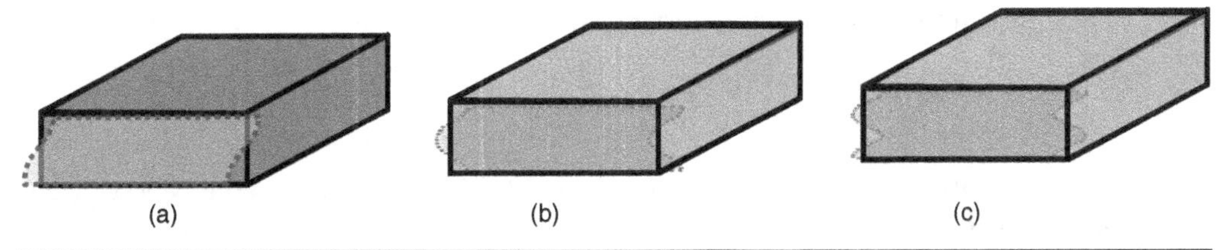

Figure 4.7 Cross-sectional views of the fundamental (a), *purported* second harmonic (b), and third harmonic (c) thickness shear vibrational mode of a plate. Mode shape of the second overtone (or harmonic) indicates why it cannot be produced through electrical voltage applied between the top and bottom faces, since the net strain from the top to bottom is zero for the second and any other even harmonics.

easiness to grow in crystalline form in large quantities at low cost and with high purity. With electrical voltage, a block of quartz can be vibrated into various modes and be made into a BAR. For example, AT-cut or SC-cut quartz can be made to resonate in thickness shear mode (Fig. 4.7a), with its resonant frequency in several to tens of several MHz (determined by the thickness). As the fundamental mode frequency f_0 is equal to one-half of the acoustic wave velocity V_a divided by the thickness d (i.e., $f_0 = V_a/2d$), for the shear wave in AT-cut quartz with its propagation velocity of 3,320 m/s the quartz thickness must be 1.66 μm for the resonant frequency to be 1 GHz.

Quartz resonator's impedance varies as a function of frequency, and the magnitude of the reactive portion of the impedance changes the largest near the fundamental resonant frequency but also at the odd harmonics (i.e., third, fifth, seventh…) as well as at the frequencies where there are spurious resonances. Even harmonics (such as second, fourth, etc.) are not generated by applying electrical voltage to produce the vibration, since the net strain energy when summed up throughout the thickness direction is zero for each of even harmonics (e.g., see the *purported* second harmonic illustrated in Fig. 4.7b), as a *nonzero* net electrical power cannot couple into *zero* net mechanical power.

4.3.3 One-Dimensional Mason's Model for Acoustic Resonator

One-Dimensional Acoustic Wave Equations

If the lateral dimensions (in x- and y directions) of an acoustic wave resonator is much larger than its thickness, the wave equation for acoustic wave traveling in z direction is

$$\frac{\partial^2 T}{\partial z^2} = \frac{\rho_{m0}}{c^E}\frac{\partial^2 T}{\partial t^2} \tag{4.3}$$

where T, c^E, and ρ_{m0} are the stress, stiffness constant (at constant or zero electric field), and mass density, respectively [11]. The solution of the wave equation is in the form of

$$T = T_0 e^{j(\omega t \pm \beta_a z)} \tag{4.4}$$

with ω and β_a being the frequency and propagation constant, respectively. By putting the solution into the wave equation, we see

$$\beta_a = \omega\sqrt{\frac{\rho_{m0}}{c^E}} = \frac{\omega}{V_a} \tag{4.5}$$

with acoustic velocity $V_a \equiv \sqrt{\dfrac{c^E}{\rho_{m0}}}$.

Acoustic impedance of a medium is defined

$$Z_o \equiv -\frac{T}{v} \tag{4.6}$$

with v being the particle velocity (equal to $-V_a T/c^E$, as can be seen in the next section). Thus, acoustic impedance is

$$Z_o \equiv -\frac{T}{v} = \frac{C^E}{V_a} = \sqrt{\rho_{m0} C^E} = V_a \rho_{m0} \tag{4.7}$$

Mason's Model for One-Dimensional Acoustic Waves

An acoustic resonator with its lateral dimension much larger than the thickness (Fig. 4.8a) can be modeled with three ports, one electrical port (having electrical voltage V and current I) and two acoustic ports (each port having mechanical force F and particle velocity v), as proposed by Mason [12] and shown in Fig. 4.8b. In the model F_1 and F_2 are the forces at $z = d/2$ and $-d/2$, respectively, while v_1 and v_2 are the velocities at $z = d/2$ and $-d/2$, respectively. Now, we just need to find the components in the 3×3 matrix below.

$$\begin{bmatrix} F_1 \\ F_2 \\ V \end{bmatrix} = \begin{bmatrix} m_{11} & m_{12} & m_{13} \\ m_{21} & m_{22} & m_{23} \\ m_{31} & m_{32} & m_{33} \end{bmatrix} \begin{bmatrix} v_1 \\ v_2 \\ I \end{bmatrix} \tag{4.8}$$

In electrical circuit with inductance L, resistance R, and capacitance C, we have the following equation that relates voltage V and current I (starting with charge Q):

$$L\frac{d^2Q}{dt} + R\frac{dQ}{dt} + \frac{Q}{C} = V = L\frac{dI}{dt} + RI + \int \frac{I}{C} dt \tag{4.9}$$

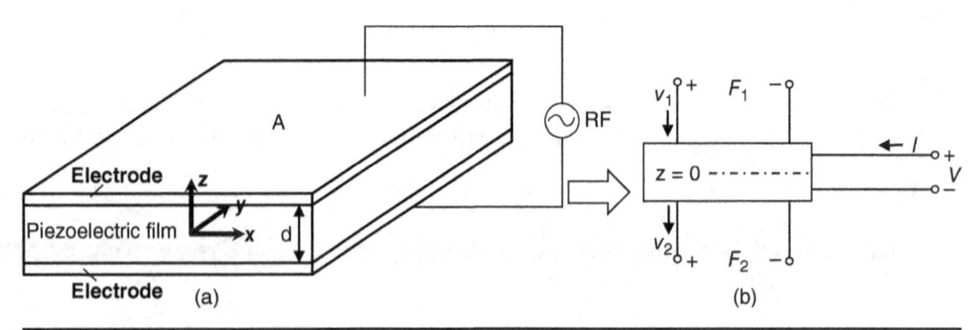

FIGURE 4.8 (a) Acoustic resonator with the lateral dimension (having an area A) much larger than the thickness d. (b) A three-port (one-dimensional) model of the resonator with one electrical port (having electrical voltage V and current I) and two acoustic ports (each port having mechanical force F and particle velocity v), if the electrodes are thin enough to be ignored.

In mechanical system with mass m, mechanical resistance r, and compliance (inverse of stiffness) s, we have the following equation that relates force F and velocity v (starting with displacement x):

$$m\frac{d^2x}{dt}+r\frac{dx}{dt}+\frac{x}{s}=F=m\frac{dv}{dt}+rv+\int\frac{v}{s}dt \tag{4.10}$$

Consequently, mass m, mechanical resistance r, and compliance s can be represented with L, R, and C in an equivalent electrical circuit where force F and velocity v are treated, as if they were voltage V and electrical current I, respectively.

For a piezoelectric medium, electric displacement D and stress T are related to electric field E and strain S as follows:

$$D=\varepsilon^S E+eS \tag{4.11}$$

$$T=-e_t E+c^E S \tag{4.12}$$

where ε^S, c^E, e, and e_t are dielectric permittivity with zero or constant strain, elastic stiffness with zero or constant electric field, piezoelectric stress constant, and transposed piezoelectric stress constant, respectively. Getting $E=(D-eS)/\varepsilon^S$ from Eq. (4.11) and inserting it into Eq. (4.12), we obtain

$$T=c^E\left(1+\frac{e_te}{c^E\varepsilon^S}\right)S-\frac{e_t}{\varepsilon^S}D=c^DS-\frac{e_t}{\varepsilon^S}D \tag{4.13}$$

where c^D (a stiffened elastic constant at $D=0$) is $c^D\equiv c^E\left(1+\frac{e_te}{c^E\varepsilon^S}\right)=c^E(1+k^2)$ with $k^2\equiv\frac{e^2}{c^E\varepsilon^S}$.

For small mechanical strain, strain along the z direction is $S=\partial u/\partial z$, where u is z-directed particle displacement [13]. For sinusoidal displacement with $u=u_oe^{j\omega t}$, by time-differentiating the equation for the strain, we get $j\omega S=\partial v/\partial z$, where v is particle velocity. Thus, we obtain the following equation that relates strain to particle velocity

$$S=\frac{1}{j\omega}\frac{\partial v}{\partial z} \tag{4.14}$$

Using Eq. (4.14) in Eqs. (4.13) and (4.11), we obtain the following equations

$$T=-\frac{jc^D}{\omega}\frac{\partial v}{\partial z}-\frac{e_t}{\varepsilon^S}D \tag{4.15}$$

$$E=\frac{je}{\omega\varepsilon^S}\frac{\partial v}{\partial z}+\frac{D}{\varepsilon^S} \tag{4.16}$$

From Newton's second law, we have

$$\frac{\partial T}{\partial z}=\rho_{m0}\frac{\partial v}{\partial t}=j\omega\rho_{m0}v \tag{4.17}$$

For an acoustic resonator with its lateral dimension much larger than the thickness, electrical displacement D is in the z direction, and $\frac{\partial D}{\partial z}=0$, since there is no free charge in the medium [13]. Differentiating Eq. (4.15) with respect to z and using Eq. (4.17) and $\partial D/\partial z = 0$, we obtain

$$\frac{\partial^2 v}{\partial z^2} = -\frac{\omega^2 \rho_{m0} v^2}{c^D}, \tag{4.18}$$

which is the wave equation for velocity v

$$\frac{\partial^2 v}{\partial z^2} = \frac{\rho_{m0}}{c^D}\frac{\partial^2 v}{\partial t^2} \tag{4.19}$$

where c^D and ρ_{m0} are stiffened elastic constant and mass density, respectively. The solution of the wave equation for velocity v is

$$v = v_F e^{j(\omega t - \beta_a z)} + v_B e^{j(\omega t + \beta_a z)} \tag{4.20}$$

where $\beta_a = \omega\sqrt{\frac{\rho_{m0}}{c^D}}$. The first component in Eq. (4.20) is for the wave propagating into the positive z direction while the second one is the one propagating in the opposite direction.

Using the boundary conditions $v(d/2) = v_1$ and $v(-d/2) = v_2$ (Fig. 4.8b), we get

$$v = \frac{v_1 \sin[\beta_a(z + d/2)] - v_2 \sin[\beta_a(z - d/2)]}{\sin \beta_a d} \tag{4.21}$$

The current I is related to electrical displacement D through $I = \frac{\partial(AD)}{\partial t} = j\omega AD$, from which we get $D = I/j\omega A$. The force F is related to stress T through $F = AT$, which gives us $T = F/A$. Using $D = I/j\omega A$ and $T = F/A$ in Eq. (4.15), we obtain

$$F = -j\frac{c^D A}{\omega}\frac{\partial v}{\partial z} + j\frac{e_t}{\omega \varepsilon^S} I, \tag{4.22}$$

which yields the following two equations at $z = d/2$ and $z = -d/2$.

$$F_1 = -j\frac{c^D \beta_a A}{\omega}\left(\frac{v_1}{\tan \beta_a d} - \frac{v_2}{\sin \beta_a d}\right) + j\frac{e_t}{\omega \varepsilon^S} I \tag{4.23}$$

$$F_2 = -j\frac{c^D \beta_a A}{\omega}\left(\frac{v_1}{\sin \beta_a d} - \frac{v_2}{\tan \beta_a d}\right) + j\frac{e_t}{\omega \varepsilon^S} I \tag{4.24}$$

Since the voltage V is related to electric field E through $V = \int_{-d/2}^{d/2} E\,dz$, integrating both sides of Eq. (4.16) from $-d/2$ to $d/2$ (with $D = I/j\omega A$) leads to

$$\begin{aligned} V &= j\frac{e}{\omega \varepsilon^S}[v(d/2) - v(-d/2)] - j\frac{d}{\omega \varepsilon^S A} I \\ &= j\frac{e}{\omega \varepsilon^S}(v_1 - v_2) - j\frac{d}{\omega \varepsilon^S A} I \end{aligned} \tag{4.25}$$

Now, using $F = AT$ and $\frac{\partial v}{\partial z} = -j\beta_a v$ (for the wave propagating into the positive z direction) in Eq. (4.15), we have

$$\begin{aligned} T &= -\frac{c^D\beta_a}{\omega} v + \frac{e_t}{\varepsilon^S} D \\ &= -\frac{c^D\beta_a}{\omega} v, \text{ for } D = 0 \end{aligned} \tag{4.26}$$

And acoustic impedance Z_o and clamp capacitance C_o are as follows.

$$Z_o \equiv -\frac{T}{v} = \frac{c^D\beta_a}{\omega} \tag{4.27}$$

$$C_o \equiv \frac{\varepsilon^S A}{d} \tag{4.28}$$

Thus, from Eqs. (4.23) to (4.28), we get the following three-port Mason's model.

$$\begin{bmatrix} F_1 \\ F_2 \\ V \end{bmatrix} = \begin{bmatrix} -j\frac{Z_o A}{\tan\beta_a d} & j\frac{Z_o A}{\sin\beta_a d} & j\frac{e_t}{\omega\varepsilon^S} \\ -j\frac{Z_o A}{\sin\beta_a d} & j\frac{Z_o A}{\tan\beta_a d} & j\frac{e_t}{\omega\varepsilon^S} \\ j\frac{e}{\omega\varepsilon^S} & -j\frac{e}{\omega\varepsilon^S} & -j\frac{1}{\omega C_o} \end{bmatrix} \begin{bmatrix} v_1 \\ v_2 \\ I \end{bmatrix} \tag{4.29}$$

For a nonpiezoelectric layer, Eq. (4.29) is simplified to

$$\begin{bmatrix} F_1 \\ F_2 \end{bmatrix} = \begin{bmatrix} -j\frac{Z_o A}{\tan\beta_a d} & j\frac{Z_o A}{\sin\beta_a d} \\ -j\frac{Z_o A}{\sin\beta_a d} & j\frac{Z_o A}{\tan\beta_a d} \end{bmatrix} \begin{bmatrix} v_1 \\ v_2 \end{bmatrix} \tag{4.30}$$

For a nonpiezoelectric layer, a transfer function from one port to another is useful, and we now obtain a so-called ABCD matrix that relates F_2 and v_2 to F_1 and v_1 (i.e., from port 1 to port 2) from Eq. (4.30) as follows.

$$\begin{bmatrix} F_2 \\ v_2 \end{bmatrix} = \begin{bmatrix} A & B \\ C & D \end{bmatrix} \begin{bmatrix} F_1 \\ v_1 \end{bmatrix} \tag{4.31}$$

where

$$\begin{aligned} A &= \left.\frac{F_2}{F_1}\right|_{v_1=0} = \cos(\beta_a d),\ B = \left.\frac{F_2}{v_1}\right|_{F_1=0} = -jZ_o A\sin(\beta_a d), \\ C &= \left.\frac{v_2}{F_1}\right|_{v_1=0} = -j\frac{\sin(\beta_a d)}{Z_o A},\ D = \left.\frac{v_2}{v_1}\right|_{F_1=0} = \cos(\beta_a d) \end{aligned} \tag{4.32}$$

The ABCD matrix can be viewed as the transfer function for an acoustic delay line (that is not piezoelectric), and we have the following transfer function from one end to the other.

$$\begin{bmatrix} F_2 \\ v_2 \end{bmatrix} = \begin{bmatrix} \cos(\beta_a d) & -jZ_o A \sin(\beta_a d) \\ -j\dfrac{\sin(\beta_a d)}{Z_o A} & \cos(\beta_a d) \end{bmatrix} \begin{bmatrix} F_1 \\ v_1 \end{bmatrix} \tag{4.33}$$

4.3.4 Using Mason's Model for Film Bulk Acoustic Resonator with Multiple Layers

An air-backed FBAR shown in Fig. 4.9a can be modeled with Mason's model as shown in Fig. 4.9b, noting that since air cannot resist force F, $F = 0$ at a solid–air interface. The model gives the following equations from Eq. (4.29) and Eq. (4.33).

$$\begin{bmatrix} F_2 \\ v_2 \end{bmatrix} = \begin{bmatrix} \cos(\beta_{a\mathrm{Al}} d_{\mathrm{Al}}) & -jZ_{o\mathrm{Al}} A \sin(\beta_{a\mathrm{Al}} d_{\mathrm{Al}}) \\ -j\dfrac{\sin(\beta_{a\mathrm{Al}} d_{\mathrm{Al}})}{Z_{o\mathrm{Al}} A} & \cos(\beta_{a\mathrm{Al}} d_{\mathrm{Al}}) \end{bmatrix} \begin{bmatrix} 0 \\ v_1 \end{bmatrix} \tag{4.34}$$

$$\begin{bmatrix} F_2 \\ F_3 \\ V \end{bmatrix} = \begin{bmatrix} -j\dfrac{Z_{o\mathrm{ZnO}} A}{\tan \beta_{a\mathrm{ZnO}} d} & j\dfrac{Z_{o\mathrm{ZnO}} A}{\sin \beta_{a\mathrm{ZnO}} d} & j\dfrac{e_t}{\omega \varepsilon^S} \\ -j\dfrac{Z_{o\mathrm{ZnO}} A}{\sin \beta_{a\mathrm{ZnO}} d} & j\dfrac{Z_{o\mathrm{ZnO}} A}{\tan \beta_{a\mathrm{ZnO}} d} & j\dfrac{e_t}{\omega \varepsilon^S} \\ j\dfrac{e}{\omega \varepsilon^S} & -j\dfrac{e}{\omega \varepsilon^S} & \dfrac{1}{j\omega C_o} \end{bmatrix} \begin{bmatrix} v_2 \\ v_3 \\ I \end{bmatrix} \tag{4.35}$$

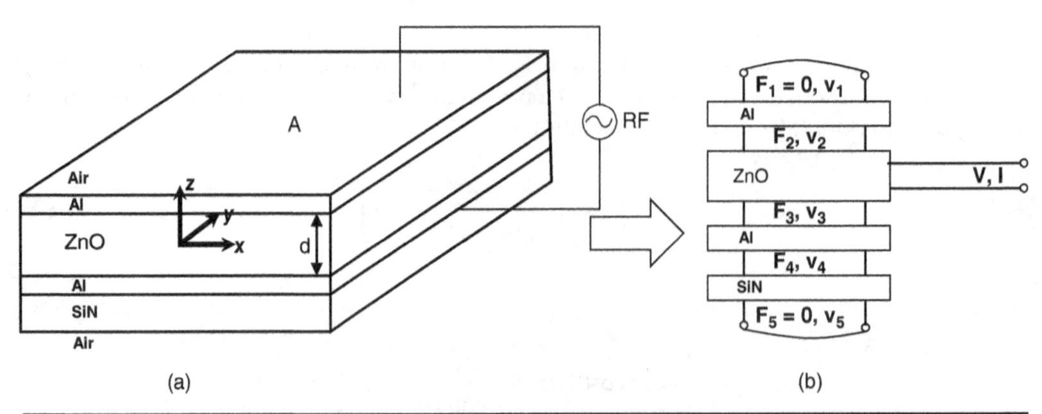

FIGURE 4.9 (a) An air-backed FBAR built on a SiN support layer with piezoelectric ZnO film. (b) A three-port (one-dimensional) model of the resonator with one electrical port (having electrical voltage and current) and two acoustic ports (each port having mechanical force and particle velocity) if the electrodes are negligible.

$$\begin{bmatrix} F_3 \\ v_3 \end{bmatrix} = \begin{bmatrix} \cos(\beta_{a\text{Al}} d_{\text{Al}}) & -jZ_{o\text{Al}} A \sin(\beta_{a\text{Al}} d_{\text{Al}}) \\ -j\dfrac{\sin(\beta_{a\text{Al}} d_{\text{Al}})}{Z_{o\text{Al}} A} & \cos(\beta_{a\text{Al}} d_{\text{Al}}) \end{bmatrix}$$

$$\begin{bmatrix} \cos(\beta_{a\text{SiN}} d_{\text{SiN}}) & -jZ_{o\text{SiN}} A \sin(\beta_{a\text{SiN}} d_{\text{SiN}}) \\ -j\dfrac{\sin(\beta_{a\text{SiN}} d_{\text{SiN}})}{Z_{o\text{SiN}} A} & \cos(\beta_{a\text{SiN}} d_{\text{SiN}}) \end{bmatrix} \begin{bmatrix} 0 \\ v_5 \end{bmatrix} \tag{4.36}$$

From Eq. (4.34) and Eq. (4.36) we see that both F_2 and v_2 are dependent on only one variable, v_1, while both F_3 and v_3 are dependent only on v_5. Consequently, Eq. (4.35) has four variables (i.e., V, I, v_1, and v_5, while it contains three separate equations and can give us V/I, the impedance of FBAR. It is messy algebraically, but a relatively simple computer programming can be used to obtain the impedance.

Mason's Model for Air-Backed, Lossless Film Bulk Acoustic Resonator with Negligible Nonpiezoelectric Layers

In a simple case of air-backed FBAR with only ZnO film (i.e., $d_{\text{Al}} = d_{\text{SiN}} = 0$), we see that $F_2 = 0$ and $v_2 = v_1$ from Eq. (4.34); $F_3 = 0$ and $v_3 = v_5$ from Eq. (4.36). Thus, from Eq. (4.35) we obtain the following equations

$$\begin{aligned} 0 &= a\frac{v_1}{I} + b\frac{v_5}{I} + c \\ 0 &= -b\frac{v_1}{I} - a\frac{v_5}{I} + c \\ \frac{V}{I} &= c\left(\frac{v_1}{I} - \frac{v_5}{I}\right) + \frac{1}{j\omega C_o} \end{aligned} \tag{4.37}$$

where $a \equiv -j\dfrac{Z_{o\text{ZnO}} A}{\tan \beta_{a\text{ZnO}} d}$, $b \equiv j\dfrac{Z_{o\text{ZnO}} A}{\sin \beta_{a\text{ZnO}} d}$, $c \equiv j\dfrac{e}{\omega \varepsilon^S}$. From the first two equations above, we get $\dfrac{v_1}{I} = -\dfrac{c}{a-b}$ and $\dfrac{v_5}{I} = \dfrac{c}{a-b}$. Substituting these two into the third equation above, we get

$$\frac{V}{I} = -\frac{2c^2}{a-b} + \frac{1}{j\omega C_o} \tag{4.38}$$

Now, putting back the definitions for a, b, and c, we get

$$\begin{aligned} \frac{V}{I} &= \frac{-2e^2}{j\omega^2 (\varepsilon^S)^2 Z_{o\text{ZnO}} A}\left(\frac{1}{1/\tan \beta_{a\text{ZnO}} d + 1/\sin \beta_{a\text{ZnO}} d}\right) + \frac{1}{j\omega C_o} \\ &= \frac{1}{j\omega C_o}\left(1 - \frac{2C_o e^2 \tan(\beta_{a\text{ZnO}} d/2)}{\omega (\varepsilon^S)^2 Z_{o\text{ZnO}} A}\right), \text{ using } \frac{1}{\tan\theta} = \frac{1}{\tan(\theta/2)} - \frac{1}{\sin\theta} \end{aligned} \tag{4.39}$$

But $\dfrac{C_o e^2}{\omega (\varepsilon^S)^2 Z_o A} = \dfrac{\varepsilon^S A e^2}{d\omega (\varepsilon^S)^2 (c^D \beta/\omega) A} = \dfrac{e^2}{\varepsilon^S c^D \beta d} = \dfrac{k_T^2}{\beta d}$, with $k_T^2 \equiv \dfrac{e^2}{\varepsilon^S c^D}$

Thus, we have the impedance of an air-backed ZnO FBAR with piezoelectric film only

$$\frac{V}{I} = \frac{1}{j\omega C_o}\left(1 - \frac{k_T^2 \tan(\beta_{aZnO}d/2)}{\beta_{aZnO}d/2}\right) \tag{4.40}$$

The impedance in Eq. (4.40) is imaginary and is either capacitive, inductive, or zero depending on the frequency; the series resonant frequency ω_s is where the impedance is zero, and the parallel resonant frequency ω_p is where the impedance approaches infinity, as shown in Fig. 4.10a.

According to Eq. (4.40), when $\beta_a d/2 = N\pi/2,\ N = 1,3,5,..,$ the impedance approaches infinity and the parallel resonant frequency ω_p is

$$\omega_p = N\pi\frac{V_a}{d}, \tag{4.41}$$

since $\beta_a = \frac{\omega}{V_a}$. Note that for this particular air-backed resonator that is composed of only piezoelectric layer, N is odd integer, which means that even harmonics are not generated. This is because for even harmonics the net strain energy (when summed up throughout the thickness direction, as illustrated in Fig. 4.7b) is zero, and consequently, *nonzero* net electrical power cannot couple into *zero* net mechanical power.

Again, referring to Eq. (4.40) we see that the series resonant frequency ω_s can be obtained from $\frac{\tan(\beta_a d/2)}{\beta_a d/2} = \frac{1}{k_T^2}$. If $k_T^2 << 1,\ \tan(\beta_a d/2) \approx \frac{4\beta_a d}{(N\pi)^2 - (\beta_a d)^2}$, and we get $(N\pi)^2 - (\beta_a d)^2 = 8k_T^2$, from which we get the series resonant frequency ω_s

$$\omega_s = \frac{V_a}{d}\sqrt{(N\pi)^2 - 8k_T^2} \tag{4.42}$$

If the difference between the two frequencies is small, then $\omega_p^2 - \omega_s^2 \approx 2\omega_p(\omega_p - \omega_s)$ and the normalized frequency difference can be approximated as follows

$$\frac{\omega_p - \omega_s}{\omega_p} \approx \frac{\omega_p^2 - \omega_s^2}{2\omega_p^2} = \frac{4k_T^2}{(N\pi)^2} \tag{4.43}$$

Thus, we see that the so-called bandwidth of FBAR depends on k_T^2, the electromechanical coupling coefficient with subscript T indicating that it is for thickness-excited (TE) vibration.

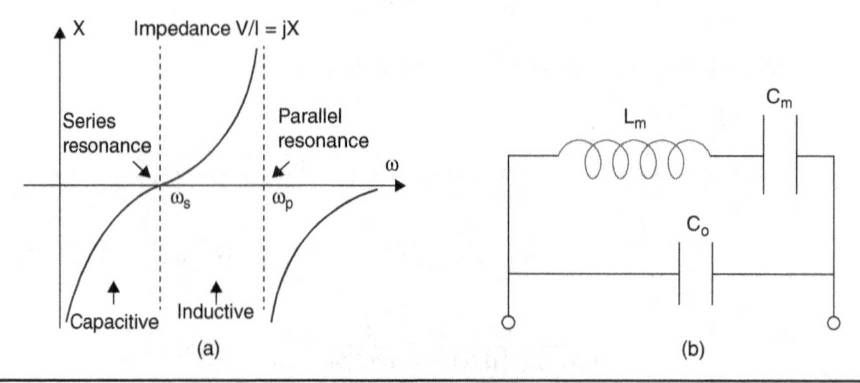

FIGURE 4.10 Impedance of Eq. (4.40) as a function of frequency.

Now, Fig. 4.10a can be modeled with an equivalent circuit shown in Fig. 4.10b. We will analyze the equivalent circuit in Section 4.3.6, where we will derive the following equation for the bandwidth

$$\omega_p - \omega_s = \frac{1}{\sqrt{L_m C_m}}\left(\sqrt{1+\frac{C_m}{C_o}} - 1\right) \approx \frac{1}{2\sqrt{L_m C_m}}\frac{C_m}{C_o}, \tag{4.44}$$

for $\frac{C_m}{C_o} << 1$ that is true for a high Q resonator. Thus, with $\omega_s = \frac{1}{\sqrt{L_m C_m}}$, the fractional bandwidth is

$$\frac{\omega_p - \omega_s}{\omega_s} \approx \frac{C_m}{2C_o} \tag{4.45}$$

Since $\frac{\omega_p - \omega_s}{\omega_s} \approx \frac{\omega_p - \omega_s}{\omega_p}$ for a high Q resonator, we get the following equation for C_m from Eqs. (4.43) and (4.45)

$$C_m = \frac{8k_T^2}{N^2\pi^2}C_o \tag{4.46}$$

Mason's Model for Piezoelectric Layer Sandwiched with Acoustic Impedances Z_1 and Z_2

In general, FBAR can be modeled with its piezoelectric layer sandwiched with layers having acoustic impedances Z_1 and Z_2, as shown in Fig. 4.11a. In that case, from Eq. (4.35) we obtain the following equations, using $Z_1 = -\frac{F_1/A}{v_1}$ and $Z_2 = \frac{F_2/A}{v_2}$

$$\begin{aligned}
-\frac{Z_1 v_1 A}{I} &= -j\frac{Z_o A}{\tan\beta_a d}\frac{v_1}{I} + j\frac{Z_o A}{\sin\beta_a d}\frac{v_2}{I} + j\frac{e_t}{\omega\varepsilon^S} \\
\frac{Z_2 v_2 A}{I} &= -j\frac{Z_o A}{\sin\beta_a d}\frac{v_1}{I} + j\frac{Z_o A}{\tan\beta_a d}\frac{v_2}{I} + j\frac{e_t}{\omega\varepsilon^S} \\
\frac{V}{I} &= j\frac{e_t}{\omega\varepsilon^S}\left(\frac{v_1}{I} - \frac{v_2}{I}\right) + \frac{1}{j\omega C_o}
\end{aligned} \tag{4.47}$$

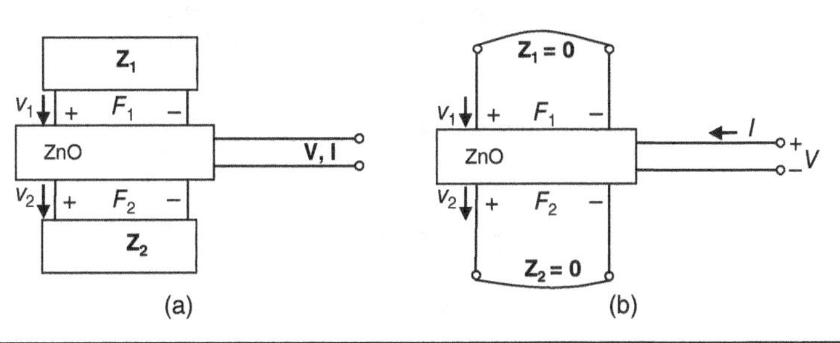

FIGURE 4.11 (a) FBAR with its piezoelectric layer sandwiched with acoustic impedances Z_1 and Z_2. (b) FBAR with its piezoelectric layer sandwiched with air.

The equations above can be converted into the following

$$0 = a\frac{v_1}{I} + b\frac{v_2}{I} + c$$

$$0 = -b\frac{v_1}{I} - d\frac{v_2}{I} + c \tag{4.48}$$

$$\frac{V}{I} = Z_o Ac\left(\frac{v_1}{I} - \frac{v_2}{I}\right) + \frac{1}{j\omega C_o}$$

where $a \equiv \frac{Z_1}{Z_o} - j\frac{1}{\tan\beta_a d}$, $b \equiv j\frac{1}{\sin\beta_a d}$, $c \equiv j\frac{e}{\omega\varepsilon^S Z_o A}$, and $d \equiv \frac{Z_2}{Z_o} - j\frac{1}{\tan\beta_a d}$. From the first two equations above, we get $\frac{v_1}{I} = \frac{c}{a} - \frac{b^2 - c}{a(b^2 - ad)}$ and $\frac{v_2}{I} = -c\left(\frac{a+b}{b^2 - ad}\right)$. Substituting these two into the third equation above, we get

$$\begin{aligned}
\frac{V}{I} &= \frac{Z_o Ac^2(a+2b+d)}{b^2 - ad} + \frac{1}{j\omega C_o} \\
&= -\frac{e^2}{\omega^2(\varepsilon^S)^2 Z_o A}\left(\frac{\frac{Z_1}{Z_o} - j\frac{2}{\tan\beta_a d} + \frac{Z_2}{Z_o} + j\frac{2}{\sin\beta_a d}}{\frac{1}{\sin^2\beta_a d} - \left(\frac{Z_1}{Z_o}\frac{Z_2}{Z_o} - \frac{1}{\tan^2\beta_a d} - j\frac{Z_1/Z_o + Z_2/Z_o}{\tan\beta_a d}\right)}\right) + \frac{1}{j\omega C_o} \\
&= \frac{1}{j\omega C_o}\left(1 - \frac{e^2}{\omega\varepsilon^S Z_o d}\frac{\left(\frac{Z_1}{Z_o} + \frac{Z_2}{Z_o}\right)\sin\beta_a d + j2(1-\cos\beta_a d)}{\left(\frac{Z_1}{Z_o} + \frac{Z_2}{Z_o}\right)\cos\beta_a d + j\left(1 + \frac{Z_1}{Z_o}\frac{Z_2}{Z_o}\right)\sin\beta_a d}\right)
\end{aligned} \tag{4.49}$$

For a special case of Z_1 and Z_2 being zero (i.e., air-backed FBAR shown in Fig. 4.11b), Eq. (4.49) becomes

$$\begin{aligned}
\frac{V}{I} &= \frac{1}{j\omega C_o}\left(1 - \frac{e^2}{\omega\varepsilon^S Z_o d}\frac{2(1-\cos\beta_a d)}{\sin\beta_a d}\right) \\
&= \frac{1}{j\omega C_o}\left(1 - \frac{e^2}{\omega\varepsilon^S Z_o d}2\tan(\beta_a d/2)\right) \\
&= \frac{1}{j\omega C_o}\left(1 - \frac{e^2}{\varepsilon^S c^D}\frac{2\tan(\beta_a d/2)}{\beta_a d}\right)
\end{aligned} \tag{4.50}$$

which is basically Eq. (4.40) that we have obtained in a different way earlier.

4.3.5 Incorporating Acoustic Loss in Mason's Model

Acoustic waves are absorbed in a medium through viscous damping and are lost as the damping converts the absorbed waves into heat. Such acoustic loss can be incorporated into Mason's model by using a complex number for β_a, that is, $\beta_a = \beta_{ar} + j\beta_{ai}$ with β_{ai} being a function of the viscosity η of a medium. Let us derive the equation for the complex β_a. When there is viscous damping in a nonpiezoelectric medium, the stress and strain are related as follows, with an additional term accounting for effect of the viscosity η.

$$T = c^E S + \eta dS/dt \tag{4.51}$$

Since strain S is related to particle displacement u as $S = \partial u/\partial z$, the above equation becomes

$$\begin{aligned} T &= c^E \frac{\partial u}{\partial z} + \eta \frac{\partial u}{\partial t \partial z} \\ &= c^E \frac{\partial u}{\partial z} + \eta \frac{\partial v}{\partial z} \end{aligned} \tag{4.52}$$

Newton's second law gives $\frac{\partial T}{\partial z} = \rho_{m0} \frac{\partial v}{\partial t}$, which becomes the following wave equation when Eq. (4.52) is used for T, and both sides of the equation are differentiated with respect to t.

$$c^E \frac{\partial^2 v}{\partial z^2} + \eta \frac{\partial^3 v}{\partial t \partial z^2} = \rho_{m0} \frac{\partial^2 v}{\partial t^2} \tag{4.53}$$

The wave equation above has a solution in the form of $v = v_0 e^{j(\omega t \pm \beta_a z)}$ with $\beta_a = \beta_{ar} + j\beta_{ai}$, and we get

$$-c^E \beta_a^2 - j\omega \beta_a^2 \eta = -\omega^2 \rho_{m0}, \tag{4.54}$$

which becomes

$$c^E \left(\beta_{ar}^2 - \beta_{ai}^2 + 2j\beta_{ar}\beta_{ai}\right) + j\omega\eta\left(\beta_{ar}^2 - \beta_{ai}^2 + 2j\beta_{ar}\beta_{ai}\right) = \omega^2 \rho_{m0} \tag{4.55}$$

From the real part of Eq. (4.55), we get

$$c^E \left(\beta_{ar}^2 - \beta_{ai}^2\right) - 2\omega\eta\beta_{ar}\beta_{ai} = \omega^2 \rho_{m0}, \tag{4.56}$$

from which we get the following equation for $\beta_{ai} << \beta_{ar}$

$$\beta_{ar} \approx \omega \sqrt{\frac{\rho_{m0}}{c^E}} = \frac{\omega}{V_a} \tag{4.57}$$

From the imaginary part of Eq. (4.55), we get

$$2c^E \beta_{ar}\beta_{ai} + \omega\eta\left(\beta_{ar}^2 - \beta_{ai}^2\right) = 0, \tag{4.58}$$

from which we get the following equation for $\beta_{ai} << \beta_{ar}$

$$\beta_{ai} \approx -\frac{\omega\eta\beta_{ar}}{2c^E} \approx -\frac{\eta\omega^2}{2V_a^3 \rho_{m0}} \tag{4.59}$$

Thus, we have

$$\beta_a \approx \frac{\omega}{V_a}\left(1 - j\frac{\eta\omega}{2V_a^2\rho_{m0}}\right) \tag{4.60}$$

Now, we can use this complex equation in any of the impedance equations that we have derived earlier such as Eqs. (4.49), (4.40), etc. in order to account for acoustic loss in a medium. For example, from Eq. (4.40), we get the following equation for the impedance (V/I)

$$\frac{V}{I} \approx \frac{1}{j\omega C_o}\left(1 - \frac{k_T^2 \tan\left(\frac{\omega d}{2V_a}\left(1 - j\frac{\eta\omega}{2V_a^2\rho_{m0}}\right)\right)}{\frac{\omega d}{2V_a}\left(1 - j\frac{\eta\omega}{2V_a^2\rho_{m0}}\right)}\right) \tag{4.61}$$

The above equation leads to an equivalent circuit shown in Fig. 4.12a, which becomes Fig. 4.12b at $\omega = 1/\sqrt{L_m C_m} \approx \omega_s$. Now, we can obtain an equation for the lossy element R_m from Eq. (4.61) by getting the impedance at $\omega \approx \omega_s = \frac{V_a}{d}\sqrt{(N\pi)^2 - 8k_T^2}$ $\left(\approx \frac{V_a \pi}{d} \text{ for } N = 1 \text{ and small } k_T^2\right)$. At $\omega \approx \omega_s$, Eq. (4.61) can be approximated as follows

$$\frac{V}{I} \approx \frac{1}{j\omega_s C_o}\left(1 - \frac{k_T^2 \tan\left(\frac{\omega_s d}{2V_a}\left(1 - j\frac{\eta\omega_s}{2V_a^2\rho_{m0}}\right)\right)}{\frac{\omega_s d}{2V_a}\left(1 - j\frac{\eta\omega_s}{2V_a^2\rho_{m0}}\right)}\right)$$

$$\approx \frac{1}{j\omega_s C_o}\left(1 - \frac{k_T^2 \tan\left(\frac{\sqrt{\pi^2 - 8k_T^2}}{2}\left(1 - j\frac{\eta\omega_s}{2V_a^2\rho_{m0}}\right)\right)}{\frac{\sqrt{\pi^2 - 8k_T^2}}{2}\left(1 - j\frac{\eta\omega_s}{2V_a^2\rho_{m0}}\right)}\right) \tag{4.62}$$

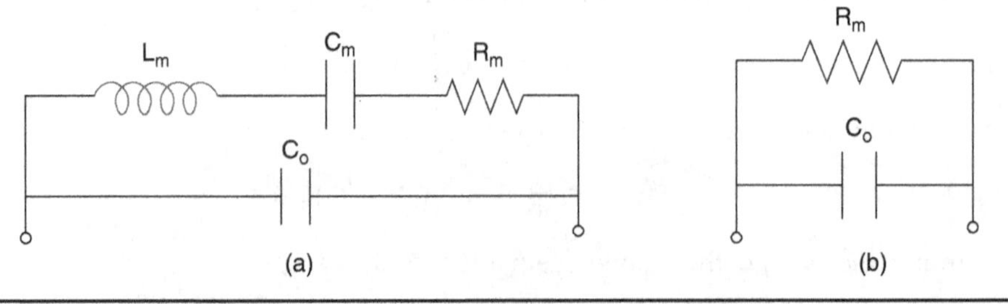

FIGURE 4.12 (a) Equivalent circuit that can be obtained from Eq. (4.61). (b) Equivalent circuit at $\omega = 1/\sqrt{L_m C_m} \approx \omega_s$.

Using $\sqrt{\pi^2 - 8k_T^2} \approx \pi$ and $\tan\left(\frac{\sqrt{\pi^2-8k_T^2}}{2}\left(1 - j\frac{\eta\omega_s}{2V_a^2\rho_{m0}}\right)\right) \approx \frac{1}{\tan\left(j\frac{\pi\eta\omega_s}{4V_a^2\rho_{m0}}\right)} \approx \frac{4V_a^2\rho_{m0}}{j\pi\eta\omega_s}$, we obtain

$$\frac{V}{I} \approx \frac{1}{j\omega_s C_o}\left(1 - \frac{4k_T^2 V_a^2 \rho_{m0}}{\left(\frac{\pi}{2} - j\frac{\pi\eta\omega_s}{4V_a^2\rho_{m0}}\right) j\pi\eta\omega_s}\right) \tag{4.63}$$

$$\approx \frac{1}{j\omega_s C_o}\left(1 - \frac{8k_T^2 V_a^2\rho_{m0}}{j\pi^2\eta\omega_s}\right) = \frac{1}{j\omega_s C_o} + \frac{8k_T^2 V_a^2 \rho_{m0} C_o}{\omega_s^2 C_o^2 \eta\pi^2}$$

Thus, we get an equation for R_m

$$R_m = \frac{\eta\pi^2}{8k_T^2 V_a^2 \rho_{m0} C_o} \tag{4.64}$$

4.3.6 BVD Equivalent Circuit for Acoustic Resonator

Near its resonant frequency, FBAR (or quartz crystal resonator) can be modeled with an equivalent circuit called Butterworth–van Dyke (BVD) model that is composed of motional inductance (L_m), motional capacitance (C_m), motional resistance (R_m), and clamp capacitance (C_o), as shown in Fig. 4.13. The three motional components that couple the mechanical motion to electrical impedance are in series to each other, while the whole motional branch is in parallel with a clamp capacitance (C_o) that accounts the fact that a dielectric (though piezoelectric) material with dielectric permittivity ε^S and thickness d is sandwiched by two electrodes over an area A and presents a capacitance equal to $\varepsilon^S A/d$.

From Eqs. (4.28), (4.46), and (4.64), we have the following

$$C_o = \varepsilon^S \frac{A}{d} \tag{4.65}$$

$$C_m = \frac{8k_T^2}{N^2\pi^2} C_o \tag{4.66}$$

$$R_m = \frac{\eta\pi^2}{8k_T^2 V_a^2 \rho_{m0} C_o} \tag{4.67}$$

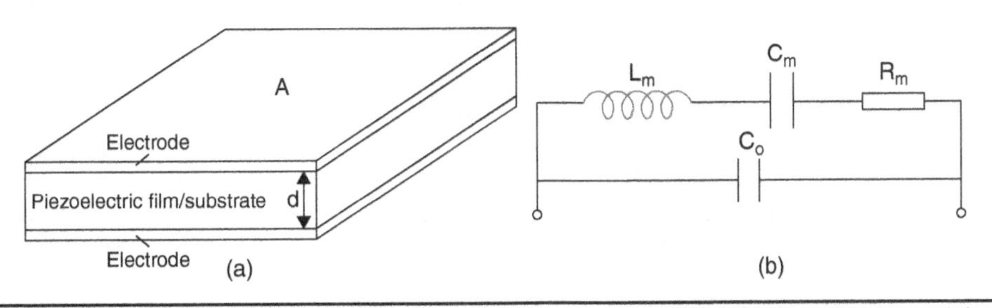

FIGURE 4.13 (a) Perspective view of a bulk acoustic resonator (BAR) defining the area *A* and the thickness *d*. (b) BVD equivalent model of BAR.

Since $1/\sqrt{L_m C_m} \approx \omega_s = 2\pi f_s$, we get

$$L_m = \frac{1}{4\pi^2 f_s^2 C_m} \tag{4.68}$$

Some exemplary values for an FBAR with 1 GHz resonant frequency are $L_m = 6.70\times10^{-7}$ H, $C_m = 2.97\times10^{-14}$ F, $R_m = 4.15\ \Omega$, and $C_o = 5.53\times10^{-12}$ F. Note the relatively large motional inductance, which stems from the high Q resonance of FBAR, since the quality factor $Q = \frac{2\pi f L_m}{R_m}$. From the BVD model, the series resonant frequency (f_s) and parallel resonant frequency (f_p) can be obtained with the following equations

$$f_s = \frac{1}{2\pi\sqrt{L_m C_m}} \tag{4.69}$$

$$f_p = \frac{1}{2\pi}\sqrt{\frac{1}{L_m C_m} + \frac{1}{L_m C_o}} \tag{4.70}$$

The insight into their physical significance can be gained through the following exercise. The imaginary part of the impedance (i.e., reactance) of the BVD circuit is sketched in Fig. 4.14b as a function of frequency. From the figure we see that there are two frequencies where the reactance is zero (or where the impedance is purely resistive). Those two frequencies are named as series resonant frequency (f_s) and parallel resonant frequency (f_p), which can be obtained from Eqs. (4.69) and (4.70). Though the equation for the series resonant frequency is an approximate equation, it yields quite accurate value as long as Q is greater than 100.

If we simplify the BVD circuit by ignoring R_m, the whole impedance becomes a reactance (i.e., an imaginary number), as shown in Fig. 4.15. The reactance versus frequency shows three distinct regions below f_s (where it is capacitive), between f_s and f_p (inductive), and above f_p (capacitive). With $R_m = 0$, the series resonant frequency f_s is exactly equal to $1/\left(2\pi\sqrt{L_m C_m}\right)$.

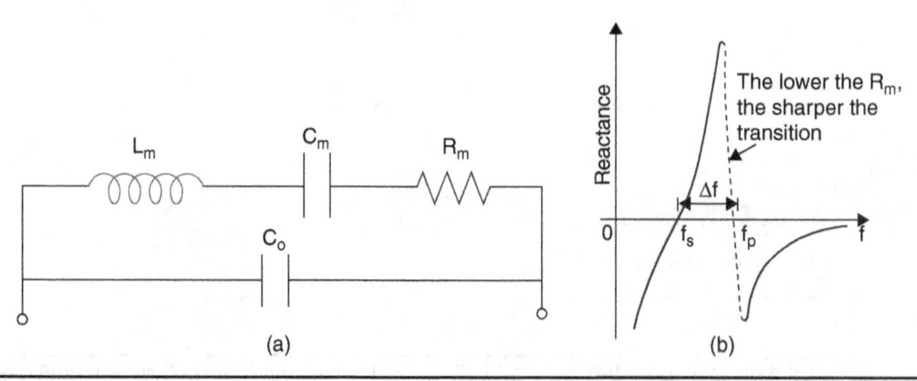

FIGURE 4.14 (a) BVD model. (b) Reactance of the BVD model versus frequency.

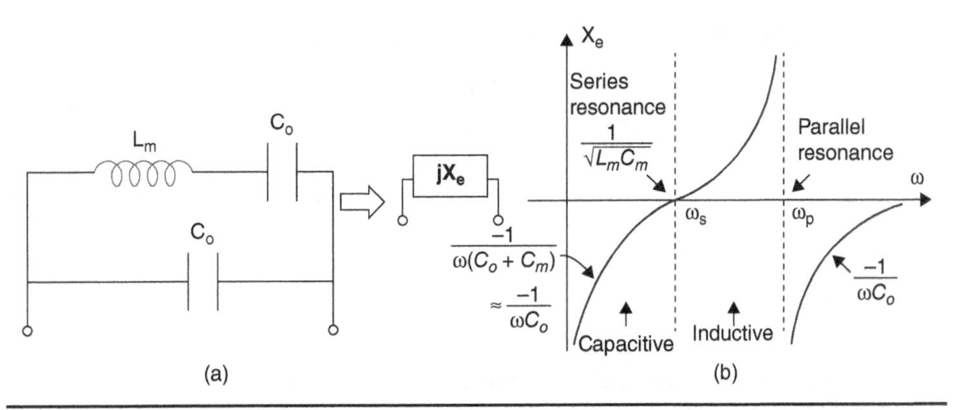

FIGURE 4.15 (a) Simplified BVD model without R_m, which has an impedance that is purely imaginary, i.e., reactance without any resistance. (b) Reactance of the simplified BVD model versus frequency.

As illustrated in Fig. 4.16, the serially connected L_m and C_m have a net impedance equal to $j\omega L_m + \frac{1}{j\omega C_m} = j\omega L_m\left(1-\frac{1}{\omega^2 L_m C_m}\right) = j\omega L_{\text{eff}}$ and can equivalently be represented with

$$L_{\text{eff}} \equiv L_m\left(1-\frac{1}{\omega^2 L_m C_m}\right), \tag{4.71}$$

which is positive for $\omega > \frac{1}{\sqrt{L_m C_m}} = \omega_s$. At the parallel resonant frequency ω_p, L_{eff} resonates with C_o, and the impedance $(=1/(j\omega C_o + 1/(j\omega L_{\text{eff}}))$ becomes infinite. Consequently, $\omega_p = \frac{1}{\sqrt{L_{\text{eff}} C_o}} = 1/\left(\sqrt{L_m\left(1-\frac{1}{\omega_p^2 L_m C_m}\right)C_o}\right)$, from which we get $\omega_p^2 L_m C_o - \frac{C_o}{C_m} = 1$ and $\omega_p^2 = \frac{1+C_o/C_m}{L_m C_o} = \frac{1}{L_m C_m} + \frac{1}{L_m C_o}$.

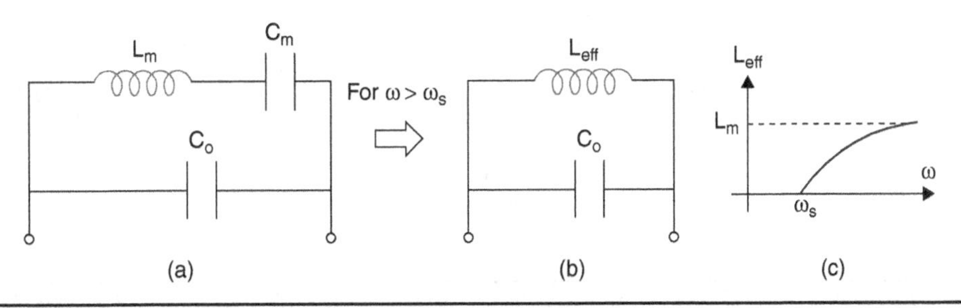

FIGURE 4.16 (a) Simplified BVD circuit without R_m. (b) Equivalent representation of the motional branch with L_{eff}. (c) L_{eff} versus frequency above the series resonant frequency.

Thus, the so-called "bandwidth" of a resonator is

$$\Delta f \equiv f_p - f_s = \frac{1}{2\pi}\left(\sqrt{\frac{1}{L_mC_m}+\frac{1}{L_mC_o}} - \sqrt{\frac{1}{L_mC_m}}\right) = \frac{1}{2\pi\sqrt{L_mC_m}}\left(\sqrt{1+\frac{C_m}{C_o}}-1\right), \tag{4.72}$$

which is $\approx \frac{1}{4\pi\sqrt{L_mC_m}}\frac{C_m}{C_o}$ for $\frac{C_m}{C_o} << 1$ that is true for a high Q resonator. In other words, the fractional bandwidth for a high Q resonator is

$$\frac{\Delta f}{f_s} \approx \frac{C_m}{2C_o} = \frac{4k_T^2}{N^2\pi^2}, \tag{4.73}$$

and consequently, the electromechanical coupling coefficient k_T^2 is critically important for filter applications of a resonator.

Now, let us put R_m back into the circuit. At the series resonant frequency $f_s = 1/\left(2\pi\sqrt{L_mC_m}\right)$, BVD circuit becomes R_m in parallel with C_o, which is approximately just R_m for a high Q resonator where R_m would be much smaller than $1/j\omega_sC_o$.

To find the resistance at the parallel resonant frequency f_p, we convert the L_{eff} and R_m in series into L'_{eff} and R_p in parallel, as shown in Fig. 4.17. Then we see that at the parallel resonant frequency $f_p = 1/\left(2\pi\sqrt{L'_{\text{eff}}C_o}\right)$, the impedance is purely resistive with resistance equal to R_p. Now, from Fig. 4.17b and Fig. 4.17a, we have

$$\frac{1}{R_p} + \frac{1}{j\omega L'_{\text{eff}}} = \frac{1}{R_m + j\omega L_{\text{eff}}} = \frac{R_m - j\omega L_{\text{eff}}}{R_m^2 + \omega^2 L_{\text{eff}}^2}, \tag{4.74}$$

which gives us the following two equations;

$$R_p = \frac{R_m^2 + \omega^2 L_{\text{eff}}^2}{R_m} = R_m + \frac{\omega^2 L_{\text{eff}}^2}{R_m} \text{ and } \frac{1}{\omega L'_{\text{eff}}} = \frac{\omega L_{\text{eff}}}{R_m^2 + \omega^2 L_{\text{eff}}^2}. \tag{4.75}$$

At $\omega = \omega_p$, we have $\frac{1}{\omega_p L'_{\text{eff}}} = \frac{\omega_p L_{\text{eff}}}{R_m^2 + \omega_p^2 L_{\text{eff}}^2} = \omega_p C_o$, from which we get $R_m^2 + \omega^2 L_{\text{eff}}^2 = \frac{L_{\text{eff}}}{C_o}$.

Thus, $R_p = \frac{L_{\text{eff}}}{R_mC_o}$, which is typically much larger than R_m for a high Q resonator.

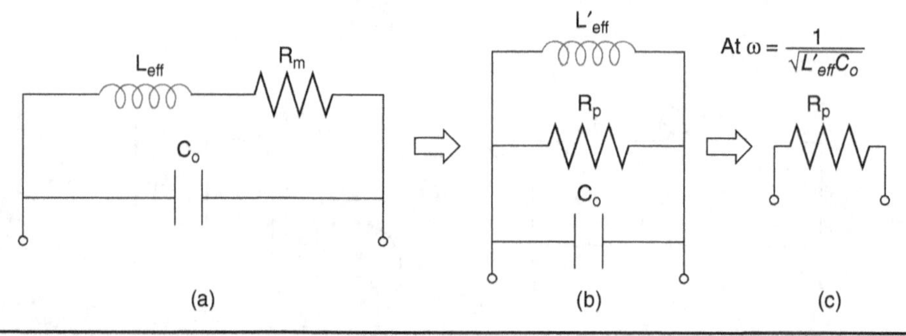

Figure 4.17 (a) BVD circuit with L_{eff} and R_m. (b) BVD circuit with L_{eff} and R_m (in series) converted into L'_{eff} and R_p (in parallel). (c) At the parallel resonant resonance where L'_{eff} and C_o *resonate each other out.*

In the later section, it will be clear that a larger R_p with respect to R_m would be better for filter applications. The ratio of the resistances at f_p and f_s is

$$\frac{R_p}{R_m} = \frac{L_{\text{eff}}}{R_m^2 C_o} = \frac{L_m\left(1-\omega_s^2/\omega_p^2\right)}{R_m^2 C_o} = \frac{L_m\left(1-\omega_s^2/\omega_p^2\right)}{R_m^2 C_o} = \frac{L_m}{R_m^2 C_o}\left(1-\frac{1}{1+C_m/C_o}\right). \tag{4.76}$$

For $C_m/C_o << 1$, the ratio is

$$\frac{R_p}{R_m} \approx \frac{L_m C_m}{R_m^2 C_o^2} = \frac{64 k_T^4 V_a^2 \rho_{m0}^2 d^2}{\pi^2 \eta^2 \varepsilon^4}, \tag{4.77}$$

which is mainly determined by the material properties of the piezoelectric layer for a given resonant frequency. Note that with Fig. 4.17b, the quality factor Q at ω_p can be obtained with $Q = \dfrac{R_p}{\omega_p L'_{\text{eff}}}$.

Example: An FBAR with series resonant frequency $f_s = 1.166$ GHz and parallel resonant frequency $f_p = 1.187$ GHz.

The impedance Z (= $R + jX$) varies as a function of frequency as shown in Fig. 4.18 over a narrow bandwidth near the resonant frequencies. The reactance X is zero at f_s and f_p, and is positive (i.e., inductive) between f_s and f_p (i.e., over Δf of 21 MHz or 1.8% of f_s), while it is negative (capacitive) in all other frequencies. The resistance R is the highest at the parallel resonant frequency and is more than 1 kΩ while the resistance at the series resonant frequency (which should be close to R_m) is about 4 Ω. The magnitude of the impedance ($\sqrt{R^2 + X^2}$) shows a dip and a peak at f_s and f_p, respectively, and shows a rapid change of more than two orders of magnitude over Δf of 21 MHz.

The impedance of FBAR (Z_{FBAR}) is typically measured with a network analyzer and with a cable having a characteristic impedance of 50 Ω. Consequently, the measured raw data is S11, a reflection coefficient, which is

$$\text{S11} = (50\ \Omega - Z_{\text{FBAR}})/(50\ \Omega + Z_{\text{FBAR}}), \tag{4.78}$$

and is a complex number varying as a function of frequency. If the S11 versus frequency is plotted on a Smith chart, it will look something like what is shown in Fig. 4.19a over a narrow-frequency range from a little lower than f_s to a little higher than f_p. The magnitude of the S11 versus frequency is shown in Fig. 4.19b.

Modified BVD Model of FBAR

The BVD model can be modified to include the series resistance of electrodes (R_s) and the dielectric loss of piezoelectric layer (R_o), as shown in Fig. 4.20. In this modified BVD model, at frequencies far from the resonant frequency the resonator behaves like a plate capacitor (C_o) in series with the two resistors R_o and R_s. The magnitude of the impedance (of the BVD model) versus frequency would still look like Fig. 4.18c.

Quality Factor of FBAR

The quality factor Q of a resonator is defined to be $Q \equiv$ stored energy/dissipated energy per cycle, and a larger Q means a lower insertion loss, lower phase noise, longer ring-down time, etc. Several different methods can be used to calculate the Q. All the methods yield somewhat different Q values for a given impedance characteristics of FBAR and must be used with caution. The one that I find most useful and accurate (especially for oscillator application) is the calculation based on the phase gradient with respect to frequency. In this method, Q is calculated by obtaining FBAR impedance (Z) versus frequency from measured S_{11} (reflection coefficient) versus frequency, followed by

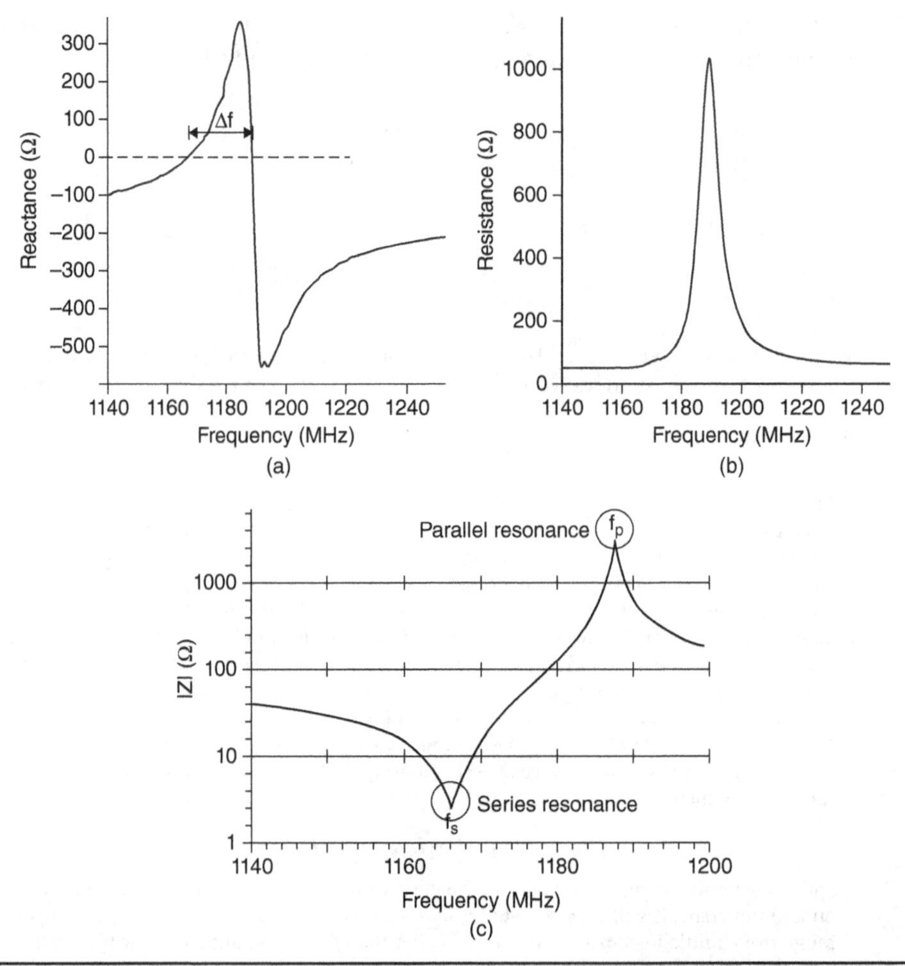

FIGURE 4.18 Impedance ($R + jX$) of an FBAR with f_s and f_p at 1.166 and 1.187 GHz, respectively: (a) reactance (X) versus frequency, (b) resistance R versus frequency, and (c) magnitude of the impedance (square root of $R^2 + X^2$) versus frequency.

calculating the slope of the phase of the FBAR impedance with respect to frequency $\left(\frac{d\angle Z}{df}\right)$ at a frequency of interest (f) and using the following formula for Q,

$$Q = \frac{f}{2}\frac{d\angle Z}{df}. \tag{4.79}$$

Though FBAR's quality factor Q can be obtained from BVD model through $Q = \frac{2\pi f L_m}{R_m}$, how fast the phase of FBAR impedance (the third plot from the top in Fig. 4.21b) changes as a function of frequency is a better indication of how well the resonator will keep the oscillation frequency when it is used in an oscillation circuit for time keeping. With this method, the faster the phase varies (as a function of frequency), the higher is the Q, and the Q depends on frequency with high values near the series and parallel frequencies.

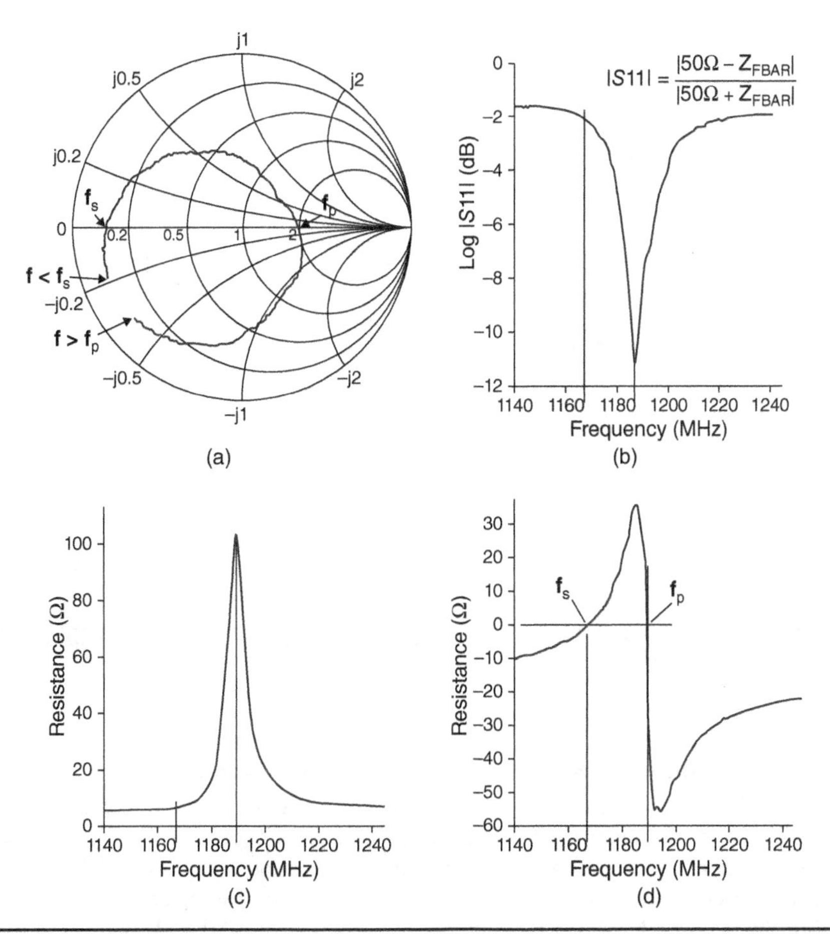

Figure 4.19 FBAR with f_s and f_p at 1.166 and 1.187 GHz, respectively: (a) S11 versus frequency plotted on a Smith chart, (b) magnitude of S11 versus frequency, (c) FBAR's resistance (R) versus frequency, and (d) FBAR's reactance X versus frequency.

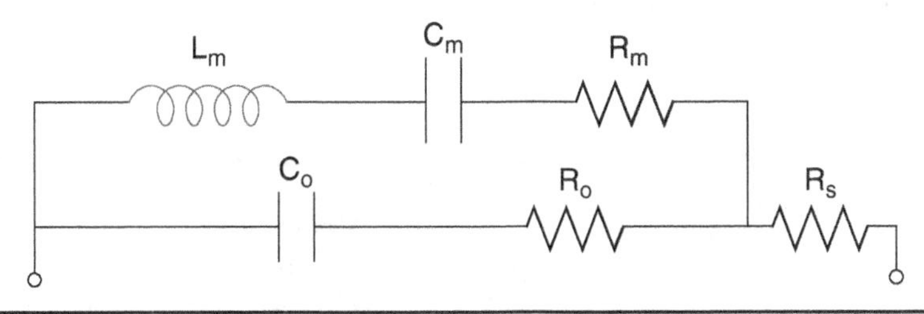

Figure 4.20 Modified BVD model of FBAR.

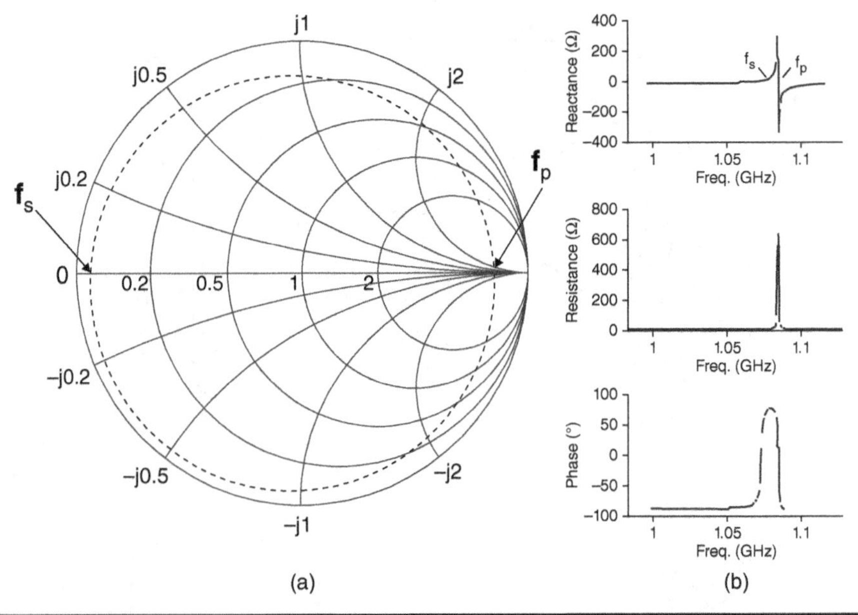

FIGURE 4.21 (a) FBAR's impedance versus frequency plotted on a Smith chart. (b) The reactance, resistance, and phase of FBAR impedance versus frequency.

4.3.7 Spurious Resonant Modes and Wave Dispersion

As bulk acoustic waves, so-called P waves, propagate and resonate in the thickness direction (z) between the top and bottom faces of a piezoelectric film (thickness d) in FBAR, there are waves propagating in-plane directions (x and y) in the piezoelectric film, which can also resonate between two edges, producing spurious modes and affecting FBAR's Q. Those in-plane waves are shear horizontal (SH) wave and the Rayleigh–Lamb (RL) wave composed of longitudinal P wave and shear vertical (SV) wave, as illustrated in Fig. 4.22 where the particle displacement directions of the waves are indicated along with the direction of the wave propagation. Since any elastic wave is either P wave, SH wave, SV waves, or any combination of P, SH, and SV waves, we see that the RL wave results from mode conversion at the solid–air interface, as either P or SV wave is reflected from the interface and is composed of P and SV waves, as shown in

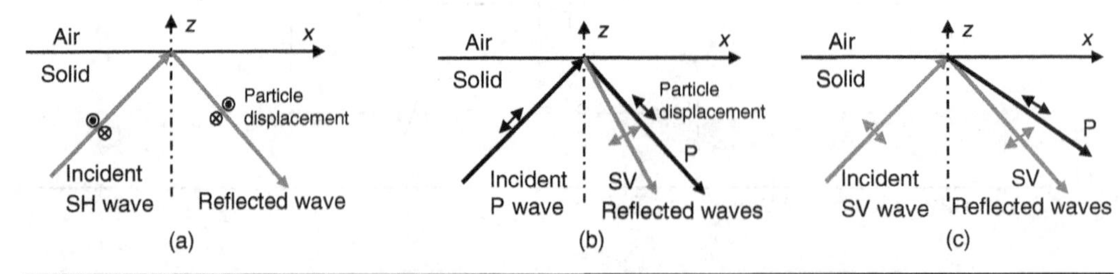

FIGURE 4.22 Reflection of elastic waves at the air–solid interface for (a) incident shear horizontal (SH) wave, (b) incident longitudinal P wave, and (c) incident shear vertical (SV) wave.

Fig. 4.22b and c. This kind of mode conversion makes sense physically if we consider the particle displacements of P and SV waves, since their particle displacements are similar. In case of SH wave, though, such mode conversion does not happen at the interface, as SH wave is reflected, since the particle displacement is along the y direction (in and out of the page), as illustrated in Fig. 4.22a, and is orthogonal to the particle displacement directions of P or SV waves. Thus, the waves in the in-plane or lateral directions of FBAR are either SH wave or the RL wave (combination of P and SV).

The SH waves propagating along the x direction of a solid piece bound by air (a representative of an FBAR) are illustrated in Fig. 4.23 for the first three modes ($n = 0$, 1, and 2). Note that the particle displacements (u_y) are along the y direction; there is no mode conversion as the wave is reflected from the solid–air interface; and there is no shape change in the x–z plane. Except the zeroth mode, the waves are dispersive, and energy is lost through heat generation. The SH waves do not produce electrical polarization across the thickness (d), if the solid is an FBAR made out of AlN or ZnO, because none of the shear stresses (T_4, T_5, and T_6) produces piezoelectric polarization in the z direction (D_3), according to the piezoelectric matrix

$$\begin{pmatrix} D_1 \\ D_2 \\ D_3 \end{pmatrix} = \begin{pmatrix} 0 & 0 & 0 & 0 & d_{15} & 0 \\ 0 & 0 & 0 & d_{15} & 0 & 0 \\ d_{31} & d_{31} & d_{33} & 0 & 0 & 0 \end{pmatrix} \begin{pmatrix} T_1 \\ T_2 \\ T_3 \\ T_4 \\ T_5 \\ T_6 \end{pmatrix}.$$

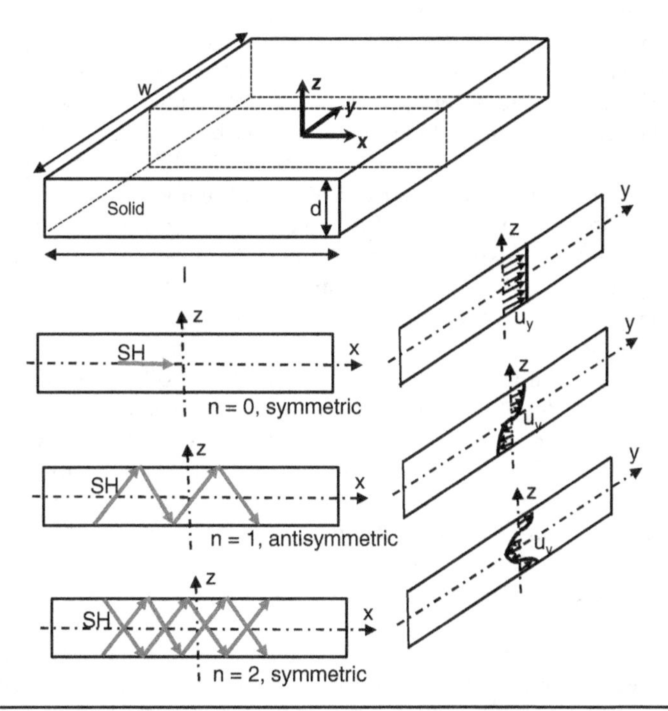

FIGURE 4.23 Shear horizontal (SH) wave propagation along the x direction and particle displacements (u_y) for the first three modes of $n = 0$, 1, and 2 in a solid piece bound by air.

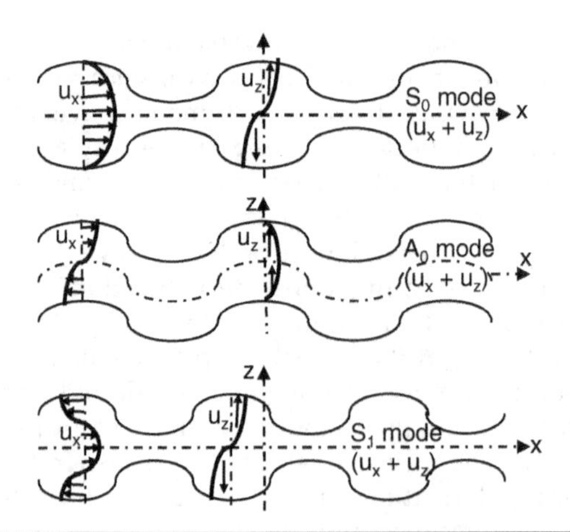

FIGURE 4.24 The first three modes of the Rayleigh–Lamb (RL) wave (composed of longitudinal *P* wave with particle displacement u_x and shear vertical [SV] wave with particle displacement u_z) propagating along the *x* direction in a solid piece shown in Fig. 4.23.

The first three modes of the RL waves propagating along the *x* direction of the same solid piece in Fig. 4.23 are illustrated in Fig. 4.24. Though the particle displacements u_x and u_z are shown separately for the sake of clarity, they are present at each and every point as the waves propagate. The particle displacements u_x and u_z can be considered as those for P and SV waves, respectively. Note that there are so-called symmetric modes (S_0, S_1, S_2 ...) and antisymmetric modes (A_0, A_1, A_2 ...), depending on the symmetricity with respect to the x–y plane. The antisymmetric modes are essentially flexural modes and produce zero net electrical polarization across the thickness direction due to the opposite signs of the induced stress above and below the neutral axis, as can be seen by combining u_x and u_z of A_0 mode in Fig. 4.24 and noting that the sign of the net stress ($= u_x + u_z$) above the neutral axis (or the mid-plane) is opposite to that of the net stress below the neutral axis. Similarly, S_0 mode produces zero net electrical polarization across the thickness direction due to the opposite signs of the induced stress ($= u_x + u_z$) above and below the neutral axis, which also can be observed from the top figure in Fig. 4.24. However, the u_x stress of S_1 mode is not symmetric with respect to the neutral axis, and the net stress ($= u_x + u_z$) across the thickness is not zero.

The RL waves and the SH waves (except the zeroth mode) are dispersive, and the dispersion characteristics depend on frequency and wave modes, as shown in Fig. 4.25. The dispersion curves offer the wave numbers (β) for the waves traveling in the in-plane directions of an FBAR, and the spurious modes due to the lateral resonances caused by those waves can be calculated. For example, if an FBAR shown in Fig. 4.23 has a series resonant frequency ω_s in thickness mode vibration, Fig. 4.25 shows that there are five to six lateral resonant modes below and above ω_s, as indicated by small circles in the figures. Among those, the only one due to the S_1 wave produces the spurious modes because the SH waves (the SH_0 and SH_1), antisymmetric RL waves (the A_0 and A_1), and the S_0 wave produce zero net electrical polarization along the thickness direction or across FBAR's top and bottom electrodes, as explained in the preceding paragraph. Shown in Fig. 4.25 are two different types of S_1 that result in the spurious modes mostly below (Fig. 4.25a) and all above (Fig. 4.25b) the series resonant

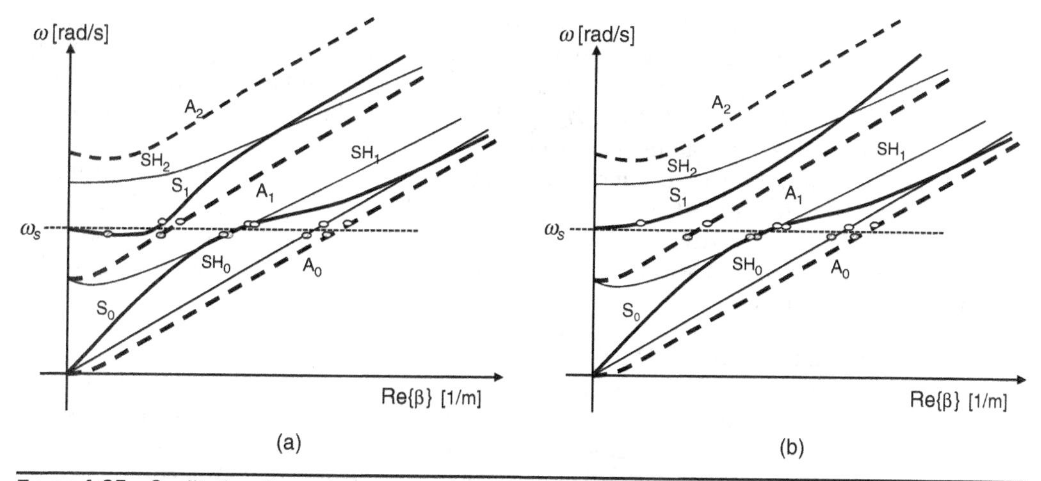

FIGURE 4.25 Qualitative dispersion curves of the Rayleigh–Lamb waves (A_0, S_0, A_1, S_1, A_2, and S_2) and the shear horizontal waves (SH_0, SH_1, and SH_2) traveling along the *x* direction in a solid piece shown in Fig. 4.23: (a) an FBAR with dominant spurious modes below its series resonant frequency ω_s and (b) a different FBAR with dominant spurious modes above ω_s.

frequency ω_s. To minimize the lateral resonances, the top-view edges (along the width and length directions in Fig. 4.23) should not be in parallel to each other but rather be nonparallel as done in each of the FBARs in a commercial FBAR filter shown in Fig. 4.30a.

More detailed studies on this topic can be done through reading pp. 311–463 in [14], pp. 67–89 and 133–145 in [15], and p. 243 in [16].

4.3.8 Film Bulk Acoustic Resonator for RF Front-End Filters

Unlike quartz crystal resonator, a film bulk acoustic wave resonator (FBAR) with a piezoelectric thin film such as AlN or ZnO can easily be made to resonate at GHz, since the fundamental resonant frequency $F_{res} = V_a/2d$, where V_a and d are the acoustic wave velocity and thickness of the piezoelectric film, respectively. For example, an FBAR

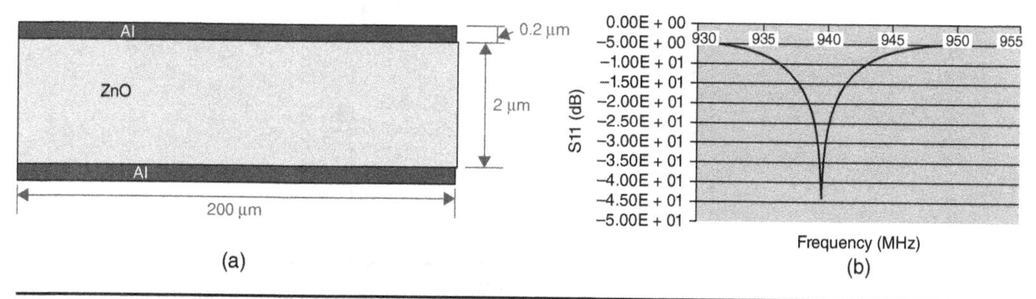

FIGURE 4.26 (a) Cross-sectional view of an FBAR with 2-μm-thick ZnO film, which has a fundamental resonant frequency of about 1 GHz (since $F_{res} = v/2d$, where $v \approx 3{,}400$ m/s and $d = 2$ μm). (b) The reflection coefficient S11 of the FBAR in (a) with impedance Z_{FBAR}. Note: $S11 = (50\ \Omega - Z_{FBAR})/(50\ \Omega + Z_{FBAR})$ when measured with 50 Ω source and line.

made of 2-μm-thick ZnO (Fig. 4.26a) would have the resonant frequency around 1 GHz, since the bulk acoustic wave velocity is ≈ 3,400 m/s. The electrical impedance of FBAR (Z_{FBAR}) varies drastically near its resonant frequency, and the FBAR's reflection coefficient, $S11 = (50\ \Omega - Z_{FBAR})/(50\ \Omega + Z_{FBAR})$, that can easily be measured with a network analyzer varies very substantially as shown in Fig. 4.26b.

In addition to FBAR's resonant frequency being 1 and 10 GHz, its lateral dimension can be down to $10 \times 10\ \mu m^2$ with Q of about 1,000 which leads to a very low insertion loss down to 1 dB. And the fabrication processes of FBARs are based on microfabrication with (or without) micromachining, and FBARs can be massively produced at a low cost.

Filter Specifications for 1.9 GHz PCS Band

As indicated in Table 4.2, personal communication system (PCS) uses 1,850–1,910 MHz for transmission (*Tx*) and 1,930–1,990 MHz for receiving (*Rx*), each band with 60 MHz bandwidth (Fig. 4.27). The specifications for filters for transmitting and for receiving (*Tx* and *Rx* filters) include < 3.5 dB of insertion loss, > 40 dB (for *Tx* filter) and > 50 dB (for *Rx* filter) of stopband-to-passband rejection ratio (stopband and passband meaning *Rx* band and *Tx* band, respectively, for *Tx* filter and vice versa for *Rx* filter), > 30 dB of second harmonic rejection, > 2.2 of voltage standing wave ratio (VSWR), a sharp roll-off (the transition from passband to stopband within the 20 MHz guardband), and operating temperature from –30 to +85°C [17, 18].

Agilent's Duplexer Built with FBARs

Since 2001, Agilent's semiconductor division (which became Avago Technologies, which in 2016 acquired Broadcom Corporation and became Broadcom Inc.) has been mass-producing FBAR-based filters for duplexers for RF front-end filtering for cell phones. The FBAR is fabricated by bulk-micromachining Si, filling the cavity with PSG, chemical mechanical polishing (CMP) of the surface, depositing and patterning of molybdenum and AlN, and removing the PSG to release the structure in a process

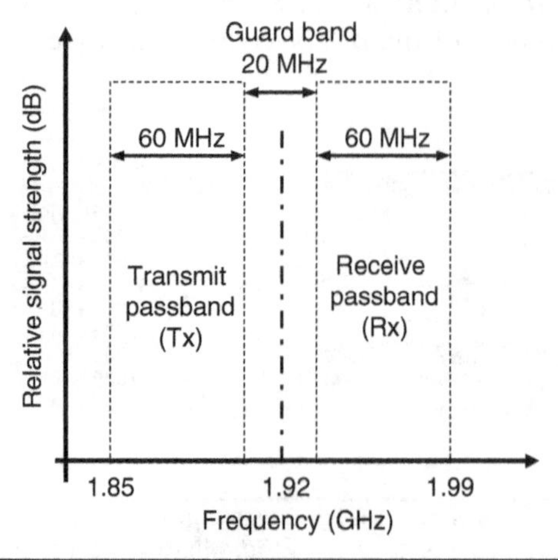

Figure 4.27 Specification for RF front-end transmission filter and receiving filter for cell phones based on personal communication system (PCS) standard. (Adapted from [19].)

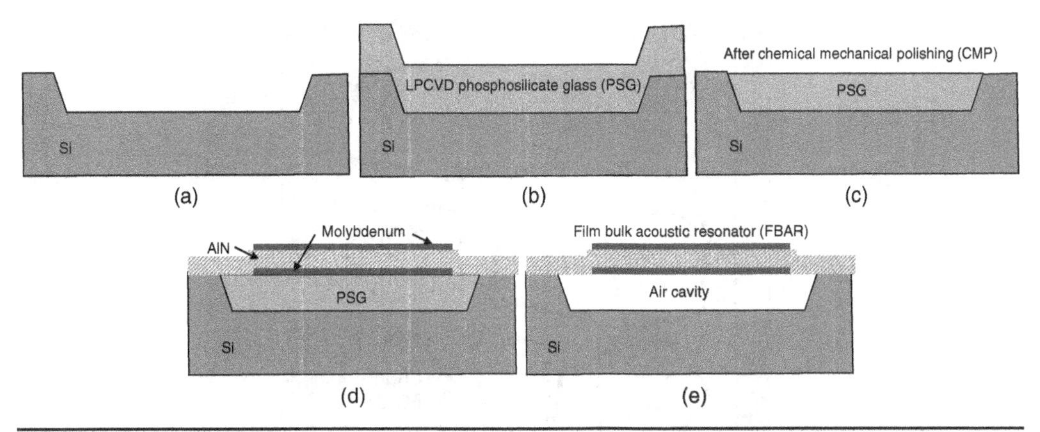

FIGURE 4.28 Brief fabrication process of an FBAR made of AlN and molybdenum on a micromachined silicon substrate.

similar to what is illustrated in Fig. 4.28. Note that in this design no support layer is used for the FBAR since AlN is mechanically sturdy enough to sustain itself and the electrodes without any support layer. This would not be the case with ZnO, which is brittle and fragile, possibly even porous.

Five to seven of the FBARs can be connected in series and parallel as shown in Fig. 4.29a to form bandpass filters for transmitting and receiving bands around 1.9 GHz. The FBARs in the serial path all have a same resonant frequency, while the FBARs in the parallel all have another resonant frequency which is lower by 60 MHz from the resonant frequency of the FBARs in the serial path. The 60-MHz shift of the resonant frequency can be obtained by adding a mass-loading layer (e.g., metal) on the FBAR, and

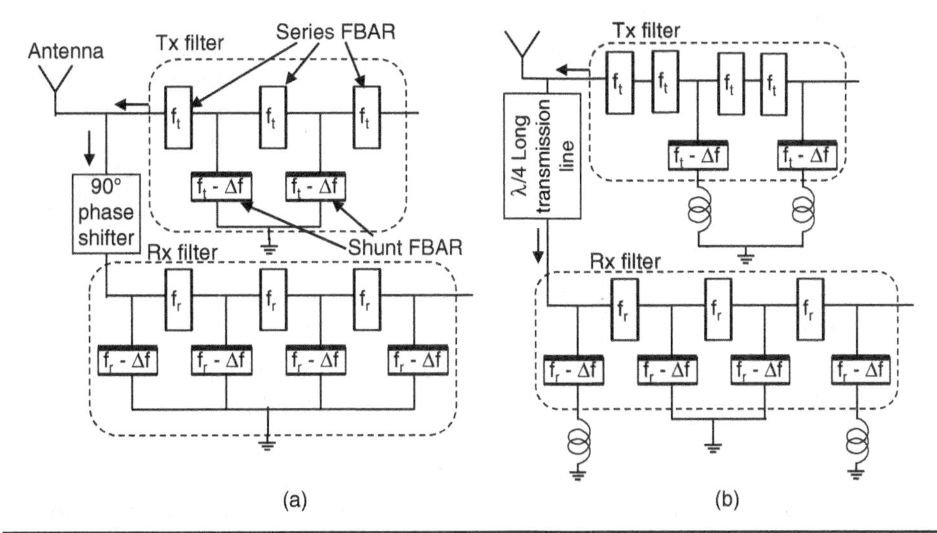

FIGURE 4.29 (a) RF front-end transmission filter and receiving filter made of FBARs. (Adapted from [19].) (b) Another configuration for the filters. (Adapted from [17].)

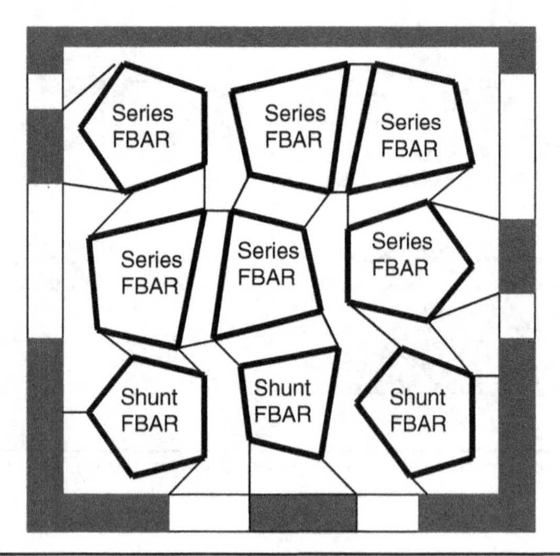

FIGURE 4.30 Top-view schematic of a transmit (*Tx*) filter composed of six series FBARs and three shunt FBARs. (Adapted from [18].)

all the FBARs for each filter are processed on a single chip. The 90° phase shifter or the quarter-wave transmission line transforms the *Rx* filter impedance from very low to very high for the passband of *Tx* filter so that the *Rx* filter may not load the *Tx* filter in the passband [17]. The small series inductance added to the two shunt FBARs (Fig. 4.29b) increases the stopband-to-passband rejection without sacrificing the insertion loss. For *Tx* filter, two series FBARs (rather than a single FBAR with area *A*) with each having an area of 2*A* can be used for each of the two inner stages in the series path in order to reduce the losses in the series path for improving the roll-off and insertion loss of the filter [18].

In one implementation, the shunt FBARs are fabricated for a resonant frequency which is about 3% lower than that of the series FBARs in order to obtain the necessary 3% bandwidth (60 MHz) required by the filter specifications [18]. The series resonant frequency (f_s) and the parallel antiresonant frequency (f_p) are designed to be separated by about 6% for a flat passband response. The chip area is efficiently used for a total of nine FBARs, as shown in Fig. 4.30a, where top view of each FBAR shows "apodizing" (i.e., no parallel lines) that minimizes the lateral resonant peaks (i.e., spurious resonances).

Question: Does the FBAR shown in Fig. 4.28 have resonance at even harmonics (i.e., the second, fourth, sixth, etc. overtones)?

Answer: No, because the even harmonics have symmetric particle displacement (with respect to the neutral plane in the diaphragm), which sums up to be zero when integrated from the bottom surface to the top.

Solid-Mounted FBAR versus Air-Backed FBAR

Though air is the best acoustic reflector for FBAR, air backing or reflecting requires micromachining to form a diaphragm as in Fig. 4.28e or Fig. 4.31a. Consequently,

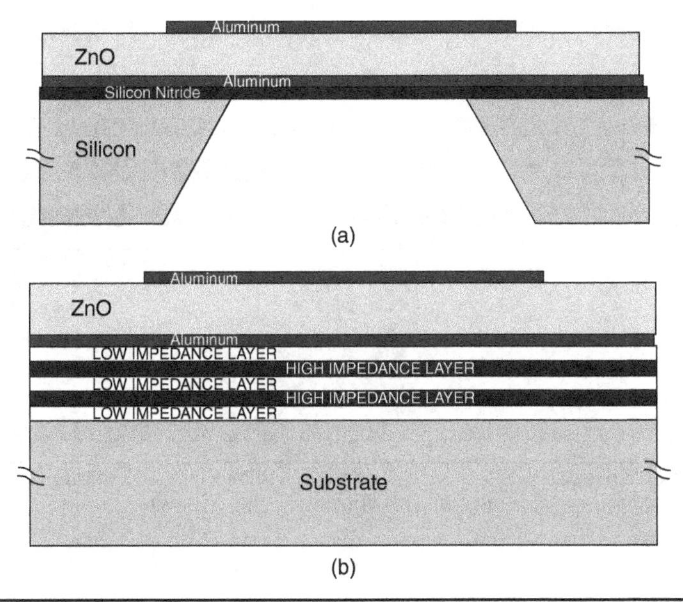

FIGURE 4.31 Two types of FBAR: (a) air-backed FBAR that requires micromachining and (b) solid-mounted FBAR with multiple layers (having high and low impedances) forming an acoustic reflector.

some commercial FBARs are fabricated without any micromachining on a solid substrate, with an embedded acoustic reflector that reflects acoustic wave into the piezoelectric film, as shown in Fig. 4.31b. Such an FBAR is commonly called solidly mounted resonator (SMR) or solid-mounted FBAR. An acoustic reflector (called Bragg reflector) can be formed with multiple numbers of quarter-wavelength matching (or mismatching) layers (less than a micron thick for each layer at GHz range). For a larger reflection (and thus a lower energy loss to the substrate), (1) the difference between the impedance of the low impedance layer and that of the high impedance layer needs to be larger, while keeping the thickness of each layer at quarter wavelength; and (2) the number of the layers needs to be larger. On the other hand, a diaphragm-supported (or an air-backed) FBAR needs only a single (or no) layer to support the piezoelectric film and electrodes, and inherently has less loss to substrate and spurious modes than SMR.

Wave Reflection from Stack of Multiple Layers

In case of an SMR with a Bragg reflector formed by SiO_2 (low impedance layer) and W (high impedance layer) shown in Fig. 4.32, the reflection coefficient R at the interface between the piezoelectric ZnO film (with impedance Z_{ZnO}) and the bottom Al is

$$R = \frac{Z_6 - Z_{ZnO}}{Z_6 + Z_{ZnO}} \tag{4.80}$$

with Z_6 being the impedance seen at the interface down toward the substrate as indicated in the figure. For longitudinal or shear waves propagating from the piezoelectric ZnO film toward the substrate (and being incident normal to the interfacial planes), the

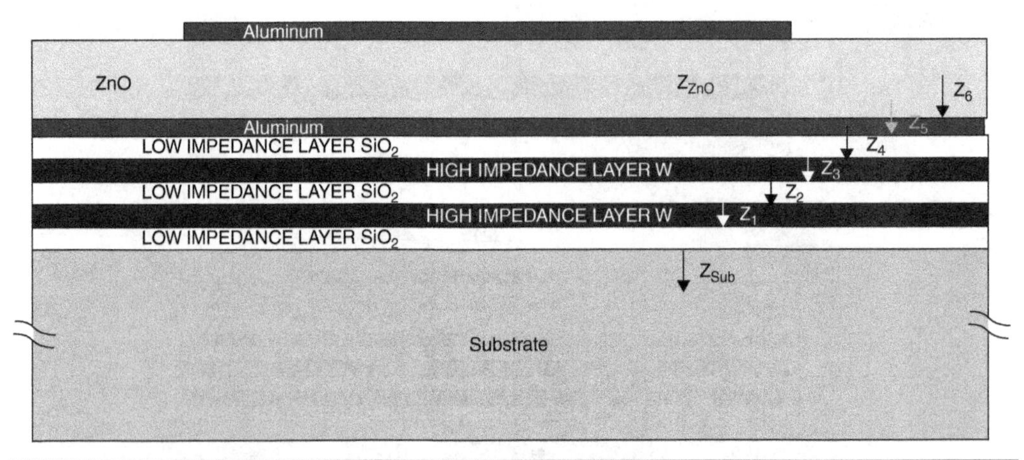

FIGURE 4.32 Impedances (Z_{sub}, Z_1, ... , and Z_6) seen at the interfaces downward for longitudinal or shear waves propagating from the piezoelectric ZnO film toward the substrate.

impedance Z_6 can be obtained by first calculating Z_1 at a frequency f with the equation below (followed by calculating Z_2 from the calculated value of Z_1, then Z_3 from Z_2, then Z_4 from Z_3, then Z_5 from Z_4, and then Z_6 from Z_5, using an equation similar to the one below) [20].

$$Z_1 = Z_{SiO_2} \frac{Z_{sub} + jZ_{SiO_2} \tan\left(\dfrac{2\pi f d_{SiO_2}}{v_{SiO_2}}\right)}{Z_{SiO_2} + jZ_{sub} \tan\left(\dfrac{2\pi f d_{SiO_2}}{v_{SiO_2}}\right)}, \tag{4.81}$$

where d_{SiO_2} and v_{SiO_2} are the SiO_2 layer thickness and acoustic velocity in SiO_2, respectively, while Z_{sub} and Z_{SiO_2} are the acoustic impedance of the substrate and SiO_2, respectively. This equation is basically the same as the equation for calculating the impedance for electromagnetic waves (or transmission line) incident normal on the interfacial plane of a layered medium with thickness d (on top of a semi-finite thick substrate), with the velocity being that of electromagnetic wave and the impedances being electromagnetic impedances. Once Z_1 is calculated, we can use the calculated Z_1 to calculate Z_2 using a similar equation as follows.

$$Z_2 = Z_W \frac{Z_1 + jZ_W \tan\left(\dfrac{2\pi f d_W}{v_W}\right)}{Z_W + jZ_1 \tan\left(\dfrac{2\pi f d_W}{v_W}\right)}, \tag{4.82}$$

where d_W and v_W are the W (tungsten) layer thickness and acoustic velocity in W, respectively, while Z_W is the acoustic impedance of W. Similarly, Z_3, Z_4, Z_5, and Z_6 can be obtained sequentially one after another.

The calculations can easily be programmed into Matlab, as shown below, and one can obtain the magnitude of the reflection coefficient $|R|$ by Bragg reflector as a function of frequency, also shown below, by running the Matlab code. The code below is for a longitudinal wave for Bragg reflector composed of three 0.2-μm-thick SiO_2 layers and two 0.4-μm-thick W layers on a silicon substrate with 0.2-μm-thick bottom Al layer and piezoelectric ZnO film, as illustrated in Fig. 4.32. A similar curve can be calculated for a shear wave by using the impedances and acoustic velocities for a shear wave in SiO_2, W, silicon substrate, and Al.

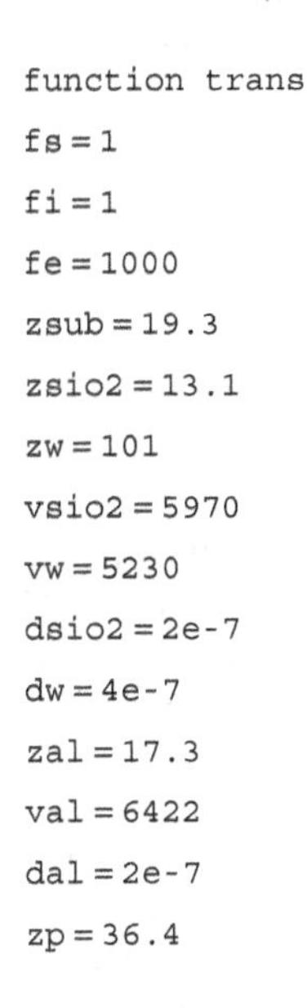

```
function trans
fs = 1
fi = 1
fe = 1000
zsub = 19.3
zsio2 = 13.1
zw = 101
vsio2 = 5970
vw = 5230
dsio2 = 2e-7
dw = 4e-7
zal = 17.3
val = 6422
dal = 2e-7
zp = 36.4
```

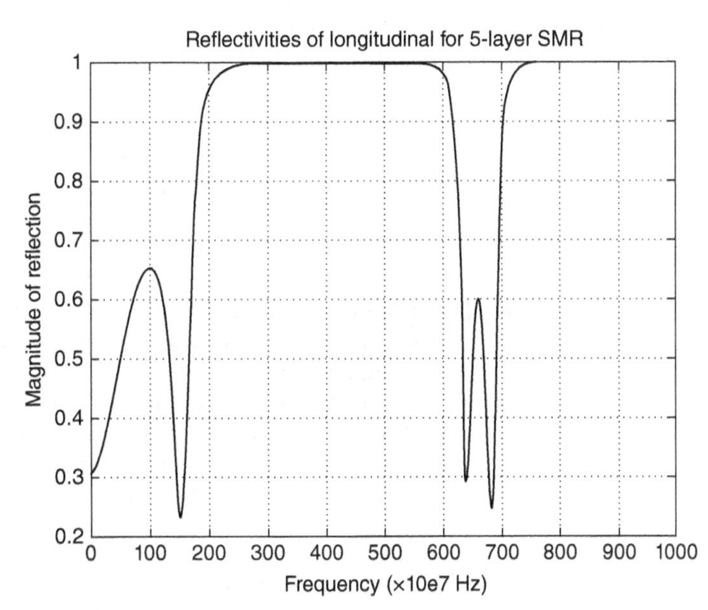


```
for f = fs:fi:fe
z(f) = zsio2*(zsub+i*zsio2*tan(2*pi*f*1e7*dsio2/vsio2))/(zsio2+i*zsub*tan
(2*pi*f*1e7*dsio2/vsio2));
z(f) = zw*(z(f)+i*zw*tan(2*pi*f*1e7*dw/vw))/(zw+i*z(f)*tan(2*pi*f*1e7*dw/
vw));
z(f) = zsio2*(z(f)+i*zsio2*tan(2*pi*f*1e7*dsio2/vsio2))/(zsio2+i*z(f)*tan
(2*pi*f*1e7*dsio2/vsio2));
z(f) = zw*(z(f)+i*zw*tan(2*pi*f*1e7*dw/vw))/(zw+i*z(f)*tan(2*pi*f*1e7*dw/
vw));
z(f) = zsio2*(z(f)+i*zsio2*tan(2*pi*f*1e7*dsio2/vsio2))/(zsio2+i*z(f)*tan
(2*pi*f*1e7*dsio2/vsio2));
r(f)=abs((z(f)-zp)/(z(f)+zp))
end
f = 1:1000
plot(f,r(f))
grid on
xlabel('Frequency (x10e7 Hz)')
ylabel('Magnitude of Reflection')
title('Reflectivities of Longitudinal for 5-Layer SMR')
```

Note: if the SiO_2 thickness is quarter wavelength (i.e., $d_{SiO_2} = \frac{\lambda_{SiO_2}}{4} = \frac{v_{SiO_2}}{4f}$), (Eq. 4.81) becomes $Z_1 = Z_{SiO_2}\frac{Z_{Sub} + jZ_{SiO_2}\tan(\pi/2)}{Z_{SiO_2} + jZ_{Sub}\tan(\pi/2)} = \frac{Z_{SiO_2}^2}{Z_{Sub}}$. And if $Z_{SiO_2}^2$ is equal to $Z_{sub}Z_W$, then $Z_1 = Z_W$, which means that the impedance seen at the interface between the first W and the first SiO_2 layers above the substrate is matched to the W impedance, through the quarter-wavelength-thick SiO_2 layer, resulting in zero reflection at that interface. This leads to a commonly used impedance matching idea based on quarter-wavelength-thick layer with its impedance being equal to $\sqrt{Z_1Z_3}$, where Z_1 and Z_3 are the impedances of the medium before and after the matching layer, respectively (Fig. 4.33b). Another note is that in an attempt to match the impedance, one may consider a d_2-thick impedance matching layer for a d_3-thick solid wall that is in between two media (e.g., water and air) having impedances Z_1 and Z_4, as illustrated in Fig. 4.33c. However, if Z_4 (e.g., 410 Pa·s/m or rayls for air) is much smaller than Z_1 (e.g., 1.5 Mrayls for water), the impedance matching layer (with Z_2) will not impact on the magnitude of the reflection coefficient ($|R|$) at the interface between Z_1 and Z_2, since $Z_{eff0} \cong jZ_3\tan\theta_3$ (where $\theta_3 = 2\pi fd_3/v_3$) for Z_3 (for solid) >> Z_4 (for air), and consequently, Z_{eff1} will be an imaginary number due to Z_{eff0} being an imaginary number (see the equations in Fig. 4.33c). With Z_{eff1} and Z_1 being pure imaginary and real numbers, respectively, the magnitude of the reflection coefficient ($|R|$) will be 1, no matter what one may choose for the thickness (d_2) and impedance (Z_2) of the impedance matching layer. This makes sense physically, since almost all of the acoustic wave is reflected from the solid–air interface and gets back to the water unless there is acoustic absorption in the layers between the water and air. If one desires very little reflection from the solid wall (with thickness d_3 and impedance Z_3) facing air (e.g., in order to minimize standing wave in water), s/he may consider (1) adding an acoustic absorber to the solid wall, (2) making the solid wall

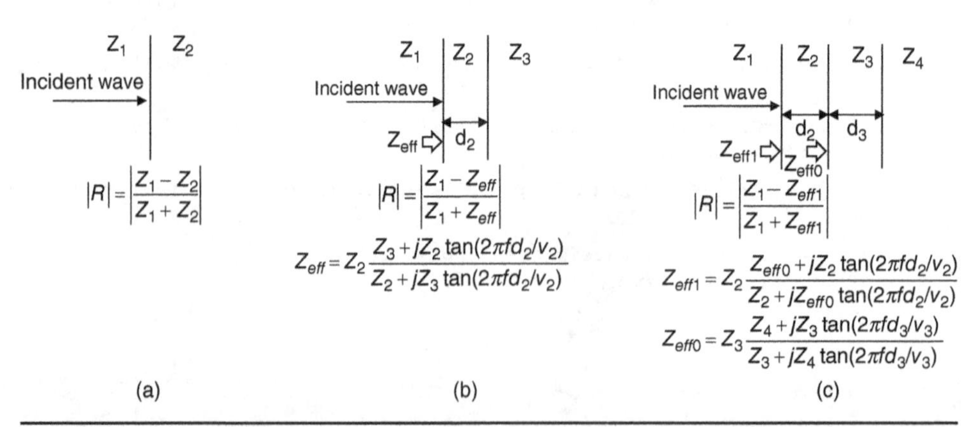

FIGURE 4.33 Wave reflection (R) at the interface between two media having impedances Z_1 and Z_2 for normally incident acoustic wave in (a) two semi-infinitely extending media (Z_1 and Z_2), (b) a similar case as (a) but with a d_2-thick layer between the two semi-infinitely extending media (Z_1 and Z_3), and (c) a similar case as (a) but with two layers of finite thicknesses (d_2 and d_3) between the two semi-infinitely extending media (Z_1 and Z_4).

with a material having a high acoustic absorption or attenuation coefficient (ideally with its impedance close to that of water or Z_1), or (3) roughening the solid surface for wave scattering.

Solid-Mounted FBARs for RF Front-End Filtering

A solid-mounted FBAR with an acoustic reflector composed of two pairs of tungsten and silicon dioxide (4 μm total thickness) was shown to have a high impedance magnitude ratio between the impedance at the series resonant frequency (f_s) and that at the parallel resonant frequency (f_p) [23]. The impedance ratio is particularly important for filter application since it determines the stopband-to-passband rejection ratio (or the stopband attenuation), as can be seen in the next section. The Q that determines the insertion loss of a filter was reported to be about 1,500 at about 2 GHz, an impressive number, as the acoustic reflector provides more than 99.99% reflection. The Q was noted to be not limited by the reflector but rather by the spurious modes due to resonances in the lateral directions. Though the acoustic reflector does a good job in reflecting acoustic energy back into the resonator, the first few layers of the reflector contain some acoustic energy, which reduces the electromechanical coupling coefficient (K^2_{eff}). Thus, SMR has inherently lower K^2_{eff} than air-backed FBAR but by only a little. Filters based on solid-mounted FBARs have been built in a similar way as the one taken with air-backed FBARs. For example, three FBARs in series and four FBARs in parallel have been used to build a filter, with the shunt FBARs being made with an additional metal layer to have their frequencies lowered by about 60 MHz from the series FBARs.

Estimating Filter Characteristics from FBAR Impedance

Transfer characteristic of FBAR-based filters such as in Fig. 4.34a can be obtained by considering the FBAR's impedance magnitude versus frequency for all the FBARs at each frequency point. For example, Fig. 4.34b (characteristic of the series FBAR) shows the impedance magnitude that is at its highest (3.4 kΩ) at 2.020 GHz and at its lowest (2.1 Ω) at 1.964 GHz (about 60 MHz below from the frequency where the highest magnitude occurs). The FBAR is designed that way because the required bandwidth for the filter for 2 GHz cell phones is 120 MHz. Each of the series FBARs

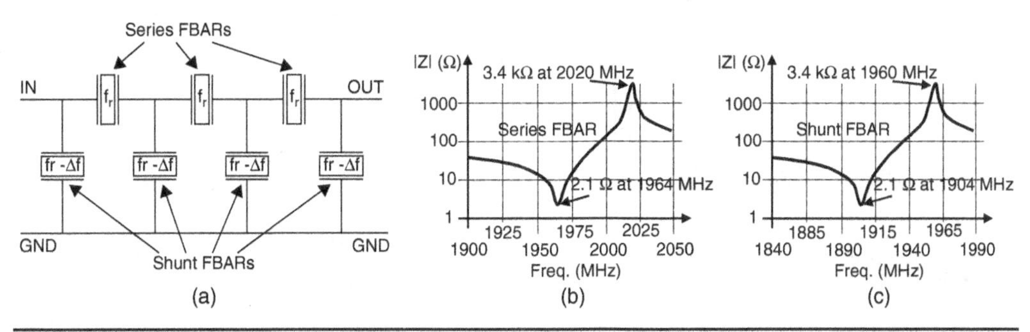

FIGURE 4.34 Ladder filter (a) made of three series FBARs and four shunt FBARs whose impedance characteristics are shown in (b) and (c) for series FBAR and shunt FBAR, respectively.

in Fig. 4.34a has its impedance magnitude as shown in Fig. 4.34b, while each of the shunt resonators has its impedance magnitude similar to that in Fig. 4.34b except that it is shifted in frequency (to lower values) by about 60 MHz, as shown in Fig. 4.34c. Thus, at 1.900 GHz, the series FBARs present 70 Ω, while the shunt FBARs present 2.1 Ω. This combination of the impedance magnitudes produces a low transfer coefficient from IN to OUT, since the shunt elements have much lower impedance values than the series elements. However, at 1.964 GHz, the series FBARs present 2.1 Ω, while the shunt FBARs present 3.4 kΩ. Now, this produces a high transfer coefficient, since the shunt elements have much higher impedance values than the series elements. One can do a similar estimate at 2.020 GHz and see a low transfer coefficient from IN to OUT.

Question: Sketch the transfer characteristic from IN to OUT between 1.8 and 2.1 GHz for the filter shown in Fig. 4.34a, assuming that each series FBAR and shunt FBAR have an impedance characteristic as shown in Fig. 4.34b and Fig. 4.34c, respectively. Please note that Fig. 4.34c is basically Fig. 4.34b shifted in frequency (to lower frequency) by about 60 MHz.

Answer: See Fig. 4.35.

Frequency Matching by Mass Loading

In Fig. 4.35 the parallel resonant frequency of the shunt FBARs is about the same as the series resonant frequency of the series FBARs. In other words, series FBARs and shunt FBARs have basically same frequency characteristics except that the frequency characteristics of shunt FBARs are the ones shifted to lower frequency by about 60 MHz. Consequently, both series and shunt FBARs can be fabricated on a same wafer with shunt FBARs receiving an additional mass loading layer (such as ≈500-Å-thick Al).

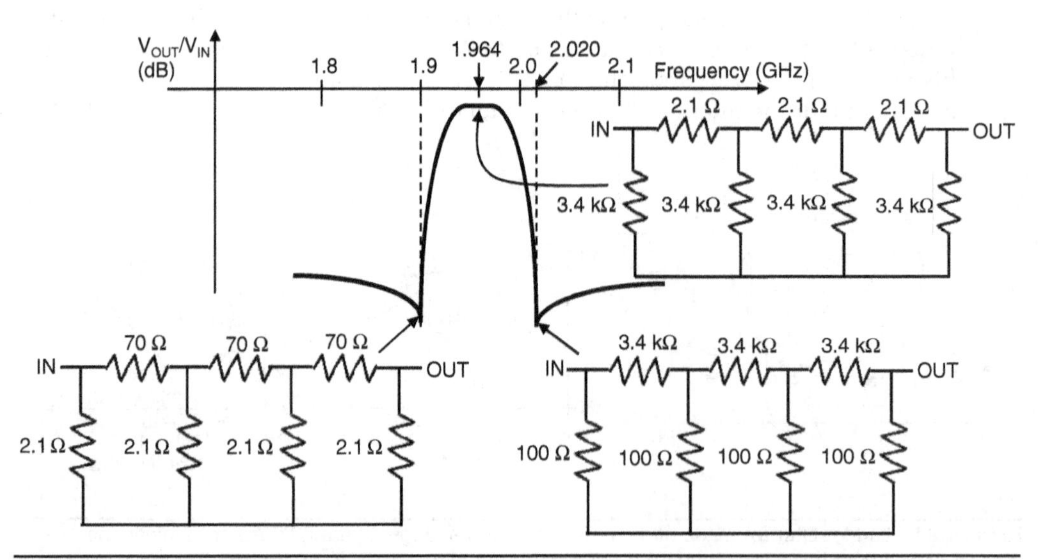

FIGURE 4.35 Transfer characteristic of the filter shown in Fig. 4.34 estimated with the impedance magnitudes of the FBARs at three different frequencies (1.900, 1.964, and 2.020 GHz).

Electromechanical Coupling Coefficient (K_t^2)

Electromechanical coupling coefficient (k_t^2) of FBAR determines the bandwidth of a FBAR-based filter, since percentage bandwidth is equal to $k_t^2/2$, and various ways to increase k_t^2 have been explored. If all the other nonpiezoelectric layers of FBAR are negligibly thin, k_t^2 is that of the piezoelectric material. Otherwise, the thicknesses and acoustic impedances of the other layers (the electrodes and support layers) impact k_t^2. Consequently, there are optimum thicknesses for those layers in relation to the thickness of the piezoelectric layer, and with a particular choice of electrode material (e.g., Al, Au, Mo, W, and Cu) and thickness ratio (between electrode and piezoelectric layer), k_t^2 can be higher than that of the piezoelectric layer [24, 25].

To calculate k_t^2 (in %) versus the thickness ratio between electrode and piezoelectric layer for an FBAR as a function of electrode material for a given piezoelectric layer (e.g., AlN), one can use the one-dimensional Mason model, particularly Eq. (4.49), from which FBAR's impedance as a function of frequency can be obtained. From the impedance, the series and parallel resonant frequencies (f_s and f_p) can be obtained as the frequencies at which the impedance magnitude is the minimum and maximum, respectively. The electromechanical coupling coefficient (k_t^2) can then be calculated with $k_T^2 = \frac{\pi^2}{4}\frac{f_s(f_p - f_s)}{f_p^2} \approx \frac{\pi^2}{4}\frac{(f_p - f_s)}{f_s}$, from Eq. (4.73), as the electrode material and thickness ratio are varied.

Number of FBARs and FBAR Size for Filter Application

With a proper design, a larger number of FBARs (used in a filter) usually translates into a steeper transition from passband to stopband and greater stopband rejection but comes at the cost of narrower bandwidth and higher insertion loss [26].

The top-view active area (A) of FBAR determines the clamp capacitance C_o. Since the capacitance ratio between shunt and series FBARs determines the stopband rejection for a given number of resonators, one will have to choose $A_{\text{series}}/A_{\text{shunt}} > 1$ for high stopband rejection. However, one will have to consider also impedance matching to 50 Ω in choosing the FBAR sizes (i.e., A_{series} and A_{shunt}).

Power Handling Capability

Power handling capability is an important issue for an RF front-end filter for transmitting (i.e., *Tx* filter). Though FBAR is inherently capable of handling high power, the reliability issue stemming from high power density needs to be well characterized and understood, especially since FBAR's physical volume is small. The small physical volume means a high power density (about 1 MW/cm^3) dissipated in the FBAR used in a *Tx* filter, resulting in high dissipative losses (and consequent heating) in the filter [27]. The heating results in resonant frequency shifts and also degrades the filter in long run. High power density may cause the strain level to approach the fracture limit of FBAR's piezoelectric film that is already under substantial residual stress.

Temperature Compensation

Since the operating temperature range for commercial wireless transceiver is to be between –30 and 85°C, the temperature coefficient of resonant frequency (TCF) needs to be minimized. One approach is to incorporate a temperature compensating layer (e.g., SiO_2 that has positive TCF unlike most other materials which have negative TCF) into FBAR layers. One can use the one-dimensional Mason model [(Eq. (4.49)] to obtain the

TCF versus SiO_2/ZnO thickness ratio by varying the material constants as a function of temperature [28].

Case for or against ZnO over AlN

The *effective* electromechanical coupling coefficients (k_t^2) for AlN and ZnO are typically 6.5% and 8.5%, respectively [29]. For a 5-GHz wireless transceiver, the required bandwidth is 200 MHz at 5.25 GHz (3.8% fractional bandwidth), though a 2-GHz wireless transceiver requires 60-MHz bandwidth at 1.88 or 1.96 GHz (3.2% or 3.1% fractional bandwidth). Thus, AlN may not offer enough electromechanical coupling coefficient (k_t^2) to meet the 3.8% fractional bandwidth for the 5-GHz wireless transceiver, since the 3.8% fractional bandwidth would require 7.6% k_t^2. This limitation of AlN has motivated recent development of scandium-doped aluminum nitride (ScAlN), of which k_t^2 depends on the doping concentration of Sc and can be more than two times that of AlN.

Though ZnO has a higher k_t^2 than AlN, the TCF of ZnO is −60 ppm/°C, more than twice of that of AlN (−25 ppm/°C). Also, with ZnO film, a support diaphragm would be needed to support ZnO and electrodes, unlike AlN. A Si or SiN diaphragm has been used to support a piezoelectric ZnO film and two Al layers (Fig. 4.36a), and an FBAR with about 2-µm-thick ZnO film has been shown to have a fundamental resonant frequency of about 1 GHz. With a nonpiezoelectric support layer, the effective coupling coefficient of ZnO-based FBAR depends on the thickness of the support layer because of the wave propagation into the support layer, and there exists an optimum thickness for the support layer (with respect to the other FBAR layers), as far as the coupling coefficient is concerned, for different harmonics including even harmonics [21]. If an FBAR is composed of nonsymmetric metals (e.g., thin metal/ZnO/thick metal) or with a support layer (e.g., metal/ZnO/metal/Si_xN_y), there are even harmonics (Fig. 4.36b) in addition to odd harmonics because of the nonsymmetric acoustic energy distribution along the thickness direction of the piezoelectric layer (i.e., nonsymmetric with respect to the mid-plane of the piezoelectric layer). The even harmonic response of an FBAR built on a support layer can be studied with Mason model, and the optimum thickness

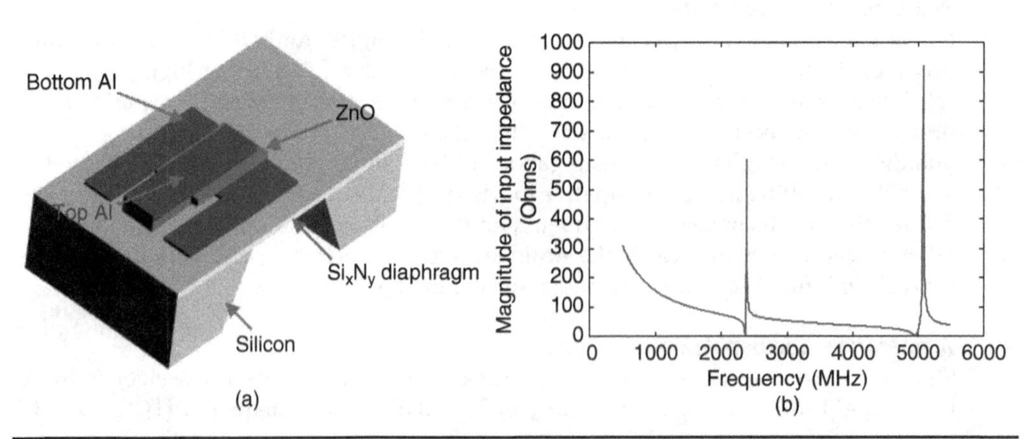

FIGURE 4.36 (a) Schematics of FBAR composed of piezoelectric ZnO film over a bulk-micromachined Si_xN_y support diaphragm. (b) Calculated broadband impedance of FBAR made of Al/ZnO/Al/Si_xN_y. (Adapted from [22].)

ratio between the piezoelectric layer and support layer can be predicted for a maximum effective electromechanical coupling coefficient. The calculation results (matched well with experimental results) for FBAR composed of $Al/ZnO/Al/Si_xN_y$ show that the maximum effective coupling coefficient for the second harmonic resonance is obtained with a thickness ratio (between the piezoelectric layer and nonpiezoelectric layer) that is close to its acoustic velocity ratio [22].

4.4 Surface Acoustic Wave Filters

Filters can be made with surface acoustic wave (SAW) transducers acting as delay lines, and various SAW filters have been commercialized for frequency from 10 MHz to 2 GHz. With a relatively slow wave velocity (especially compared to electromagnetic wave), the dimension needed for delay lines with SAW (Table 4.3) is small enough for SAW filters to offer the following advantages: (1) monolithic fabrication, similar to IC manufacturing process that allows easy mass production with low cost and (2) high performing filter characteristics. Lately, SAW filters have successfully been commercialized for RF front-end filtering (especially for *Rx* filters) for cell phones as they can be manufactured cheaper than FBAR filters. However, power handling capability of SAW filters at GHz has not been high enough to compete with FBAR filters for *Tx* filters.

4.4.1 SAW Generation by Interdigitated Electrodes over Piezoelectric Substrate

The voltage difference between the two interdigitated electrodes produces electrical fields in the gaps (between the electrodes), as shown in Fig. 4.37. The electrical fields (most of which are directed laterally) produce acoustic waves (at the surface since the electrical field is the strongest at the surface) that are sinusoidally varying in space if the electrodes are sitting on top of a piezoelectric substrate. The SAW generated under an electrical impulse applied to the electrodes is assumed to be sinusoidal in time when bandlimited near the transducer's center frequency. This kind of SAW transducer is called interdigitated transducer (IDT), since it is based on interdigitated electrodes. An electrical voltage (V_o) applied between two IDT electrodes produces the electrical field lines in the piezoelectric substrate, as indicated in Fig. 4.37. Thus, the electrical fields are large *between* the IDT fingers and near the surface of the piezoelectric substrate, while the electrical fields just underneath the IDT fingers are small. The high and low electrical

	Surface Acoustic Wave	Electromagnetic Wave
Wave velocity	3×10^3 m/s	3×10^8 m/s
Wavelength	300 μm for 10 MHz	30 m for 10 MHz
	3 μm for 1 GHz	0.3 m for 1 GHz
Length for 0.1 μs delay	0.3 mm	30 m

Table 4.3 Surface Acoustic Wave versus Electromagnetic Wave

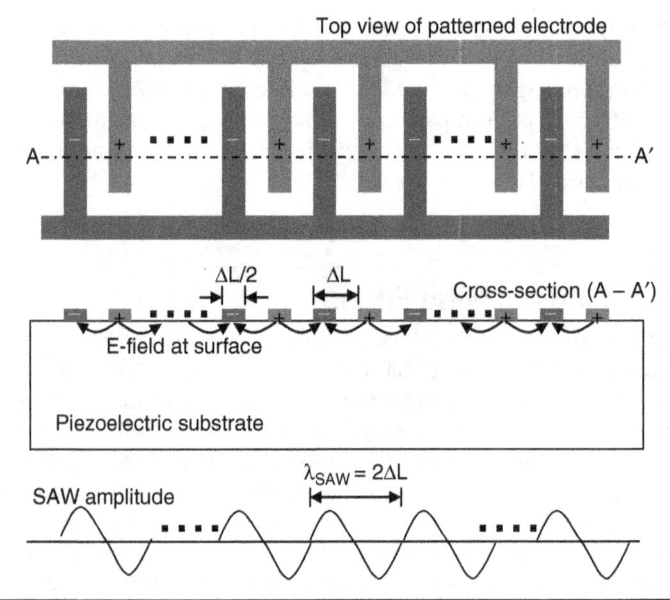

FIGURE 4.37 Top view and cross-sectional (or side) views of a SAW transducer based on interdigitated electrode pattern sitting on top of a piezoelectric substrate. The top view shows the patterned electrode, while the side views show the patterned electrode along with the voltage signs, electric field lines, and the amplitude of SAW generated on the surface of the piezoelectric substrate.

fields near the substrate surface produce SAW with its amplitude varying as shown just below the side view in Fig. 4.37. And we see that the wavelength (λ) of the SAW is equal to $2\Delta L$, where ΔL = the IDT finger width + the IDT finger spacing. Since the finger width is usually the same as the spacing, the SAW wavelength λ is equal to four times the IDT finger width or spacing.

Consequently, to generate 2 GHz SAW (having a wavelength $\lambda \approx 3{,}000$ m/s ÷ 2 GHz = 1.5 μm) for a duplexer, the spacing between the IDT fingers (and the IDT finger width) must be 1.5/4 = 0.375 μm. This means not only submicron lithography for the electrode but also a very large electrical field between the IDT fingers near the surface of the piezoelectric substrate. To appreciate how large the electrical field can be for RF transmit filters in a cell phone, we calculate the voltage V_o to be about 10 V ($= \sqrt{1\text{W} \cdot 2 \cdot 50\Omega}$, since power = $V_o^2/2$ R) when the RF power is 1 W. Thus, the largest electrical field near the surface of a piezoelectric substrate is about 27 MV/m (= 10 V/0.375 μm), which exceeds the reported coercive field of lithium niobate [30], a common piezoelectric substrate used for SAW filters. Another common piezoelectric material used for SAW filters is ZnO thin film, which is known to have a breakdown field of 10 MV/m. Another piezoelectric material, quartz, has a much higher breakdown field of 1,000 MV/m but has a very low electromechanical coupling coefficient, which makes very hard for a quarz-resonator-based filter to meet the bandwidth requirement.

Question: For a SAW generator described in Fig. 4.37, where will the piezoelectric material fail when an applied RF power is large? Indicate the locations on both the top and side views.

Answer:

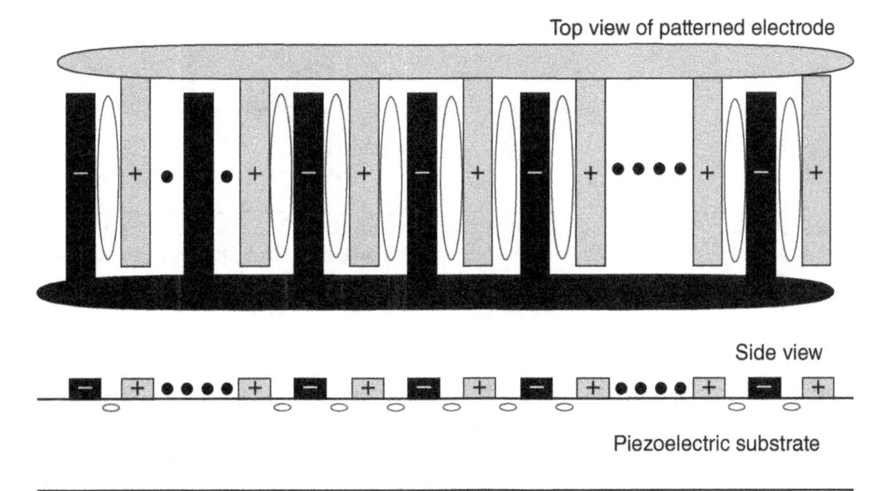

4.4.2 SAW Filter Components

A SAW launched with one set of IDT fingers can be picked up by another set of nearby IDT fingers, and the time it takes for the wave to travel from one IDT finger to another is a delay time that can be used in a transversal filter shown in Fig. 4.38. The filter topology shown in Fig. 4.38 can easily be implemented with an IDT shown in Fig. 4.39 for the output IDT that picks up the SAW generated by the input IDT. The delays are obtained as the wave propagates from the input IDT to the output IDT, while the weights are obtained through varying the overlapping regions between two adjacent electrodes in the output IDT (Fig. 4.39). Thus, a filter can easily be made by depositing and patterning an electrode over a piezoelectric substrate, and the number of the fabrication steps is relatively small.

A SAW-based filter needs at least two IDTs on a piezoelectric substrate: one acting as a SAW generator and the other acting as a receiver of the SAW. Consequently, the SAW typically travels a significant amount of distance on a surface of the substrate, losing its energy as it propagates. Thus, the Q of SAW filter is inherently low (about 200 at

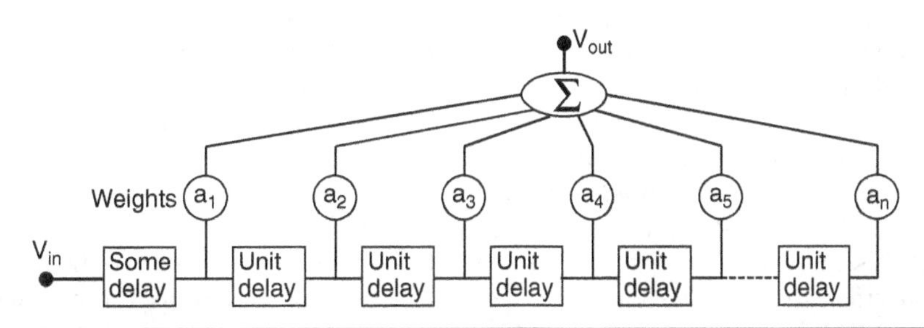

Figure 4.38 Schematic of a "digital" filter formed by summing weighted signals after delays. The weights in SAW filter can be obtained by varying lengths of the IDT electrodes for different overlapping distances between two adjacent electrodes, as done for the output IDT in Fig. 4.39.

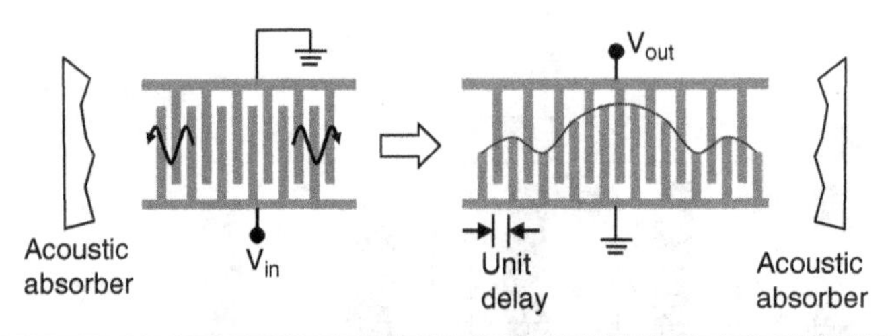

FIGURE 4.39 Top view of a SAW filter consisting of two IDTs: one to generate SAW and the other to pick up the SAW with built-in weighting that along with other parameters determines the filter characteristics.

2 GHz), unless the surface is ensured to have very low defect and not to scatter waves, but SAW transducers built on lithium niobate ($LiNbO_3$) have very large electromechanical coupling coefficient ($k_t^2 \approx 14\%$). Thus, the insertion loss of a SAW-based filter can be made small. However, the sharpness of the roll-off at the filter edge still suffers due to the low Q. Since high Q is needed to meet the roll-off requirements of the 20-MHz guard band for PCS duplexer, the SAW filter faces uphill battle against duplexers based on FBARs or SMRs which have Q of about 1,000 at 2 GHz.

4.5 Tunable Capacitors

Tunable capacitors can be used for impedance matching for wide-bandwidth amplifier, voltage-controlled oscillators (VCOs), tunable filters, tunable oscillators, etc. Conventional tunable capacitors are based on varactor diodes, which have limited tuning range with relatively high insertion loss and poor electrical isolation, nonlinearity, and substantial power consumption. On the other hand, MEMS-based tunable capacitors have been reported to have a very large tuning range and low insertion loss. For example, the tunable capacitor based on a silicon comb drive is reported to have 300% tunability with base capacitance variation of 1–5 pF by < 15 V drive voltage and also a high quality factor [31]. Since silicon is semiconducting, a tunable capacitor made on top of Si can lose RF energy into the substrate and may have high insertion loss. Thus, micromachining can be used to remove the silicon under a tunable capacitor, or a glass substrate can be used.

Since its debut in 1991, MEMS tunable capacitor has been thought as a potential alternative for varactor because of its large tuning ratio and high quality factor. Basic MEMS capacitor is composed of two parallel plates (one of which is movable) with air in between, and the air gap is varied (through moving the movable plate) to change the capacitance. Electrostatic actuation has most actively been used to move the movable (or deflectable) plate. Advantage for electrostatically actuated tunable filter is that it can be made out of commonly used semiconductor materials. However, electrostatic actuation is only unidirectional (not bidirectional) due to extreme difficulty of obtaining a repulsive force and is nonlinear. Also, the controllable range of the air gap is only one-third of the initial gap due to the pull-in effect explained in Chap. 3. Thermal actuation also is unidirectional (as heat produces thermal expansion not contraction), though it provides strong actuation force. Unlike electrostatic actuation, thermal actuation consumes high power for the needed heat and is slow as heat transfer is inherently slow.

Piezoelectric actuation does not have all the above disadvantages of electrostatic and thermal actuations. However, it requires piezoelectric thin film such as AlN, ZnO,

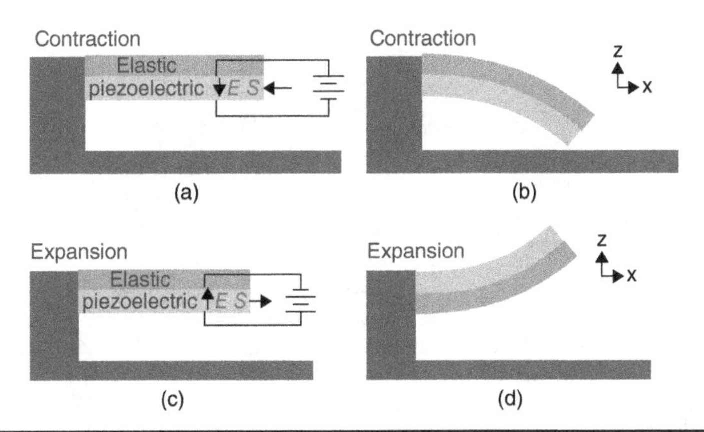

FIGURE 4.40 Piezoelectric unimorph cantilever: (a) and (b) when the applied voltage produces in-plane contraction on the piezoelectric film, which makes the cantilever bend downward; (c) and (d) when the applied voltage is reversed and generates in-plane expansion on the piezoelectric film, which in turn produces upward motion of the cantilever.

PZT, etc. For actuation, PZT would be best with its large piezoelectric coefficients, but its sputter deposition requires tight process control for good quality, while sol-gel PZT has large tensile stress. Though d_{31} of ZnO is 5 pC/N, about 40 times lower than that of PZT (but about twice that of AlN), piezoelectric actuation with ZnO is large enough for tunable capacitor application. With a piezoelectric layer and a nonpiezoelectric layer(s), a piezoelectric unimorph cantilever such as the one shown in Fig. 4.40 can be formed. When an electric field E_z is applied across the thickness of the piezoelectric layer (i.e., in the z direction), a strain developed in the x and y directions is $S_x = S_y = d_{31}E_z$, which can be positive or negative, depending on the field direction. If it is negative, the piezoelectric film contracts in in-plane directions and makes the cantilever bend downward, as shown in Fig. 4.40b (resulting in increased air-gap capacitance). If the in-plane strain is positive, the cantilever bends upward (Fig. 4.40d), and the air-gap capacitance decreases. This kind of bidirectional actuation is not possible with electrostatic or thermal actuation.

4.5.1 Bulk-Micromachined Silicon-Supported Tunable Capacitor with Mass Structure

A silicon mass structure formed with bulk micromachining has been used to form an air-gap capacitor over a glass substrate, as shown in Fig. 4.41. The mass structure is suspended and actuated by a piezoelectric beam, and follows the movement of the piezoelectrically actuated beam. The thickness of the silicon beam in the piezoelectric actuator is critical for optimum actuation and structure sturdiness, and is chosen to be 25 μm for robustness and fabrication easiness, which leads to 3-μm-thick ZnO for constant voltage optimization [32]. The lateral dimension of the capacitor is 3×0.3 mm^2, and the initial air gap is 1.5–2 μm.

The deflection of the actuator beam is bidirectional and highly linear, with a downward displacement observed to be around 0.086 μm/V until the 2-μm-wide initial gap is closed, and with an equally large upward response from which a 3-μm upward

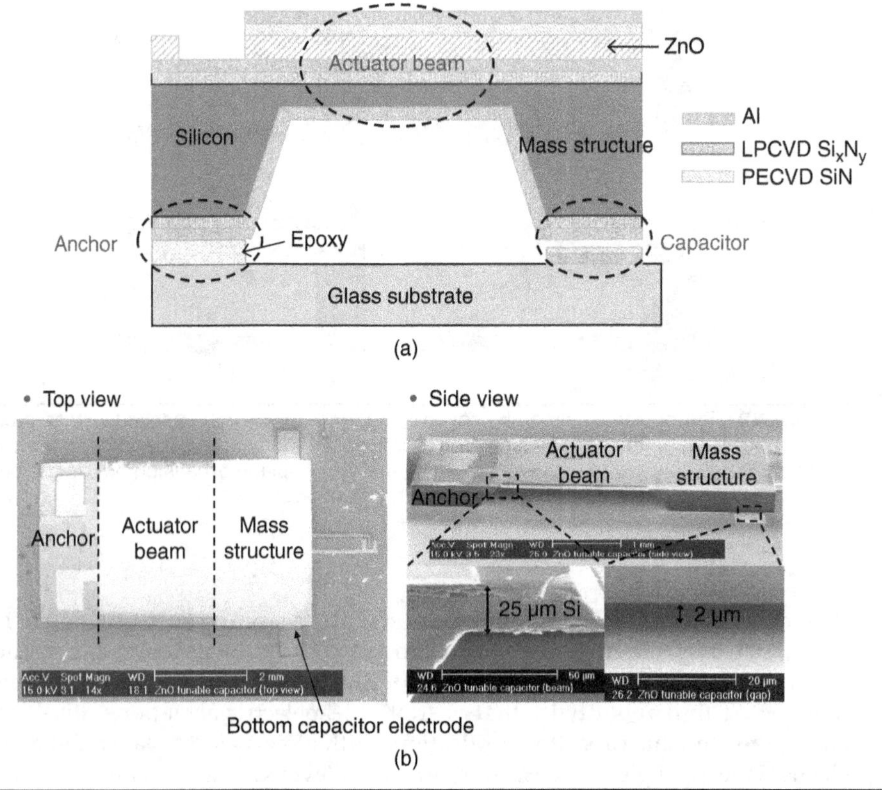

FIGURE 4.41 Bulk-micromachined silicon-supported tunable capacitor with mass structure: (a) cross-sectional schematic and (b) SEM photo. (Adapted from [32].)

displacement can be obtained at –35 V (Fig. 4.42a). Theoretically, one can expect a complete gap closure to get a very large capacitance but not experimentally because of unavoidable particles in an open test environment as well as surface roughness. With the parasitic capacitance (0.35 pF) de-embedded, the measured capacitance is C_0 = 1.23 pF, C_{max} = 10.02 pF at +25 V, and C_{min} = 0.46 pF at –35 V, resulting in a tuning ratio of 22:1, as shown in Fig. 4.42b.

4.5.2 Bridge-Type Surface-Micromachined Tunable Capacitor

In order to reduce the device size, one would want to use a short cantilever that can be made with surface micromachining, similar to the one shown in Fig. 4.40. However, the air-gap distance between a bending cantilever and the substrate is not constant along the length direction. The largest air-gap capacitance is obtained when the cantilever's free end touches the substrate, at which point, though, the other portion of the cantilever still has an air gap. Consequently, the capacitance tuning ratio is limited by the inherent incomplete closure of the gap.

On the other hand, a constant gap distance variation and complete gap closure can be attained with a bridge structure. However, bridge-based MEMS structure usually suffers from the tension that gets developed as the bridge is deflected, not to mention

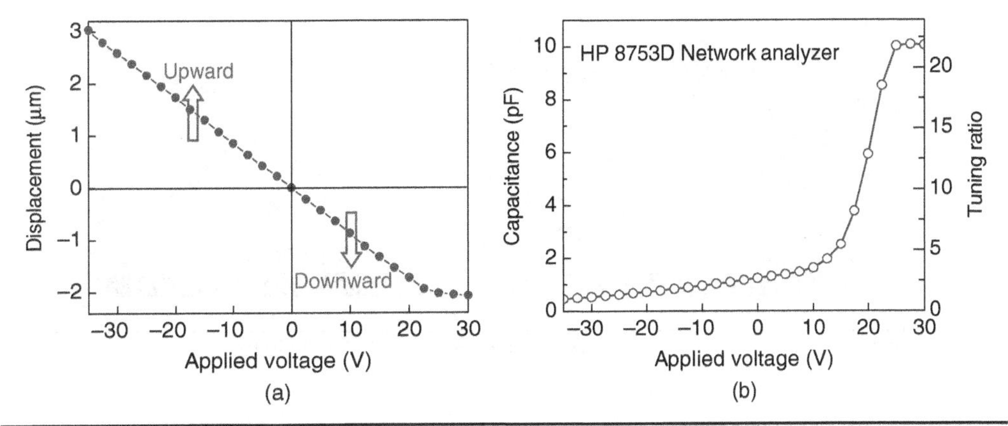

Figure 4.42 (a) Measured displacement of the silicon-supported tunable capacitor with mass structure. (b) Measured capacitance tuning at 2 GHz for the silicon-supported tunable capacitor. (Adapted from [32].)

the effect of the residual stress on the deflection. As a result, the tension hinders the deflection and relatively high voltage is required for the bridge deflection. To minimize the tension developed in the bridge, a bridge structure can be simply supported (rather than clamped) at the anchors. These points are summarized in Fig. 4.43a.

A released bridge (simply supported by two cantilevers) shown in Fig. 4.43b is free to move without any residual stress. The SU-8 confinement posts and confinement bridge are used to enclose the simply supported bridge in space as shown in Fig. 4.43b and Fig. 4.44. By varying the uniform gap of the parallel-plate tunable capacitor, a large capacitance tuning without the cost of high voltage for large displacement can be obtained.

The simply supported bridge is completely released right after the fabrication, that is, it floats over the two piezoelectric cantilevers, confined by the confinement posts and bridge. Simple supports happen when the bridge is electrostatically pulled down to the

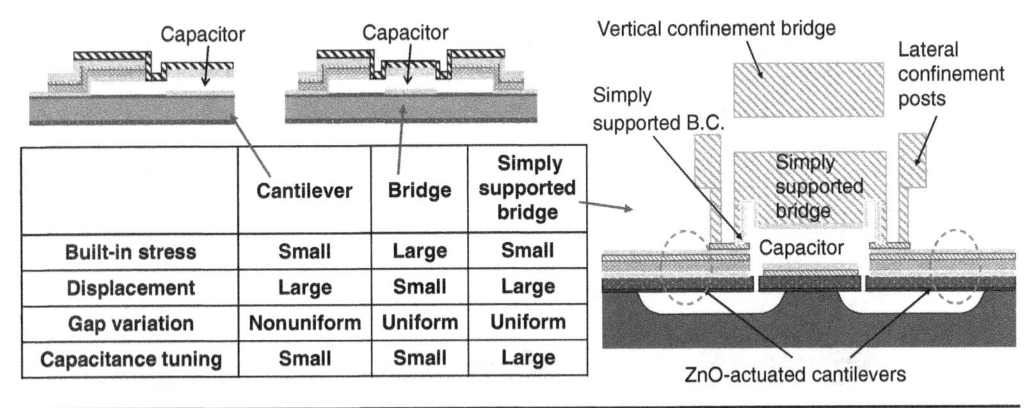

	Cantilever	Bridge	Simply supported bridge
Built-in stress	Small	Large	Small
Displacement	Large	Small	Large
Gap variation	Nonuniform	Uniform	Uniform
Capacitance tuning	Small	Small	Large

Figure 4.43 (a) Comparison of the three types of structures for tunable capacitor. (b) Schematic illustration of a piezoelectrically actuated tunable capacitor with simply supported bridge structure. (Adapted from [33].)

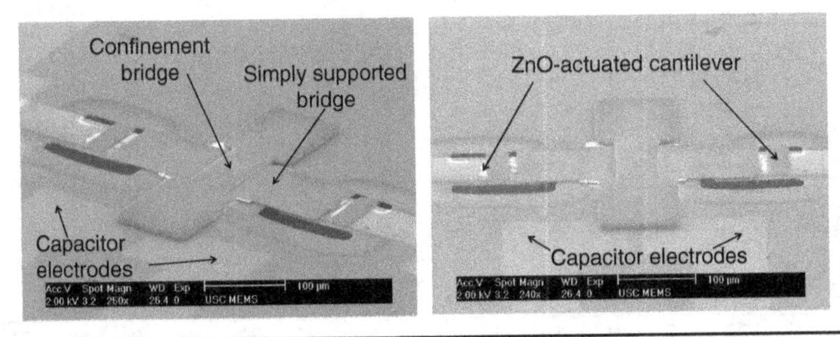

Figure 4.44 SEM photos of bridge-type tunable capacitor. (Adapted from [33].)

support cantilever. The bridge is flat, even if the two released support cantilevers are warped due to the symmetry of the two cantilevers. Since the air gap within parallel-plate tunable capacitor is uniform, a large capacitance tuning can be obtained. Also, the simply supported bridge without built-in stress allows a large change of the gap distance with a relatively low voltage.

As the measured data in Fig. 4.45 show, the piezoelectric actuation is bidirectional and highly linear at the tip of the actuating cantilever (with and without the bridge) and at the center of the bridge. At the center of the simply supported bridge, the displacement is 50% of that at the actuating cantilever and is continuously downward for ~ 0.35 μm until the gap is closed.

The measured capacitances (Fig. 4.46) at 3 GHz show that $C_0 = 0.23$ pF; $C_{max} = 1.82$ pF (at +40 V); $C_{min} = 0.13$ pF (at −30 V); and tuning ratio = 14:1, with the parasitic capacitance (0.2 pF) de-embedded. The theoretical plot is made with 0.008 μm/V gap variation. The measured quality factor of the capacitor (0.42 pF) at 2GHz is 20.

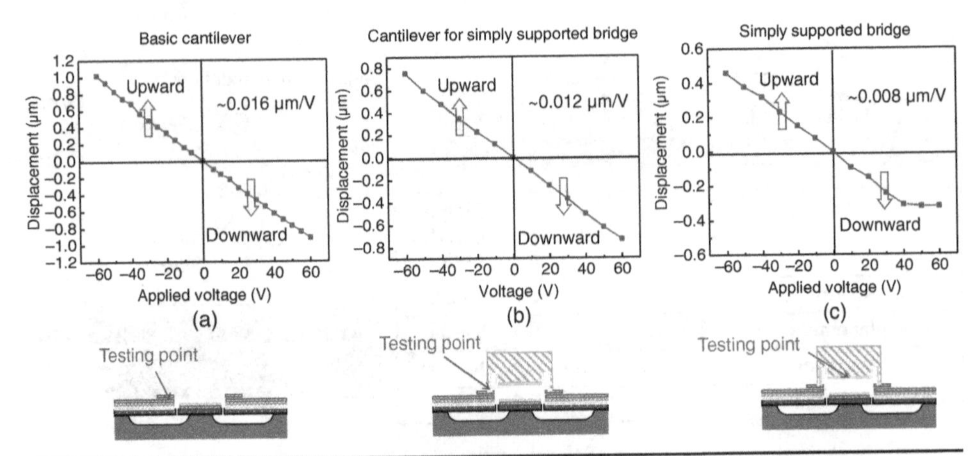

Figure 4.45 Measured displacements of the bridge-type tunable capacitor: piezoelectric displacement versus actuation voltage at the actuating cantilever without (a) and with (b) the bridge and at the center of the simply supported bridge (c). (Adapted from [33].)

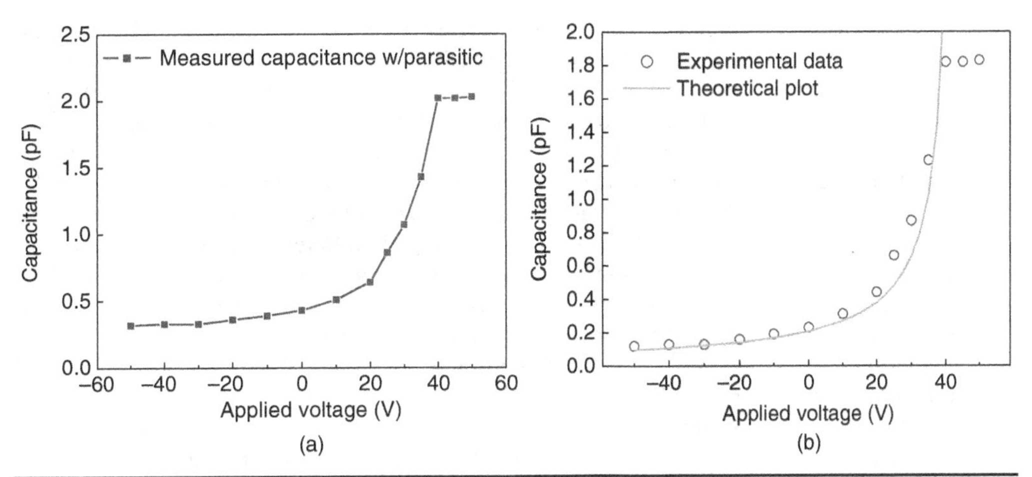

Figure 4.46 Measured capacitances at 3 GHz versus applied voltage for the bridge-type tunable capacitor (a) as measured with parasitic capacitance and (b) after the parasitic capacitance (0.2 pF) is de-embedded. (Adapted from [33].)

4.6 RF MEMS Switches

Since mid-90s, MEMS switches have been explored for DC to RF applications including measurement instruments, phased array antenna, cell phones, etc. The main advantage that MEMS switches offer over transistor-based switches is lower insertion loss and higher power isolation ratio. MEMS switches have been considered for signal routing, phase shifter for phased-array antenna, broadband amplifier that is formed by many amplifiers connected by switches, adjustable gain microwave amplifier, active impedance matching, etc. However, one major concern with MEMS switches has been long-term reliability as most of them are based on electrostatic actuation and operate with physical contact between two pieces of materials. The physical contacts happen millions to billions times during the lifetime of a switch, and will cause wear and tear as well as other electrically induced effects such as dielectric charging.

One notable example of an electrostatically actuated MEMS switch (or relay) for RF measurement systems is illustrated in Fig. 4.47 along with coplanar waveguide (CPW).

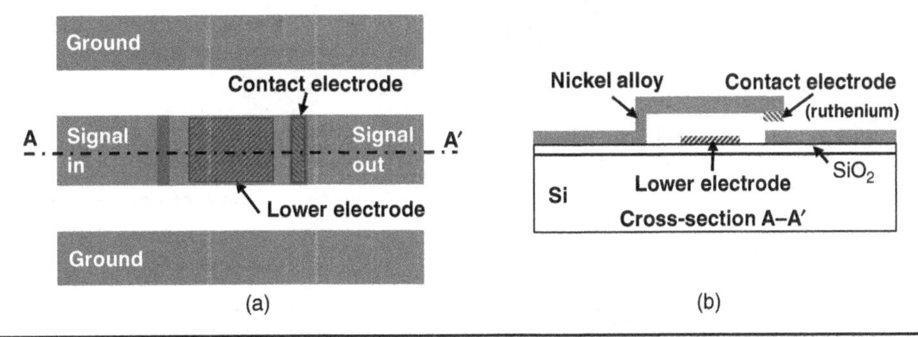

Figure 4.47 MEMS relay switch for DC to RF signals: (a) top view with coplanar waveguide (CPW) and (b) cross-sectional view across A–A′ clearly showing the contact and lower electrodes.

Here an electrical contact is made between ruthenium and nickel alloy [34] when the nickel alloy cantilever is electrostatically pulled down to the silicon substrate by the voltage applied between the nickel alloy and the lower electrode (Fig. 4.47b). Since the contact is between two conductors, the switch is good for DC switching as well as RF, as long as the metals are highly conductive and the contact resistance is sufficiently low. However, the alloy needs to be mechanically sturdier by a factor of 10 than a single-element metal so that the cantilever may go through many cycles of bending. Metal-to-metal contact (which is needed for DC switching) is inherently prone to mechanical failure, worse than metal-to-dielectric contact. That is why many macroscale relays had used mercury (i.e., liquid metal), until mercury was found to be hazardous to human health and phase-out or phase-down of mercury from products was initiated. Alternative liquid metal such as gallium (with melting temperature of 30°C) has been explored for MEMS relay switch [35].

Without a metal-to-metal contact, a switch can still be made for RF applications as a large capacitance difference between *on* and *off* states can provide a large difference on the amount of signal passing through the switch. A notable example of such a capacitive MEMS switch is shown in Fig. 4.48 along with CPW. Here the switch is composed of a perforated metal bridge which is suspended over a lower electrode covered with an insulating layer. An RF signal passes through the switch with very little loss when the perforated metal bridge is not deflected (with no DC voltage applied between the lower electrode and the bridge), since the capacitance between the lower electrode and the bridge (which is electrically and mechanically connected to the ground electrodes) is very small due to the large air gap [36, 37]. However, when the bridge is electrostatically pulled down with a DC voltage applied between the metal bridge and the lower electrode, it makes contact with the insulating dielectric layer over the lower electrode, and

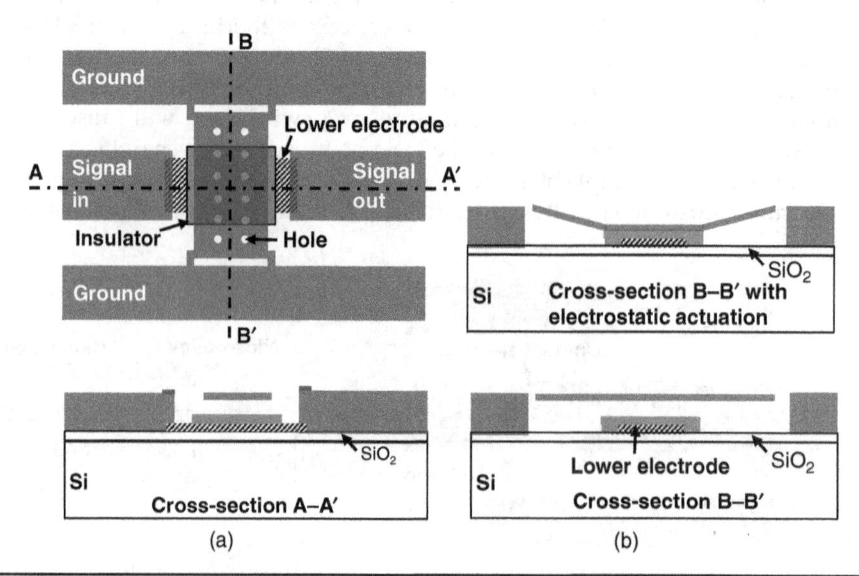

FIGURE 4.48 Capacitive RF MEMS switch: (a) top view and cross-sectional view across A–A′ and (b) cross-sectional view across B–B′ with (top) and without (bottom) any electrostatic voltage between the deflectable metal bridge and the lower electrode.

thus, the capacitance between the two metals can be more than 100 times larger than when the bridge is under zero applied voltage. This large capacitance has a similar effect of shorting the input signal to the ground, and very little of the input signal passes through the switch. The larger the capacitance ratio between on and off states, the larger the difference will be for the signal passing through the switch. Consequently, a large contact area is desirable. However, a larger contact area has more areas to fail as the physical contacts go on many times. Also, after many cycles of physical contacts with pulses of applied voltage, the dielectric insulating layer between two metals may be charged up with mobile charges, and electrostatic actuation may stop. This dielectric charging problem can be solved with bipolar pulses but indicates difficulty of making the switch reliable for many cycles.

After decades of research and development, long-term reliability of a capacitive RF MEMS switch is no longer considered to be a major issue. However, high power handling and hot-switching capabilities remain to be elusive.

Wireless systems handle many frequency bands with many filters, which are selected with switches, and the number of antennas inside a smartphone is growing to accommodate ever increasing frequency bands, especially for 5G. Currently, semiconductor switches are used to configure antennas and capacitors (or inductors) for optimum performance of antennas for their allocated frequency bands. MEMS switches offering lower insertion loss and higher isolation ratio may be an option to replace the semiconductor switches.

References

1. S. Sethi, "FBAR gets excellent reception," The Linley Group, Mobile Chip Report, July 27, 2015.
2. "SiTime's MEMS First™ Process," SiT-AN20001 Review, 1.7, February 17, 2009.
3. L. Lin, C.T.C. Nguyen, R. T. Howe, and A. P. Pisano, "Micro electromechanical filters for signal processing," Technical Digest, IEEE Micro Electromechanical Systems Workshop, Travemunde, Germany, February 4–7, 1992, pp. 226–231.
4. K. Wang, F.D. Bannon, J.R. Clark, and C.T.C. Nguyen, "Q-enhancement of micromechanical filters via low-velocity spring coupling," IEEE Ultrasonics Symposium, Toronto, Canada, October 5–8, 1998, pp. 323–327.
5. K. Wang, A.C. Wong, and C.T.C. Nguyen, "VHF free-free beam high-Q micromechanical resonators," Journal of Microelectromechanical Systems, 9(3), September 2000, pp. 347–360.
6. S.S. Li, Y.W. Lin, Y. Xie, Z. Ren, and C.T.C. Nguyen, "Micromechanical "hollow-disk" ring resonators," 17th IEEE International Conference on Micro Electro Mechanical Systems, Maastricht MEMS 2004 Technical Digest, Maastricht, Netherlands, 2004, pp. 821–824.
7. J. Wang, Z. Ren, and C.T.C. Nguyen, "Self-aligned 1.14-GHz vibrating radial-mode disk resonators," Transducers '03, 12th International Conference on Solid-State Sensors, Actuators and Microsystems, Boston, MA, 2003, pp. 947–950.
8. J. Wang, J.E. Butler, T. Feygelson, and C.T.C. Nguyen, "1.51-GHz nanocrystalline diamond micromechanical disk resonator with material-mismatched isolating support," 17th IEEE International Conference on Micro Electro Mechanical Systems, Maastricht MEMS 2004 Technical Digest, Maastricht, Netherlands, 2004, pp. 641–644.

9. W.T. Hsu, J.R. Clark, and C.T.C. Nguyen, "A sub-micron capacitive gap process for multiple-metal-electrode lateral micromechanical resonators," Technical Digest, MEMS 2001, 14th IEEE International Conference on Micro Electro Mechanical Systems, Interlaken, Switzerland, 2001, pp. 349–352.
10. M.U. Demirci, M.A. Abdelmoneum, and C.T.C. Nguyen, "Mechanically corner-coupled square microresonator array for reduced series motional resistance," Transducers '03, 12th International Conference on Solid-State Sensors, Actuators and Microsystems, Boston, MA, 2003, pp. 955–958.
11. J.F. Rosenbaum, Bulk Acoustic Wave Theory and Devices, Artech House, Norwood, MA, 1988.
12. W.P. Mason, Electronical Transducers and Wave Filters, Van Nostrand, New York, NY, 1948.
13. G.S. Kino, Acoustic Waves: Devices, Imaging & Analog Signal Processing, Prentice-Hall, 1987.
14. K.F. Graff, Wave Motion in Elastic Solids, Ohio State University Press, Ohio, OH, 1975.
15. K. Hashimoto, Editor, RF Bulk Acoustic Wave Filters for Communications, Artech House, Boston, MA, 2009.
16. C.M. Grünsteidl, I.A. Veres, and T.W. Murray, "Experimental and numerical study of the excitability of zero group velocity Lamb waves by laser-ultrasound," The Journal of the Acoustical Society of America, 138, 2015, 242; doi: 10.1121/1.4922701.
17. P. Bradley, et al. "A film bulk acoustic resonator (FBAR) duplexer for USPCS handset applications," IEEE MTT-S Digest, 2001, 367.
18. D. Feld, K. Wang, P. Bradley, A. Barfknecht, B. Ly, and R. Ruby, "A high performance 3.0 mm × 3.0 mm × 1.1 mm FBAR full band Tx filter for U.S. PCS handsets," IEEE Ultrasonics Symposium, Munich, Germany, 2002, pp. 913–918 vol.1.
19. US Patent 6,262,637, July 17, 2001.
20. K. M. Lakin, G.R. Kline, and K.T. McCarron, "Development of miniature filters for wireless applications," IEEE Transactions on Microwave Theory and Techniques, 43(12), 1995, pp. 2933–2939.
21. K. M. Lakin and J. S. Wang, "Acoustic bulk wave composite resonators," Applied Physics Letters, 38(125), 1981.
22. W. Pang, H. Zhang, and E.S. Kim, "Analytical and experimental study on second harmonic response of FBAR for oscillator application," IEEE International Ultrasonics Symposium, Rotterdam, The Netherlands, September 18–21, 2005, pp. 2136–2139.
23. R. Aigner, et al., "RF-filters in mobile phone applications," Transducers '03, 12th International Conference on Solid-State Sensors, Actuators and Microsystems, Boston, MA, 2003, pp. 891–894.
24. K.M. Lakin, J. Belsick, J.F. McDonald, and K.T. McCarron, "Improved bulk wave resonator coupling coefficient for wide bandwidth filters," IEEE Ultrasonics Symposium, Atlanta, GA, 2001, pp. 827–831.
25. J.D. Larson and Y. Oshmyansky, "Measurement of effective kt^2, Q, R_p, R_s vs. temperature for Mo/AlN FBAR resonators," IEEE Ultrasonics Symposium, Munich, Germany, 2002, pp. 939–943.
26. Ken-Ya Hashimoto, RF Bulk Acoustic Wave Filters for Communications, Artech House, Boston, MA, 2009, pp. 35–36.

27. J.D. Larson, J. D. Ruby, R. C. Bradley, J. Wen, S.L. Kok, and A. Chien, "Power handling and temperature coefficient studies in FBAR duplexers for the 1900 MHz PCS band," IEEE Ultrasonics Symposium, San Juan, Puerto Rico, 1, 2000, pp. 869–874.
28. S.V. Krishnaswamy, J. Rosenbaum, S. Horwitz, C. Vale, and R.A. Moore, "Film bulk acoustic wave resonator technology," IEEE Ultrasonics Symposium, Honolulu, HI, 1990, pp. 529–536.
29. T. Nishihara, T. Yokoyama, T. Miyashita, and Y. Satoh, "High performance and miniature thin film bulk acoustic wave filters for 5 GHz," IEEE Ultrasonics Symposium, Munich, Germany, 1, 2002, pp. 969–972.
30. S. Kim and V. Gopalan, "Coercive fields in ferroelectrics: A case study in lithium niobate and lithium tantalite," Applied Physics Letters, 80, 2002, pp. 2740–2742.
31. J.J. Yao, et al., "High tuning-ratio MEMS-based tunable capacitors for RF communications applications," Solid-State Sensor and Actuator Workshop, Hilton Head Island, SC, 1998, pp. 124–127.
32. C.Y. Lee and E.S. Kim, "Piezoelectrically actuated tunable capacitor," IEEE/ASME Journal of Microelectromechanical Systems, 15(4), 2006, pp. 745–755.
33. C.Y. Lee, et al, "Surface micromachined, complementary-metal-oxide-semiconductor compatible tunable capacitor with 14:1 continuous tuning range," Applied Physics Letters, 92(044103), 2008.
34. T. Moran, C. Keimel, and T. Miller, "Advances in MEMS switches for RF test applications," Proceedings of the 11th European Microwave Integrated Circuits Conference, London, UK, October 2016, pp. 440–443.
35. Q. Liu and N.C. Tien, "Design and modeling of liquid gallium contact RF MEMS switch," Proceedings of the ASME Second International Conference on Micro/Nanoscale Heat and Mass Transfer, Volume 3, Shanghai, China, December 18–21, 2009, pp. 191–198.
36. Z.J. Yao, S. Chen, S. Eshelman, D. Denniston, and C. Goldsmith, "Micromachined low-loss microwave switches," Journal of Microelectromechanical Systems, 8(2), 1999, pp. 129–134.
37. S. Shekhar, K.J. Vinoy, and G.K. Ananthasuresh, "Design, fabrication and characterization of capacitive RF MEMS switches with low pull-in voltage," 2014 IEEE International Microwave and RF Conference (IMaRC), pp. 182–185.

Questions

Question 4.1 For a film bulk acoustic wave resonator (FBAR) at 1 GHz, answer the questions listed in the leftmost column of the table below for ZnO and AlN films by circling your answers in the table.

	ZnO film	AlN film
Electromechanical coupling (k_t^2)?	8.5% or 6.5%	8.5% or 6.5%
Film thickness?	2.5 μm or 5 μm	2.5 μm or 5 μm
FBAR without any support layer?	Yes or No	Yes or No
Typical quality factor?	1,000 or 500	1,000 or 500

Question 4.2 Currently, 4G cell phones use RF front-end duplexers based on air-backed or solid-mounted film bulk acoustic wave resonator (FBAR) or surface acoustic wave (SAW) resonator. Rank the three types 1 through 3 (with 1 being the best) for the issues listed on the first column.

	Solid-mounted FBAR	**Air-backed FBAR**	**SAW**
Quality factor			
Overall filter performance			
Power handling capability			
Manufacturing easiness			

Question 4.3 (a) Sketch, on the figure below, how the impedance characteristics of a film bulk acoustic resonator (FBAR) shown below will change if FBAR's quality factor (Q) is increased, while all other parameters remain the same. (b) Sketch, on the figure below, how the impedance characteristics of an FBAR shown below will change if FBAR's electromechanical coupling coefficient (K_t^2) is increased, while all other parameters remain the same.

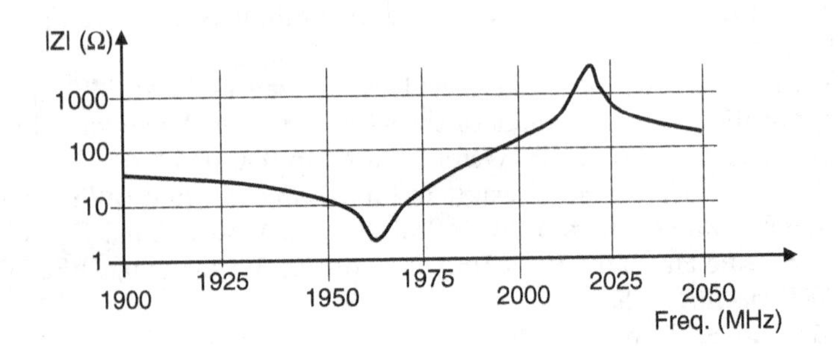

Question 4.4 Approximately estimate $|V_{out}|/|V_{in}|$ at 1.964 and 2.020 GHz for the filter made of one series FBAR and two shunt FBARs below, assuming that the series FBAR and shunt FBARs have impedance characteristics as shown below.

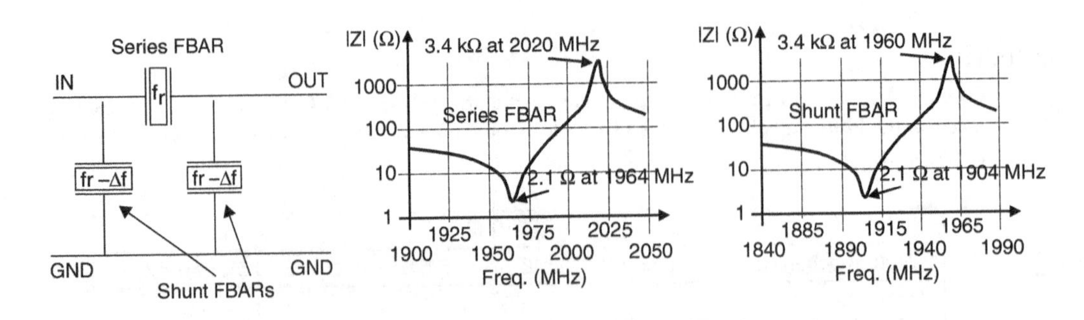

Question 4.5 For an RF front-end filter at 1–2 GHz, which of the two FBARs shown above would you choose if your *only* concern is the issue listed in the leftmost column of the following table? Please indicate your choice by putting a check mark in the table with a short explanation.

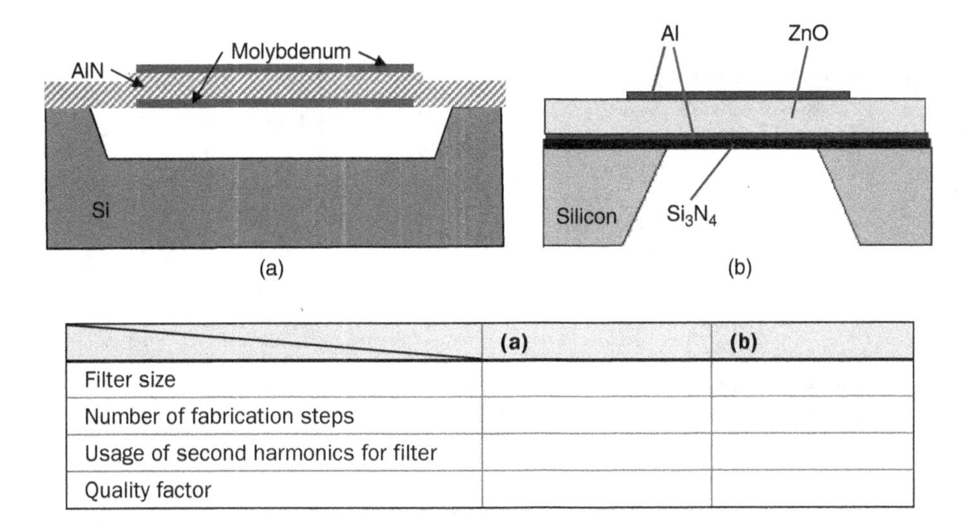

	(a)	(b)
Filter size		
Number of fabrication steps		
Usage of second harmonics for filter		
Quality factor		

Question 4.6 Electrical impedance of a film bulk acoustic resonator (FBAR) near its resonance was measured and plotted as below. Each of the figures (a)–(c) must be any of the real part, imaginary part, and phase of the impedance. First, indicate (at the left of each of the three figures) which part of the impedance it is showing (e.g., the real part, imaginary part, or phase?). Then, sketch the magnitude of the impedance as a function of the frequency over 1,000–2,500 MHz.

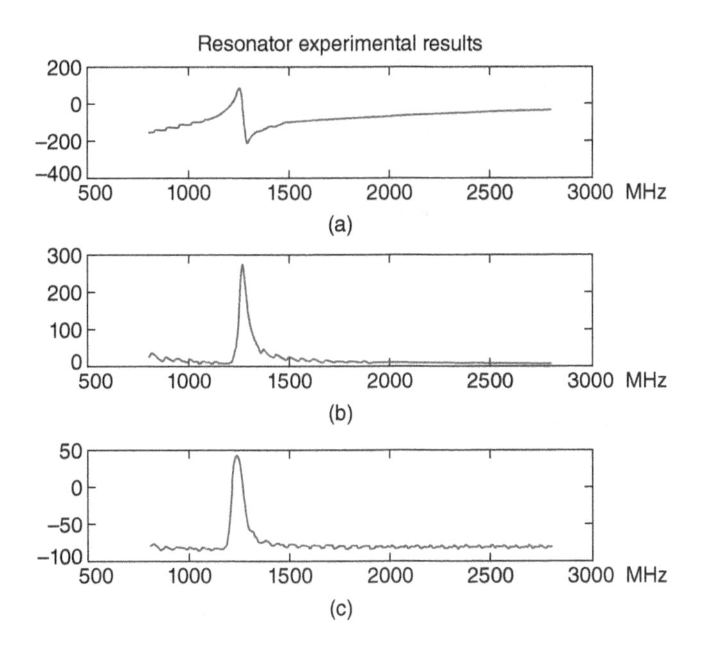

Question 4.7 For an FBAR made with 2.5-μm-thick AlN as shown below, indicate the resonant frequencies and spacing between the frequencies on the figure at the right. Assume that the acoustic velocity in AlN is 10,000 m/s and the top-view of the patterned molybdenum is a square having $100 \times 100\ \mu m^2$ area.

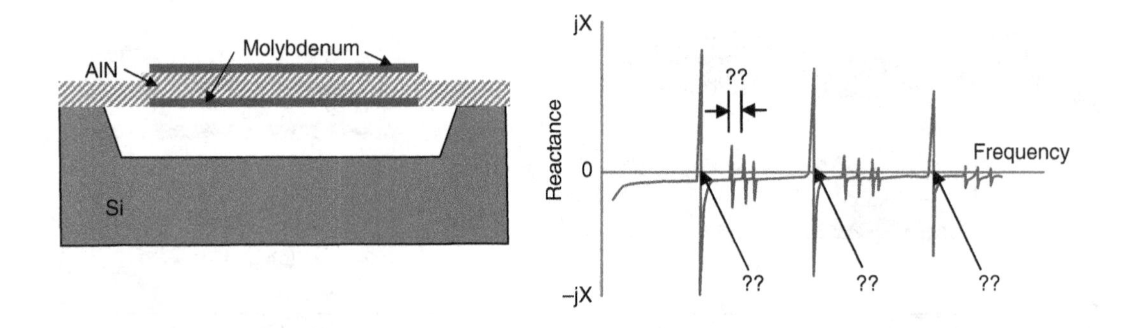

Question 4.8 For an extensional mode resonator with 400-μm-thick quartz as shown below, what are the resonant frequencies? Assume that the longitudinal–acoustic wave velocity in quartz is 6,000 m/s.

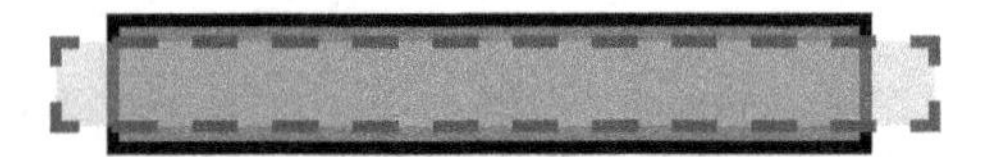

Question 4.9 For a disk resonator with submicron air gap (Fig. 4.4), (a) will there be any measurable electromechanical resonance, if V_p is equal to 0 V and (b) what will happen if V_p is increased higher than 300 V?

Question 4.10 How could the stem post be perfectly centered in the process (b) shown below? In other words, what is the key idea behind centering the stem in the self-aligned disk resonator described in Fig. 4.5? For the disk resonators (Fig. 4.4) shown below, draw the photomask patterns for the disk and the stem post for both (a) and (b).

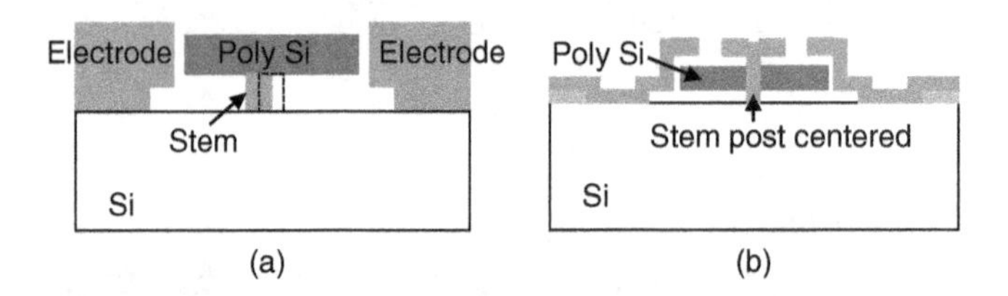

Question 4.11 For the disk resonator **(Fig. 4.4)** with 0.3-μm air gap, what is the *maximum* DC bias voltage V_p that can be applied in ambient pressure (i.e., 1 bar)? If the disk resonator has a pull-in voltage of 100 V, what is the maximum RF power that can pass through the resonator without pulling in the deformable disk to the electrode? Assume that the RF frequency is 50 MHz.

Question 4.12 One issue with electrostatically actuated resonators is that the equivalent resistance is much higher than 50 Ω. Thus, one would try to reduce the resistance by reducing the air gap, increasing the coupling area, and/or increasing the DC bias voltage. However, there are disadvantages associated with each of the three approaches. Indicate on the table below by check marks whether any of the disadvantages listed in the first column are applicable to each of the three approaches.

	Reduce the air gap	**Increase the coupling area**	**Increase the DC bias voltage**
Reduced power handling capability			
Increased resonator dimension			
Increased likelihood of air breakdown			
Reduced manufacturing yield			

Question 4.13 For each of the issues listed on the first column below, rank the three different RF front-end filter types by writing 1, 2, or 3 for the best, second best, and third best, respectively.

	Filter based on air-backed FBAR	**Filter based on solid-mounted resonator (SMR)**	**SAW filter built on lithium niobate ($LiNbO_3$)**
Power handling capability			
Filter insertion loss			
Fabrication easiness			

CHAPTER 5
Optical MEMS

There have been active research activities in optical microelectromechanical systems (MEMS) including MEMS for free space optics. However, in this chapter, we will first study micromirror arrays for projection display (as TI has been successfully commercializing such mirrors) and for fiber optic communication, followed by micro-opto-mechanical modulators. We will also study micromachined Fabry–Perot photonic crystal, filter, and interferometer built with Bragg reflectors.

5.1 Micromirror Array for Projection Display

Projection displays have been one notable application area of MEMS technology [1], though projection displays can be made with liquid crystal display (LCD) valves, laser, etc. In this section, we will study three MEMS-based approaches: a commercially successful digital light processing (DLP) from Texas Instruments Inc., grating light valve (GLV), and thin-film micromirror array (TMA).

5.1.1 Digital Light Processing

The DLP mirrors can be microfabricated with a process, briefly illustrated in Fig. 5.1, on a complementary metal–oxide semiconductor (CMOS) wafer with static random access memory (SRAM) after planarizing the surface with chemical mechanical polishing (CMP). A key idea is to use polymers for the spacers 1 and 2, and then to remove them with oxygen plasma to release the Al mirrors. The Al mirrors are doped with a few percent of silicon for mechanical sturdiness.

The cross-sectional views (along with the top view) of a DLP mirror after the spacer layers are removed are shown in Fig. 5.2. Each of such as Al mirrors can be positioned into any of two bistable positions by an electrostatic voltage, as illustrated in Fig. 5.2e and Fig. 5.2f. When the electrostatic voltage is removed, the mirror rotates to its flat state due to the spring action by the torsional hinges. The actuating mechanism is buried under the mirror so that the fill factor (the ratio of the reflective surface area over the total area) can be maximized. Light reflection is controlled by tilting the electrostatically actuated bistable mirror, as torsional hinges suspend a yoke that supports the mirror through a post. The yoke is in its flat state when there is no voltage applied, but is rotated by an electrostatic voltage into any of the two bistable positions that are different by 20° (i.e., ±10°). The yoke is designed to have "landing tips" to minimize the contact area when the yoke is pulled down to the substrate. Dicing a fabricated wafer into chips is challenging when the wafer contains released structures with air gaps, because the dicing saw usually requires water and also produces dusts during the dicing. Texas Instruments dealt with this issue by partially dicing a wafer

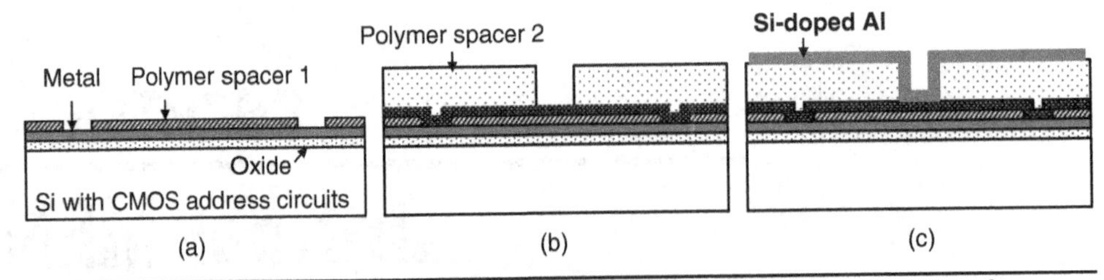

FIGURE 5.1 Simplified fabrication process of digital light processing (DLP) mirrors: (a) after depositing and patterning polymer spacer layer 1 on top of metal/oxide layers on top of planarized surface of CMOS wafer, (b) after depositing and patterning hinge and yoke layer, followed by deposition and patterning of polymer space layer 2, and (c) after deposition and patterning of silicon-doped Al, before removing polymer spacer layers 1 and 2 with oxygen plasma.

before releasing the mirrors and then completing the dicing through a "wafer break" with a dome-shaped piece of block, without using a dicing saw, after the release of the mirrors.

Optical Switching Principle of DLP

Static random access memory (SRAM) cell is used to store the information for the position of the yoke (either +10° or –10°), as shown in Fig. 5.3a. The SRAM cell has two complementary sides where the voltage levels are opposite to each other. That is, if one of them is a logic "high" (e.g., about 5 V), the other is a logic "low" (e.g., about 0 V).

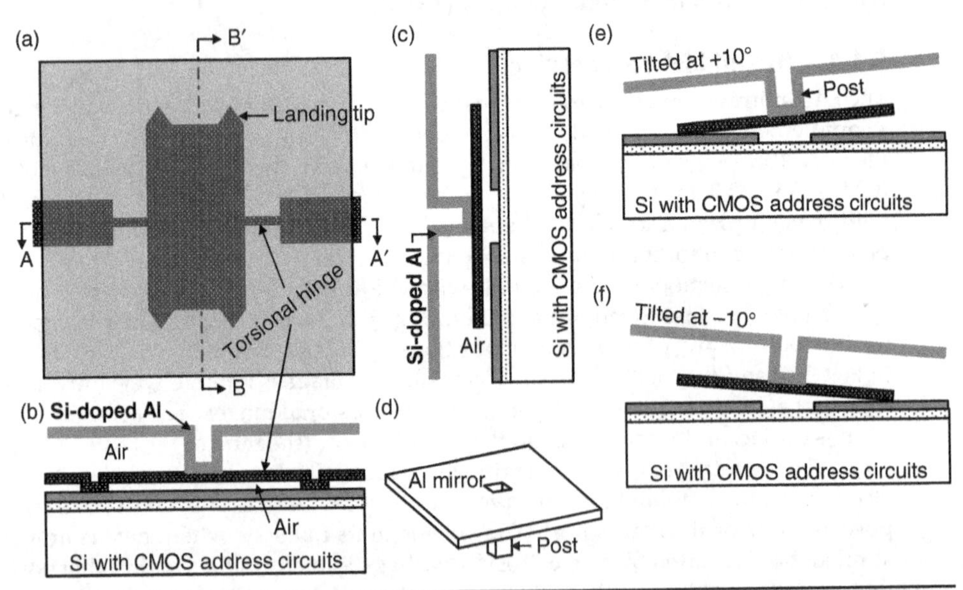

FIGURE 5.2 Simplified schematics of released DLP mirror: (a) top view, (b) cross-sectional view across A–A′, (c) cross-sectional view across B–B′, (d) perspective view of the release mirror, and (e) and (f) the Al mirror tilted at +10° and –10° by electrostatically pulling down the mirror-support platform to the left and right, respectively. The mirror-support platform is encapsulated with insulating layer so that there may be no electrical shortage between it and the electrode to which it makes physical contact.

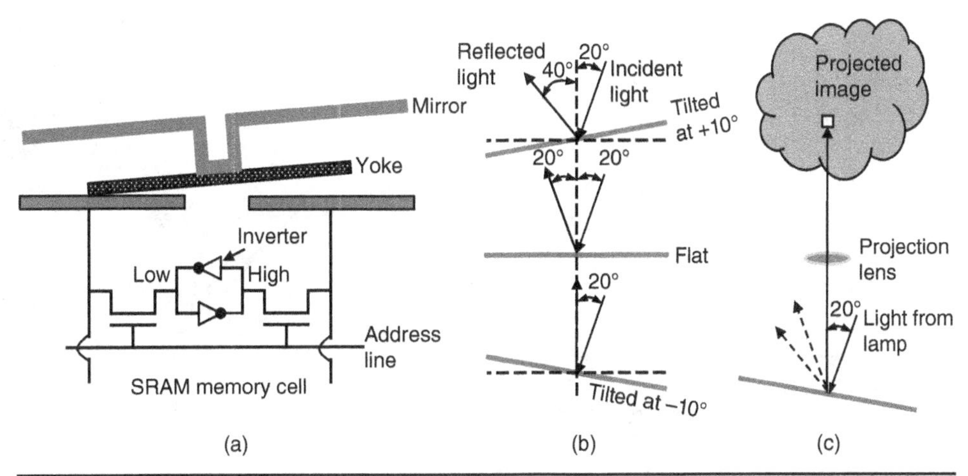

Figure 5.3 (a) DLP mirror and its drive electronics, (b) schematics showing how the mirror's tilting angle determines the direction to which a light with an incident angle of 20° is reflected, and (c) schematic showing how a pixel on a projected image is brightened by the reflected light when the mirror is tilted at –10°, while the mirror at flat or +10° reflects the light into a direction off from the projection lens.

The 0 V, for instance, is low enough to pull the yoke to that side, when (1) there is a bias voltage (e.g., 15 V) on the yoke via a bias-reset bus and (2) the yoke is positioned in its flat state. This way the yoke can be latched into either of the two complementary sides. And once the yoke is latched, a change of the voltage level in the SRAM cell does not affect the yoke position, since the bias voltage is sufficient to keep the yoke electrostatically "pulled in." The yoke comes out of the latched position only when the bias voltage on the yoke is removed. And when the bias voltage is removed, the yoke returns to its flat state and is now ready to be latched into any of the two bistable positions, according to the bit information in the SRAM, when the bias voltage is applied.

Light incident on a mirror is reflected by the mirror either into a projection lens or into an area outside the lens depending on the position of the mirror, as illustrated in Fig. 5.3. Thus, the amount of the light intensity (that is projected into a spot on a screen by the mirror) depends on the relative amount of time the mirror is positioned at –10° position per frame (usually about 30 ms). State of the SRAM cell (1,0) determines which mirror rotation angle is selected.

Address Sequence of DLP

Grayscale is digitally controlled with a word containing a certain number of bits. For example, if a 4-bit word is used for the grayscale control, the time intervals for each bit are divided as shown in Fig. 5.4a with the most significant bit (MSB) being allocated eight times longer duration than the least significant bit (LSB) in order to obtain 2^4 equally spaced gray levels (0, 1/15, 2/15, …., 15/15). And for each bit time, the following steps (1–5) are sequenced to position the mirror into a desired angle: (1) reset all the mirrors in the array by turning off the bias voltage to allow the mirrors to rotate to flat state, (2) turn the bias voltage on to enable mirrors to rotate to address states (±10°), (3) keep the bias voltage on to latch mirrors (which will not respond to new

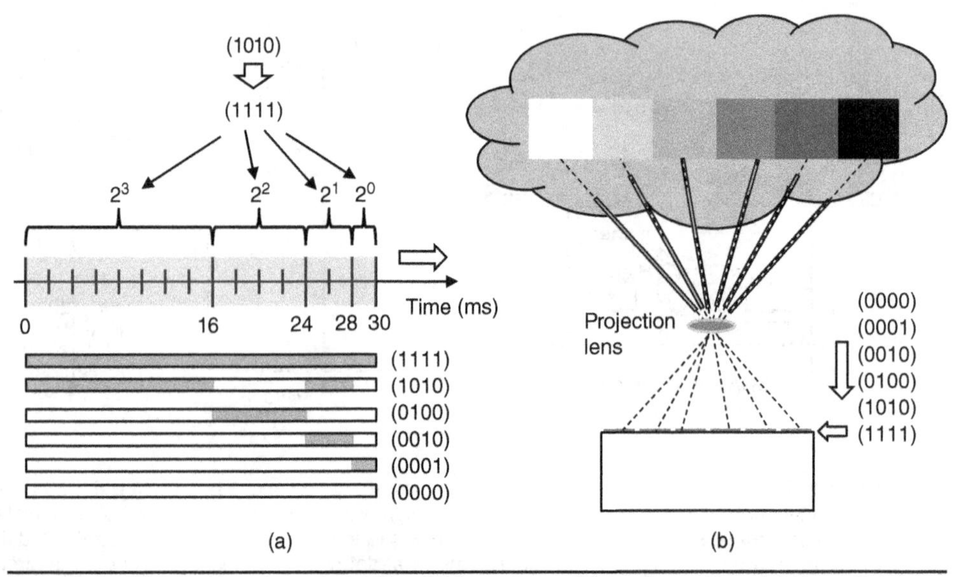

FIGURE 5.4 Digital grayscale control with a 4-bit word over 30 ms frame time: (a) binary time intervals for a 4-bit grayscale and (b) a 4-bit word input into the memory element of each DLP mirror, 1 bit at a time, beginning with the most significant bit (MSB) for each word.

address states), (4) address SRAM array under the mirrors, one line at a time, and (5) repeat the sequence beginning at step 1 at a new bit time.

Different grayscales obtained with a 4-bit word are illustrated in Fig. 5.4. Note how the video field time (30 ms) is divided into four binary time intervals and how the six different words affect the light intensity in each of the time intervals. Color display can be obtained by using a rotating color filter that lets red, green, and blue lights to pass through one-third time of each frame. In this case, a single DLP chip handles three different color components at three different time periods per frame in order to produce a color image. An alternative is to use three different DLP chips to handle three different color components at all time.

Question: Show a few cross-sectional views of the process steps that will produce the mirror and mirror support post (shown in Fig. 5.2d and repeated below).

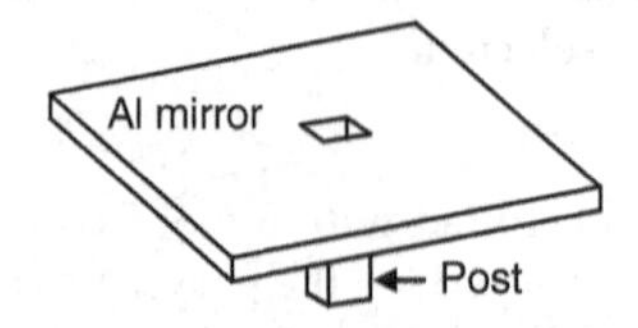

Answer:

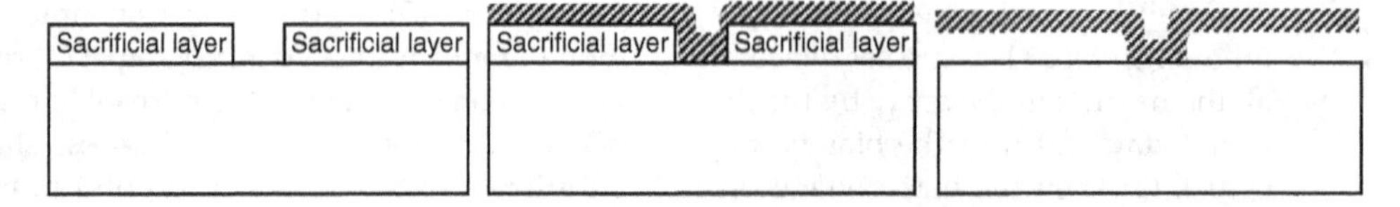

Question: For TI's DLP shown in Fig. 5.3a, the state of the SRAM cell (1,0) determines which mirror rotation angle is selected. When and for what is a voltage applied to the bias-reset bus?

Answer: The voltage (e.g., 15 V) is applied when the mirrors need to be rotated to the desired angles and to keep them latched on those angles.

5.1.2 Grating Light Valve

Another MEMS approach for a projection display is to use diffractive interference of light to discriminate between on and off pixel states. Pixels are identical movable structures, which are pulled to the substrate to modulate reflected light through controlled diffraction of incident light. The ribbons (pixels) shown in Fig. 5.5 are made of stoichiometric silicon nitride (Si_3N_4), known for its good tensile strength and durability, and can be fabricated with a simple two-mask process. Stoichiometric silicon nitride (Si_3N_4) has a large tensile stress and returns to its flat state vary fast upon the removal of an applied voltage (within 20 ns for the fabricated ribbon [2]). Aluminum is deposited on top of the silicon nitride for an electrical connection and optical throughput.

Pulling down (by electrostatic actuation) every other ribbon produces a diffraction grating, as shown in Fig. 5.5f. Thus, the ribbons can be either in reflective state or diffractive state depending on the position of every other ribbon. If one looks at an angle θ_1 from the normal direction to the surface (Fig. 5.5f), the light intensity can be either very small (in the reflective case) or very large (in the diffractive case).

In Fig. 5.5, reflection and ± first-order diffraction with grating period Λ and grating groove depth Δ are illustrated with the Al-coated silicon nitride ribbons. As can be seen in the calculated light throughputs from Al in Fig. 5.6, the specular reflection (i.e., when viewed along the direction normal to the substrate surface) is close to 80% when all the

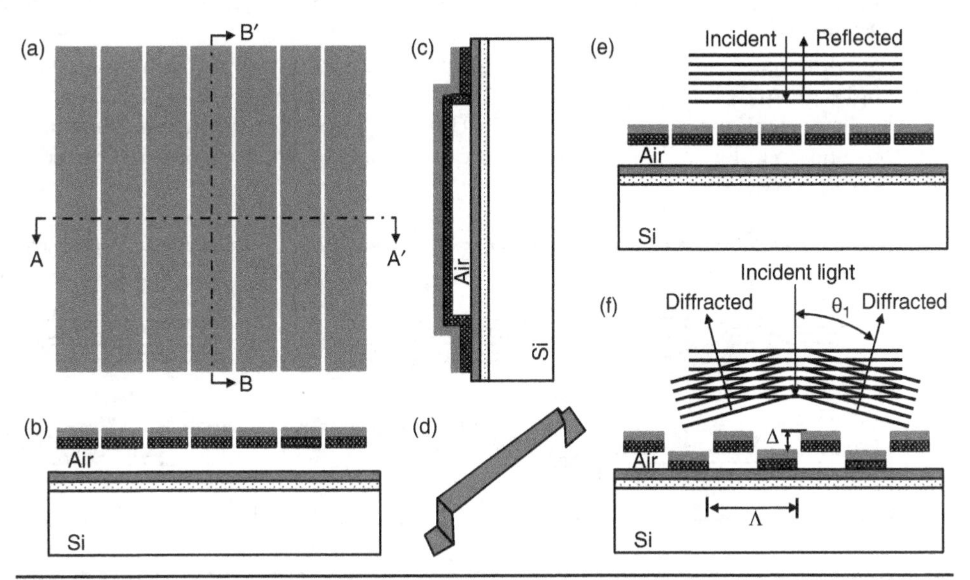

FIGURE 5.5 Simplified schematics of a released array of surface-micromachined ribbons (or bridges) for optical diffraction grating: (a) top view, (b) cross-sectional view across A–A′, (c) cross-sectional view across B–B′, (d) perspective view of a release ribbon, and (e) and (f) before and after electrostatically pulling down every other ribbon to form a diffraction grating. (Adapted from [2].)

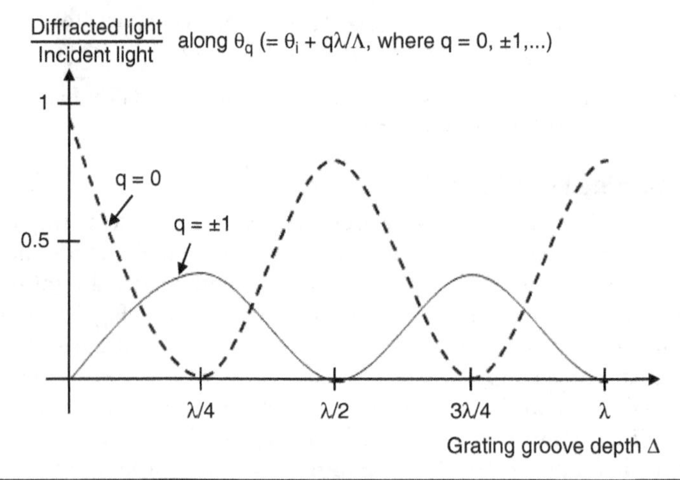

Figure 5.6 Calculated diffraction coefficients versus grating groove depth Δ for aluminum grating. (Adapted from [2].)

Al/Si_3N_4 beams provide a total of $2\cdot\lambda/2$ traveling distance for an incident light with wavelength λ, but is close to zero when the beams presents a $2\cdot\lambda/4$ traveling distance to the incident light. On the other hand, the light amount along the ± first-order diffraction angle is close to zero when all the beams provide a total of $2\cdot\lambda/2$ traveling distance for the incident light, but is close to 40% when the beams present a $2\cdot\lambda/4$ traveling distance to the incident light. Thus, one can use a digital mode between the 80% and zero (or between the 40% and zero) for controlling grayscale for display application. Or linear grayscale control can be obtained by varying the air gap a finite amount near the 97% reflectivity region.

Illuminated single-element gratings produce a row of diffraction orders spaced at angles given by $\theta_q = \theta_i + q\lambda/\Lambda$; $q = 0, \pm1, \pm2, \ldots$, where θ_i = angle of incident beam relative to the normal and Λ = grating period. Thus, for a constant incident angle (such as the one shown in Fig. 5.7), the grating period determines which color is diffracted normal

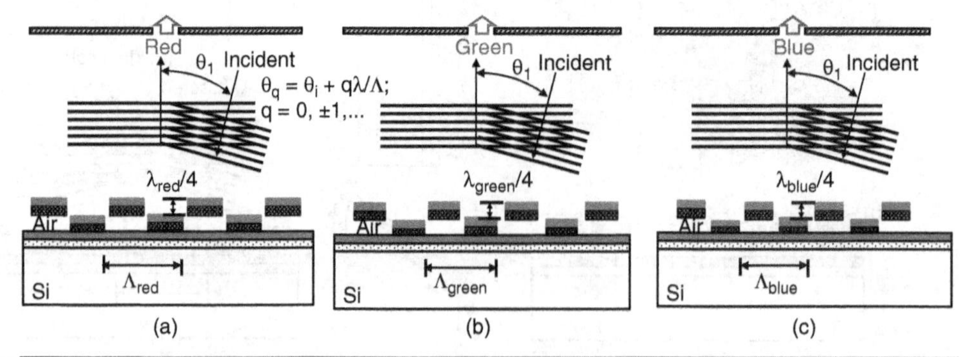

Figure 5.7 Diffraction gratings for red (a), green (b), and blue (c) that diffract a light with a constant incident angle such that the diffracted light normal to the grating may be red, green, and blue, respectively. Light diffracted perpendicular to the diffraction grating depends on the grating period Λ. (Adapted from [2].)

to the surface. Consequently, a larger grating period (Λ) diffracts a larger wavelength light (more reddish) into the normal direction for a constant incident angle. For color display one can use three different diffraction grating periods for red, blue, and green, as shown in Fig. 5.7. Note that for GLV the light source must be monowavelength, and only LED or laser can be used. This puts burden on the system cost since LED or laser is more expensive than halogen light bulbs.

Question: For the grating light valve shown in Fig. 5.5, how do the grating period (Λ) and groove depth (Δ) affect the diffraction angle and efficiency? Indicate your answer on the table below by filling the blanks with increase, decrease, or no effect.

	Diffraction Angle (θ)	Diffraction Efficiency (or Intensity of Diffracted Light)
Increase Λ		
Increase Δ		

Answer:

	Diffraction Angle (θ)	Diffraction Efficiency (or Intensity of Diffracted Light)
Increase Λ	Decrease	No effect
Increase Δ	No effect	Increase or decrease

Question: Which of the three (shown in Fig. 5.7) has the largest grating period Λ?

Answer: The one for red light (Fig. 5.7a), since red light has the largest wavelength λ among the three.

5.1.3 Thin-Film Micromirror Array

The third MEMS approach for a projection display is to use a so-called thin-film micromirror array (TMA) based on a piezoelectrically actuated cantilever that tilts a mirror linearly. In fabricating TMA (Fig. 5.8) polysilicon is used for the spacer layer 1, since the deposition of LPCVD SiN and the annealing of sol–gel-deposited PZT require a high temperature, much higher than a polymer can withstand, though the spacer layer 2 that

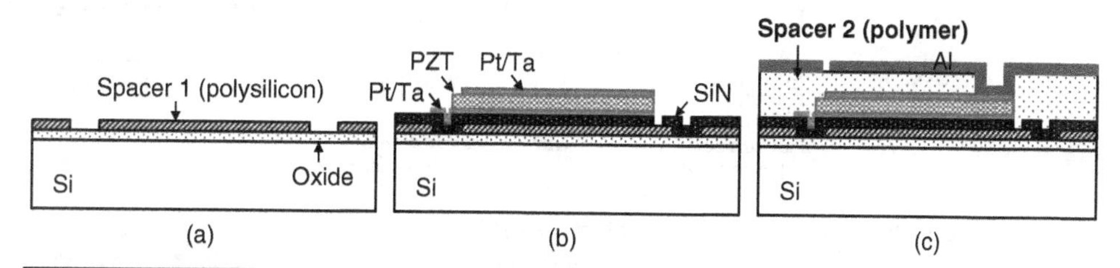

FIGURE 5.8 Brief fabrication steps for TMA mirror: (a) polysilicon deposition and patterning for spacer layer 1, (b) depositions and patterning of low-pressure chemical vapor deposition (LPCVD) SiN, bottom electrode, PZT, and top electrodes, in the listed order, and (c) polymer deposition and patterning for spacer layer 2, and then Al layer deposition and patterning for mirrors, before dry etching of the polymer spacer (with O_2 plasma) and polysilicon spacer (with XeF_2) to release the mirrors. (Adapted from [3].)

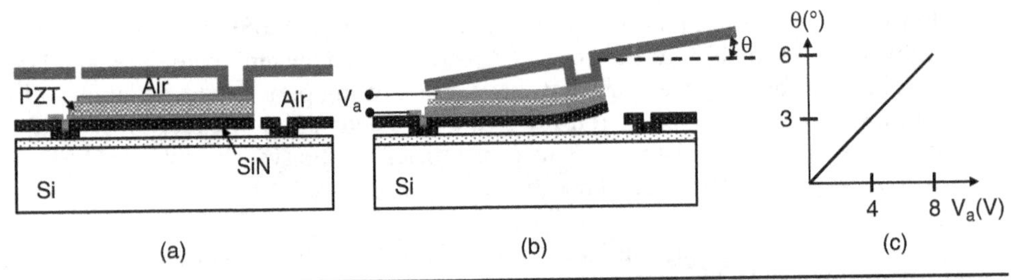

FIGURE 5.9 Cross-sectional views of released TMA mirror without (a) and with (b) applied voltage V_a and the mirror's tilting angle θ versus applied voltage V_a (c). (Adapted from [3].)

is deposited and patterned before Al mirror is composed of a polymer [3]. Thus, the spacer layers are dry etched with oxygen plasma etching of the spacer 2 followed by XeF_2 dry etching of the spacer 1 in order to release the Al mirror (Fig. 5.9a).

When a voltage is applied between the top and bottom electrodes (Pt) across the PZT, the Pt-PZT-Pt-SiN unimorph bends at an angle θ proportional to the applied voltage V_a, as indicated in Fig. 5.9b, since the PZT contracts in-plane directions (while expanding in the thickness direction) due to the applied voltage. The bending angle is linearly proportional to the applied voltage for a relatively small tilting angle (Fig. 5.9c).

As illustrated in Fig. 5.10, the grayscale control is obtained by tilting the mirror at an angle from 0 to 6° and using a knife filter to block out a portion of the light that the mirror reflects into the opaque region of the filter [4]. As shown in Fig. 5.10, the amount of the light that passes through the opening of the knife filter can be from 0 to 100% depending on the tilting angle of the piezoelectrically actuated cantilever, which was shown to be about 0.75°/V, resulting in 6° tilting with 8 V applied across the PZT film [3]. Though the applied voltage of 0–8 V appears to be low, the electrical field

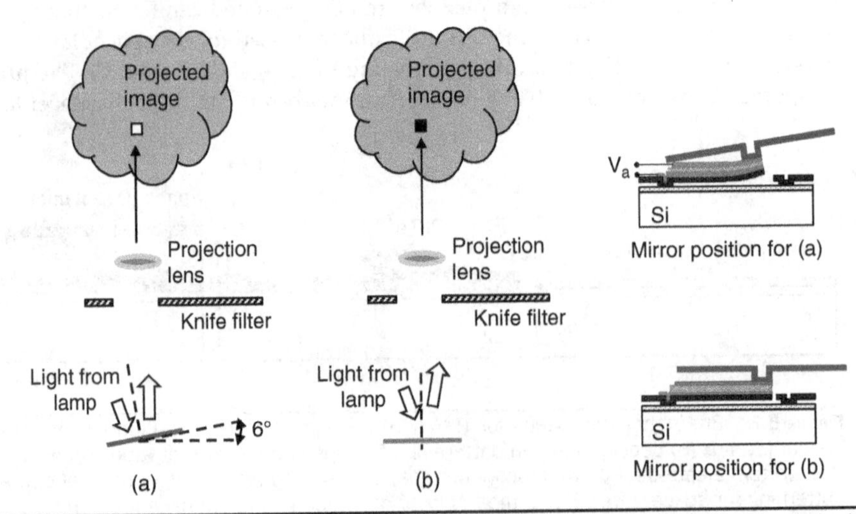

FIGURE 5.10 Light modulation through a knife filter: (a) when a piezoelectrically actuated mirror is tilted 6° and (b) when the mirror is not tilted.

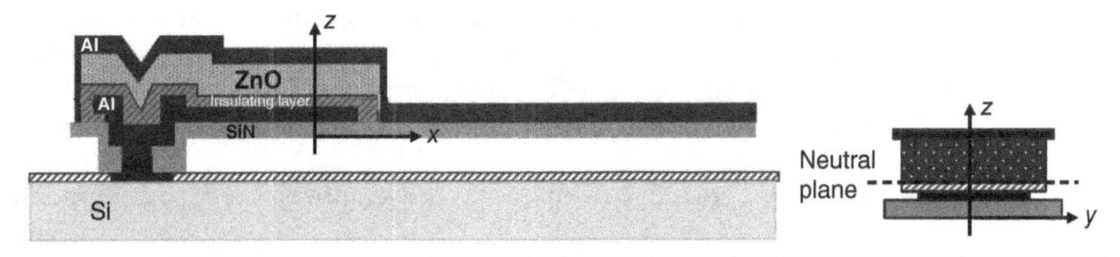

Figure 5.11 Cross-sectional view of a piezoelectrically actuated cantilever with piezoelectric ZnO layer [5].

across a thin PZT film (e.g., 1 μm thick) is at MV/m level, which make the PZT go through a large ferroelectric hysteresis loop, potentially affecting reliability.

Question: For the thin-film micromirror array (TMA) shown in Fig. 5.9b, will the thickness of the piezoelectric PZT layer increase, decrease, or remain the same, when the voltage is applied to bend the cantilever as shown in Fig. 5.9b (i.e., from the flat to the bent cantilever)?

Answer: Increases, since the lateral dimension (i.e., in the x- and y-directions) is decreased to produce the bending shape shown in Fig. 5.10b.

Design Issues with Piezoelectrically Actuated Mirror Array

In designing a piezoelectrically actuated cantilever such as the one shown in Fig. 5.11, we need to consider materials selection (piezoelectric layer, support layer, etc.), thickness selection, thin insulation layer to remove thermal response, etc. For a multilayer cantilever actuated by a piezoelectric film, the neutral plane is typically designed to be at the bottom plane of the piezoelectric film in order to induce maximum bending stress in the piezoelectric film, as shown at the right in Fig. 5.11 (the cross-section of a cantilever viewed from the free end). For a large deflection, the cantilever should be designed as thin as possible, and we can first set the thickness of each layer to its minimum thickness limited and/or allowed in a given fabrication process. Then, based on the thickness of each layer, we calculate the thickness of ZnO to make the neutral plane of the cantilever coincide with the bottom surface of the ZnO layer [5].

As illustrated in Fig. 5.12, calculation of the deflection and tip displacement of the ZnO-actuated cantilever can be treated as a pure bending with an effective momentum M^* and moment of inertia I^* [6]. Once the M^* and I^* are obtained in terms of the force

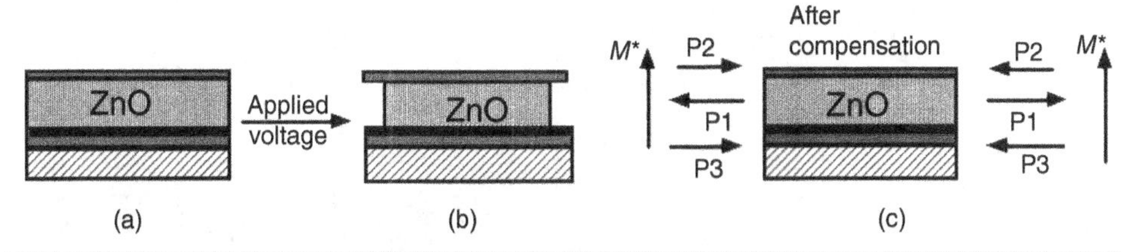

Figure 5.12 A segment of the cantilever along the length direction (a) under no applied voltage, (b) under an applied voltage which produces a piezoelectric strain for a hypothetical case of the layers not attached to each other, and (c) under an applied voltage for a real case where all the layers are attached to each other, and the piezoelectrically induced strain can be equivalently modeled with the forces applied to the layers as shown.

densities (P in N/m^2) and elastic moduli (and geometrical dimensions) of the layers, the deflection curvature of the cantilever bending can be calculated as $M^*/E_{SiN}I^*$, where E_{SiN} is Young's modulus for SiN, as we will see in Chapter 6.

The fundamental resonant frequency of the cantilever can be estimated with $f_r = \frac{1.875^2}{2\pi l^2}\sqrt{\left(\frac{EI}{\rho A}\right)_{\text{eff}}}$, where l, E, I, ρ, and A are the length, elastic modulus, moment of inertia, mass density, and cross-sectional area of the composite cantilever, respectively. The tip displacement of a piezoelectrically actuated cantilever can be measured with a laser interferometer. In order to distinguish piezoelectric response from thermal response, an electrical input of an unbiased square wave can be used to deflect the cantilever. As shown in Fig. 5.13, the tip of a 150-μm-long cantilever (Fig. 5.11) was observed to deflect about 1.9 μm vertically in piezoelectric response to a 10-$V_{\text{zero-to-peak}}$ square wave of 50 Hz. Due to the relative high resistance between the top and bottom electrodes of the cantilever (above 100 MΩ), the thermal deflection is about two or three orders of magnitude smaller than the piezoelectric response according to a theoretical calculation and experimental measurements.

Due to the relative narrow air gap underneath the cantilever (Fig. 5.11), there can be a substantial amount of "squeeze film damping," which increases as the air gap is reduced and/or the cantilever's lateral dimension increases, since the air in the gap will have to be pushed in and out sideways as the cantilever bends upward and downward. Squeeze film damping can be much stronger than air damping, though both are due to the air molecules, and can reduce the vibrational amplitude of the cantilever significantly. For a cantilever with about 1-μm air gap, the vibrational amplitude in the upward direction is observed to be almost twice larger than that in the downward direction. Indeed, the frequency response of the cantilever is dominated by the air-gap distance, which is determined by the thickness of the sacrificial layer. Typical frequency responses to a square wave with a 5-V amplitude are shown in Fig. 5.14a and 5.14b for the cantilevers with the air gap of about 0.5 and 1 μm, respectively; the bending

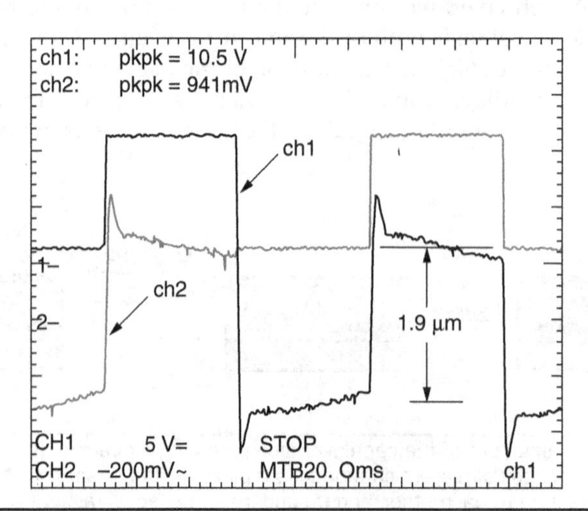

FIGURE 5.13 The measured cantilever-tip displacement (Chap. 2) for a 50-Hz, 10-$V_{\text{zero-to-peak}}$ square wave input (Chap. 1) [5].

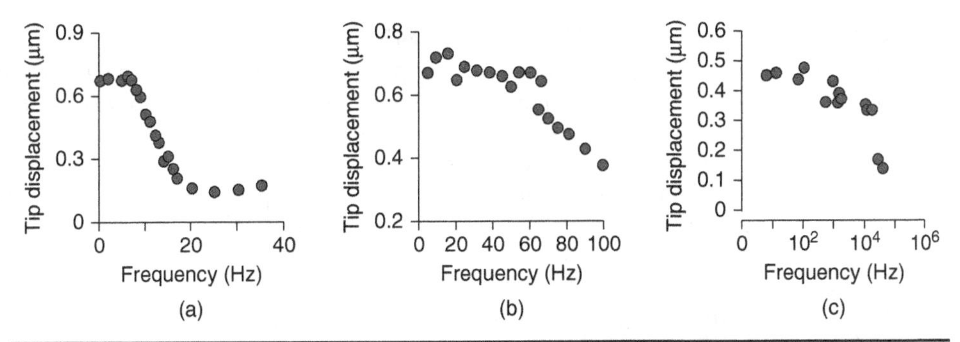

FIGURE 5.14 Measured frequency responses of (a) a piezocantilever with 0.5-µm gap that has a large squeeze film damping, (b) a piezocantilever with 1-µm gap, which still has substantial squeeze film damping, though less, and (c) a piezocantilever that is initially bent up by 30° and has very little squeeze film damping [5].

displacements start dropping when the frequency is increased beyond 10 and 60 Hz for the two cases, as squeeze film damping dominates the displacements. However, if the cantilevers are initially bent by about 30°, there is very little squeeze film damping, and the roll-off frequency can be as high as 10 kHz as shown in Fig. 5.14c, in which case just an air damping affects the frequency response.

5.2 Micromirrors for Optical Fiber Communication

Data transfer through optical fiber has been growing exponentially ever since wavelength division multiplexing (WDM) overtook time division multiplexing (TDM). Data transmission through optical fiber requires a set of optical components every certain distance (say every kilometer or so) to amplify the light and relay the amplified light further along, since light attenuates in a fiber as it travels at 0.2–0.6 dB/km for 1.1–1.7 µm wavelength. Erbium-doped fiber (EDF) provides optical amplification, when it is "pumped" by a laser, albeit nonuniformly in wavelength. After an optical amplification, the amplified signal must go through gain equalizing filter (GEF) to make the amplified light have a uniform spectrum over a wavelength bandwidth. An optical isolator (OI) is used to isolate the laser from affecting anything other than the EDF, while a variable attenuator (VA) is used to reduce the light intensity over the whole bandwidth (after GEF) to a proper level [7, 8]. Some of the components, particularly GEF and VA, have been explored to be replaced by MEMS-based devices.

Question: The light used in a WDM contains a multiple number of wavelengths and needs to be separated into lights, in spatial domain, according to the wavelengths. Give two possible methods one can use to obtain the spatial separation.

Answer: Prism and diffraction grating.

Question: If an EDF can amplify optical signal in the wavelength between 1,530 and 1,562 nm, how much bandwidth (in Hz) can the fiber carry? Note that the speed of light is 0.3 Gm/s.

Answer: $$\frac{3\times10^{8}\ \text{m/s}}{1530\times10^{-9}\ \text{m}}-\frac{3\times10^{8}\ \text{m/s}}{1562\times10^{-9}\ \text{m}}=4\times10^{12}\ \text{Hz}$$

5.2.1 Mechanical Reflection Optical Switch

An electrostatically actuated beam (e.g., SiN_x) can be an optical switch, as the SiN_x beam reflects light when it is undeflected (Fig. 5.15a), while the beam lets light to pass through when it is pulled down to the silicon substrate through a voltage applied between the Al and doped polysilicon, as shown in Fig. 5.15b [9]. An electromagnetic wave normally incident into the interface between two media having impedances Z_1 and Z_2, as shown Fig. 5.15c, experiences zero reflection if $Z_2 = \sqrt{Z_1 Z_3}$ and the thickness of medium 2 is a quarter wavelength, because the quarter-wavelength layer makes the impedance seen by the wave at the interface (between Z_1 and Z_2) matched to Z_1, as can easily be derived from Eq. (4.81) for a semi-infinitely thick Z_3 medium (i.e., thick Si substrate), so that the reflection coefficient at that interface [Eq. (4.2)] is zero. This is commonly called a quarter-wavelength impedance matching, and has been used for eye glasses for antireflection (with limited success since sun light has multiple wavelengths). As the characteristic impedance of a dielectric material is $\sqrt{\mu/\varepsilon}$ (where μ and ε are permeability and permittivity, respectively) and refractive index n is $\sqrt{\varepsilon_r}$ (where ε_r is relative permittivity), the impedance requirement for the quarter-wavelength matching can be written as $n_2 = \sqrt{n_1 n_3}$.

For an electrostatically actuated optical switch with a large air gap (Fig. 5.16a), the air-gap distance affects the reflection coefficient at the interface between air and the membrane, as shown in Fig. 5.16b. Thus, two totally different reflection states (i.e., reflective vs. antireflective) can be obtained through varying the air gap by integer number of the quarter wavelength. Operating the membrane between the two states for an incident light allows a light modulation in time domain. One main requirement for such light modulator is the speed of the membrane deflection, which depends on the membrane's resonant frequency as well as squeeze film damping effect. Thus, the size of the membrane should be small or highly tensile-stressed for high resonant frequency, and the membrane should be perforated for low squeeze film damping. The optical switch or modulator shown in Fig. 5.16 is made with a membrane (SiN_x and doped polysilicon) designed to have an equivalent refractive index of $\sqrt{1 \cdot n_{Si}}$ and thickness of a quarter wavelength of the incident light.

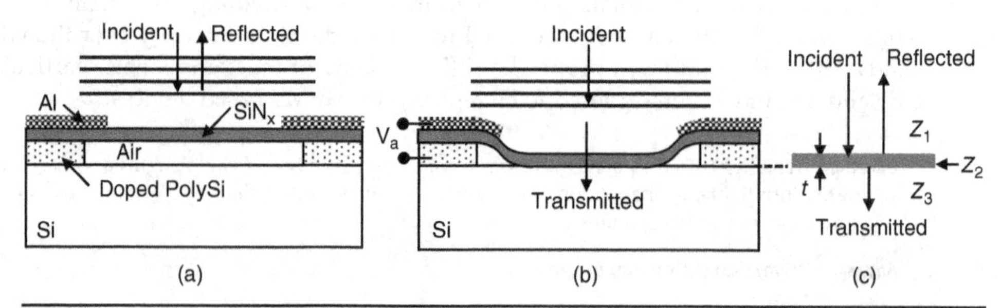

FIGURE 5.15 Electrostatically actuated optical reflective switch: (a) and (b) cross-sectional views without and with an applied electrostatic voltage V_a between the Al and doped polysilicon, respectively, and (c) light wave normally incident on an interface between two media with two different impedances (Z_1 and Z_2), as the thickness t and impedance Z_2 may be chosen to make the impedance (seen by the light at the interface) be matched to Z_3 (of a semi-infinite-thick substrate).

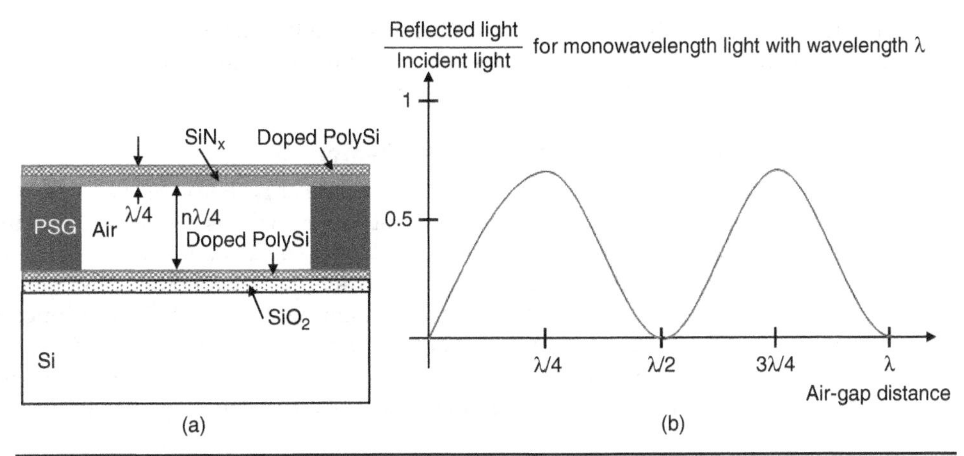

FIGURE 5.16 Electrostatically actuated optical reflective switch with a SiN_x/PolySi membrane and a large air gap: (a) cross-sectional view that shows the membrane and air gap over a silicon substrate and (b) reflectivity of the membrane versus the air-gap distance. (Adapted from [10].)

Question: In the optical modulator shown in Fig. 5.16, what percentage of normal incident light (with a single wavelength such as 1.425 μm) is reflected (a) when the air gap is one-fourth of 1.425 μm and (b) when the air gap is zero.

Answer: (a) 70% and (b) 0%

Variable Attenuator with Electrostatically Actuated Optical Switch

A variable attenuator (VA) can be made with a device like the one shown in Fig. 5.16a as it presents varying reflectivity to incident light as the membrane is electrostatically deflected without being pulled down to the substrate. Varying the air gap between the membrane and substrate affects the reflectivity of the membrane, as explained in the previous section, because the impedance presented to the incident light by the membrane is a function of the air-gap distance, as commonly encountered with an electromagnetic wave incident on a medium that has different impedance than the one in which the wave has traveled. A uniform attenuation (with < 1-dB variation and attenuation level up to 15 dB) over 1.53–1.57-μm-wavelength range with an applied voltage between 0 and 48 V was obtained with a composite p-Si/nitride/p-Si membrane [11].

Question: For a variable attenuator based on Fig. 5.16a, what is the minimum air-gap distance if linear variable gain is to be obtained for a wavelength of 1.55 μm?

Answer: The desirable control range is 1.55 μm ÷ 4 = 0.3875 μm. Thus, the minimum air-gap distance is 0.3875 μm × 3 = 1.16 μm, since electrostatic actuation has the pull-in effect.

Gain Equalization with Electrostatically Actuated Optical Switch

A contiguous membrane can electrostatically be deformed to form a partially deformed surface and be made to be a gain equalization filter (GEF) for light wave that is spatially spread through a prism or diffraction grating [12]. The particular area (where the air gap is different from the rest of the contiguous membrane) presents a different impedance to the incident light and reflects the light with a different reflection coefficient. Consequently, a spatially spread spectrum of wavelengths can be reflected with different reflection coefficients from the various points over the partially deformed contiguous membrane. This kind of GEF can be used to flatten out a nonuniform spectrum of light (e.g., after an amplification with Er-doped fiber).

5.2.2 Fabry–Perot Opto-Mechanical Modulator

Another approach to modulate reflectance is based on Fabry–Perot resonance, as shown in Fig. 5.17. Light in the air gap between two reflective surfaces resonates when the air gap is an integer multiple of the half wavelength ($m_{gap} \cdot \lambda_c/2$), according to Fabry–Perot optical resonance. When this resonance happens, there is large amount of light intensity in the air gap. Thus, the reflectivity R of the incident light in Fig. 5.17a is a function of the air-gap distance and can vary between 100 and 0%, as shown in Fig. 5.17b. From Eq. (4.81), for a lossless mirror, the impedance Z_1 seen at the left end of the fixed mirror in Fig. 5.17a with Z_0 and Z_{Si} being the impedances of air and Si, respectively, is

$$Z_1 = Z_{Si}\frac{Z_0 + jZ_{Si}\tan\left(\dfrac{2\pi f d_{Si}}{v_{Si}}\right)}{Z_{Si} + jZ_0\tan\left(\dfrac{2\pi f d_{Si}}{v_{Si}}\right)} = \frac{Z_{Si}^2}{Z_0} \quad \text{for} \quad d_{Si} = \frac{\lambda_c}{4n_{Si}} \text{ or quarter-wavelength-thick fixed}$$

Si mirror with n_{Si} being a refractive index of silicon. With the Z_1 obtained above, we get

$$Z_2 = Z_0\frac{Z_1 + jZ_0\tan\left(\dfrac{2\pi f d_{gap}}{v_{air}}\right)}{Z_0 + jZ_1\tan\left(\dfrac{2\pi f d_{gap}}{v_{air}}\right)} = \begin{cases} Z_1 \left(= \dfrac{Z_{Si}^2}{Z_0}\right) & \text{for } d_{gap} = \dfrac{\lambda_c}{2} \\ \dfrac{Z_0^2}{Z_1}\left(= \dfrac{Z_0^3}{Z_{Si}^2}\right) & \text{for } d_{gap} = \dfrac{\lambda_c}{4} \end{cases} \tag{5.1}$$

And with the Z_2 above, for $d_{Si} = \dfrac{\lambda_c}{4n_{Si}}$ for the movable Si mirror, we get

$$Z_3 = Z_{Si}\frac{Z_2 + jZ_{Si}\tan\left(\dfrac{2\pi f d_{Si}}{v_{Si}}\right)}{Z_{Si} + jZ_2\tan\left(\dfrac{2\pi f d_{Si}}{v_{Si}}\right)} = \frac{Z_{Si}^2}{Z_2} = \begin{cases} Z_0 & \text{for } d_{gap} = \dfrac{\lambda_c}{2} \\ \dfrac{Z_{Si}^4}{Z_0^3} & \text{for } d_{gap} = \dfrac{\lambda_c}{4} \end{cases} \tag{5.2}$$

Thus, the reflection coefficient of the incident light is

$$R = \frac{Z_0 - Z_3}{Z_0 + Z_3} = \begin{cases} 0 & \text{for } d_{gap} = \dfrac{\lambda_c}{2} \\ \dfrac{Z_0^4 - Z_{Si}^4}{Z_0^4 + Z_{Si}^4}\left(= \dfrac{n_{Si}^4 - n_o^4}{n_{Si}^4 + n_o^4} \text{ for lossless Si}\right) & \text{for } d_{gap} = \dfrac{\lambda_c}{4} \end{cases} \tag{5.3}$$

The reflection is zero when the air gap is equal to $m_{gap} \cdot \lambda_c/2$ with m_{gap} being an integer, and is independent of the mirror's reflectivity, because the optical resonance in the air gap between the mirrors takes in all the incident light (from the left on Fig. 5.17a) and transmits it out to the right (with some optical energy stored in the air gap, which does not impact the steady-state transmissivity). The reflection coefficient R is highest when the air gap is equal to $m_{gap} \cdot \lambda_c/4$ with integer m_{gap}, and is equal to 0.99 for λ_c = 1.3 μm (at which silicon is lossless), from Eq. (5.3) with $n_{Si} = 3.52$ and $n_0 = 1.00$.

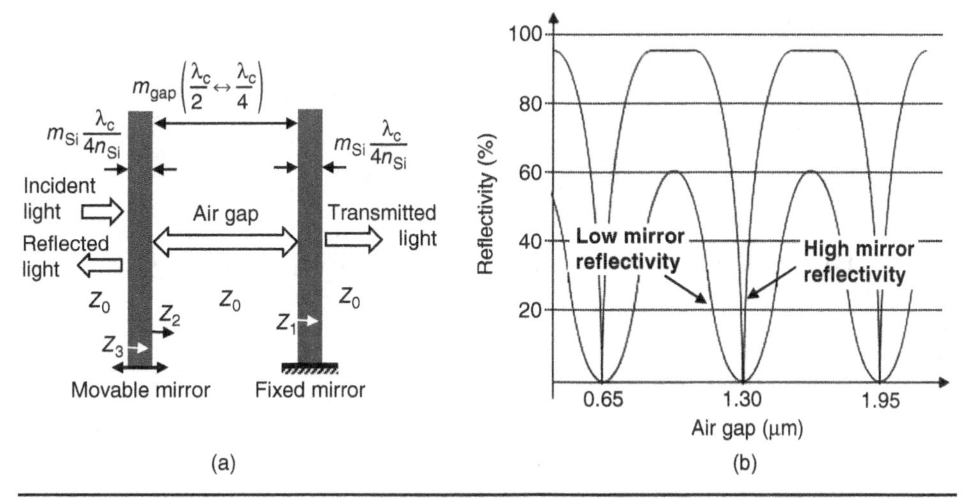

FIGURE 5.17 (a) Tunable Fabry–Perot cavity with a movable mirror and (b) reflectivities of a light (with wavelength $\lambda_c = 1.3$ μm in air) for mirrors with low- and high reflectivities versus the air-gap distance.

The idea illustrated in Fig. 5.17a can be implemented for a Fabry–Perot opto-mechanical light modulator with a double-supported silicon nitride (SiN) beam separated from an underlying polysilicon fixed electrode by an air gap, as shown in Fig. 5.18. Since the polysilicon is on top of an antireflection layer (which can be treated as having an impedance same as that of polysilicon), the impedance Z_1 seen from the top surface of the polysilicon to the Si substrate is

$$Z_1 = Z_{\text{PolySi}} \frac{Z_{\text{PolySi}} + jZ_{\text{PolySi}} \tan\left(\dfrac{2\pi f d_{\text{PolySi}}}{v_{\text{PolySi}}}\right)}{Z_{\text{PolySi}} + jZ_{\text{PolySi}} \tan\left(\dfrac{2\pi f d_{\text{PolySi}}}{v_{\text{PolySi}}}\right)} = Z_{\text{PolySi}} \tag{5.4}$$

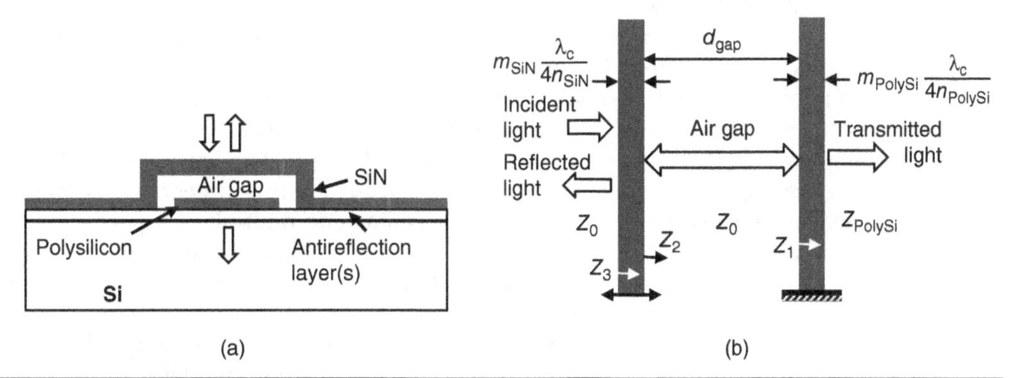

FIGURE 5.18 (a) Surface-micromachined electrostatically tunable Fabry–Perot cavity with SiN and polysilicon acting as movable and fixed mirrors, respectively and (b) equivalent representation of the Fabry–Perot cavity treating the antireflection layer(s) with impedance matched to the polysilicon fixed mirror.

And we get

$$Z_2 = Z_0 \frac{Z_1 + jZ_0 \tan\left(\frac{2\pi f d_{gap}}{v_{air}}\right)}{Z_0 + jZ_1 \tan\left(\frac{2\pi f d_{gap}}{v_{air}}\right)} = \begin{cases} Z_1 \ (= Z_{PolySi}) & \text{for } d_{gap} = \frac{\lambda_c}{2} \\ \frac{Z_0^2}{Z_1}\left(= \frac{Z_0^2}{Z_{PolySi}}\right) & \text{for } d_{gap} = \frac{\lambda_c}{4} \end{cases} \tag{5.5}$$

With the Z_2 above, for $d_{SiN} = \frac{\lambda_c}{4n_{SiN}}$ for the movable SiN mirror, we get

$$Z_3 = Z_{SiN} \frac{Z_2 + jZ_{SiN} \tan\left(\frac{2\pi f d_{SiN}}{v_{SiN}}\right)}{Z_{SiN} + jZ_2 \tan\left(\frac{2\pi f d_{SiN}}{v_{SiN}}\right)} = \frac{Z_{SiN}^2}{Z_2} = \begin{cases} \frac{Z_{SiN}^2}{Z_{PolySi}} & \text{for } d_{gap} = \frac{\lambda_c}{2} \\ \frac{Z_{SiN}^2 Z_{PolySi}}{Z_0^2} & \text{for } d_{gap} = \frac{\lambda_c}{4} \end{cases} \tag{5.6}$$

Thus, the reflection coefficient of the incident light with $n_{PolySi} \approx n_{Si}$ is

$$R = \frac{Z_0 - Z_3}{Z_0 + Z_3} = \begin{cases} 0, \text{ if } Z_{SiN}^2 = Z_{PolySi} Z_0 \ (\text{or } n_{SiN} = \sqrt{n_{PolySi} n_0}), & \text{for } d_{gap} = \frac{\lambda_c}{2} \\ \frac{Z_0^2 - Z_{PolySi}^2}{Z_0^2 + Z_{PolySi}^2}\left(= \frac{n_{Si}^2 - n_o^2}{n_{Si}^2 + n_o^2} \text{ for lossless PolySi}\right), \text{ if } Z_{SiN}^2 = Z_{PolySi} Z_0, & \text{for } d_{gap} = \frac{\lambda_c}{4} \end{cases} \tag{5.7}$$

A structure similar to Fig. 5.18a was implemented with two polysilicon layers (acting as the mirrors) and was shown to be able to modulate light up to a rate ≈ 2.8 Mb/s for light with $\lambda_c = 1.3\ \mu m$ [13].

5.2.3 Fabry–Perot Photonic Crystal, Filter, and Interferometer Built with Bragg Reflectors

As can be seen in Fig. 5.17b, the reflectivity (and the transmitivity) and its sharpness of Fabry–Perot cavity depends on the mirror's reflectivity. For many applications including photonic crystal, photonic bandpass filter and sensors based on Fabry–Perot interferometer, high reflectivity of the mirrors is desired. Consequently, silicon micromachining has been used to microfabricate Bragg mirrors that are composed of layers alternating between high impedance layer (e.g., silicon) and low impedance layer (e.g., air), each layer being a quarter-wavelength thick (or an odd multiple of a quarter wavelength, m_{Si} and m_{air}), similar to what are shown in Fig. 5.19 [14–17]. Very smooth sidewalls of (111) silicon planes that are perpendicular to a wafer surface can be made on a (110) silicon wafer through KOH wet etching (as shown in Fig. 2.5b), which produces nonperpendicular (111) planes that can limit the structural shape. To avoid such a structural limitation, one can first use deep reactive ion etching (DRIE) of silicon to delineate the structure on a (110) silicon wafer followed by a short etching in KOH to make the vertical sidewalls be of very smooth (111) silicon plane (Fig. 5.19b).

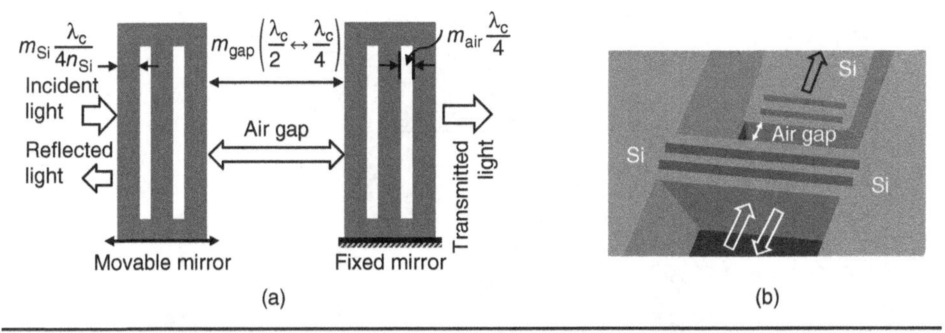

Figure 5.19 (a) Tunable Fabry–Perot cavity with Bragg mirrors and (b) its implementation on (110) silicon substrate with a combination of deep reactive ion etching and KOH wet etching.

Question: Which of the two approaches shown in Figs. 5.16 and 5.18 would you choose if you care about the ratio between the highest and lowest reflectivity?

Answer: Figure 5.18, which has a reflectivity as high as 85% which is equal to $(3.52^2 - 1)/(3.52^2 + 1)$, compared to 70% for Fig. 5.16.

5.2.4 Micromirrors for Optical Cross-Connect

Optical fibers carrying data need an optical cross-connect (OXC) to route N optical input channels (or input fibers) to N output channels (or output fibers). At first, in 1998–99, pop-up MEMS mirrors made on a silicon wafer were considered for such a cross-connect, with a pop-up mirror being electrostatically pulled down and being popped up by a built-in spring mechanism [18]. However, when a mirror was popped up to route one fiber to another, the mirror could/would be in a path for the other fibers, blocking the optical path(s) for those. Thus, a better solution was needed, and routing optical signal in 3-D space with a 2-D mirror array was conceived to solve the blocking problem encountered with the pop-up mirrors. An array of 2N MEMS mirrors can route N input fibers into N output fibers, as illustrated in Fig. 5.20. Each MEMS mirror needs to provide linearly controllable tilting on each of the two orthogonal axes with fine angular accuracy at a reasonably high speed.

Question: In an optical cross-connect (with MEMS mirror array) shown Fig. 5.20 for fiber signal routing, which one of the following gives the MEMS mirrors the information regarding where to route the fibers signals? Choose one: (1) the address embedded in the signal carried by the fibers or (2) the network manager.

Answer: Since the routing is done without deciphering the address information embedded in the signal carried by the fibers in a typical optical cross-connect, only the network manager (2) knows which output fibers will take signals carried by which input fibers.

Electrostatic Mirror Array for Fiber Optic Cross-Connect

Electrostatic actuation is used in the mirror shown in Fig. 5.21, which is a part of 2N mirror array for an $N \times N$ fiber optic cross-connect. Two sets of torsional beams are used to support a mirror and an outer frame, and make the mirror to tilt in two orthogonal axes. The mirror was formed in a silicon substrate with deep reactive ion etching (DRIE) and was released after the silicon substrate is "screen-print" bonded to Pyrex glass [19]. Due to the large air gap of 300 μm, the required voltage for 2° tilting was about 100 V.

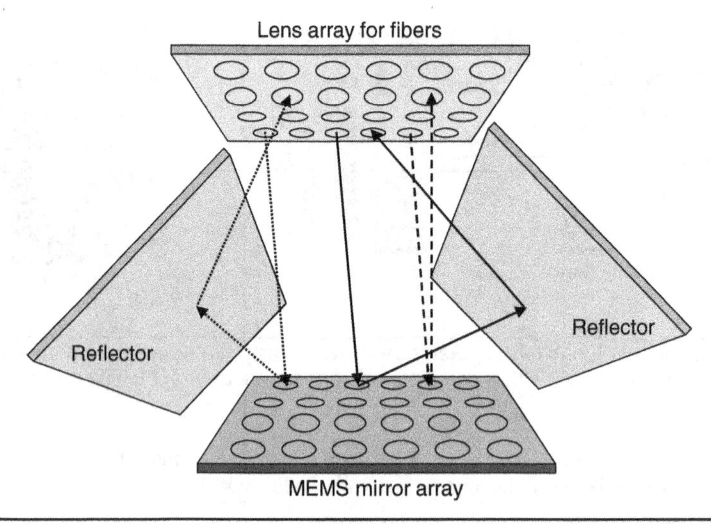

Figure 5.20 MEMS mirror array for optical cross-connect (OXC) for fiber optic communication.

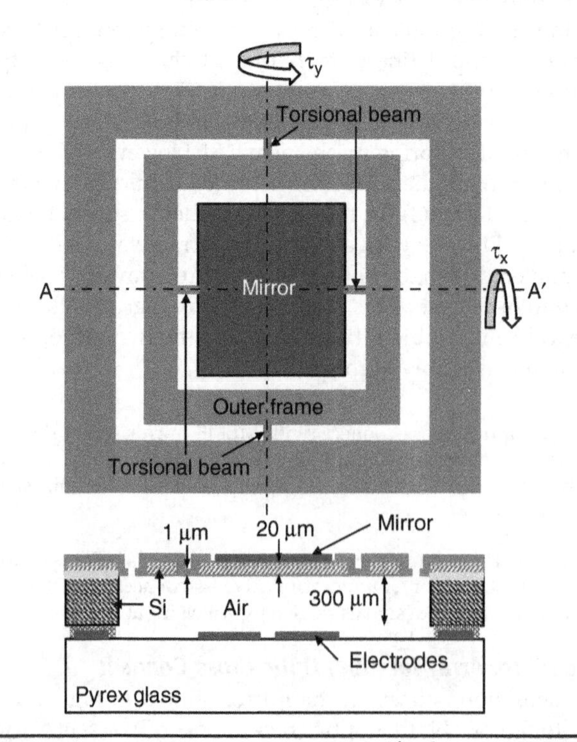

Figure 5.21 Top view and cross-sectional view (across A–A′) of an electrostatically actuated MEMS mirror for N × N single-mode fiber optical cross-connect. (Adapted from [19].)

Electromagnetic Mirror Array for Fiber Optic Cross-Connect

Electromagnetically actuated 2N MEMS mirrors (similar to Fig. 3.13) also were developed for an N × N single-mode fiber optical cross-connect with a current of 0.5 mA inducing a magnetic moment in the coil that worked with the external magnetic field to produce 6° tilting [20]. Electromagnetic actuation with coils on the mirror has several advantages over electrostatically actuated mirrors such as large torques over a large range of tilting angles, linear torque versus current and angle, drive currents of a few mA and 1–2 V (unlike electrostatically actuated mirrors needing voltages of 10–200 V), no electrostatic pull-in effect, no dielectric charging and discharging, etc. However, electromagnetic actuation needs external magnets; suffers from the embedded coils' residual stress which warps the mirrors (reducing radius of curvature); and dissipates electrical power on the coils through I^2R heating.

Question: For a deformable mirror array for optical cross-connect application, which of the two actuation methods (between electrostatic and electromagnetic actuations) would you choose if your *only* concern is the issue listed in the leftmost column of the following table? Please indicate your choice by putting a check mark in the table.

	Electrostatic	Electromagnetic
Power consumption		
Required voltage level		
Mirror flatness		
Manufacturing cost		

Answer:

	Electrostatic	Electromagnetic
Power consumption	√	
Required voltage level		√
Mirror flatness	√	
Manufacturing cost	√	

References

1. P.F. van Kessel, et al., "A MEMS-based projection display," Proceedings of the IEEE, 86(8), August 1998.
2. R.B. Apte, F.S.A. Sandejas, W.C. Banyai, and D.M. Bloom, "Deformable grating light valves for high resolution displays," Solid State Sensor and Actuator Workshop, Hilton Head Island, SC, June 1994, pp. 1–6.

3. S.G. Kim and K.H. Hwang, Euro Display '99 in Berlin, Paper 9.1.
4. J.A. van Raalte, "A new Schlieren light valve for television projection," Applied Optics, 9(10), 1970, pp. 2225–2230.
5. Y.L. Huang, H. Zhang, E.S. Kim, S.G. Kim, and Y.B. Jeon, "Piezoelectrically Actuated microcantilever for actuated mirror array application," Solid State Sensor and Actuator Workshop, Hilton Head Island, SC, June 2–6, 1996, pp. 191–195.
6. F.R. Shanley, "Strength of Materials," McGraw-Hill, New York, NY, 1957, pp. 312–316.
7. J. Zyskind, et al, "Fast power transients in optically amplified multiwavelength optical networks," Proceeding of the Optical Fiber Communication Conference '96—postdeadline session, Washington, DC, paper PD31.
8. Y. Sun, A. Srivastava, J. Zhou, and J. Sulhoff, "Optical fiber amplifiers for WDM optical networks," Bell Labs Technical Journal, 4, 1999, pp. 187–206.
9. J. Walker, K. Goossen, and S. Arney, "Fabrication of a mechanical antireflection switch for fiber-to-the-home systems," Journal of Microelectromechanical Systems, 5(1), March 1996, pp. 45–51.
10. J. A. Walker, K. W. Goossen, S. C. Arneyt, N. J. Frigo, and P. P. Iannone, "A silicon optical modulator with 5 MHz operation for fiber-in-the-loop applications," Proceedings of the International Solid State Sensors and Actuators Conference—Transducers '95, Stockholm, Sweden, 1995, pp. 285–288.
11. J. Ford and J. Walker, "Micromechanical fiber-optic attenuator with 3 ms response," Journal of Lightwave Technology, 16, 1998, pp. 1663–1670.
12. J.E. Ford and J.A. Walker, "Dynamic spectral power equalization using micro-opto-mechanics," IEEE Photonics Technology Letters, 10, 1998, pp. 1440–1442.
13. C. Marxer et al., "MHz opto-mechanical modulator," Proceedings of the International Solid State Sensors and Actuators Conference—Transducers '95, Stockholm, Sweden, 1995, pp. 289–292.
14. D.S. Greywall, "Micromechanical light modulators, pressure gauges, and thermometers attached to optical fibers," Journal of Micromechanics and Microengineering, 7, 1997, pp. 343–352.
15. A. Lipson and E.M. Yeatman, "Low-loss one-dimensional photonic bandgap filter in (110) silicon," Optics Letters, 31(3), February 2006, pp. 395–397.
16. V.A. Tolmachev, T.S. Perova, E.V. Astrova, B.Z. Volchek, and J.K. Vij, "Vertically etched silicon as 1D photonic crystal," Physica Status Solidi (a), 197(2), 2003, pp. 544–548.
17. O. Solgaard, A.A. Godil, R.T. Howe, L.P. Lee, Y. Peter, and Hans Zappe, "Optical MEMS: From micromirrors to complex systems," Journal of Microelectromechanical Systems, 23(3), June 2014, pp. 517–538.
18. L.Y. Lin and E.L. Goldstein, "Lightwave micromachines for optical crossconnects," Proceedings of the 1999 European Conference on Optical Communications—ECOC '99, Nice, France, September 26–30, 1999, pp. I-114–I-115.
19. T. D. Kudrle et al., "Electrostatic micromirror arrays fabricated with bulk and surface micromachining techniques," The Sixteenth Annual International Conference on Micro Electro Mechanical Systems, MEMS-03 Kyoto IEEE, Kyoto, Japan, 2003, pp. 267–270.
20. J. Bernstein et al., "Two axis-of-rotation mirror array using electromagnetic MEMS," The Sixteenth Annual International Conference on Micro Electro Mechanical Systems, MEMS-03 Kyoto IEEE, Kyoto, Japan, 2003, pp. 275–278.

Questions

Question 5.1 For a wave normally incident into the interface between two media having impedances Z_1 and Z_2, as shown below, calculate the magnitudes of the reflection coefficients at 1.0 and 1.1 GHz, if $Z_1 = 50\ \Omega$ and $Z_2 = 2 + j3 \times 10^{-7} \times (f - 10^9)\ \Omega$, where f is frequency in Hz.

Z_1 | Z_2

incident wave →

Note that the reflection coefficient $R = (Z_1 - Z_2)/(Z_1 + Z_2)$.

Question 5.2 For an array of vertically deflectable cantilevers (i.e., cantilevers that can be pulled down [or pushed up] to/fro the substrate), rate the following three actuation techniques between "good" and "no good" for the issues listed in the first row of the table below.

	Electrical Power Consumption	**Bidirectional Motion Generation**	**Dynamic Range (i.e., Linear Control Range)**	**Required Voltage Level for a Large Deflection**	**IC Process Compatibility**
Electrostatic					
Electromagnetic					
Piezoelectric					

Question 5.3 With TI's DLP mirrors, how fast the mirror has to be rotated for the projection display to have a grayscale of 255 levels (that would require an 8-bit word)? Assume that each frame is refreshed every 30 ms.

Question 5.4 If we fit 256 wavelength division multiplexing (WDM) channels between 1.3 and 1.7 μm in optical wavelength, how much frequency bandwidth is each channel capable of handling?

Question 5.5 For a diffraction grating shown in Fig. 5.5f, calculate the diffraction angles (θ_1 and θ_3) in degree (°) for the first- and third-order diffractions when the wavelength of the incident light is λ, if the grating period Λ is equal to λ.

Question 5.6 What percentage of the incident light intensity is diffracted into the angle equal to $\theta_i + \lambda/\Lambda$ (where θ_i = angle of incident beam relative to the normal and Λ = grating period) by an aluminum diffraction grating with 2.0-μm period, when light having a wavelength (λ) of 550 nm is illuminated at normal incidence. [Hint: See Fig. 5.6.]

Question 5.7 For the electrostatically actuated optical reflective switch (Fig. 5.15), what is the desired refractive index for the SiN_x membrane that is built on a silicon substrate that has a refractive index of 3.4?

Question 5.8 If the electrostatically actuated optical reflective switch shown in Fig. 5.15 has the light reflectivity as shown in Fig. 5.16b, what are the thickness and refractive index of the SiN_x membrane?

Question 5.9 For each of the issues listed on the first column below, rank DLP, GLV, and TMA with short comments.

	DLP	GLV Operating in Digital Mode with Diffraction	GLV Operating in Analog Mode with Reflection	TMA
Fabrication easiness				
In-operation stiction				
Light throughput (or brightness for a given light source)				
Performance drift				

Question 5.10 Shown below is a cross-sectional view of a piezoelectrically actuated cantilever made for a projection display application.

(a) Which of the following *cannot* be used for a sacrificial layer in fabricating the cantilever: LPCVD polysilicon, evaporated silicon, silicon oxide, and photoresist? (b) Right after the cantilever fabrication, you notice that the cantilever is warped (or bent) in the length direction. State two potential sources for this initial warping. (c) What is the minimum thickness for the Si_xN_y layer, if we want the neutral plane (for cantilever bending) outside ZnO (i.e., not within ZnO)? Assume that the ZnO (0.3 μm thick) and the insulating layer (0.2 μm thick) have Young's modulus of 210 and 70 GPa, respectively, while the top and bottom Al films (0.2 μm thick each) have Young's modulus of 70 GPa. Also, assume that Young's modulus of Si_xN_y is 270 GPa.

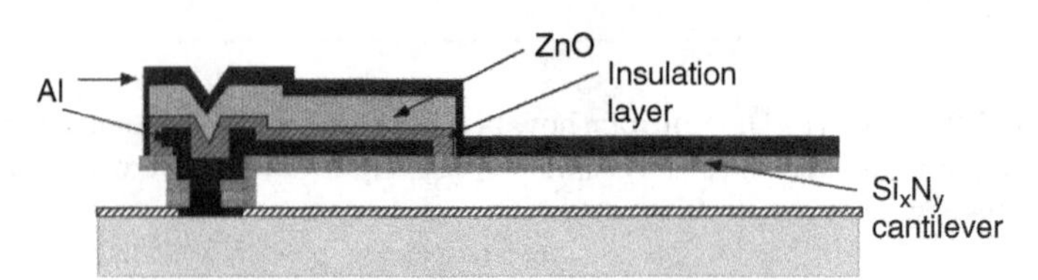

Question 5.11 Shown below is a piezoelectrically actuated cantilever made for a projection display application. For the following questions assume that the piezoelectric layer (0.3 μm thick) has Young's modulus of 2 10 GPa while the top and bottom electrodes (0.2 μm thick each) have Young's modulus of 70 GPa. Also, assume that Young's modulus of the support layer is 270 GPa.

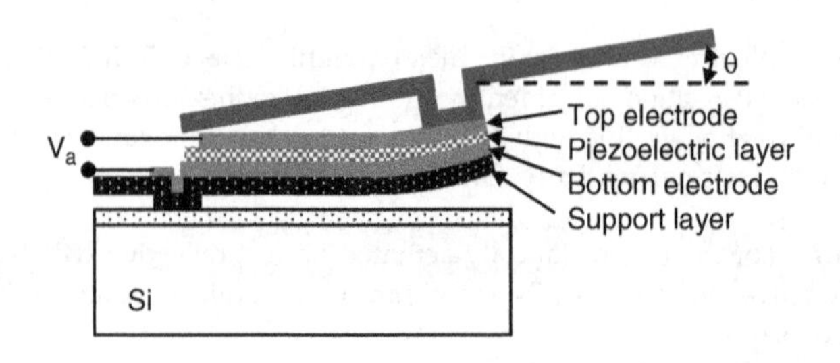

(a) What is the minimum thickness for the supporting layer, if we want the neutral plane (for cantilever bending) outside the piezoelectric layer (i.e., not within the piezoelectric layer)? (b) If PZT thin film is used for the piezoelectric layer with the PZT poled through the top and bottom electrodes, which piezoelectric coefficient(s) among d_{33}, d_{31}, and d_{15} will directly impact the cantilever bending? (c) Write down an equation for the deflection angle θ at the free end of the cantilever in terms of the curvature of the cantilever bending ($1/R$) and the cantilever length (L).

Question 5.12 Answer the following questions on the DLP mirror shown at Fig. 5.2. (a) As the state of the SRAM cell (1 or 0, i.e., 5 or 0 V) determines which rotation angle (+ or −10°) is selected, what prevents the mirror from rotating during the time period when the SRAM cells are updated with new values? (b) To obtain the desired tilting angle (±10°), which two dimensions will have to be designed properly?

Question 5.13 For the DLP operation explained in Fig. 5.4b, if one of the 4-bit words is changed from (0000) to (1001), how will the brightness in each of the video fields change?

CHAPTER 6

Mechanics and Inertial Sensors

Unless we know the mechanics of various structures such as cantilever, bridge, plate, etc., we will not be able to analyze and/or design microelectromechanical systems (MEMS) structures. In this chapter, we will study both statics and dynamics of structures followed by MEMS accelerometers and gyroscopes. Though this chapter contains many equations, it is written such that those with understanding of college freshmen physics may follow and become proficient in the mechanics.

6.1 Statics

6.1.1 Stress and Strain as Single Indexed Matrix Elements

You may want to review what is covered with Fig. 3.19 on stress. The units for stress or pressure are many, and the following covers most of the pressure units.

$$1 \text{ atmosphere} = 760 \text{ Torr} = 1.013 \times 10^5 \text{ Pa}$$
$$1 \text{ Torr} = 1 \text{ mmHg} = 133.3 \text{ Pa}$$
$$1 \text{ mTorr} = 1 \text{ μHg}$$
$$1 \text{ Pa} = 1 \text{ N/m}^2 = 10 \text{ dyne/cm}^2 = 7.5 \text{ mTorr}$$
$$1 \text{ bar} = 1 \times 10^5 \text{ Pa} = 750 \text{ Torr}$$
$$1 \text{ μbar} = 0.1 \text{ Pa} = 1 \text{ dyne/cm}^2$$
$$1 \text{ psi (lb/in}^2) = 6.89 \times 10^3 \text{ Pa} = 51.9 \text{ Torr}$$

Since there are only six independent variables for stresses, we can use $T_1 - T_6$ to represent the three normal stresses (i.e., $T_1 \equiv \sigma_x$, $T_2 \equiv \sigma_y$, and $T_3 \equiv \sigma_z$) and three shear stresses ($T_4 \equiv \sigma_{xy}$, $T_5 \equiv \sigma_{yz}$, and $T_6 \equiv \sigma_{zx}$). Then strain (which is supposed to be a second rank tensor, similar to stress) also can be represented with a single indexed parameter S_i, with $i = 1, 2, 3, 4, 5,$ and 6. For anisotropic solid, strain S is related to stress T as in $S_i = s_{ij} T_j$, while stress T is related to strain S as $T_i = c_{ij} S_j$, where c_{ij} and s_{ij} are the elastic stiffness coefficients and the elastic compliance coefficients, respectively, with many of them being zero and only several independent variables due to crystal symmetry. These equations come from general Hook's law for elastic medium. The equations

$S_i = s_{ij}T_j$ and $T_i = c_{ij}S_j$ use repeated indices (with i and j = 1, 2, 3, 4, 5, and 6) to shorten the following matrices.

$$\begin{pmatrix} S_1 \\ S_2 \\ S_3 \\ S_4 \\ S_5 \\ S_6 \end{pmatrix} = \begin{pmatrix} s_{11} & s_{12} & s_{13} & s_{14} & s_{15} & s_{16} \\ s_{21} & s_{22} & s_{23} & s_{24} & s_{25} & s_{26} \\ s_{31} & s_{32} & s_{33} & s_{34} & s_{35} & s_{36} \\ s_{41} & s_{42} & s_{43} & s_{44} & s_{45} & s_{46} \\ s_{51} & s_{52} & s_{53} & s_{54} & s_{55} & s_{56} \\ s_{61} & s_{62} & s_{63} & s_{64} & s_{65} & s_{66} \end{pmatrix} \begin{pmatrix} T_1 \\ T_2 \\ T_3 \\ T_4 \\ T_5 \\ T_6 \end{pmatrix}$$

$$\begin{pmatrix} T_1 \\ T_2 \\ T_3 \\ T_4 \\ T_5 \\ t_6 \end{pmatrix} = \begin{pmatrix} c_{11} & c_{12} & c_{13} & c_{14} & c_{15} & c_{16} \\ c_{21} & c_{22} & c_{23} & c_{24} & c_{25} & c_{26} \\ c_{31} & c_{32} & c_{33} & c_{34} & c_{35} & c_{36} \\ c_{41} & c_{42} & c_{43} & c_{44} & c_{45} & c_{46} \\ c_{51} & c_{52} & c_{53} & c_{54} & c_{55} & c_{56} \\ c_{61} & c_{62} & c_{63} & c_{64} & c_{65} & c_{66} \end{pmatrix} \begin{pmatrix} S_1 \\ S_2 \\ S_3 \\ S_4 \\ S_5 \\ S_6 \end{pmatrix} \tag{6.1}$$

Repeated Indices

Equations can be shortened through using repeated indices. Whenever an index is repeated in a product between two or more element, the repeated index is taken to indicate summation of the products as all the allowed numbers for the index are taken. For example, a product $e_{1i}E_i$ has the repeated index i and is a shortened version of $e_{11}E_1 + e_{12}E_2 + e_{13}E_3$ for $i = 1, 2$, and 3.

Example: For i and $k = 1, 2, 3$; $j = 1, 2, 3, 4, 5, 6$, the equation $D_i = d_{ij}T_j + e_{ik}E_k$ is equivalent to the following matrix equation.

$$\begin{pmatrix} D_1 \\ D_2 \\ D_3 \end{pmatrix} = \begin{pmatrix} d_{11} & d_{12} & d_{13} & d_{14} & d_{15} & d_{16} \\ d_{21} & d_{22} & d_{23} & d_{24} & d_{25} & d_{26} \\ d_{31} & d_{32} & d_{33} & d_{34} & d_{35} & d_{36} \end{pmatrix} \begin{pmatrix} T_1 \\ T_2 \\ T_3 \\ T_4 \\ T_5 \\ T_6 \end{pmatrix} + \begin{pmatrix} \varepsilon_{11} & \varepsilon_{12} & \varepsilon_{13} \\ \varepsilon_{21} & \varepsilon_{22} & \varepsilon_{23} \\ \varepsilon_{31} & \varepsilon_{32} & \varepsilon_{33} \end{pmatrix} \begin{pmatrix} E_1 \\ E_2 \\ E_3 \end{pmatrix} \tag{6.2}$$

6.1.2 Isotropic Media

In case of isotropic medium, the 36 components of the stiffness coefficients (c_{ij}) reduce to only two independent coefficients: Young's modulus E and Poison's ratio ν. Consequently, Hooke's law for isotropic, nonpiezoelectric solid gives the following equations for normal strains and stresses:

$\varepsilon_x = +\frac{\sigma_x}{E} - \nu\frac{\sigma_y}{E} - \nu\frac{\sigma_z}{E}$, $\varepsilon_y = -\nu\frac{\sigma_x}{E} + \frac{\sigma_y}{E} - \nu\frac{\sigma_z}{E}$, and $\varepsilon_z = -\nu\frac{\sigma_x}{E} - \nu\frac{\sigma_y}{E} + \frac{\sigma_z}{E}$, with the normal strains defined as $\varepsilon_x = \Delta_x/L_x$, $\varepsilon_y = \Delta_y/L_y$, and $\varepsilon_z = \Delta_z/L_z$, where Δ and L are length change and original length, respectively.

The plus or minus signs in the equations above reflect the physical nature of solid deformation in three dimensions. When a tensile stress (positive stress) is applied in x direction, a positive strain (extension in length) in x direction is produced, while negative strains (contraction in length) are produced in y- and z directions. Young's modulus (E) is an indication of how stiff a material is, and a material with a higher Young's modulus deforms less for a given stress. Poisson's ratio (ν) gives a measure on how

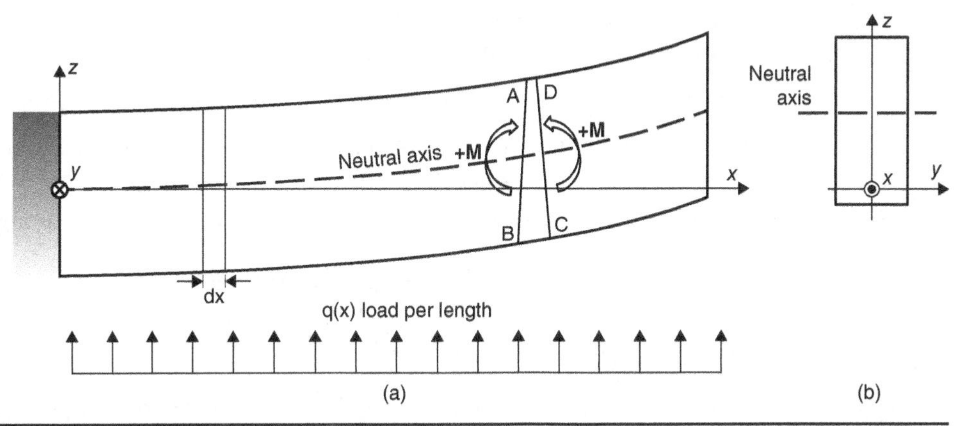

FIGURE 6.1 Cross-sectional views of a cantilever under bending due to an applied load: (a) along the beam axis, *x*, and (b) looking into the beam along the beam axis.

much strain is produced along the other two directions (e.g., *y*- and *z* directions) for a given strain in one direction (e.g., *x* direction).

Shear strains (γ_{xy}, γ_{yz}, and γ_{zx}) and shear stresses are related through $\gamma_{xy} = \tau_{xy}/G$, $\gamma_{yz} = \tau_{yz}/G$, and $\gamma_{zx} = \tau_{zx}/G$, where shear modulus G is related to Young's modulus E and Poisson's ratio ν, as $G = \frac{E}{2(1+\nu)}$ A shear strain is produced by a pair of shear stresses and is a measure of angular displacement.

6.1.3 Bending of Isotropic Cantilever

A cantilever is a special form of a beam that is defined to be a solid with its length much longer than the other dimensions, such as the one shown in Fig. 6.1, and goes through a bending when there is a bending moment *M* along the beam length due to an applied load. The bending is near "pure" when the cantilever is an isotropic medium. The strain due to such a pure bending varies *linearly* in the beam thickness direction starting from zero at the beam axis (or the neural axis that goes through the centroid of the beam cross-section). Note that the strain is opposite in sign above and below the beam axis (Fig. 6.2b).

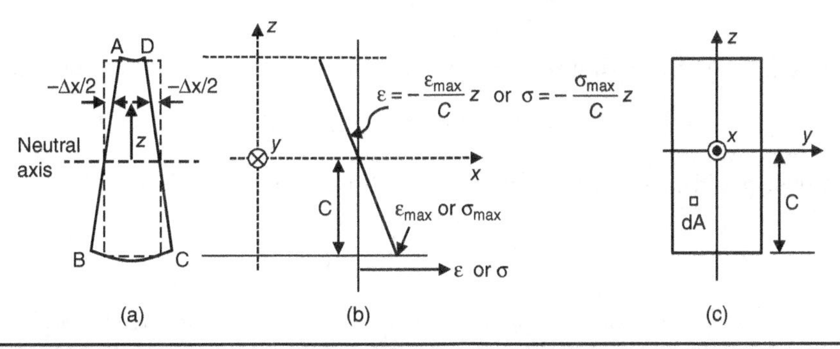

FIGURE 6.2 Bending of an isotropic cantilever (due to bending moment *M*) of which the bending shape along the beam axis, *x*, is shown in Fig. 6.1: (a) a differential element ABCD showing the length change Δx as a function of the distance from the neutral axis, *z*, (b) strain ε and stress σ developed in the beam along the beam thickness direction, *z*, and (c) differential area element *dA* on the plane perpendicular to the beam axis.

Relationship between Stress and Bending Moment

An applied moment M is resisted by an internal bending moment developed by the flexural stresses, and $M = \int_A \left(-\frac{z}{C}\sigma_{max}\right)\cdot dA \cdot z = -\frac{\sigma_{max}}{C}\int_A z^2 dA$, noting that $\left(-\frac{z}{C}\sigma_{max}\right)\cdot dA$ is the normal force on the differential area element dA, according to Fig. 6.2, and also noting that moment is obtained when the arm, z, is multiplied to the force. Since moment of inertia $I = \int_A z^2 dA$, we obtain $M = -\frac{\sigma_{max}}{C} I$ or $\sigma_{max} = -\frac{MC}{I}$. Thus, noting that the stress σ on any point of a cross-section is $-\frac{z}{C}\sigma_{max}$, we obtain

$$\sigma = -\frac{My}{I} \tag{6.3}$$

The Eq. (6.3) shows an important relationship between stress σ (on any point of the beam cross-section, directed toward the beam axis, i.e., x) and the bending moment M, when a beam goes through pure bending. The stress σ is dependent on M, y, and I (moment of inertia), and we need to know M to obtain stress distribution.

Question: What is the moment of inertia for a cantilever with a rectangular cross-section with b and h being the width and height (or thickness) of the beam, respectively?

Answer: $I = \int_A z^2 dA = \int_{-b/2}^{b/2}\int_{-h/2}^{h/2} z^2 dz\,dy = \frac{bh^3}{12}$.

Question: If a beam (with a moment of inertia equal to 10^{-20} m^4) goes through a pure bending by a bending moment equal to 10^{-10} N·m, what is the stress at a point 1 μm away (in the thickness direction) from the beam axis?

Answer: $\sigma = \left|\frac{My}{I}\right| = \frac{(10^{-10}\,\text{N}\cdot\text{m})\times(10^{-6}\,\text{m})}{10^{-20}\,\text{m}^4} = 10^{-4}\,\text{N/m}^2$

Relationship among Applied Load, Shear Force, and Bending Moment

Let us consider a cantilever under a load $q(x)$, as shown in Fig. 6.3.

Summing the vertical forces and setting the sum to zero for equilibrium, we obtain $0 = V - (-qdx) - (V + dV)$, which leads to

$$q = \frac{dV}{dx} \tag{6.4}$$

At equilibrium, the summation of moments around the point A must be zero and we get

$$0 = (M + dM) - M - (V + dV)dx + qdx\frac{dx}{2} \approx dM - Vdx \tag{6.5}$$

Consequently, we have the following relationship between M and V:

$$V = \frac{dM}{dx} \tag{6.6}$$

Thus, we see that bending moment M is related to shear force through a spatial differentiation (or integration) as seen in Eq. (6.6), while spatial differentiation of shear force is equal to applied load as in Eq. (6.4). In other words, integration of an applied load $q(x)$ along the beam length gives the shear force distribution $V(x)$ in the beam, and

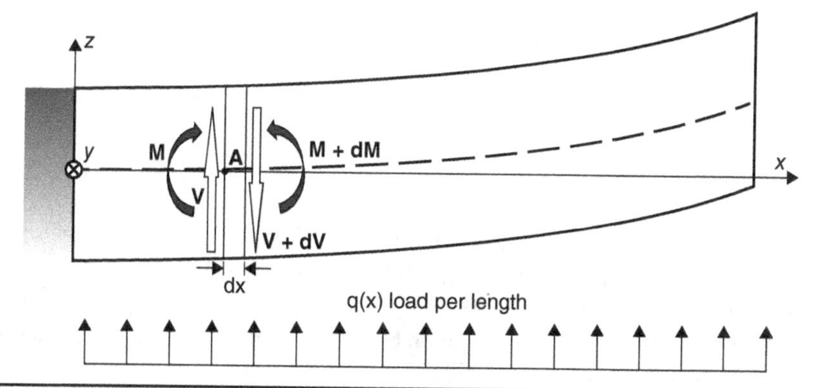

Figure 6.3 A cantilever under a load $q(x)$ applied along the thickness direction, z, and a differential area element dx along the beam axis.

integration of the shear force $V(x)$ along the beam length yields the bending moment distribution $M(x)$ in the beam.

Relationship among Bending Moment, Slope of Bending Curve, and Bending Curve

Referring to Fig. 6.4 for pure bending of a beam, we see that $-\Delta u = z\Delta\theta$, from which we obtain $\frac{du}{dx} = -z\frac{d\theta}{dx}$, since $\lim_{\Delta x \to 0} \frac{\Delta u}{\Delta x} = -z \lim_{\Delta x \to 0} \frac{\Delta\theta}{\Delta x}$. However, $\frac{du}{dx}$ is the linear strain ε in a bending beam at a distance z from the neutral axis or $\frac{du}{dx} = \varepsilon$. Thus, $\varepsilon = -z\frac{d\theta}{dx}$. Now

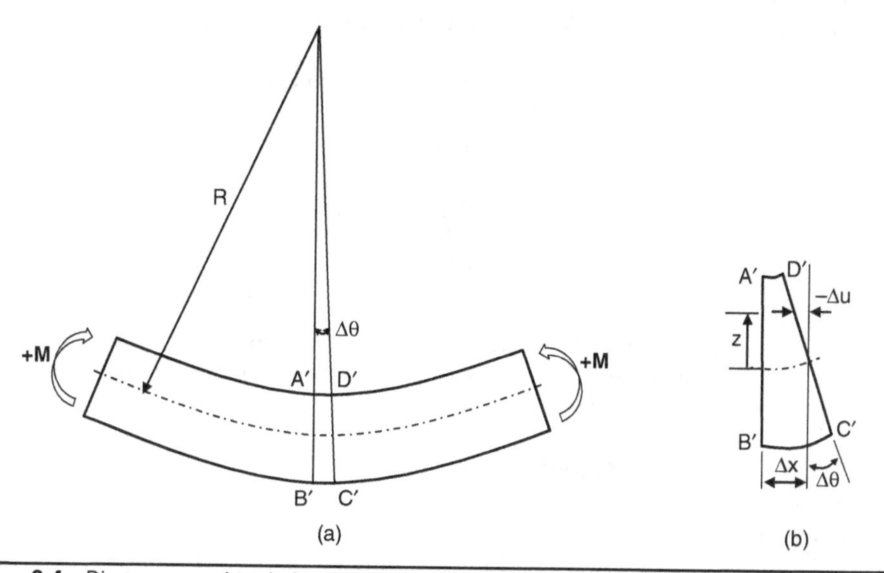

Figure 6.4 Diagrams used to derive the relationship between strain-curvature and moment-curvature in pure bending of a beam: (a) bending curve and (b) a differential element A′B′C′D′ along the beam axis.

noting $\Delta x = R\Delta\theta$, we see $\lim_{\Delta x \to 0} \frac{\Delta\theta}{\Delta x} = \frac{d\theta}{dx} = \frac{1}{R} = \kappa$, with κ being defined to be the curvature of the bending curve, and obtain $\frac{1}{R} = \kappa = -\frac{\varepsilon}{z}$. Finally, since $\varepsilon = \varepsilon_x = \frac{\sigma_x}{E}$ and $\sigma_x = -\frac{Mz}{I}$, we obtain

$$\frac{1}{R} = \frac{M}{EI} \tag{6.7}$$

Consequently, in case of pure bending of a beam, we see that the curvature of the bending shape is equal to bending moment M divided by EI (Young's modulus times moment of inertia). The EI is often called flexural rigidity (D), since it determines how much bending there will be for a given bending moment. Since $\frac{1}{R} \approx \frac{d^2w}{dx^2}$ with w = deflection of the bending curve, Eq. (6.7) is

$$\frac{d^2w}{dx^2} = \frac{M}{EI} \tag{6.8}$$

Also, from $\frac{1}{R} = -\frac{\varepsilon}{z}$, we get

$$\varepsilon = -z\frac{d^2w}{dx^2} \tag{6.9}$$

Question: If a beam going through a pure bending has a radius of curvature equal to 1 m, how much bending moment is being applied? Assume that the Young's modulus and moment of inertia are equal to 200 GPa and 10^{-22} m^4, respectively.

Answer: Since $\frac{1}{R} = \frac{M}{EI}$, $M = EI/R = (200 \times 10^9 \text{N/m}^2) \times (10^{-22} \text{m}^4)/(1 \text{ m}) = 2 \times 10^{-11} \text{N} \cdot \text{m}$

Differential Equations for Deflection of Elastic Beams

The equations in the previous sections lead to the following set of equations:

$$\frac{dw}{dx} = \text{slope of bending curve} \tag{6.10}$$

$$M(x) = EI\frac{d^2w}{dx^2} \tag{6.11}$$

$$V(x) = \frac{dM}{dx} = EI\frac{d^3w}{dx^3} \tag{6.12}$$

$$q(x) = \frac{dV}{dx} = EI\frac{d^4w}{dx^4} \tag{6.13}$$

Thus, for a given loading $q(x)$, we need to solve a fourth-order differential equation to obtain the elastic deflection curve $w(x)$ using the boundary conditions to evaluate the four integration constants.

6.1.4 Boundary Conditions

There are mainly three types of boundaries for mechanical support: fixed support, simple support, and free end, as illustrated in Fig. 6.5. At fixed and simple supports, the displacement w must be zero, since the supports do not allow any such displacement. The main difference between fixed support and simple support is whether the support can sustain bending moment or not. Fixed support can resist bending moment, while

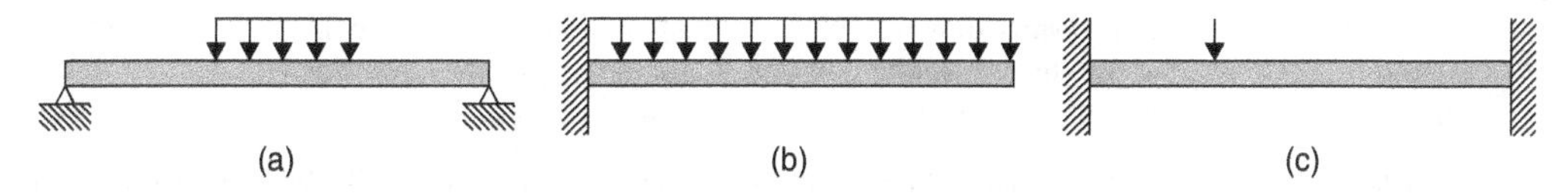

Figure 6.5 Cross-sectional views of beams under various loading: (a) a beam with its two edges simply supported with uniform loading over a portion of the beam, (b) a beam with its one end clamped and the other end free with uniform loading over the whole beam, and (c) a beam with its two edges clamped with a point loading on the beam.

simple support cannot. Thus, we have the following boundary conditions (two equations at each support), which need to be used to evaluate the integrating constants when the fourth-order differential equation is solved for the displacement.

1. *Clamped or fixed or built-in support*: displacement w and slope dw/dx must be zero.
2. *Simple support*: displacement w and bending moment M ($\propto d^2w/dx^2$) must be zero.
3. *Free end*: bending moment M ($\propto d^2w/dx^2$) and shear force V ($\propto d^3w/dx^3$) must be zero.

Cantilever is in general defined to be a beam with one edge fixed (or clamped) and the other edge free (such as Fig. 6.5b), while bridge is usually defined to be a beam with its two edges fixed (such as Fig. 6.5c). Because of different boundary conditions, beam bending shapes (and the stresses induced by the bending) are quite different for cantilever and bridge even under a same uniform load, as illustrated in Fig. 6.6. A bridge (or a diaphragm with its four edges fixed) goes through bending according to the shape shown in Fig. 6.6c. Note the zero slope of the deflection curve at the fixed edges. The slope would not have been zero, if the edges are simply supported (Fig. 6.6a) or free.

Because of the zero slope requirement at the fixed edges, an applied load produces such a bending shape that the bending-induced stress changes its sign (from compression to tension and back to compression above the beam's neutral plane) from one edge to the other for a bridge. This kind of sign change in the stress does not happen with simply supported edges nor with a cantilever.

Also, note the sign change of the stress from compression to tension (or vice versa), along the beam's thickness direction from top to bottom surface. This is exactly according to Eq. (6.3). The stress variation along the thickness direction is linear, though the stress variation along the beam's length direction is not.

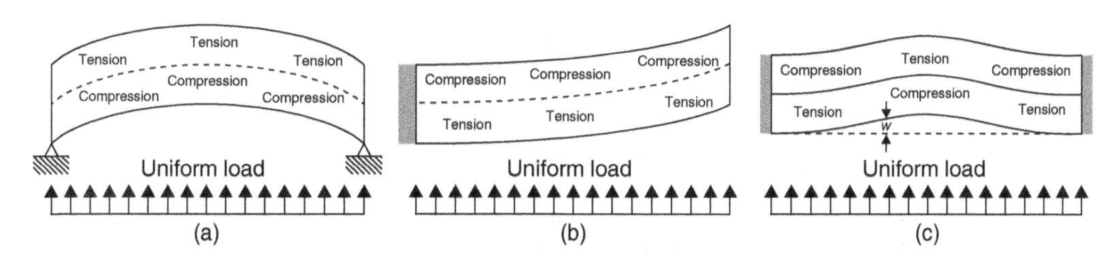

Figure 6.6 Cross-sectional views of beam bending under uniform loading: (a) a beam with its two edges simply supported, (b) a cantilever, and (c) a bridge.

Question: For isotropic beams shown in Fig. 6.6, what is the net stress if we integrate the bending-induced stress along the thickness direction (i.e., along the z direction) from the very bottom to the very top? Compressive, tensile, or zero?

Answer: It depends on the shape of the beam's cross-section, and is zero, if the beam's cross-section is rectangular, as can be understood from Fig. 6.2.

Example 6.1: Cantilever under Static Point Loading

Problem: For a cantilever deflected by a static point force (applied at its free end, as shown below), plot the shear force, bending moment, and deflection as a function of the position along the length direction.

Answer: Due to the applied point force, there is a reactive force at the fixed end in the opposite direction with the same magnitude. Shear force is $V = \int q dx = \int -p\delta(x-l)dx = p$ for $0 < x < l$.

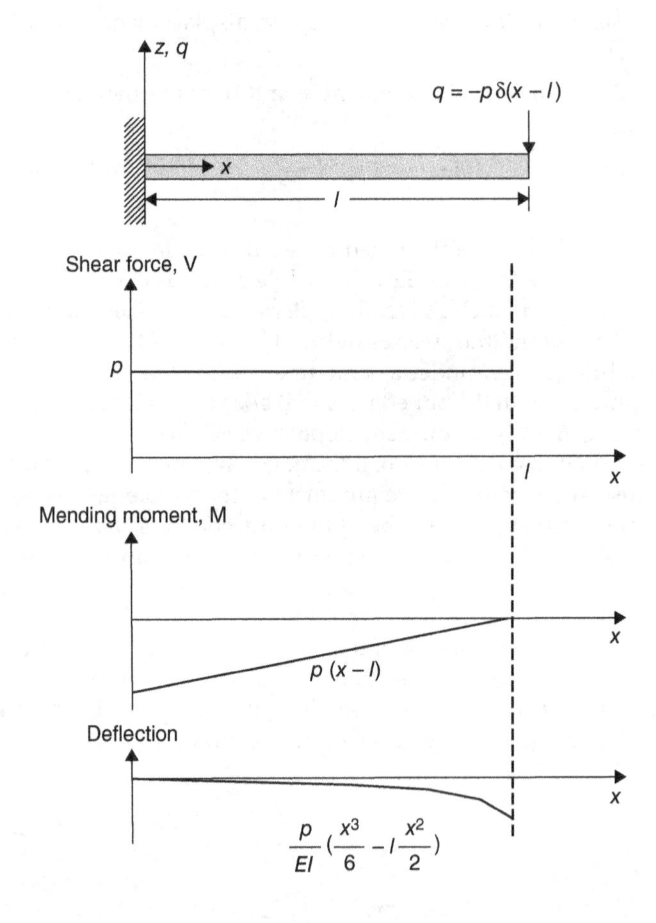

The sign of the shear force can best be obtained through physical reasoning. However, one can use the conventional definitions of positive shear force and bending moment as illustrated in Fig. 6.3. In other words, the particular set of the applied point force at the free end and the reactive force at the fixed end produces a positively signed shear stress between the two forces.

Integrating the shear force, we obtain the bending moment, $M = \int V dx = \int p dx = px + C_1$, where $C_1 = -pl$, since M must be zero at $x = l$, due to the *free* boundary condition.

From Eq. (6.8), we obtain the slope of the deflection curve, $\frac{dw}{dx} = \frac{1}{EI}\int M dx = \frac{p}{EI}\left(\frac{x^2}{2} - lx + C_2\right)$, where $C_2 = 0$, since dw/dx must be zero at $x = 0$, due to the *fixed* boundary condition.

Integrating dw/dx, we obtain the deflection $w = \int \frac{dw}{dx}dx = \frac{p}{EI}\left(\frac{x^3}{6} - \frac{lx^2}{2} + C_3\right)$, where $C_3 = 0$, since w must be zero at $x = 0$, due to the *fixed* boundary condition.

6.1.5 Equivalent Spring Constants for Common MEMS Structures

For an applied uniform pressure P (in N/m^2) and an applied electrostatic voltage V, an equivalent spring constant k' (in N/m^3) can be used to calculate air-gap distance at the free end of a cantilever, at the center of a bridge, or at the center of a clamped circular plate (Fig. 6.7). The k' values for a cantilever, a bridge, and a clamped circular plate are listed in Table 6.1 and can be used to calculate air-gap closing (i.e., $g_o - w$, where g_o and w are the initial gap and the displacement, respectively, as illustrated in Fig. 6.8) at the free end of a cantilever, at the center of a bridge, or at the center of a clamped circular plate. Note that residual stress σ affects the k' for a bridge or a clamped plate significantly but not for a cantilever, because a cantilever releases residual stress through its free end. Tensile residual stress can severely limit the bending displacement of a bridge or a clamped plate in response to an applied pressure or voltage.

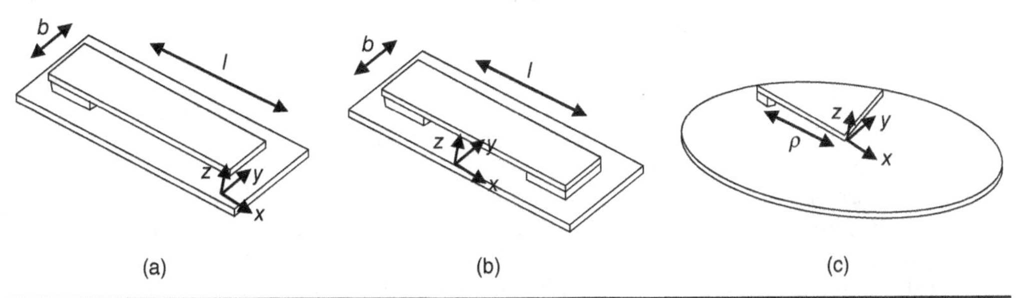

Figure 6.7 Perspective views of (a) a cantilever, (b) a bridge with its two edges clamped, and (c) a clamped circular plate, with b = beam width, t = beam or plate thickness, l = beam length, ρ = circular plate radius.

Structure	Governing Equations	Equivalent Spring Constant k' (N/m^3)	Buckling Stress σ_c (N/m^2)
Cantilever	$EI\frac{d^4w}{dx^4} = Pb - \frac{\varepsilon_o bV^2}{2g^2}$	$\frac{2Et^3}{3l^4}$	NA
Bridge (two edges clamped)	$EI\frac{d^4w}{dx^4} + \sigma(1-\nu)bt\frac{d^2w}{dx^2} = Pb - \frac{\varepsilon_o bV^2}{2g^2}$	$\frac{32Et^3}{l^4} + \frac{8\sigma(1-\nu)t}{l^2}$	$\frac{4Et^2}{3l^2(1-\nu)}$
Clamped circular plate	$D\nabla^4 w + \sigma t\nabla^2 w = P - \frac{\varepsilon_o V^2}{2g^2}$	$\frac{16Et^3}{3\rho^4(1-\nu^2)} + \frac{4\sigma t}{\rho^2}$	$\frac{4Et^2}{3\rho^2(1-\nu^2)}$

Table 6.1 Governing Equations and Equivalent Spring Constant k's for Cantilever, Bridge, and Clamped Circular Plate: $I = bt^3/12$, $D = Et^3/[12(1-\nu^2)]$, σ = biaxial residual stress [1]

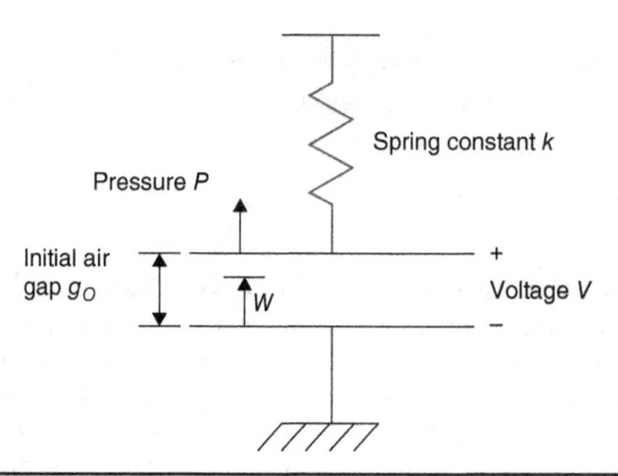

Figure 6.8 Equivalent model to calculate the largest displacement of a MEMS structure (e.g., any of those shown in Fig. 6.7) under a uniform pressure P (in N/m^2) and an electrostatic voltage.

The governing differential equation for a cantilever in Table 6.1 is basically the equation that we saw earlier, i.e., $EI\frac{d^4w}{dx^4} = q(x)$, with the loads $q(x)$ coming from uniform pressure P and electrostatic force. Referring to Figs. 6.7 and 6.8, we can derive the electrostatic force by differentiating the energy stored in a capacitor $= CV^2/2 = \varepsilon_o AV^2/2g$, where $A = bl$. Consequently, we have

$$|\text{Force}| = |dE/dg| = \varepsilon_o AV^2/2g^2 \tag{6.14}$$

$$|\text{Force}| \text{ per length} = \varepsilon_o bV^2/2g^2 \tag{6.15}$$

In case of a bridge and clamped circular plate, the effect of biaxial residual stress σ is taken care by adding an additional differential term as shown in Table 6.1, and the equivalent spring constant k' is increased substantially for tensile residual stress for a bridge and a clamped circular plate. If the residual stress is compressive, the sign of the residual stress σ will be negative, and the equivalent spring constant k' will decrease, as the compressive residual stress increases in magnitude. At a large enough compressive residual stress, the equivalent spring constant k' will become zero, at which point the bridge or clamped circular plate is buckled or warped. Thus, one can obtain an equation for the buckling residual stress σ_c for the bridge and clamped circular plate from the equivalent spring constant k' by setting $k' = 0$.

Note that the equivalent spring constant for a bridge is about 5 times larger than that for a cantilever when $\sigma = 0$. Furthermore, even if there is zero residual stress initially, bending displacement of a bridge is limited by induced tension in the bridge as the bridge undergoes bending due to the fact that the two edges are fixed. The more a bridge is bent, the larger is the induced tension, which acts like a tensile residual stress and limits the bending. Thus, if one desires a large bending displacement, a cantilever would be a preferred structural choice over a bridge. However, a cantilever can easily be warped, both along the length and width directions, if the residual stress varies along the thickness direction, as shown in Chap. 7 with Fig. 7.7. Consequently, the equivalent spring constant of a cantilever may turn out to be much larger than what the equation in Table 6.1 predicts due to the warping of the structure.

Question: By how much is the equivalent stiffness constant for a clamped/clamped beam (Fig. 6.7b) larger than that for a cantilever beam (Fig. 6.7a) when the deflection is small and when the residual stress is negligible. Assume that everything is same for the two structures except the boundary condition.

Answer: 5.3 times, according to Table 6.1.

Example 6.2: Cantilever under Static Point Loading

Problem: The figures below are the top and cross-sectional views of a polysilicon cantilever (100 μm long, 2 μm thick, and 6 μm wide) with one edge clamped. The polysilicon cantilever is separated by 1-μm air gap from the underlying substrate and deflected by an electrostatic voltage applied between the cantilever and the substrate.

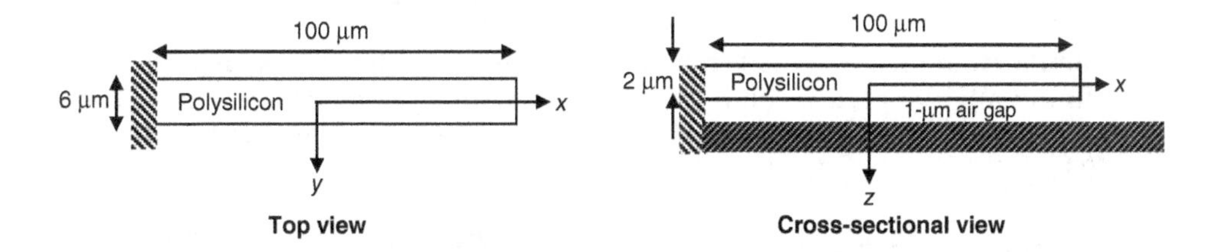

Calculate the cantilever displacement at its free end if a 1 GHz RF signal with a power level of 1 μW is applied between the cantilever and the substrate. For this calculation, take the Young's modulus of the polysilicon to be 150 GPa.

Answer: First, we calculate how much electrostatic force (per unit area) the polysilicon cantilever experiences due to the 1 GHz RF signal at 1 μW. Across the air gap of an air-filled parallel capacitor (C), there exists the following electrostatic attractive force F due to voltage V owing to RF power (P_{RF}) passing through the capacitor:

$$F = \frac{\varepsilon_o AV^2}{2g^2} = \frac{CV^2}{2g} = \frac{\omega CV^2}{2g\omega} = \frac{P_{RF}}{2g\omega} \tag{6.16}$$

where A, g, ε_o, and ω are the capacitor area, the air gap of the capacitor, vacuum permittivity, and RF frequency, respectively. In deriving the equation above, we use Eq. (6.14) and $P_{RF} = V^2/|Z_c|$ with $|Z_c| = 1/\omega C$. Note that the equation shows an electrostatic attractive force due to RF power passing through an air-gap capacitor; the narrower the gap is, the higher the attractive force for a given power. Consequently, the electrostatic force per unit area is

$$\frac{F}{A} = \frac{P_{RF}}{2g\omega A} = \frac{10^{-6}}{2\times 10^{-6}\times 2\pi\times 10^{9}\times 100\times 6\times 10^{-12}} = 0.13\ \text{N/m}^2 \tag{6.17}$$

Now, from Table 6.1, the equivalent spring constant $k' = \dfrac{2Et^3}{3l^4}$ and we calculate

$$\frac{2Et^3}{3l^4} = \frac{2\times 150\times 10^{9}(2\times 10^{-6})^3}{3\times(100\times 10^{-6})^4} = 8\times 10^{9}\ \text{N/m}^3. \tag{6.18}$$

Thus, the cantilever displacement at its free end is

$$\Delta z = \frac{F/A}{k'} = \frac{0.13\ \text{N/m}^2}{8\times 10^{9}\text{N/m}^3} = 1.7\times 10^{-11}\text{m} \tag{6.19}$$

6.1.6 Plate Bending

Just as a thin membrane is a two-dimensional string, so a plate may be considered as a two-dimensional "beam." The governing equation for plate bending is similar to the one for beam bending, except that the derivatives now involve two axes. In this section, we will go through the equations for plate bending, without actually deriving them, just enough for us to be able to grasp the similarity between beam and plate bendings. Exact derivations can be found in [2].

In case of beam bending, from Table 6.1, the governing equation for a bridge under lateral loading P (with no applied electrostatic voltage) is

$$EI\frac{d^4w}{dx^4} = Pb - \sigma(1-\nu)bt\frac{d^2w}{dx^2} \tag{6.20}$$

Similarly, in case of plate bending, the governing equation for small deflections of an initially flat plate of uniform thickness made of homogenous isotropic material (subjected to normal and shear forces in the plane of the plate) under lateral loading P is

$$D\nabla^4 w = D\left(\frac{\partial^4 w}{\partial x^4} + 2\frac{\partial^4 w}{\partial x^2 \partial y^2} + \frac{\partial^4 w}{\partial y^4}\right) = P + N_x\frac{\partial^2 w}{\partial x^2} + 2N_{xy}\frac{\partial^2 w}{\partial x \partial y} + N_y\frac{\partial^2 w}{\partial y^2}, \tag{6.21}$$

where $D = Et^3/12(1-\nu^2)$. Included in the governing equation is the effect of the in-plane normal and shear forces N_x, N_y, and N_{xy}, as defined in [2], which depend not only on the residual stress but also on the stretching of the plate due to bending.

The governing differential equation will have to be solved with the plate's edges satisfying the boundary conditions. For an edge parallel to x axis, we have the following boundary conditions:

for fixed edge, $w = 0$ and $\dfrac{\partial w}{\partial y} = 0$

for simply supported edge, $w = 0$ and $M_{1y} = -D\left(\dfrac{\partial^2 w}{\partial y^2} + \nu\dfrac{\partial^2 w}{\partial x^2}\right) = 0$

for free edge, $M_{1y} = -D\left(\dfrac{\partial^2 w}{\partial y^2} + \nu\dfrac{\partial^2 w}{\partial x^2}\right) = 0$ and $\dfrac{\partial}{\partial y}\left(\dfrac{\partial^2 w}{\partial y^2} + (2-\nu)\dfrac{\partial^2 w}{\partial x^2}\right) = 0$

from $T_{1xy} = 0$ and $S_{1y} = 0$

Bending Stress

When the governing equation is solved for the deflection w, we can calculate stresses induced by plate bending from the deflection w. First, recalling the bending moment for beam is $M = EI\dfrac{d^2w}{dx^2}$, we can see the reasonableness of the following equations for M_{1x} and M_{1y}, the exact derivations of which are shown in [2].

$$M_{1x} = -D\left(\frac{\partial^2 w}{\partial x^2} + \nu\frac{\partial^2 w}{\partial y^2}\right),\quad M_{1y} = -D\left(\frac{\partial^2 w}{\partial y^2} + \nu\frac{\partial^2 w}{\partial x^2}\right),\quad T_{1xy} = D(1-\nu)\frac{\partial^2 w}{\partial x \partial y} \tag{6.22}$$

Similarly, from the equation for shear force for beam, $V(x) = \frac{dM}{dx} = EI\frac{d^3w}{dx^3}$, we can see the reasonableness of the following equations for the shear stresses S_{1x} and S_{1y}

$$S_{1x} = -D\left(\frac{\partial^3 w}{\partial x^3} + \frac{\partial^3 w}{\partial x\,\partial y^2}\right), \quad S_{1y} = -D\left(\frac{\partial^3 w}{\partial y^3} + \frac{\partial^3 w}{\partial x^2\,\partial y}\right) \tag{6.23}$$

The strain-curvature relation [Eq. (6.9)] derived earlier for beam bending is a good starting point for strain-curvature relations for plate bending, and we can see the reasonableness of the following equations of the strains by plate bending, without actually deriving them

$$\varepsilon_x = -z\frac{\partial^2 w}{\partial x^2} \qquad \varepsilon_y = -z\frac{\partial^2 w}{\partial y^2} \qquad \gamma_{yx} = 2z\frac{\partial^2 w}{\partial x\,\partial y} \tag{6.24}$$

where z is the distance from the mid-plane of the plate.

In case of plate bending, normal stress in z direction (σ_z) is zero, since any stress in the z direction is released through shape change in that direction due to the fact that only air is resisting such shape change. Thus, the Hooke's law for isotropic medium is a bit simplified, and we obtain the equations that relate stresses to strains.

$$\begin{gathered}\varepsilon_x = \frac{1}{E}(\sigma_x - \nu\sigma_y) \quad \varepsilon_y = \frac{1}{E}(\sigma_y - \nu\sigma_x) \quad \gamma_{xy} = \frac{\tau_{xy}}{G} \quad (\sigma_z = 0) \\ \sigma_x = \frac{E}{1-\nu^2}(\varepsilon_x + \nu\varepsilon_y) \quad \sigma_y = \frac{E}{1-\nu^2}(\varepsilon_y + \nu\varepsilon_x) \quad \tau_{xy} = G\gamma_{xy}\end{gathered} \tag{6.25}$$

Now, plugging the strain-curvature equations [Eq. (6.24)] into Eq. (6.25), we obtain the following equations that relate stresses to bending curvature (and bending moment).

$$\sigma_x = -\frac{Ez}{1-\nu^2}\left(\frac{\partial^2 w}{\partial x^2} + \nu\frac{\partial^2 w}{\partial y^2}\right) = \frac{12M_{1x}}{h^3}z \tag{6.26}$$

$$\sigma_y = -\frac{Ez}{1-\nu^2}\left(\frac{\partial^2 w}{\partial y^2} + \nu\frac{\partial^2 w}{\partial x^2}\right) = \frac{12M_{1y}}{h^3}z \tag{6.27}$$

$$\tau_{xy} = 2G\frac{\partial^2 w}{\partial x\,\partial y}z = \frac{12T_{1xz}}{h^3}z$$

Simply Supported Rectangular Plate with Zero N_x, N_y, and N_{xy}

For a rectangular plate with its four edges simply supported, the deflection w along the edges must be zero. At the same time, these edges can rotate freely with respect to the edge line; i.e., there are no bending moments along these edges. Thus, for the coordinate axes shown at right, Eq. (6.21) should be solved with the following boundary conditions:

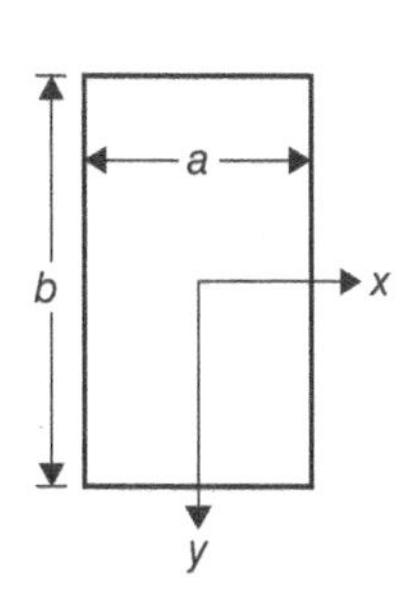

$$w = 0 \text{ and } \left(\frac{\partial^2 w}{\partial x^2} + \nu\frac{\partial^2 w}{\partial y^2}\right) = 0 \text{ at } x = \pm a/2$$

$$w = 0 \text{ and } \left(\frac{\partial^2 w}{\partial y^2} + \nu\frac{\partial^2 w}{\partial x^2}\right) = 0 \text{ at } y = \pm b/2$$

Note that $\frac{\partial^2 w}{\partial y^2} = 0$ at $x = \pm a/2$ and also $\frac{\partial^2 w}{\partial x^2} = 0$ at $y = \pm b/2$ since $w = 0$ at the edges. Thus, we can represent the boundary conditions as follows:

$$w = 0 \text{ and } \frac{\partial^2 w}{\partial x^2} = 0 \text{ at } x = \pm a/2$$

$$w = 0 \text{ and } \frac{\partial^2 w}{\partial y^2} = 0 \text{ at } y = \pm b/2$$

First, let us assume that (1) N_x, N_y, and N_{xy} are zero and (2) the load distributed over the surface of the plate is given by the expression

$$P = a_{mn} \cos\frac{m\pi x}{a} \cos\frac{n\pi y}{b} \tag{6.28}$$

In this case, Eq. (6.21) becomes

$$\frac{\partial^4 w}{\partial x^4} + 2\frac{\partial^4 w}{\partial x^2 \partial y^2} + \frac{\partial^4 w}{\partial y^4} = \frac{a_{mn}}{D} \cos\frac{m\pi x}{a} \cos\frac{n\pi y}{b} \tag{6.29}$$

It may be seen that all the boundary conditions are satisfied, if we take for deflections the expression

$$w = A\cos\frac{m\pi x}{a} \cos\frac{n\pi y}{b} \tag{6.30}$$

Substituting Eq. (6.30) into Eq. (6.29), we get

$$A = \frac{a_{mn}}{\pi^4 D\left(\frac{m^2}{a^2} + \frac{n^2}{b^2}\right)^2} \tag{6.31}$$

Hence,

$$w = \frac{a_{mn}}{\pi^4 D\left(\frac{m^2}{a^2} + \frac{n^2}{b^2}\right)^2} \cos\frac{m\pi x}{a} \cos\frac{n\pi y}{b} \tag{6.32}$$

This solution can be used for calculating deflections produced in a simply supported rectangular plate by any kind of loading given by the equation $P = f(x,y)$. Using Fourier series, we represent the function $f(x,y)$ in the form of a double trigonometric series:

$$f(x,y) = \sum_{m=1}^{\infty}\sum_{n=1}^{\infty} a_{mn} \cos\frac{m\pi x}{a} \cos\frac{n\pi y}{b}$$

where $a_{mn} = \frac{4}{ab} \int_{-b/2}^{b/2} \int_{-a/2}^{a/2} f(x,y) \cos\frac{m\pi x}{a} \cos\frac{n\pi y}{b}\, dx dy$

The total deflection due to $P = f(x,y)$ can be obtained by superposing such terms as the ones given by Eq. (6.32). Thus,

$$w = \frac{1}{\pi^4 D}\sum_{m=1}^{\infty}\sum_{n=1}^{\infty}\frac{a_{mn}}{\left(\frac{m^2}{a^2}+\frac{n^2}{b^2}\right)^2}\cos\frac{m\pi x}{a}\cos\frac{n\pi y}{b}. \tag{6.33}$$

For example, let us take a load uniformly distributed over the entire surface of the plate represented by $f(x,y) = P_o$. Then,

$$a_{mn} = \frac{4P_0}{ab}\int_{-\frac{b}{2}}^{\frac{b}{2}}\int_{-\frac{a}{2}}^{\frac{a}{2}}\cos\frac{m\pi x}{a}\cos\frac{n\pi y}{b}dxdy$$

$$= \begin{cases} \frac{16P_0}{\pi^2 mn}\sin\frac{m\pi}{2}\sin\frac{n\pi}{2}, & \text{for } m \text{ and } n \text{ odd integers} \\ 0, & \text{otherwise} \end{cases}$$

Substituting these into Eq. (6.33), we get

$$w = \frac{16P_0}{\pi^6 D}\sum_{m=1}^{\infty}\sum_{n=1}^{\infty}\frac{\sin\frac{m\pi}{2}\cos\frac{m\pi x}{a}\sin\frac{n\pi}{2}\cos\frac{n\pi y}{b}}{mn\left(\frac{m^2}{a^2}+\frac{n^2}{b^2}\right)^2}$$

$$= \frac{16P_0}{\pi^6 D}\sum_{m=1}^{\infty}\sum_{n=1}^{\infty}\frac{\sin\left[\frac{m\pi}{a}\left(x+\frac{a}{2}\right)\right]\sin\left[\frac{n\pi}{b}\left(y+\frac{b}{2}\right)\right]}{mn\left(\frac{m^2}{a^2}+\frac{n^2}{b^2}\right)^2} \tag{6.34}$$

where m and n are odd integers.

Rectangular Plate with All Edges Fixed and Zero N_x, N_y, and N_{xy}

For the coordinate axes shown at right, we have the following boundary conditions for rectangular plates with all edges fixed:

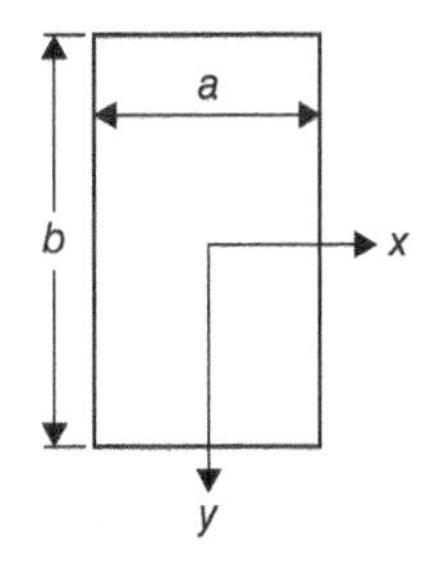

$$w = 0 \text{ and } \partial w/\partial x = 0 \text{ at } x = \pm a/2$$
$$w = 0 \text{ and } \partial w/\partial y = 0 \text{ at } y = \pm b/2.$$

To obtain an equation for the deflection (of a rectangular plate with its four edges fixed) under a load uniformly distributed over the entire surface of the plate represented by $f(x,y) = P_o$, we can employ the equation for the deflection of a simply supported rectangular plate, and superpose on it the equation for the deflection of a plate driven by the moments distributed along the edges. These moments are adjusted in such a manner as to satisfy the conditions $\partial w/\partial x = 0$ at $x = \pm a/2$ and $\partial w/\partial y = 0$ at $y = \pm b/2$. The actual derivation [2] is beyond the scope of this book, and only the final equation for a **square plate** with side a [3] is presented below:

$$w = \frac{2P_0a^4}{\pi^5 D} \sum_{m=1,3,5\ldots}^{\infty} \frac{(-1)^{\frac{m-1}{2}}}{m^2\cosh\frac{m\pi}{2}}$$

$$\left[\cos\frac{m\pi x}{a}\left\{\frac{2}{m^3}\cosh\frac{m\pi}{2} - \left(\left\langle\frac{1}{m^3}+e_m\right\rangle\frac{m\pi}{2}\tanh\frac{m\pi}{2}+\frac{2}{m^3}\right)\cosh\frac{m\pi y}{a}\right.\right.$$

$$\left. + \left(\frac{1}{m^3}+e_m\right)\frac{m\pi y}{a}\sinh\frac{m\pi y}{a}\right\}$$

$$\left. + e_m\cos\frac{m\pi y}{a}\left\{\frac{m\pi x}{a}\sinh\frac{m\pi x}{a} - \frac{m\pi}{2}\tanh\frac{m\pi}{2}\cosh\frac{m\pi x}{a}\right\}\right] \tag{6.35}$$

where $D = Et^3/12(1-\nu^2)$ and $e_1 = 0.3721$, $e_3 = -0.0380$, $e_5 = -0.0177$, $e_7 = -0.0085\ldots$.

Using Eqs. (6.26) and (6.27), we obtain the following equations for the normal stresses:

$$\sigma_x = \frac{24P_0a^2z}{\pi^3t^3} \sum_{m=1,3,5\ldots}^{\infty} \frac{(-1)^{(m-1)/2}}{\cosh\frac{m\pi}{2}}$$

$$\left[\cos\frac{m\pi x}{a}\left\{\frac{2}{m^3}\cosh\frac{m\pi}{2} - \left(\left\langle\frac{1}{m^3}+e_m\right\rangle\frac{m\pi}{2}\tanh\frac{m\pi}{2}\langle 1-\nu\rangle+\frac{2}{m^3}+2\nu e_m\right)\cosh\frac{m\pi y}{a}\right.\right.$$

$$\left. + \left(\frac{1}{m^3}+e_m\right)(1-\nu)\frac{m\pi y}{a}\sinh\frac{m\pi y}{a}\right\}\Bigg\}$$

$$\left. - e_m\cos\frac{m\pi y}{a}\left\{\langle 1-\nu\rangle\frac{m\pi x}{a}\sinh\frac{m\pi x}{a} + \left(2+\langle\nu-1\rangle\frac{m\pi}{2}\tanh\frac{m\pi}{2}\right)\cosh\frac{m\pi x}{a}\right\}\right] \tag{6.36}$$

$$\sigma_y = \frac{24P_0a^2z}{\pi^3t^3} \sum_{m=1,3,5\ldots}^{\infty} \frac{(-1)^{\frac{m-1}{2}}}{\cosh\frac{m\pi}{2}}$$

$$\left[\cos\frac{m\pi x}{a}\left\{\frac{2\nu}{m^3}\cosh\frac{m\pi}{2} - \left(\left\langle\frac{1}{m^3}+e_m\right\rangle\frac{m\pi}{2}\tanh\frac{m\pi}{2}\langle\nu-1\rangle+\frac{2\nu}{m^3}+2e_m\right)\cosh\frac{m\pi y}{a}\right.\right.$$

$$\left. + \left\langle\frac{1}{m^3}+e_m\right\rangle\langle\nu-1\rangle\frac{m\pi y}{a}\sinh\frac{m\pi y}{a}\right\}$$

$$\left. - e_m\cos\frac{m\pi y}{a}\left\{\langle\nu-1\rangle\frac{m\pi x}{a}\sinh\frac{m\pi x}{a} + \left(2\nu+\langle 1-\nu\rangle\frac{m\pi}{2}\tanh\frac{m\pi}{2}\right)\cosh\frac{m\pi x}{a}\right\}\right] \tag{6.37}$$

Note that the stresses depend on the geometry of the diaphragm and Poisson's ratio, but not on Young's modulus. Also, the stresses are linearly proportional to the applied pressure and to z direction.

The series for σ_x and σ_y [Eqs. (6.36) and (6.37)] converge rather fast as m increases. Thus, for less than 10% error, we can approximate the two equations as follows, for $\nu = 0.3$ [3]:

$$\sigma_x \approx \frac{24P_0a^2z}{\pi^3t^3}\left[\cos\frac{\pi x}{a}\left\{2-1.4375\cosh\frac{\pi y}{a}+0.3828\frac{\pi y}{a}\sinh\frac{\pi y}{a}\right\}\right.$$
$$-\cos\frac{3\pi x}{a}\left\{\frac{2}{27}-0.0008638\cosh\frac{3\pi y}{a}-0.0000363\frac{\pi y}{a}\sinh\frac{3\pi y}{a}\right\}$$
$$\left.-\cos\frac{\pi y}{a}\left\{0.147\sinh\frac{\pi x}{a}+0.1038\frac{\pi x}{a}\sinh\frac{\pi x}{a}\right\}\right] \tag{6.38}$$

$$\sigma_y \approx \frac{24P_0a^2z}{\pi^3t^3}\left[\cos\frac{\pi x}{a}\left\{0.6+0.01573\cosh\frac{\pi y}{a}-0.3828\frac{\pi y}{a}\sinh\frac{\pi y}{a}\right\}\right.$$
$$\left.-\cos\frac{\pi y}{a}\left\{0.2385\cosh\frac{\pi x}{a}-0.1038\frac{\pi x}{a}\sinh\frac{\pi x}{a}\right\}\right] \tag{6.39}$$

To show that the errors are less than 10% when Eqs. (6.38) and (6.39) are used, the normalized stresses (i.e., $\bar{\sigma}_x = \frac{\pi^3t^3}{24P_0a^2z}\sigma_x$ and $\bar{\sigma}_y = \frac{\pi^3t^3}{24P_0a^2z}\sigma_y$) calculated by using different number of terms in Eqs. (6.36) and (6.37) with $\nu = 0.3$ are tabulated as follows:

		x = 0, y = 0	x = 0, y = a/2	x = a/2, y = 0	x = a/2, y = a/2
$\bar{\sigma}_x = \frac{\pi^3t^3}{24P_0a^2z}\sigma_x$	**Eq. (6.38)**	**0.3423**	**−0.2232**	**−0.7441**	**0**
	Up to m = 3	0.3432	−0.2460	−0.8201	0
	Up to m = 5	0.3592	−0.2354	0.7847	0
	Up to m = 7	0.3533	−0.2405	0.8017	0
$\bar{\sigma}_y = \frac{\pi^3t^3}{24P_0a^2z}\sigma_y$	**Eq. (6.39)**	**0.3772**	**−0.7443**	**−0.2232**	**0**
	Up to m = 3	0.3514	−0.8203	−0.2459	0
	Up to m = 5	0.3563	−0.7849	−0.2352	0
	Up to m = 7	0.3545	−0.8019	−0.2403	0

Contour plots for the normalized stresses (i.e., $\bar{\sigma}_x = \frac{\pi^3t^3}{24P_0a^2z}\sigma_x$ and $\bar{\sigma}_y = \frac{\pi^3t^3}{24P_0a^2z}\sigma_y$) are plotted in Figs. 6.9 and 6.10. The values plotted in Fig. 6.9 are calculated by using terms up to $m = 7$ in Eqs. (6.36) and (6.37), while values in Fig. 6.10 are plots calculated with Eqs. (6.38) and (6.39). It can be seen that Eqs. (6.38) and (6.39) give reasonable approximations to Eqs. (6.36) and (6.37), respectively.

The contour plots in Figs. 6.9 and 6.10 are for the normal stresses (σ_x and σ_y) over uniformly loaded square diaphragm with its four edges clamped and with t, a, z, and P_o

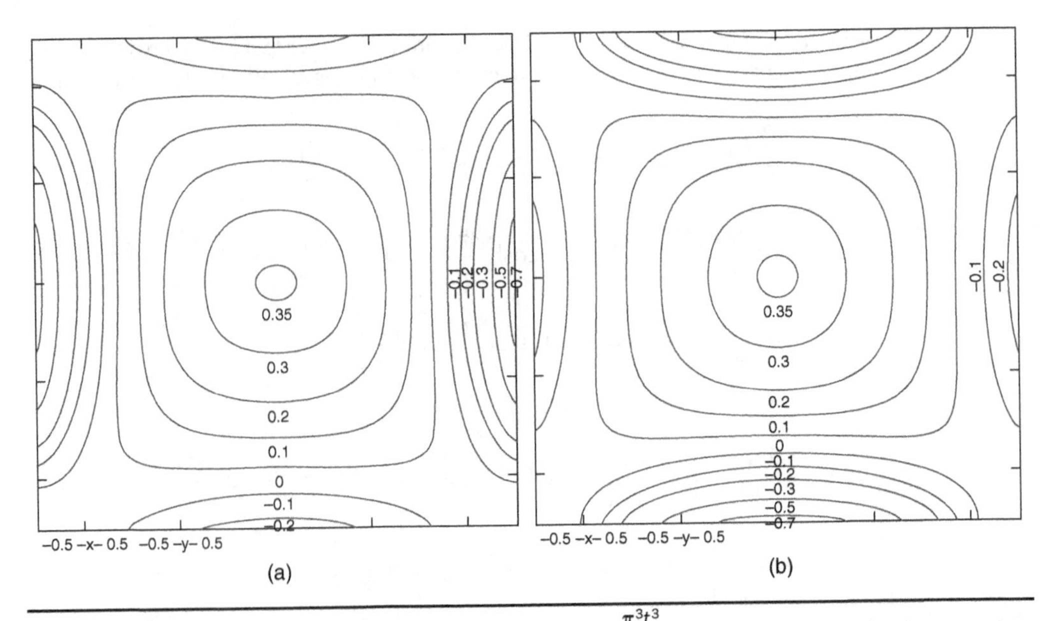

FIGURE 6.9 Contour plots of (a) the normalized stress $\bar{\sigma}_x = \dfrac{\pi^3 t^3}{24P_0 a^2 z}\sigma_x$ calculated by using terms up to $m = 7$ in Eq. (6.36) with $\nu = 0.3$ and (b) the normalized stress $\bar{\sigma}_y = \dfrac{\pi^3 t^3}{24P_0 a^2 z}\sigma_y$ calculated by using terms up to $m = 7$ in Eq. (6.37) with $\nu = 0.3$ [3].

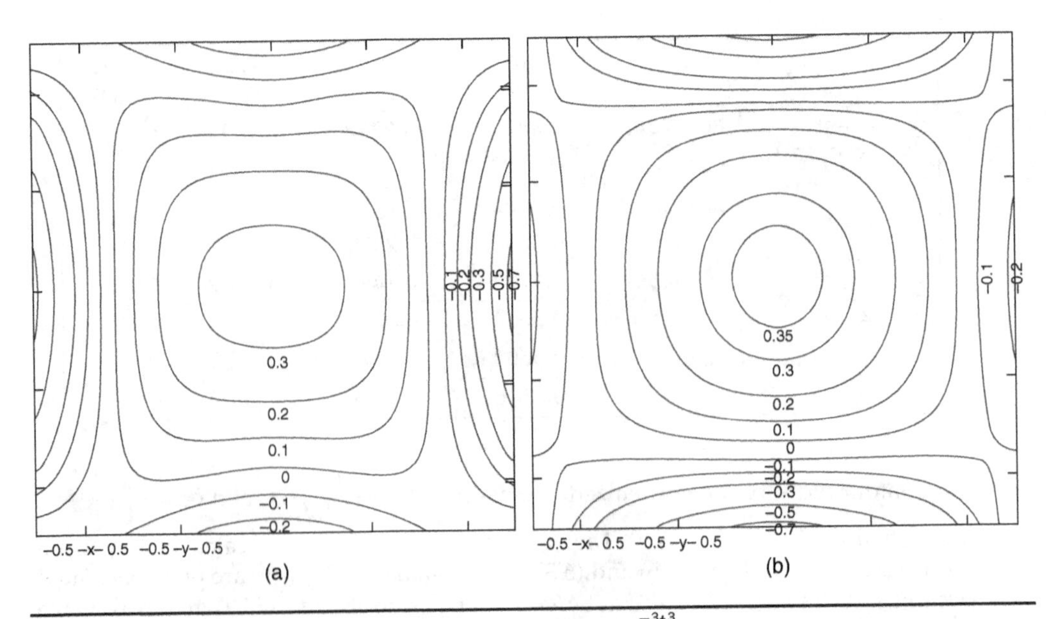

FIGURE 6.10 Contour plots of (a) the normalized stress $\bar{\sigma}_x = \dfrac{\pi^3 t^3}{24P_0 a^2 z}\sigma_x$ calculated by using the approximate equation, Eq. (6.38), with $\nu = 0.3$ and (b) the normalized stress $\bar{\sigma}_y = \dfrac{\pi^3 t^3}{24P_0 a^2 z}\sigma_y$ calculated by using the approximate equation, Eq. (6.39), with $\nu = 0.3$ [3].

being the thickness, length of a side, distance from the neutral axis in the thickness direction, and applied pressure, respectively. From the plots, note the following:

1. σ_x and σ_y are symmetrical with respect to x and y axes, and thus only one quadrant of a square diaphragm is needed to describe the stress distributions.
2. $\sigma_x(x,y) = \sigma_y(y,x)$, at the same z. In other words, σ_x is same as σ_y when folded with respect to the line $y = x$.
3. $\sigma_x(x,y) \neq \sigma_y(x,y)$ in most of the places due to how the normal stresses are developed as the diaphragm bends.
4. σ_x is largest at $x = a/2$ and $y = 0$, while σ_y is largest at $x = 0$ and $y = a/2$. In other words, the points where the normal stresses are the largest are at the centers of the edges.
5. σ_x and σ_y are zero at the four corners of the diaphragm.

Useful Table for Plates

Table 6.2 provides the following information for square- and circular plates with their edges either simply supported or clamped (fixed or built-in) under a uniform loading over the plates: (1) bending displacements at the centers of the plates and (2) bending moments at selected points, from which bending stresses can be calculated with

w_{max} = maximum deflection, P = uniform pressure, t = plate thickness, E = Young's modulus		Simply supported edges			Built-in edges		
		At center	At edge		At center	At edge	
		$x = 0$, $y = 0$	$x = \pm a/2$, $y = 0$	$x = 0$, $y = \pm a/2$	$x = 0$, $y = 0$	$x = \pm a/2$, $y = 0$	$x = 0$, $y = \pm a/2$
Square plates ($\nu = 0.3$)	$\frac{w_{max}}{Pa^4/Et^3}$	0.044	0	0	0.014	0	0
	$\frac{M_{1x}}{Pa^2}$	0.048	0	0	0.026	−0.051	−0.015
	$\frac{M_{1y}}{Pa^2}$	0.048	0	0	0.026	−0.015	−0.051
Circular plates ($\nu = 0.3$)	$\frac{w_{max}}{Pr^4/Et^3}$	0.695	0		0.171	0	
	$\frac{M_{radial}}{Pr^2}$	0.206	0		0.0813	−0.125	
	$\frac{M_{tangential}}{Pr^2}$	0.206	0		0.0813	−0.0375	

Adapted from [4].

TABLE 6.2 Maximum Static Deflections and Bending Moments in Uniformly Loaded Plates for Small Deflections and No In-Plane Normal nor Shear Forces

Eqs. (6.26) and (6.27) for rectangular plates. For rectangular plates with various aspect ratios, please refer to Table 7.6 in [4].

For a square plate with its four edges clamped, note that $(M_{1x})_{x=a/2,\,y=0}$ is exactly same as $(M_{1y})_{x=0,\,y=a/2}$, while $(M_{1x})_{x=a/2,\,y=0}$ and $(M_{1y})_{x=a/2,\,y=0}$ are not same as M_{1x} and M_{1y} are directly related to stresses as shown in Eqs. (6.26) and (6.27). For a square plate under a uniform loading, the M_{1x} at $x = a/2$ and $y = 0$ is the same with the M_{1y} at $x = 0$ and $y = a/2$, because the square plate is 90° symmetrical, in that if the square plate is rotated 90°, all parameters do not change except the 90° rotated vectors. Thus, $(M_{1x})_{x=a/2,y=0}$ is exactly same as $(M_{1y})_{x=0,\,y=a/2}$, and the normal stresses σ_{xx} and σ_{yy} (that are directly related to the bending moments M_{1x} and M_{1y}, respectively) are 90° symmetrical, too. At a same location, say at $x = a/2$ and $y = 0$, σ_{xx} and σ_{yy} are usually not same, because, by definition, σ_{xx} is the normal stress directed in the x direction working on a plane that is cutting the x axis perpendicularly, while σ_{yy} is the normal stress directed in the y direction working on a plane that is cutting the x axis perpendicularly.

Question: Calculate the maximum displacement (i.e., the displacement at the center) of a 1×1 mm^2 square diaphragm (1 μm thick) with its four edges built-in for an applied pressure of 1 Pa, assuming $E = 200$ GPa and Poisson's ratio $\nu = 0.3$.

Answer: $w_{max} = 0.014Pa^4/Et^3 = 0.014 \times 1 \times (10^{-3})^4/(200 \times 10^9 \times (10^{-6})^3) = 0.07$ μm

Example 6.3: Square Plate with Edges Clamped under Static Uniform Loading

Problem: The figures below are the top- and cross-sectional views of a micromachined silicon nitride square diaphragm with its four edges clamped on a silicon substrate. Assume that $a = 1$ mm, $t = 1$ μm, $E = 300$ GPa, and $\nu = 0.3$ for the diaphragm and also that the diaphragm is uniformly loaded with 1 Pa of static pressure (as shown in the figure). Assuming that the diaphragm is free of any residual stress, answer the following: (a) what is the displacement at the center of the diaphragm (i.e., at $x = y = 0$) due to the 1 Pa pressure? (b) At which points in the diaphragm is σ_{xx} largest? At which points in the diaphragm is σ_{yy} largest? What is the largest σ_{xx}? (c) How does σ_{xx} vary as a function of z from $-t/2$ to $+t/2$ at $x = 0$ and $y = 0$? (d) How do σ_{xx} and σ_{yy} vary as a function of x from $-a/2$ to $+a/2$ at $y = +a/2$ and $z = t/2$ (and at $y = 0$ and $z = t/2$)?

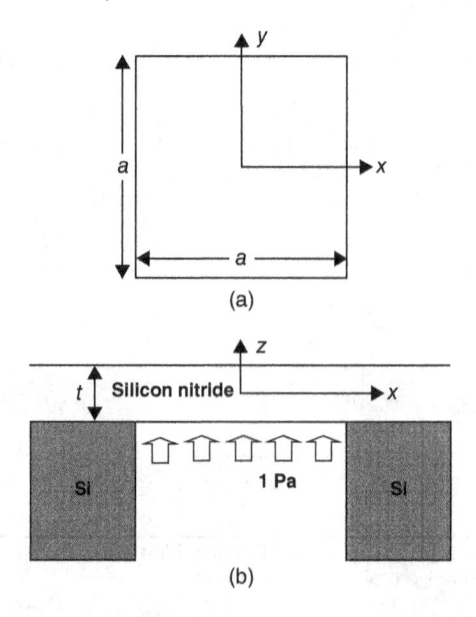

(a)
(b)

Answers: (a) From Table 6.2, $w_{max} = \frac{\alpha Pa^4}{Et^3} = \frac{0.014 \times 1\text{Pa} \times (1\times10^{-3}\text{m})^4}{300 \times 10^9\text{Pa} \times (1\times10^{-6}\text{m})^3} = 0.05\ \mu\text{m}$

(b) From Fig. 6.9a, σ_{xx} is the largest at ($\pm a/2$, 0, $\pm t/2$), while σ_{yy} is the largest at (0, $\pm a/2$, $\pm t/2$).

$$\sigma_{x,max} = 0.7\frac{24a^2(t/2)P}{\pi^3 t^3} = \frac{8.4(10^{-3})^2 \times 1}{\pi^3 \times (10^{-6})^2} = 0.27 \times 10^6\,\text{Pa}$$

(c)

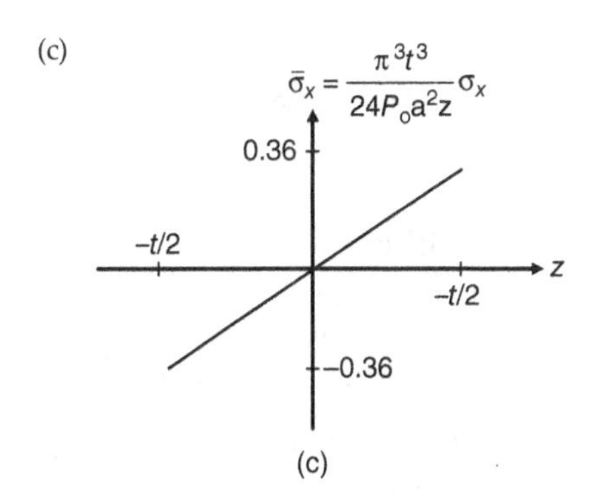

(c)

(d)

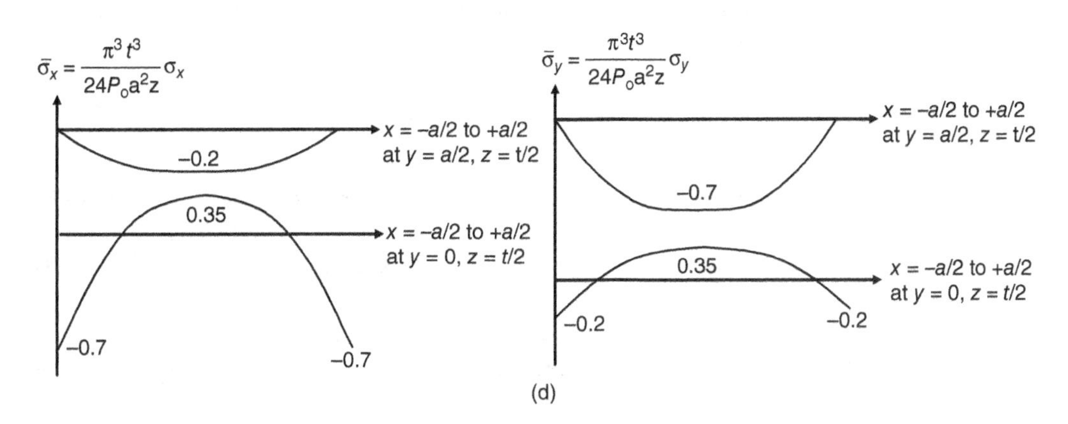

(d)

6.2 Dynamics

In case of a simple mass-spring system, we can derive an equation for the resonant frequency of its mechanical vibration very easily as shown below. However, for an elastic body with distributed mass and elasticity, the resonant frequency calculation is usually not easy because of infinite number of vibration modes and infinite degrees of freedom. Two methods exist for the resonant frequency calculation: (1) classical method through writing the differential equation of motion from Newton's second law and (2) energy method for approximate solution. As we will see in next several sections, the energy method is usually much easier to use for calculating the fundamental resonant frequency.

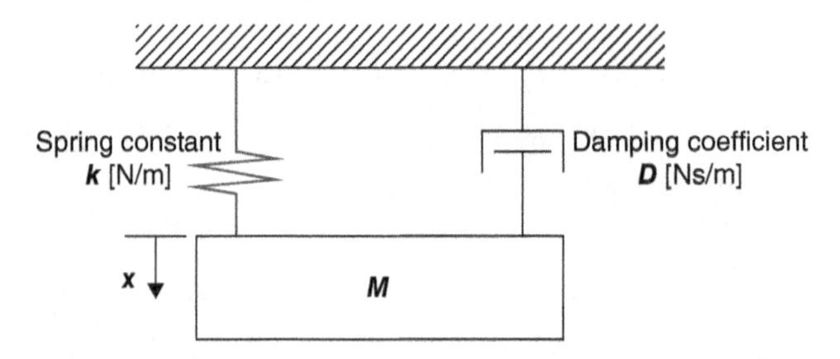

Figure 6.11 Spring-mass-dashpot with spring constant k, damping coefficient D, and mass M.

6.2.1 Mass-Spring-Dashpot and Tensioned String

Newton's law for free vibration of a mass-spring-dashpot (with spring constant k, damping coefficient D, and mass M) shown in Fig. 6.11 is $M \cdot d^2x/dt^2 + D \cdot dx/dt + k \cdot x = 0$. For harmonic motion, $x(t) = Xe^{j\omega t}$, and the time differential equation becomes $-\omega^2 M + j\omega D + k = 0$, from which we can obtain ω. If $D = 0$ (i.e., no damping), $\omega_R = \sqrt{\frac{k}{M}}$. We will study more on this later when we go through accelerometers.

Consider a statically loaded string of length L with no flexural stiffness (similar to a guitar string). Let μ = mass/length = ρA, where ρ = volume density and A = cross-sectional area of a string, and $z(x,t)$ = deflection curve of vibrating string. Applying Newton's law in Fig. 6.12, we obtain

$$\mu dx \frac{\partial^2 z}{\partial t^2} = T\frac{\partial z}{\partial x} + T\frac{\partial^2 z}{\partial x^2}dx - T\frac{\partial z}{\partial x} \quad \Rightarrow \mu\frac{\partial^2 z}{\partial t^2} = T\frac{\partial^2 z}{\partial x^2} \tag{6.40}$$

Assuming harmonic vibration of string, $z(x,t) = z(x)\sin\omega t$ or $z(x)e^{j\omega t}$. And we see

$$-\mu\omega^2 z(x) = T\frac{\partial^2 z}{\partial x^2} \quad \Rightarrow \quad \frac{\partial^2 z}{\partial x^2} + \frac{\mu\omega^2}{T}z(x) = 0 \tag{6.41}$$

General solution for the spatial differential equation above is $z(x) = C_1 \sin\sqrt{\frac{\mu\omega^2}{T}}x + C_2\cos\sqrt{\frac{\mu\omega^2}{T}}x$. From the boundary conditions $z(0) = z(L) = 0$, we get $z(x) = C_1\sin\sqrt{\frac{\mu\omega^2}{T}}x$

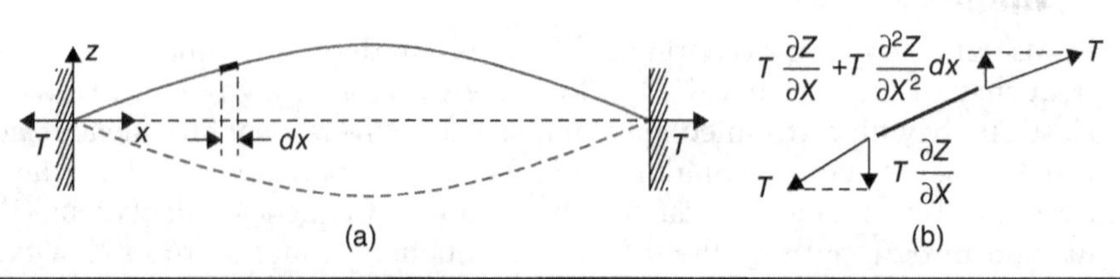

Figure 6.12 A string in free vibration: (a) at a particular point in time and (b) expanded view of a differential element dx with the vertical force components derived from the tension T (induced by the string deflection).

and $L\sqrt{\frac{\mu\omega^2}{T}} = n\pi$, where n = 1, 2, 3, ... Thus, the first resonant frequency and mode shape are $\omega_1 = \frac{\pi}{L}\sqrt{\frac{T}{\mu}}$ and $z(x) = C_1 \sin\frac{\pi x}{L}$, respectively.

So far, we have derived the equations for the first resonant frequency and mode shape for a tensioned string (such as a guitar string), using the classical method. Note that the resonant frequency is inversely proportional to the string length and proportional to the square of the tension in the string. Now, we will derive the equations using the energy method (Rayleigh's method).

6.2.2 Energy Method (Rayleigh's Method)

Using the fact that the maximum potential energy due to string stretching (which occurs at the string's maximum deflection) is equal to the maximum kinetic energy (which occurs as the string goes through the center line), we can obtain an equation for the fundamental resonant frequency. In the case of the tensioned string with a constant tension T, the potential energy due to stretching is equal to $\int_0^L T\,\delta l$, integrated along the string, with δl being the incremental increase of the string length.

Referring to a diagram below,

$$dl = \sqrt{dx^2 + dz^2} = dx\sqrt{1+\left(\frac{dz}{dx}\right)^2} = dx\left(1+\frac{1}{2}\left(\frac{dz}{dx}\right)^2\right) \quad \text{if } dz \ll dx$$

$$\delta l = dl - dx = \frac{1}{2}\left(\frac{dz}{dx}\right)^2 dx$$

dl
dz
dx

Thus, the maximum potential energy due to string stretch is

$$\text{P.E.}_{\max} = \int_0^L T\frac{1}{2}\left(\frac{dz}{dx}\right)^2 dx = \frac{T}{2}\int_0^L \left(\frac{dz}{dx}\right)^2 dx$$

Equating $\text{P.E.}_{\max}$ and $\text{K.E.}_{\max}$ $(= mv^2/2 = \frac{\mu}{2}\int_0^L \omega^2 z^2\, dx)$, we obtain

$$\omega^2 = \frac{\frac{T}{2}\int_0^L \left(\frac{dz}{dx}\right)^2 dx}{\frac{\mu}{2}\int_0^L z^2\, dx}$$

The power of this technique called Rayleigh's method is in that virtually any reasonable assumption for mode shape will give a good estimate of the resonant frequency. For example, using the exact mode shape, $z(x) = C_1 \sin\frac{\pi x}{L}$, we obtain

$$\omega^2 = \left(\frac{\pi}{L}\right)^2 \frac{T}{\mu} \frac{\int_0^L C_1^2 \cos^2\left(\frac{\pi x}{L}\right) dx}{\int_0^L C_1^2 \sin^2\left(\frac{\pi x}{L}\right) dx} \quad \Rightarrow \quad \omega = \frac{\pi}{L}\sqrt{\frac{T}{\mu}} \tag{6.42}$$

Using an approximate mode shape shown below, $y(x) = \begin{cases} 2C_1x/L & x \le L/2 \\ C_1(2-2x/L) & x \ge L/2 \end{cases}$, we obtain

$$\int_0^L \left(\frac{dy}{dx}\right)^2 dx = \int_0^L \left(\frac{2C_1}{L}\right)^2 dx = \frac{(2C_1)^2}{L}$$

$$\int_0^L y^2\, dx = 2\int_0^{L/2} \left(\frac{2C_1}{L}\right)^2 x^2\, dx = \frac{(2C_1)^2 L}{12}$$

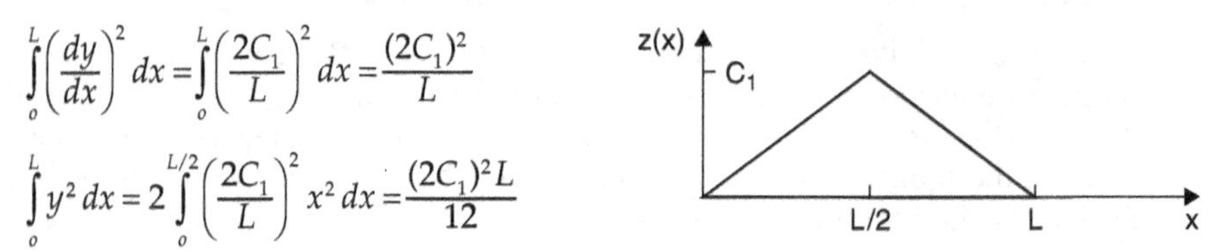

$\omega^2 = \dfrac{T}{\mu}\dfrac{12}{L^2} \quad \Rightarrow \quad \omega = \dfrac{3.464}{L}\sqrt{\dfrac{T}{\mu}}$, which is 10% greater than the exact value.

It has been mathematically proven that Rayleigh's approximation (generalized energy method) always gives for the lowest natural frequency a value which is greater than the exact value. Consequently, among a number of approximate results found in Rayleigh's methods, the smallest is always the best one.

Consequently, to use the energy method, (1) assume any approximate mode shape, (2) calculate the maximum potential (strain) energy and kinetic energy, deriving the formulas for them or simply using the formulas that have already been derived such as those listed in Table 6.3, (3) calculate the fundamental resonant frequency, (4) choose a different mode shape, (5) repeat steps 2 and 3, (6) repeat steps 4 and 5 as many times as you want, and (7) and choose the smallest resonant frequency value for the fundamental frequency.

Question: For a relatively complex structure, we *approximately* calculate the fundamental resonant frequency by using Rayleigh's method, for which we need to choose a mode shape that looks similar to the exact mode shape and satisfies the boundary conditions. Will the actual resonant frequency be always smaller than, larger than, or anywhere near the calculated frequency?

Answer: Always smaller.

		Kinetic Energy	
	Strain Energy	**General**	**Harmonic Motion**
Beam in bending	$\frac{EI}{2}\int_0^L \left(\frac{\partial^2 w}{\partial x^2}\right)^2 dx$	$\frac{A\rho}{2}\int_0^L \left(\frac{\partial w}{\partial t}\right)^2 dx$	$\frac{A\rho\omega^2}{2}\int_0^L W^2 dx$
Rectangular plate in bending	$\frac{D}{2}\iint \left\{\left(\frac{\partial^2 w}{\partial x^2}+\frac{\partial^2 w}{\partial y^2}\right)^2 - 2(1-\nu)\left[\frac{\partial^2 w}{\partial x^2}\frac{\partial^2 w}{\partial y^2} - \left(\frac{\partial^2 w}{\partial x\, \partial y}\right)^2\right]\right\} dxdy$	$\frac{\rho t}{2}\iint \left(\frac{\partial w}{\partial t}\right)^2 dx\, dy$	$\frac{\rho t\omega^2}{2}\iint W^2\, dx\, dy$
Circular plate in bending	$\frac{D}{2}\int_0^a \left\{\left(\frac{\partial^2 w}{\partial r^2}+\frac{1}{r}\frac{\partial w}{\partial r}\right)^2 - 2(1-\nu)\frac{\partial^2 w}{\partial r^2}\frac{1}{r}\frac{\partial w}{\partial r}\right\} rdr$	$\pi\rho t\int_0^a \left(\frac{\partial w}{\partial t}\right)^2 rdr$	$\pi\rho t\omega^2 \int_0^a W^2 rdr$

A = cross-sectional area of beam, a = radius of circular plate, L = beam length, t = plate thickness, ρ = mass density, I = moment of inertia of beam, E = Young's modulus, ν = Poisson's ratio, $D = \dfrac{Et^3}{12(1-\nu^2)}$.

Adapted from [4].

TABLE 6.3 Strain and Kinetic Energies in Bending of Beam and Plates

Example 6.4: Fundamental Resonant Frequency of Cantilever with Proof Mass

Problem: The figures below are the top- and cross-sectional views of a micromachined silicon cantilever supporting a silicon proof mass. For an acceleration applied to the proof mass producing a force F_i at the center of gravity of the proof mass as shown below, derive the equations for the cantilever's displacement and fundamental resonant frequency. Assume that $w_2 >> w_1$, $l_2 >> l_1$, and $t_2 >> t_1$.

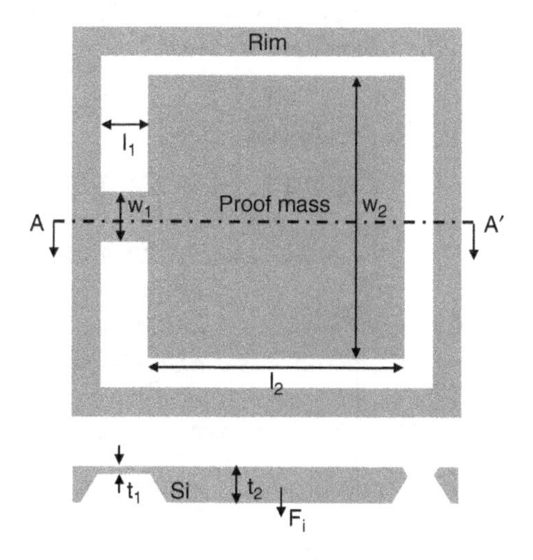

Answer: The cantilever's free body diagram is shown below.

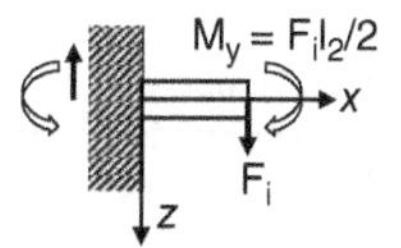

Deflection (w) in response to F_i: $EI_y \dfrac{d^2w}{dx^2} = -M_y(x) = F_{iz}(l_1 - x) + F_i l_2/2$, where $I_y = w_1 t_1^3/12$.

$$\text{Let } y = w'. \Rightarrow y' = \frac{-F_i x}{EI_y} + C_1 \Rightarrow y = \int\left(\frac{-F_i x}{EI_y} + C_1\right)dx = \frac{-F_i x^2}{2EI_y} + C_1 x + C_2 = w'$$

$$\text{B.C.: } w(0) = w'(0) = 0. \Rightarrow w = \int\left(\frac{-F_i x^2}{2EI_y} + C_1 x\right)dx = \frac{-F_i x^3}{6EI_y} + \frac{C_1 x^2}{2} + C_3$$

$\therefore w = \dfrac{F_i x^2}{6EI_y}[3(l_1 + l_2/2) - x]$, where l_2 is the proof-mass length, and typically $>> l_1$.

Rayleigh's method for the resonant frequency: kinetic energy is mainly by proof mass, while potential energy comes from cantilever bending.

$$d(\text{PE}_{\text{bend}}) = \frac{1}{2} EI_y \left(\frac{d^2w}{dx^2}\right)^2 dx, \quad \text{KE}_{\max} \approx \frac{M_p}{2}\omega_1^2[w(L)]^2 = \omega_1^2 \text{KE}'_{\max}$$

$$\omega_1^2 = \frac{\text{PE}_{\max}}{\text{KE}'_{\max}} = \frac{Ew_1 t_1^3}{M_p} \frac{4l_1^3 + 6l_1^2 l_2 + 3l_1 l_2^2}{(4l_1^3 + 3l_1^2 l_2)^2}$$

6.2.3 Vibrations of Beams

The figure below shows the relationships among *static* load, shear force, bending moment, deflection-curve slope, and deflection curve for a beam with its two edges simply supported. Those are all related to each other through spatial differentiation or integration, stemming from the governing equation, static load $= EI\dfrac{d^4w}{dx^4}$, which is solved according to the boundary conditions. Both the bending moment and the displacement are *zero* at the two edges because they are simply supported.

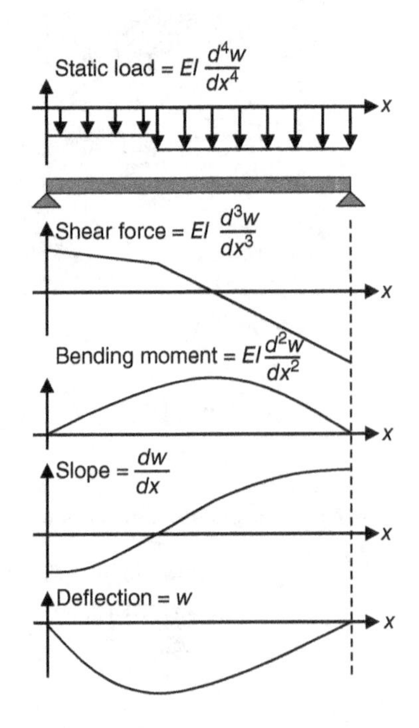

Now, if the beam is naturally vibrating (without any forced input, i.e., zero static load), there is an alternating inertia load due to the acceleration/deceleration of the beam going through vibration, and the governing equation is $EI\dfrac{\partial^4 w}{\partial x^4} = -\rho_1\dfrac{\partial^2 w}{\partial t^2}$, where ρ_1 is the beam mass per unit length. Assuming a sustained free vibration ($w(x, t) = w(x)\sin\omega t$ or $w(x)e^{j\omega t}$), the governing equation becomes $\rho_1\omega^2 w = EI\dfrac{\partial^4 w}{\partial x^4}$.

General solution is $w(x) = C_1e^{ax} + C_2e^{-ax} + C_3\sin ax + C_4\cos ax$, where $a = \sqrt[4]{\dfrac{\rho_1\omega^2}{EI}}$. From B.C.'s for simply supported ends, $w(0) = w(l) = 0$ and $w''(0) = w''(l) = 0$, we obtain $w(x) = C\sin ax$ and $al = l\sqrt[4]{\dfrac{\rho_1\omega^2}{EI}} = n\pi$, where $n = 1,2,3,\ldots$ Thus, $\omega_1 = \dfrac{\pi^2}{l^2}\sqrt{\dfrac{EI}{\rho_1}}$, $\omega_2 = \dfrac{4\pi^2}{l^2}\sqrt{\dfrac{EI}{\rho_1}}, \ldots, \omega_n = \dfrac{n^2\pi^2}{l^2}\sqrt{\dfrac{EI}{\rho_1}}$. So, we see that the solution of the differential equation along with the resonant frequencies can easily be obtained for simply supported edges. Other boundary conditions yield more complex solutions, which are tabulated in Table 6.4 that shows the resonant frequencies and mode shapes for the first, second, third, and

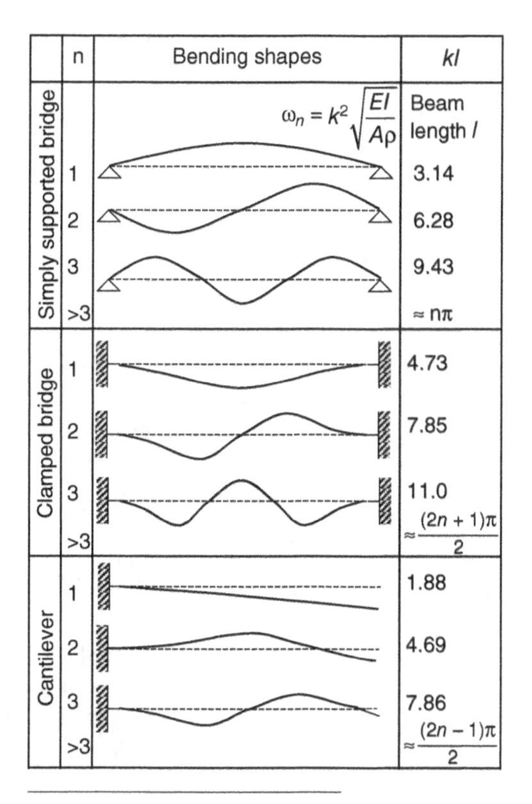

	n	Bending shapes	kl
		$\omega_n = k^2\sqrt{\frac{EI}{A\rho}}$	Beam length l
Simply supported bridge	1		3.14
	2		6.28
	3		9.43
	>3		$\approx n\pi$
Clamped bridge	1		4.73
	2		7.85
	3		11.0
	>3		$\approx \frac{(2n+1)\pi}{2}$
Cantilever	1		1.88
	2		4.69
	3		7.86
	>3		$\approx \frac{(2n-1)\pi}{2}$

Adapted from [4].

Table 6.4 Beam's Resonant Frequencies and Bending Shapes at Fundamental and Harmonic Frequencies

fourth harmonic resonances for beams with various boundary conditions at their two supports. The resonant frequencies for those modes can be obtained from the fourth column that lists the numerical values for kl. For a given beam length l, the k value can be calculated from the listed kl and can be plugged into $\omega_n = k^2\sqrt{\frac{EI}{A\rho}}$, where E, I, A, and ρ are Young's modulus, moment of inertia, beam cross-sectional area, and mass density (kg/m^3), respectively, in order to calculate the resonant frequency.

Question: Calculate the fundamental resonant frequency of a cantilever that is 100 μm long and has its cross-sectional area equal to 100 μm^2. Assume that the Young's modulus and moment of inertia are equal to 100 GPa and 10^{-22} m^4, respectively. Also, assume that the mass density is equal to 1,000 kg/m^3.

Answer: From Table 6.4, $k = \frac{1.88}{l} = 1.875 \times 10^4 \text{m}^{-1}$

$$\omega_1 = (1.875 \times 10^4 \text{m}^{-1})^2 \sqrt{\frac{100 \times 10^9 \text{N/m}^2 \times 10^{-22} \times \text{m}^4}{100 \times 10^{-12} \text{m}^2 \times 1{,}000 \text{ kg/m}^3}} = 3.5 \times 10^6 \text{rad/s}$$

Question: For a cantilever, by what factor will the fundamental resonant frequency be changed, if its length is reduced by a factor of 2, while all the other parameters remain same?

Answer: According to Table 6.4, the resonant frequency will be increased by a factor of 4.

Question: If a cantilever beam has the fundamental resonant frequency equal to 2 kHz, what is its second harmonic resonant frequency?

Answer: $f_2 = 2\text{ kHz}\left(\frac{4.69}{1.88}\right)^2 = 12.5\text{ kHz}$

Effects of Axial Loads

Just as tension in a guitar string affects the string's resonant frequency, a tensile or compressive stress along the axial direction of a beam affects the beam's resonant frequency. The higher the tensile stress, the higher is the resonant frequency. Specifically, when axial load F acts on a beam, the resonant frequencies for a beam with simply supported edges are

$$\omega_n = \frac{n^2\pi^2}{l^2}\sqrt{\frac{EI}{A\rho}}\sqrt{1\pm\frac{\alpha^2}{n^2}} = \omega_0\sqrt{1\pm\frac{\alpha^2}{n^2}},\ \text{where } \alpha = \frac{Fl^2}{EI\pi^2} \text{ and } n = \text{mode number.} \tag{6.43}$$

The plus sign in the equation above is for a tensile axial load, while the minus sign is for a compressive axial load. In case of compressive axial load, the frequency becomes imaginary number, if the load F is greater than $nEI\pi^2/l^2$. What physically happens is that for a simply supported beam, the beam buckles when the compressive axial load $F = EI\pi^2/l^2$.

Axial loads exist due to residual stress or large vibration amplitude. When the amplitude of vibration is large, an axial load is induced in a beam with its two edges simply supported (or clamped), because the two supports limit the axial distance for the beam and make the beam to bend in vertical direction without any length change in the axial direction. In other words, when a beam with its two ends restrained (so that the distance between the two ends is fixed) vibrates *laterally*, there is stretching of the midline, which increases the resonant frequency. For example, if both ends of a beam are simply supported and cannot move axially, tensile force is induced as the beam deflects. The induced axial force (and so the natural resonant frequency) depends on the amplitude of vibration [4].

Similarly, in case of a clamped–clamped beam, as the applied load ($F_o\cos\omega t$) increases, the vibration amplitude (and induced axial tension) increases, and so does the resonant frequency [5]. This effect is shown in the amplitude versus frequency curve for a clamped–clamped beam in Fig. 6.13, and is analyzed with Duffing equation (which is a nonlinear second-order differential equation). And the fundamental resonant frequency ω_1 is [5]:

$$\omega_1^2 = \omega_o^2\left(1+0.53(1-\nu^2)(w_{max}/t)^2\right) \tag{6.44}$$

where ω_o = the resonant frequency at small vibration amplitude, ν = Poisson's ratio, w_{max} = maximum vibration amplitude, and t = beam thickness. Interestingly, for a given F_o, the amplitude curve as a function of frequency follows different lines, when the frequency is increased from low to high and when it is decreased from high to low, in

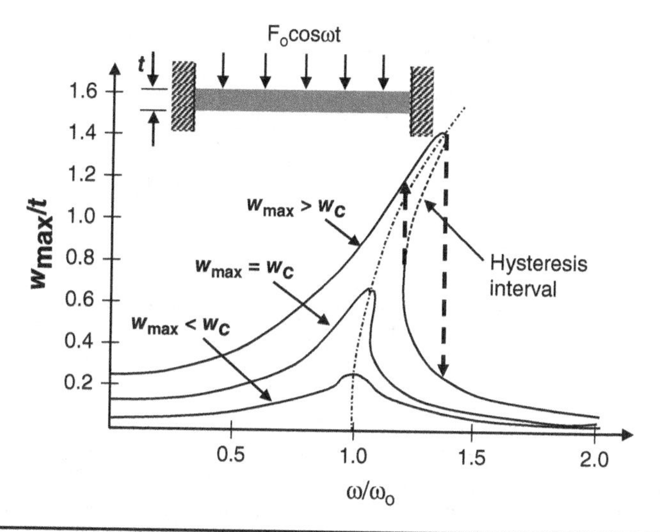

FIGURE 6.13 Normalized amplitude versus normalized frequency for a clamped–clamped beam with rectangular cross-section and zero built-in axial strain. (Adapted from [5].)

the region denoted as "hysteresis interval" in Fig. 6.13, if $w_{max} > w_c$, where the critical amplitude w_c (beyond which the curve becomes tripled valued) is $w_c = \dfrac{t}{\sqrt{0.53Q_1(1-\nu^2)}}$, with Q_1 being the quality factor of the fundamental mode.

6.2.4 Vibrations of Plates

As a string (e.g., guitar string) is an extreme case of a beam having no flexural rigidity, a membrane (e.g., drum skin) is an extreme case of a plate having no flexural rigidity (i.e., no bending stiffness). The resonant frequency of a membrane depends on its built-in tension per running length (T) as well as its area (A) and mass per unit area (ρ_1), and the fundamental natural frequency of a circular membrane with its radius R is

$$\omega_1 = 2.40\sqrt{\frac{T}{\mu_1 R^2}} = 4.26\sqrt{\frac{T}{\mu_1 A}}. \tag{6.45}$$

A plate with its edges fixed/clamped goes through bending according to the shape shown in Fig. 6.14. Note the zero slope of the deflection curve at the edges. The slope would not have been zero, if the edges are simply supported, and also is different from a membrane bending. The stress variation along the thickness direction is linearly dependent on the distance (z), though the stress variation along the length direction is not linearly dependent on the distance (x or y).

The governing equation for free vibration of a plate is

$$D\nabla^4 w = -\rho t\frac{\partial^2 w}{\partial t^2} + \left(N_x\frac{\partial^2 w}{\partial x^2} + 2N_{xy}\frac{\partial^2 w}{\partial x\partial y} + N_y\frac{\partial^2 w}{\partial y^2}\right). \tag{6.46}$$

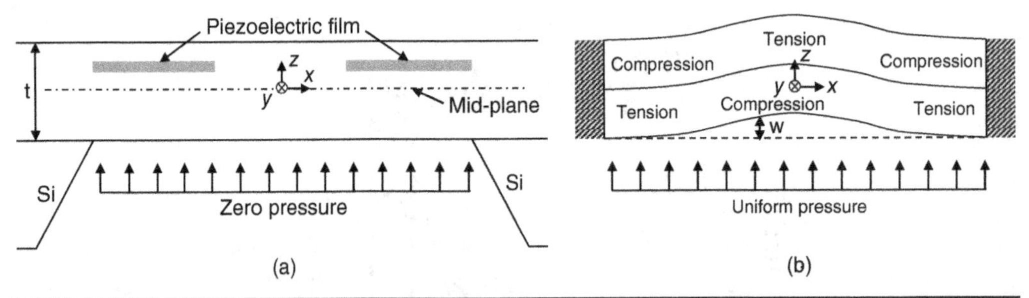

FIGURE 6.14 Cross-sectional views of (a) a bridge or diaphragm with its two edges fixed or clamped and (b) its bending shape under a uniform load.

This equation can be solved according to the boundary conditions for various plates to obtain the resonant frequencies, mode shapes, and nodal lines. Solving the differential equation for free vibration of a plate is relatively easy for a rectangular plate with its edges simply supported, as shown below. However, a plate with its four edges clamped would require an infinite sum of sinusoidal equations to have the deflection curve equation satisfy the boundary conditions. For $w = AW(x,y)\cos(\omega_{mn}t+\theta)$, Eq. (6.46) becomes the following spatial differential equation:

$$D\nabla^4 W = \rho t\omega_{mn}^2 W + \left(N_x \frac{\partial^2 W}{\partial x^2} + 2N_{xy}\frac{\partial^2 W}{\partial x \partial y} + N_y \frac{\partial^2 W}{\partial y^2}\right). \tag{6.47}$$

For a simply supported rectangular plate of length a, width b, and thickness t, $W = A\sin\frac{m\pi x}{a}\sin\frac{n\pi y}{b}$ satisfies the boundary conditions. Substituting the derivatives of W into Eq. (6.47), we obtain

$$D\left[\left(\frac{m}{a}\right)^4 + 2\left(\frac{m}{a}\right)^2\left(\frac{n}{b}\right)^2 + \left(\frac{n}{b}\right)^4\right]\pi^4 \sin\frac{m\pi x}{a}\sin\frac{n\pi y}{b}$$

$$= \rho t\omega_{mn}^2 \sin\frac{m\pi x}{a}\sin\frac{n\pi y}{b} - \pi^2\left[N_x\left(\frac{m}{a}\right)^2 + N_y\left(\frac{n}{b}\right)^2\right]\sin\frac{m\pi x}{a}\sin\frac{n\pi y}{b} \tag{6.48}$$

Solving for ω_{mn}^2, we obtain $\omega_{mn}^2 = \frac{1}{\rho t}\left\{\pi^4 D\left[\left(\frac{m}{a}\right)^2 + \left(\frac{n}{b}\right)^2\right]^2 + \pi^2\left[N_x\left(\frac{m}{a}\right)^2 + N_y\left(\frac{n}{b}\right)^2\right]\right\}$, and for a square plate, we have

$$\omega_{mn}^2 = \frac{1}{\rho t}\left(\pi^4 D\frac{\eta^2}{a^4} + \pi^2 N\frac{\eta}{a^2}\right) = \frac{\pi^4 D\eta^2}{\rho t a^4}\left(1 + \frac{Na^2}{\pi^2 D\eta}\right) = \frac{\pi^4 D\eta^2}{\rho t a^4}\left(1 + \frac{12(1-\nu^2)Na^2}{\pi^2 E\eta t^3}\right) \tag{6.49}$$

where $\eta = m^2 + n^2$, $D = Et^3/12(1-\nu^2)$, and ν = Poisson's ratio.

Thus, for a square plate with its four edges simply supported and with no in-plane stress (i.e., $N = 0$), the fundamental and harmonic resonant frequencies are $\omega_{11} = \frac{2\pi^2}{a^2}\sqrt{\frac{D}{\rho t}}$, $\omega_{12} = \frac{5\pi^2}{a^2}\sqrt{\frac{D}{\rho t}}$, and $\omega_{22} = \frac{8\pi^2}{a^2}\sqrt{\frac{D}{\rho t}}$ .....

Question: For a 1×1 mm^2, 1-μm-thick square diaphragm with its four edges simply supported, how much will the fundamental resonant frequency be changed due to a residual in-plane tensile stress of 10 kPa·m compared to a zero residual stress case. Assume that Young's modulus and Poisson's ratio are 200 GPa and 0.3, respectively.

Answer:

$$\omega_{11}^2 = \frac{\pi^4 D\eta^2}{\rho t a^4}\left(1+\frac{12(1-\nu^2)Na^2}{\pi^2 E\eta t^3}\right) = \frac{\pi^4 D\eta^2}{\rho t a^4}\left(1+\frac{12(1-0.3^2)10^4\cdot 10^{-6}}{\pi^2 2\cdot 10^{11}\cdot 2\cdot 10^{-18}}\right) = \frac{\pi^4 D\eta^2}{\rho t a^4}(1+2{,}736).$$

Thus, the resonant frequency will be increased by $\sqrt{2{,}736}$ times or 52 times.

Unlike a plate with its four edges simply supported, there are no simple analytical equations for the resonant frequencies and mode shapes for the plates with other boundary conditions. Consequently, for a quick reference, one can refer to tabulated values such as Table 6.5. One particular structure of interest in MEMS is the plate with its four edges clamped, which is commonly used for pressure sensing, and one would want to know how to calculate the resonant frequencies for different harmonics. Note that the bending displacement is zero on the nodal lines.

Just as an axial load affects the resonant frequency of a beam, an in-plane load due to (1) residual stress and/or (2) stretching of the neutral plane (owing to large displacement) affects the resonant frequency of a plate. As the displacement increases, the plate stretching effect contributes to the total induced stress more [5].

For a square diaphragm with its four edges clamped, the fundamental resonant frequency is linearly proportional to the thickness and inversely proportional to the area. For example, by reducing the diaphragm thickness by a factor of 2 (say, from 3 to 1.5 μm) and increasing the area by a factor of 4 (say, from 3×3 to 6×6 mm^2), one can lower the resonant frequency by a factor of 8 (say, from 8 to 1 kHz). By adding a compressive stress in the diaphragm, one can lower the resonant frequency further, but not much, since the diaphragm buckles under such a small compressive stress that it is hard to control the stress level accurately from run to run, as can be understood better in the following analysis.

For example, for a square plate with its four edges *simply* supported, the fundamental resonant frequency can be obtained from Eq. (6.49) to be

$$\omega_{11}^2 = \frac{4\pi^4 D}{\rho t a^4}\left(1+\frac{12(1-\nu^2)Na^2}{2\pi^2 E t^3}\right) \tag{6.50}$$

	$n = 1$	$n = 2$	$n = 3$	$n = 1$	$n = 2$	$n = 3$	$n = 4$	$n = 5$	$n = 6$
ω_n	$3.49\sqrt{\frac{D}{\rho t a^4}}$	$8.55\sqrt{\frac{D}{\rho t a^4}}$	$21.4\sqrt{\frac{D}{\rho t a^4}}$	$36.0\sqrt{\frac{D}{\rho t a^4}}$	$73.4\sqrt{\frac{D}{\rho t a^4}}$	$108.3\sqrt{\frac{D}{\rho t a^4}}$	$131.6\sqrt{\frac{D}{\rho t a^4}}$	$132.3\sqrt{\frac{D}{\rho t a^4}}$	$161.2\sqrt{\frac{D}{\rho t a^4}}$
Nodal lines									

$D = Et^3/[12(1-\nu^2)]$, t = plate thickness, a = plate length, ρ = mass density

Adapted from [4].

TABLE 6.5 Resonant Frequencies and Nodal lines of Square Plates with Various Edge Conditions

Thus, one can intentionally introduce compressive stress in the diaphragm (e.g., through a compressive silicon nitride diaphragm) to reduce the resonant frequency and also to improve the sensitivities of diaphragm-based transducers. However, as can be seen in Fig. 6.15 showing the calculated resonant frequencies versus residual stress for a square diaphragm with its four edges simply supported, the buckling compressive stress that causes the diaphragm to buckle is quite small (< 1 MPa, which is very difficult to control in thin-film deposition for this particular case). Moreover, the resonant frequency drops rapidly over a very narrow range of the compressive stress near the buckling stress. A diaphragm with different boundary conditions (such as clamped edges) behaves similarly under a compressive residual stress.

Bulk silicon micromachining on (100) Si wafer with a wet anisotropic etchant produces sloped (111) planes on the edge of a micromachined cantilever, bridge, or diaphragm, as shown at the top of Fig. 6.16b, and the edge boundary condition is a little different from clamped edge (Fig. 6.16a). However, as long as the micromachined structure is much thinner than the silicon substrate, the clamped boundary condition is almost as good as the more accurate modeling of the boundary condition, as indicated by the calculated fundamental resonant frequencies for the two cases in Fig. 6.16.

Question: If a square, isotropic diaphragm with its four edges clamped has the fundamental resonant frequency equal to 7 kHz, what will be the fundamental resonant frequency if the diaphragm thickness is increased by a factor of 2, while all the other parameters remain the same?

Answer: From Table 6.5, we get $\omega_1 = 35.99\sqrt{\frac{D}{\rho t a^4}} = 35.99\sqrt{\frac{Et^3/12(1-\nu^2)}{\rho t a^4}} = 10.39\frac{t}{a^2}\sqrt{\frac{E}{\rho(1-\nu^2)}}$.

Consequently, the resonant frequency will increase by a factor of 2 to 14 kHz.

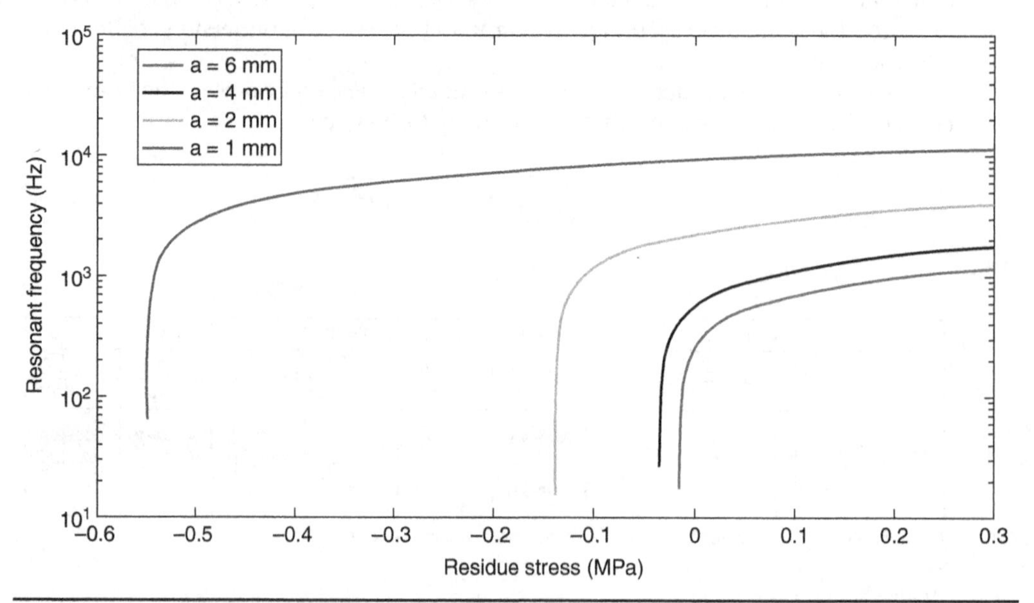

FIGURE 6.15 Fundamental resonant frequency versus residual stress for square 5-μm-thick diaphragms of various sizes (from 1 × 1 to 6 × 6 mm²) with its four edges simply supported.

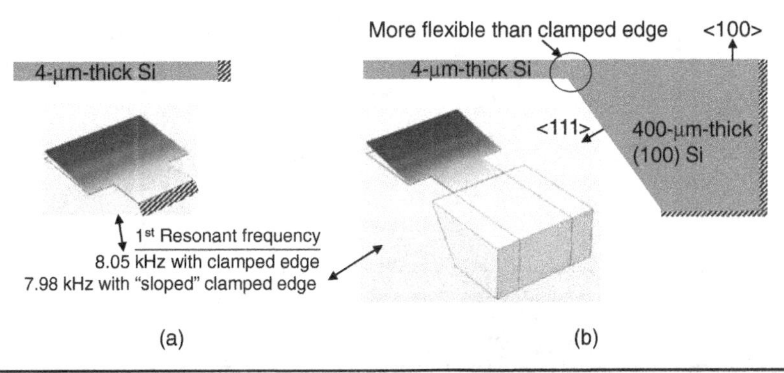

FIGURE 6.16 Cross-sectional (top) and perspective (bottom) views of a paddle-shaped cantilever with clamped edge (a) and "sloped" clamped edge (b) which is more accurate boundary condition for a cantilever formed on a (100) silicon substrate with an anisotropic etching in which (111) planes limit the etching. Calculated fundamental resonant frequencies for both cases differ by less than 1%, though the boundary conditions are substantially different, since the 4-μm-thick cantilever is much thinner than the 400-μm-thick silicon substrate.

6.3 MEMS Accelerometers

6.3.1 Spring-Mass-Dashpot as Accelerometer

Let us consider a simplest accelerometer composed of spring with spring constant k, mass M_p, and dashpot with damping coefficient D, shown in Fig. 6.17. For an applied acceleration $a(t)$ that produces an inertial force F_i $(=M_p a(t))$, the equation of motion to be solved is

$$M_p a(t) = M_p x'' + Dx' + kx. \tag{6.51}$$

For a sinusoidal acceleration $a(t) = A\cos\omega t = \mathrm{Re}\{Ae^{j\omega t}\}$, the solution of Eq. (6.51) for the displacement $x(t)$ will be $x(t) = |X|\cos(\omega t + \theta) = \mathrm{Re}\{Xe^{j\omega t}\}$. Substituting the expected solution into Eq. (6.51), we get $M_p A = -\omega^2 M_p X + j\omega X + kX$, from which we obtain the

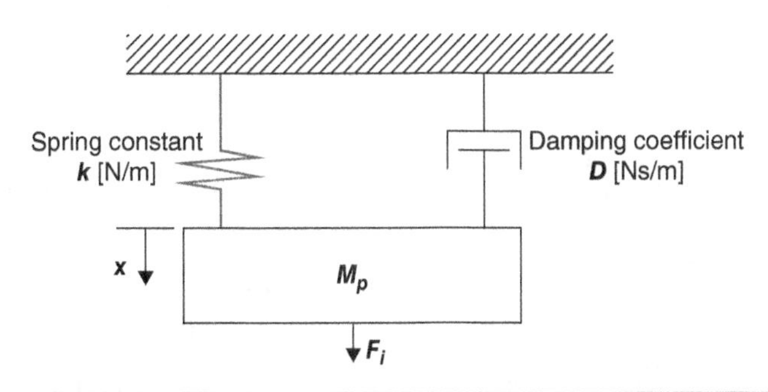

FIGURE 6.17 Spring-mass-dashpot as an accelerometer.

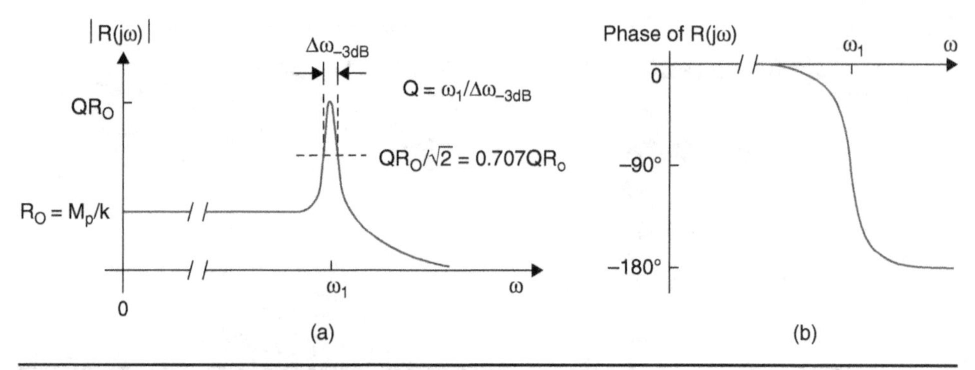

FIGURE 6.18 The magnitude (a) and phase (b) of the sinusoidal response of the simple spring-mass-dashpot system as a function of frequency.

following sinusoidal steady-state response, $R(j\omega)$, of the spring-mass-dashpot, which is a complex number having a magnitude and a phase.

$$R(j\omega) = \frac{X}{A} = \frac{M_p}{-\omega^2 M_p + j\omega D + k} = \frac{R_o}{1-(\omega/\omega_1)^2 + j\omega/(Q\omega_1)}, \tag{6.52}$$

where $R_o \equiv M_p/k$ which is the response when the frequency ω is close to zero, $\omega_1 \equiv \sqrt{k/M_p}$ which is the resonant frequency, and $Q \equiv \dfrac{k}{\omega_1 D}$ which we call the quality factor.

The magnitude and phase of the response, $R(j\omega)$, are plotted in Fig. 6.18 as a function of frequency. As can be seen in Fig. 6.18a, the response (governed by Eq. (6.52)), a second order equation) shows its magnitude increasing by a factor of Q at the fundamental resonant frequency (ω_1) from its low-frequency response (R_o). The response versus frequency is uniform up to about the resonant frequency, and the sharpness of the peak at the resonant frequency is determined by Q which gives $\omega_1/\Delta\omega_{-3\text{dB}}$ (where $\Delta\omega_{-3\text{dB}}$ is the bandwidth between the two frequencies at which the magnitude is 3 dB lower than that at the resonant frequency) for a second-order system. The phase goes through a total of 180° shift due to the two poles of a second-order system, as shown in Fig. 6.18b. Note that the phase lags by 90° at the resonant frequency compared to that at a very low frequency.

Because of the peaking at the resonant frequency (i.e., $Q > 1$ for underdamped resonance), the spring-mass will respond to a step acceleration as shown in Fig. 6.19b and c in time domain, depending on how high the Q is. The response has ringing because the spring-mass responds very wildly to the frequency component (out of so many

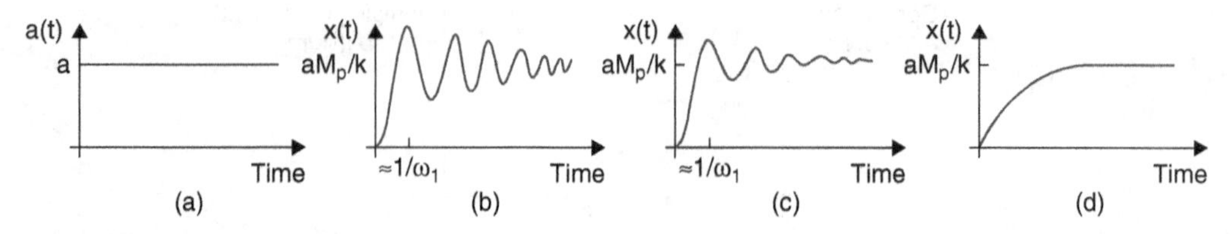

FIGURE 6.19 For a step input (a), qualitative time responses for varying degrees of Q: $Q >> 1$ (b), $Q > 1$ (c) and $Q < 1$ (d).

frequency components contained in a step input) that corresponds to its resonant frequency, and the main frequency component in the ringing is basically the fundamental resonant frequency (ω_1). The higher the Q, the larger is the ringing amplitude and the longer the ring-down time (τ, defined to be the time it takes for the amplitude to drop to 36.8%, or $1/e$, of the initial amplitude), which is about $2Q/\omega_1$. Thus, to avoid such ringing, one would have (1) to intentionally lower the Q by increasing damping or (2) to restrict the bandwidth of input signal. If the Q is less than 1 (i.e., overdamped), the response to a step acceleration will look like Fig. 6.19d.

Question: For a simple spring-mass system shown in Fig. 6.17, sketch the magnitude and phase responses for two cases of the proof mass M_p being different by a factor of 2, while the spring and damping constants remain same.

Answer: In the figures below, $R_o' = 2R_o$, while $\omega_o' = \omega_o/\sqrt{2} = 0.707\omega_o$ and $Q' = k/(\omega_1' D) = \sqrt{2} \cdot Q$.

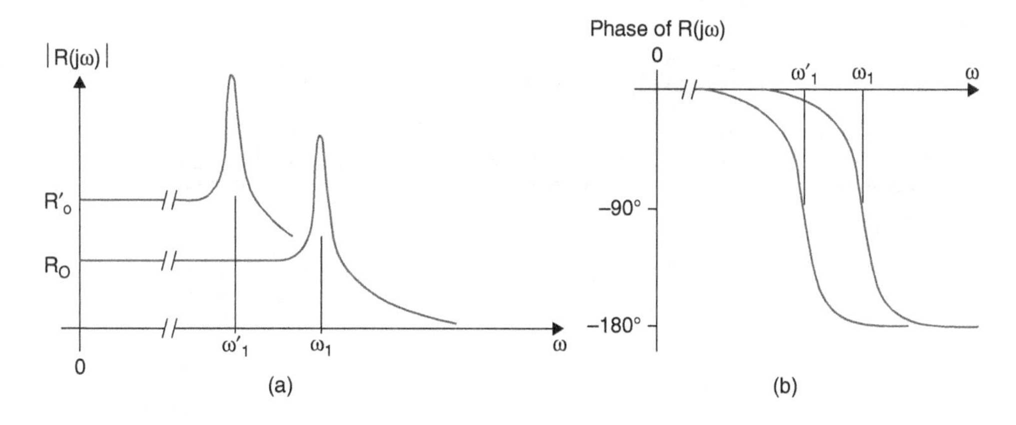

Trade-Off between Sensitivity and Bandwidth

Noting what factors affect the low-frequency response (R_o), fundamental resonant frequency (ω_1), and quality factor (Q), we see that there is a trade-off between the low-frequency response (or the sensitivity to acceleration) and the fundamental resonant frequency (or the bandwidth in case of an accelerometer, since the accelerometer would have uniform response only up to the fundamental resonant frequency). Since $R_o = \dfrac{M_p}{k}$ and $\omega_1 = \sqrt{\dfrac{k}{M_p}}$ for a spring-mass system, one cannot increase both the sensitivity (R_o) and the bandwidth (ω_1) by varying M_p and/or k. The product of the sensitivity and the square of the bandwidth (i.e., $R_o\omega_1^2$) is equal to 1 in the case of a simple spring-mass system and is commonly defined to be the figure of merit (FOM) in comparing different types of accelerometers, rather than just the sensitivity or just the bandwidth.

Question: If the FOM of an accelerometer is equal to 1, as in a simple spring-mass system, what will be the DC response (in terms of μm per 9.8 m/s²), if the fundamental resonant frequency is equal to 365 Hz (= 2,293 rad/s).

Answer: For FOM $\equiv R_o\omega_1^2 = 1$, $R_o = 1/\omega_1^2 = 1/(2{,}293)^2 = 1.9 \times 10^{-7}\text{s}^2/\text{rad}^2$. Consequently, there will be $(1.9 \times 10^{-7}\ \text{s}^2/\text{rad}^2)(9.8\ \text{m/s}^2) = 1.9\ \mu\text{m}$ response for 1 g acceleration.

6.3.2 A Bulk-Micromachined Silicon Accelerometer

Accelerometer uses a proof mass for converting acceleration into a measurable displacement or stress. For example, silicon can be bulk-micromachined to produce an accelerometer with a proof mass suspended by a beam ($l \times w \times t$) shown in Fig. 6.20. The proof mass responds to accelerations in all three axes, as indicated with F_{ix}, F_{iy}, and F_{iz} (acting at the mass' center of gravity) in Fig. 6.20b, though the accelerometer is designed to sense the acceleration in the z direction (a_z) through the displacement of the proof mass in the z direction, d. If we define the three axial responses to an input acceleration of $a(x,y,z,t)$ as $R_x \equiv d/a_x$, $R_y \equiv d/a_y$, and $R_z \equiv d/a_z$, the cross-axis responsivities would be R_x and R_y, which need to be minimized for an accurate single-axis accelerometer.

The response, R_z, can be obtained through solving the following differential equation, referring to the bottom figure of Fig. 6.20b (and ignoring $F_{ix}t_p$, since we are trying to obtain the response due to F_{iz}),

$$EI_y \frac{d^2 d}{dx^2} = -M_y(x) = F_{iz}(l-x) + F_{iz}l_p, \text{ where } I_y = wt^3/12. \tag{6.53}$$

$$\text{Let } y = d'. \Rightarrow y' = \frac{-F_{iz}x}{EI_y} + C_1 \text{ with } C_1 \equiv \frac{F_{iz}(l+l_p)}{EI_y} \Rightarrow y = \int\left(\frac{-F_{iz}x}{EI_y} + C_1\right)dx = \frac{-F_{iz}x^2}{2EI_y} + C_1x + C_2 = d'$$

$$\text{From the B.C. of } d'(0) = 0, C_2 = 0. \quad \Rightarrow d = \int\left(\frac{-F_{iz}x^2}{2EI_y} + C_1x\right)dx = \frac{-F_{iz}x^3}{6EI_y} + \frac{C_1x^2}{2} + C_3$$

$$\text{From another B.C. of } d(0) = 0, C_3 = 0. \Rightarrow d = \frac{F_{iz}x^2}{6EI_y}[3(l+l_p) - x].$$

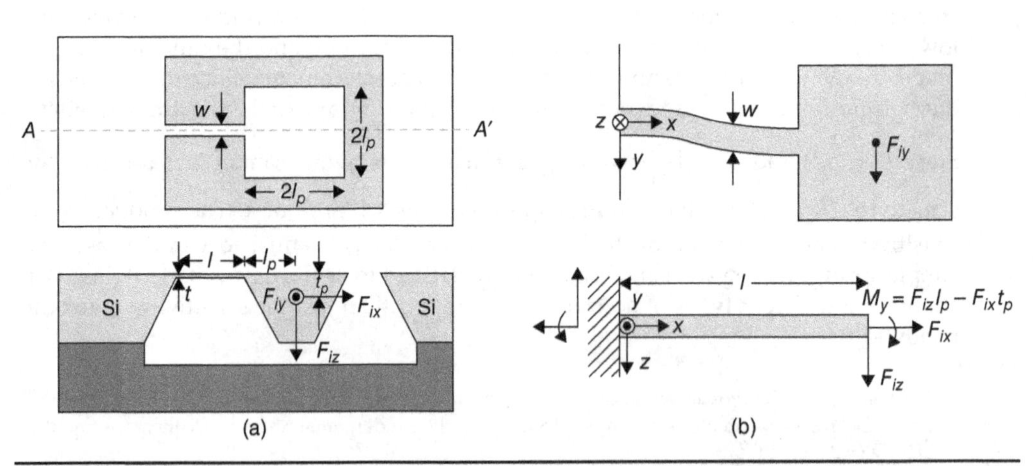

FIGURE 6.20 (a) Top view and cross-sectional (across A–A') views of a bulk-micromachined silicon accelerometer. (b) Bottom: Cross-sectional-view free body diagram of the beam due to a z-directed acceleration. Top: Top-view diagram of the beam deflection due to a y-directed acceleration.

Thus, $R_z = d(l)/a_z = \frac{2M_p l^2(2l+3l_p)}{Ewt^3}$, since $F_{iz} = M_p a_z$ with M_p being the mass of the proof mass. We can define effective spring constant, $k_z = \frac{F_{iz}}{d(l)} = \frac{Ewt^3}{2l^2(2l+3l_p)}$, and use it to obtain the response

$$R_z = \frac{M_p}{k_z} = \frac{2M_p l^2(2l+3l_p)}{Ewt^3}. \tag{6.54}$$

To obtain the bandwidth of the accelerometer, we need to calculate the fundamental resonant frequency using either the effective spring constant or Rayleigh's method. Using the effective spring constant, ignoring the mass of the beam, we get $\omega_1^2 \approx \frac{k_z}{M_p} = \frac{Ewt^3}{M_p}\frac{1}{2l^2(2l+3l_p)}$. For Rayleigh's method, we notice that the kinetic energy is mainly due to moving of the proof mass, while the potential energy is from the beam bending (Table 6.3): $d(\text{PE}_{\text{bend}}) = \frac{1}{2}EI_y\left(\frac{d^2d}{dx^2}\right)^2 dx$, $\text{KE}_{\max} \approx \frac{M_p}{2}\omega_1^2[d(l)]^2 = \omega_1^2\text{KE}'_{\max}$. After the integration of the potential energy and some algebraic manipulation, we get $\omega_1^2 \approx \frac{\text{PE}_{\max}}{\text{KE}'_{\max}} = \frac{Ewt^3}{M_p}\frac{l^3+3l^2l_p+3ll_p^2}{(2l^3+3l^2l_p)^2}$. In this case, Rayleigh method has taken more steps, but yields a more accurate equation for the fundamental resonant frequency, since it does not ignore the effect of the beam mass. Figure of merit (FOM) for this particular accelerometer is $\text{FOM} \equiv R_z\omega_1^2 \approx \frac{2l^2(2l+3l_p)(l^3+3l^2l_p+3ll_p^2)}{(2l^3+3l^2l_p)^2}$. For $l >> l_p$, $l = l_p$, and $l << l_p$, FOM is ≈ 1, 2.8, and $\approx 2l_p/l$, respectively.

Cross-Axis Response due to Axial Acceleration a_x

Acceleration in the x direction also produces a signal in the accelerometer that is designed to detect acceleration in the z direction, mainly because the proof mass' center of mass is not at the same plane as the beam, as illustrated in Fig. 6.20. The force directed in the x direction produces a bending moment ($F_{ix}t_p$) on the proof mass through the arm between the beam axis and the center of mass. The bending moment produces a spurious signal equal to $\tilde{d}(x) = -\frac{F_{ix}t_p}{2EI_y}x^2$, which needs to be minimized, if the accelerometer is to detect the acceleration in the z direction accurately. The cross-axis sensitivity is $\left|\frac{R_x}{R_z}\right| = \frac{-(M_p/F_{ix})\tilde{d}(l)}{(M_p/F_{iz})d(l)} = \frac{3t_p}{2l+3l_p}$.

To suppress the response to acceleration in the x direction, one can (1) plate heavy metal on top of the proof mass to move the center of mass upward so that $t_p \approx 0$, (2) add another beam (to the other side of the proof mass) to constrain rotation, which limits, though, the motion of the proof mass even in the sensitive axis due to the fact that the proof mass is now clamped on both sides, or (3) use skew-symmetric proof mass [6].

Cross-Axis Response due to Axial Acceleration a_y

Acceleration in the y direction also produces a lateral displacement and torsion in the beam, as illustrated in the top figure of Fig. 6.20b. The lateral displacement is not along the sensitive axis (i.e., z axis). However, the torsion produces a displacement in z direction and needs to be minimized with any of the techniques (mentioned in the previous section) that brings the center of mass (of the proof mass) in the same plane as the beam, such as the skew-symmetric proof mass [6].

If the beam displacement is converted to electrical signal through piezoresistive effect of silicon (i.e., through implanting a resistor in silicon via doping the silicon with impurities), the lateral displacement in the y direction due to a_y may be made to produce little signal by integrating the induced tension and compression in the beam. The piezoresistors in the tensioned and compressed regions can appropriately arranged and/or connected such that the readout is insensitive to the lateral displacement because tension and compression are summed up to be near zero.

6.3.3 Piezoresistive Readout

For an accelerometer shown in Fig. 6.20, the axial stress is maximum at the "base" (i.e., the clamped edge) on the top (or bottom) surface of the beam where we would want to implant a piezoresistor to convert the stress (or strain) into electrical resistance change. To take advantage of the axial stress, the piezoresistor is usually laid out in axial direction (i.e., a long, narrow line along the beam axis). The axial stress due to the z-directed acceleration is

$$\sigma_x(x,z) = -\frac{M_y(x)z}{I_y} = \frac{zF_{iz}(l-x+l_p)}{I_y}, \tag{6.55}$$

which is maximum at the top of beam ($z = t/2$) at the base ($x = 0$). Thus, using $I_y = wt^3/12$, we obtain the maximum stress

$$\sigma_{x,\max} = \frac{6F_{iz}(l+l_p)}{wt^2}. \tag{6.56}$$

The maximum axial stress produces a maximum axial stain, $\varepsilon_{x,\max} = \sigma_{x,\max}/E$.

Question: A piezoresistive silicon accelerometer shown in Fig. 6.20 is made on an (100) n-type silicon and has the following dimensions: t = 10 μm (the beam thickness), w = 50 μm, l = 400 μm, M_p = 1 mg, and l_p = 1,000 μm. Calculate the normalized resistance changes (i.e., $\Delta R/R$) due to 1 g acceleration (i.e., 9.8 m/s²) directed in the z direction assuming that (1) the piezoresistive element is placed at $x = 0$ and $z = t/2$ along the length direction; and (2) $\Delta R/R$ is approximately equal to $\pi_l\sigma_x$, where π_l is the relevant piezoresistive constant and is equal to -3.12×10^{-10} Pa⁻¹. [Hint: Due to 1 g acceleration, there will be a force (i.e., F_{iz} directed in the z direction in Fig. 6.20) equal to 9.8×10^{-6} N.]

Answer: $\sigma_x(x=0, z=t/2) = \dfrac{6F_{iz}(l+l_p)}{wt^2} = \dfrac{6\times 9.8\times 10^{-6}\times(4\times 10^{-4}+2\times 10^{-3}/2)}{5\times 10^{-5}\times(10^{-5})^2} = 1.65\times 10^7\ \text{N/m}^2$

$\dfrac{\Delta R}{R} = \pi_l\sigma_x = -3.12\times 10^{-10}\times 1.65\times 10^7 = -5.1\times 10^{-3}$

MEMS Example: Quality factor (Q) can be about 100 in air (with its viscous damping) for accelerometers having its fundamental resonant frequency around 1 kHz. That kind of Q can cause much ringing in the accelerometer's step response. To reduce the ringing, squeeze-film damping can be used for a critical damping or overdamping. Squeeze-film damping can easily be obtained by suspending a proof mass above a support wafer with a narrow air gap [7]. In the piezoresistive silicon accelerometer with built-in damping [7], a proof mass is suspended by four beams, each of which has two piezoresistors, one at the edge next to the proof mass and the other at the edge next to a frame anchored to a substrate. The eight piezoresistors change their resistance values differently for accelerations in three different directions (x, y, and z), as the proof mass moves in response to acceleration in each of the three directions differently and since the proof mass movement is accompanied by unique bending shapes of the four beams. By arranging the eight piezoresistors in a Wheatstone bridge configuration, the net resistance change can be made to be nonzero only for the z-directed acceleration, thus, minimizing the cross-axis sensitivities.

6.3.4 Piezoelectric Readout

Piezoelectric material can be used to convert mechanical strain (caused by acceleration) into an electrical polarization and has been used in commercial accelerometers with a bulk piezoelectric element in contact with a proof mass [8]. These accelerometers are simple and rugged; yet have a very good sensitivity, shock survivability, and dynamic range. However, they are bulky and heavy, especially compared to MEMS counterparts. Also, unique to piezoelectric sensing (as covered in detail in Chap. 3 under piezoelectric transduction), a static response is inherently not possible due to transfer of electrons from one electrode to the other electrode (usually through finite resistance at the input of a preamp), and the lowest frequency that the accelerometers can detect is typically about 0.1 Hz.

MEMS Example 1: Piezoelectric thin film such as ZnO, AlN, and PZT can be used on a micromachined structure on a silicon substrate to produce a piezoelectric MEMS accelerometer. One example is a piezoelectric accelerometer integrated with metal–oxide–semiconductor (MOS) field-effect transistor (FET) on a single chip [9]. Achievable sensitivity (and minimum detectable signal level) as well as power consumption would be better with piezoelectric detection than with piezoresistive one. However, piezoelectric sensing is not amenable for sensing static DC acceleration. One can consider to encapsulate the piezoelectric layer with an electrically insulating layer (e.g., silicon dioxide) to cut out any electrical path for electrons to move between the top and bottom electrodes (sandwiching a piezoelectric film). Still, the best reported charge retention time (measured with an electrometer) is about 30 days [10]. Also, due to such high resistance at the input of a preamp, piezoelectric sensing for low frequency measurand inherently suffers from electromagnetic interference.

MEMS Example 2: A triaxis piezoelectric bimorph accelerometer can be built on four beams suspending a single proof mass, as shown in Fig. 6.21a [11]. Each of the suspension beam is composed of Parylene/Al/Si_xN_y/ZnO/Si_xN_y/Al/Si_xN_y/ZnO/Si_xN_y/Al/Parylene along the thickness direction from top to bottom (or from bottom to top). The neutral plane is designed to be just at the middle of the symmetric bimorph beam. Parylene is used because (1) it has a very small Young's modulus (~3.2 Gpa) and adds negligible stiffness to the sensing structure; and (2) it is a non-brittle plastic material with a very large linear elastic range (its yield strain being ~3%). When z-axis acceleration is applied, the proof mass moves vertically, while the four suspension beams deflect with induced stresses, as shown in Fig. 6.21b. Each suspension beam is designed to have two sensing electrodes (exploiting the different induced stress types): one for z-axis acceleration sensing and the other for x-axis or y-axis acceleration sensing.

As the x-axis and y-axis sensitivities of the triaxis accelerometer are same due to the symmetric structure, we just need to consider the x-axis sensitivity. The cross-sectional view of the triaxis accelerometer when x-axis acceleration is applied is shown in Fig. 6.22. The proof mass rotates around the y-axis beam, because the center of the

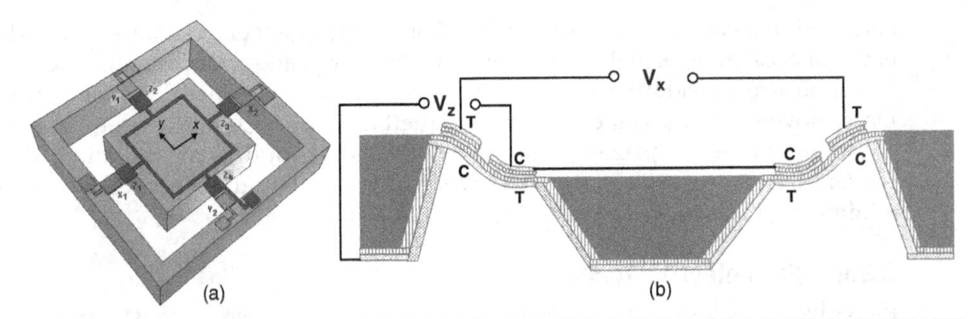

FIGURE 6.21 Schematic view (a) and operating principle (b) of the triaxis piezoelectric bimorph accelerometer when the applied acceleration is along the *z* axis [11]. Note how the proof mass is deflected and how the supporting beams are deflected, particularly the stress types, tension (*T*) versus compression (*C*) on the eight portions of the two beams.

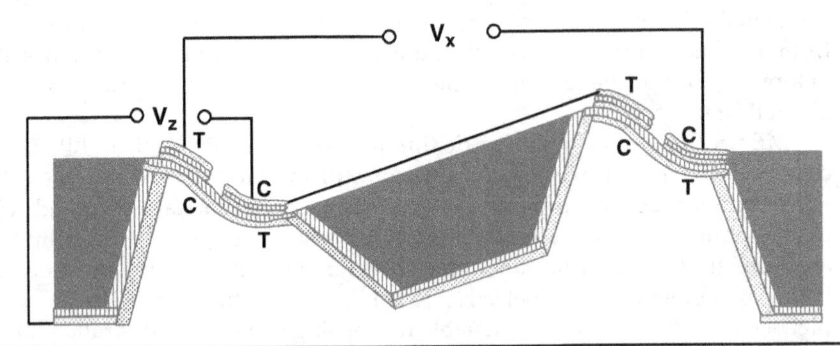

FIGURE 6.22 Operating principle of the triaxis piezoelectric bimorph accelerometer when lateral acceleration (either along the *x*- or *y* axis) is applied [11]. Compare the stress types on the eight portions of the two beams with those in Fig. 6.21b.

proof mass is not in the same plane as that of the suspension beam, resulting in a rotation angle θ (around the two *y*-axis beams) and bending moments (on the two *x*-axis beams). The unamplified sensitivities of the *x*-, *y*-, and *z*-axis electrodes (of the triaxis accelerometer) in response to accelerations in *x*-, *y*-, and *z*-axis were measured to be 0.93, 1.13, and 0.88 mV/g, respectively [11]. The worst-case minimum detectable signal of the triaxis accelerometer was measured to be 0.04 g over a sub-Hz to 100 Hz bandwidth. The cross-axial sensitivity among the *x*-, *y*-, and *z*-axis electrodes was less than 15% in the triaxis accelerometer.

6.3.5 Capacitive Readout

Capacitive detection is another method to convert mechanical motion (due to an applied acceleration) into an electrical signal. For capacitive readout, an air gap is needed between a proof mass and a stationary electrode; the smaller the air gap is, the higher the sensitivity at the cost of lower dynamic range. To increase the dynamic range, one can use an electrostatic voltage across the air gap to balance the force due to acceleration and to maintain the proof mass at one location [12]. The force-balancing voltage is directly related to the position of the proof mass and used to electrostatically move the

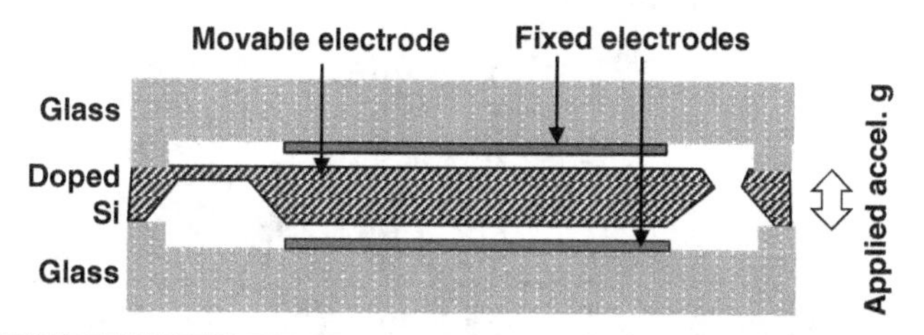

Figure 6.23 A micromachined force-balance capacitive accelerometer with a pulse width modulation (PWM) electrostatic actuation that allows high sensitivity and durability with limited air gaps between the movable and fixed electrodes. (Adapted from [13].)

proof mass. This way, no matter how large an applied acceleration is, the proof mass can be maintained at one constant location, and the range of detectable acceleration (i.e., dynamic range) can be very high. In this case, the force-balancing voltage needed to maintain the proof mass at a constant location is directly related to the applied acceleration.

MEMS Example 1: A bulk-micromachined force-balanced accelerometer whose proof mass is held electrostatically in its neutral position is shown in Fig. 6.23 [13]. In this design, the movable electrode at the end of the cantilever acts as a proof mass, responding to an applied acceleration g. The movable electrode's position can be turned into a pulse width modulation (PWM) signal by detecting the difference in the capacitances between the movable electrode and the two fixed electrodes (ΔC), through complementary metal–oxide–semiconductor (CMOS) switched capacitor technique. The PWM signal and its inverted signal are applied to the top and bottom fixed electrodes, so that the net amount of the electrostatic forces to the movable electrode is controlled by ΔC. This way the movable electrode is maintained exactly halfway between the fixed electrodes for any acceleration. The pulse width is proportional to the acceleration g, and the low-pass-filtered signal of the PWM signal yields an analog voltage that is proportional to g.

MEMS Example 2: An MEMS capacitive accelerometer shown Fig. 6.24 is connected to an electronic circuit chip on a single package, via a wire bonded to the anchor, for air bag deployment for automobiles in late 90s [14]. This hybrid approach of integrating an accelerometer with electronics did not work out well in competition against monolithic integration, due to higher manufacturing cost. In response to an applied acceleration,

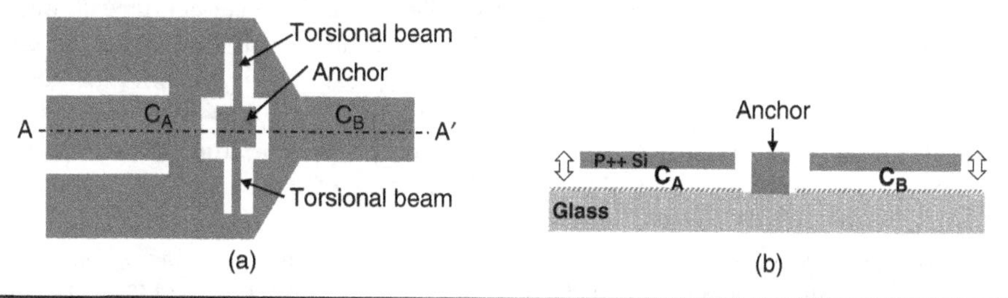

Figure 6.24 An MEMS capacitive accelerometer for air bag deployment of automobile: (a) top-view schematic, (b) cross-sectional view across A–A′ showing anodic bonding between silicon and glass, with bulk of the silicon being dissolved to form heavily-boron-doped, thin silicon structure. (Adapted from [14].)

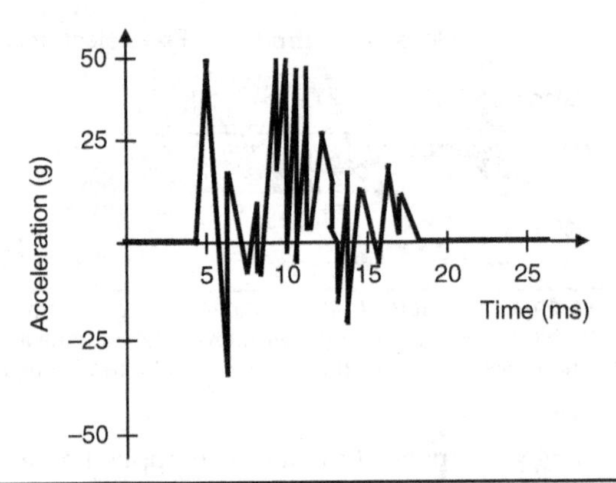

Figure 6.25 A typical acceleration in time when a car crashes from which air bag deployment has to be determined through analysis of acceleration, velocity (i.e., time integration of acceleration), and distance (time integration of velocity) profile.

the capacitances (C_A and C_B) between the released thin silicon plate and the substrate change due to rotation of the plate through the torsional beams, and the $(C_A - C_B)/(C_A + C_B)$ varies linearly in response to an applied acceleration. The static capacitance is designed to be about 150 fF, which varies by about 15 fF due to acceleration caused by car crash (which produces acceleration variation typically between −30 and +50 g within 15 ms, as shown in Fig. 6.25). As the thin, heavy-boron-doped silicon plate (Fig. 6.24) is large in lateral dimension compared to the air gap between the plate and glass substrate, squeeze-film damping overdamps the vibration to a level to bring the dominant pole to eight times less than the fundamental resonant frequency.

MEMS Example 3: The most successful MEMS accelerometers so far have been surface-micromachined capacitive accelerometers that are monolithically integrated with BiCMOS circuits on a single chip from Analog Devices, Inc. The sensing structure looks like a comb drive with its proof mass being the center beam from which the comb fingers branch out, as shown in Fig. 6.26. The proof mass is extremely small and light due to its thin thickness and small size, and its displacement due to an applied acceleration also is very small. The folded tether (Fig. 6.26) releases residual stress and also is much more compliant than the straight tether. Even with the folded tether, though, a typical capacitance change due to ±5 g is less than 1 fF, a very small value. However, because circuitry is integrated with the sensor on a single chip, the parasitic capacitance can be small and predictable. Thus, the smallest detectable capacitance change has been claimed to be as small as zepto F (10^{-21}F). Also, monolithic integration of circuits and sensor(s) on a single chip lowers manufacturing cost from hybrid integration of circuits and sensor(s), as long as the sensor size is small (e.g., less than 100 × 100 μm²). In addition to signal detection, amplification, and processing, the circuits can take care of temperature compensation, self-calibration, etc.

Question: When an automobile crashes, the acceleration varies in *time* as shown in Fig. 6.25. The Fourier analysis of the acceleration shows that the highest frequency component of the acceleration is about 10 kHz. What will the minimum value for the fundamental resonant frequency of an accelerometer have to be for it to be used as an air bag deployment sensor for automobiles?

Answer: Greater than 10 kHz.

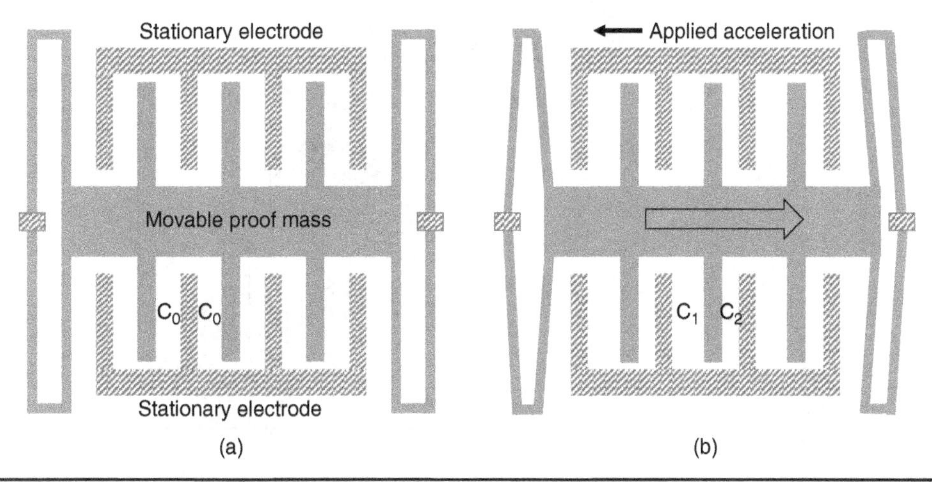

FIGURE 6.26 Top-view illustrations of how surface-micromachined capacitive accelerometers work: a folded tether supporting a movable comb-like structure (a) without and (b) under applied acceleration.

6.4 Vibratory Gyroscopes

6.4.1 Working Principle

For an object (such as a beam) moving with linear velocity equal to v_r, Coriolis force equal to $2\Omega \times v_r$ is induced when an angular rotation (with a rate Ω) is applied to the moving object, as illustrated in Fig. 6.27. The Coriolis force moves the linearly vibrating object into a direction that is perpendicular to both the vibrating direction and the angular rotation axis, as the cross-product ($\times$) of the two vectors Ω and v_r implies. Sensing this movement can yield how much angular rate (i.e, rotational velocity) is

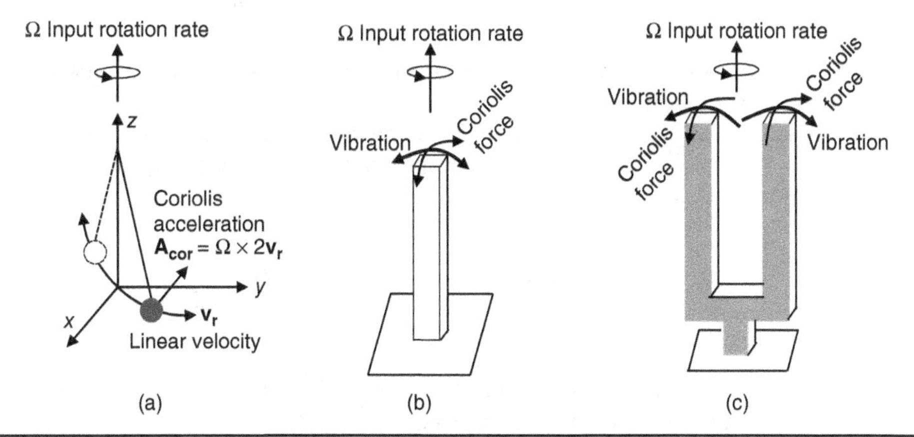

FIGURE 6.27 Three examples of Coriolis force produced by an input angular rate on a linearly vibration object: (a) a pendulum, (b) a cantilever beam, and (c) a tuning fork.

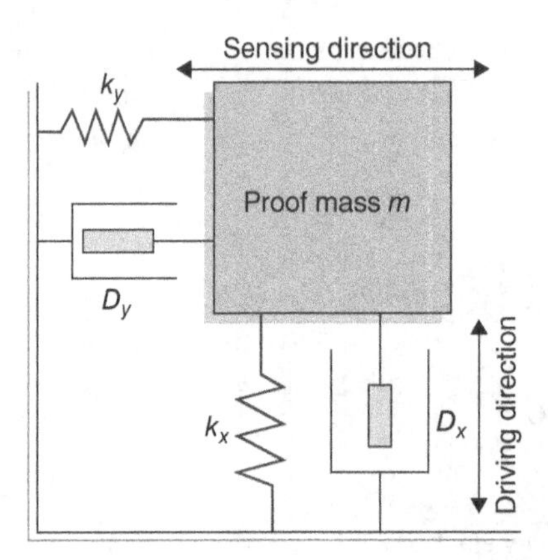

FIGURE 6.28 A model of a simple vibratory gyroscope.

acting on the object, and a gyroscope based on this principle is called a vibratory gyroscope. In such a gyroscope, there are drive mode and sensing mode that determine the vibrational amplitudes for the linear vibration (usually driven by electrical signal) and the Coriolis vibration (that can be sensed in various ways), respectively.

For a proof mass driven only along the x axis with a sinusoidal force $F_d = F_0 \sin(\omega_d t)$, the driving-mode and sensing-mode vibrations of a vibratory gyroscope can be modeled as shown in Fig. 6.28, if the input angular rotation is around the z axis with its angular rate equal to Ω_z. For a constant input angular rate $\left(\text{i.e., } \frac{d\Omega_z}{dt} = 0\right)$, the vibration of the proof mass is governed by the following two equations:

$$m\frac{d^2x}{dt^2} + D_x\frac{dx}{dt} + k_x x = F_0 \sin(\omega_d t) \tag{6.57}$$

$$m\frac{d^2y}{dt^2} + D_y\frac{dy}{dt} + k_y y = -2m\Omega_z\frac{dx}{dt} \tag{6.58}$$

The first equation is for the driving-mode vibration, while the second equation is for the sensing-mode vibration.

Assuming the solution of Eq. (6.57) to be $x(t) = X_0\cos(\omega_d t + \theta_x) = \mathrm{Re}\{Xe^{j\omega_d t}\}$, we get $F_0 = -\omega_d^2 mX + j\omega_d D_x X + k_x X$ from the differential equation. Consequently, the displacement along the x axis is

$$X = \frac{F_0}{-\omega_d^2 m + j\omega_d D_x + k_x} = \frac{F_0/k_x}{1-(\omega_d/\omega_x)^2 + j2\zeta_x\omega_d/\omega_x} \tag{6.59}$$

where F_0 is the input drive amplitude; $\omega_x (= \sqrt{k_x/m})$ is the resonant frequency in the drive direction (x); ω_d is the driving frequency; ζ_x $(= \omega_x D_x/2k_x)$ is the damping ratio. Since the quality factor Q_x is equal to $k_x/\omega_x D_x$ [see the text following Eq. (6.51)], the damping ratio ζ_x is equal to $1/2Q_x$. The magnitude of the displacement X_o and the phase difference θ (between $x(t)$ and $F_0 \sin(\omega_d t)$) are then

$$X_0 = \frac{F_0/k_x}{\sqrt{\left(1-\left(\frac{\omega_d}{\omega_x}\right)^2\right)^2 + \left(2\xi_x \frac{\omega_d}{\omega_x}\right)^2}} = \frac{F_0/k_x}{\sqrt{\left(1-\left(\frac{\omega_d}{\omega_x}\right)^2\right)^2 + \frac{1}{Q_x^2}\left(\frac{\omega_d}{\omega_x}\right)^2}} \quad (6.60)$$

$$\text{and } \theta_x = -\tan^{-1}\frac{2\zeta_x \omega_d/\omega_x}{1-(\omega_d/\omega_x)^2}. \quad (6.61)$$

Thus, the relationship between X_0 and F_0 is dependent on the frequency ω_d and damping ratio ζ_x; the vibrational amplitude X_0 is the largest when $\omega_d = \omega_x$ and is equal to $X_0 = Q_x F_0/k_x$ (which is increased by a factor of Q from a displacement at a low frequency). When the drive frequency ω_d is close to the resonant frequency ω_x along the x axis (the drive-mode axis), the vibration amplitude X_0 increases as the damping ratio ζ_x decreases (or the Q increases).

If the gyroscope is rotated around the z axis with an angular rate, Ω_z, there will be a Coriolis force F_c on the mass in the y direction, $F_c = 2m\Omega_z \frac{dx}{dt} = 2mX_0\omega_d\Omega_z \cos(\omega_d t + \theta_x)$. Consequently, the differential equation for the mass movement along the y axis (the sensing-mode axis) is

$$m\frac{d^2y}{dt^2} + D_y\frac{dy}{dt} + k_y y = -2mX_0\Omega_z\omega_d \cos(\omega_d t + \theta_x) \quad (6.62)$$

Assuming the solution to be $y = Y_0 \cos(\omega_d t + \theta_x + \theta_y)$ and plugging y, dy/dt, and d^2y/dt^2 into the differential equation, we can obtain the following vibrational amplitude and phase along the y axis,

$$Y_0 = \frac{2X_0\omega_d\Omega_z}{\omega_y^2\sqrt{\left(1-\frac{\omega_d^2}{\omega_y^2}\right)^2 + 4\zeta_y^2\frac{\omega_d^2}{\omega_y^2}}} = \frac{2X_0\omega_d\Omega_z}{\omega_y^2\sqrt{\left(1-\frac{\omega_d^2}{\omega_y^2}\right)^2 + \frac{1}{Q_y^2}\frac{\omega_d^2}{\omega_y^2}}} \text{ and } \theta_y = -\tan^{-1}\frac{2\zeta_y\omega_d\omega_y}{\omega_y^2-\omega_d^2} \quad (6.63)$$

where $\zeta_y \equiv \omega_y D_y/2k_y$, which is equal to $1/2Q_y$, since the quality factor $Q_y = k_y/\omega_y D_y$. Thus, Y_0 is proportional to the input angular rate Ω_z (as well as the x axis vibrational amplitude X_0) and can be used for sensing an angular rate. To maximize the angular sensitivity, the driving frequency ω_d should be equal to the resonant frequency of the driving mode along the x axis ω_x for the largest x axis vibrational amplitude X_0.

The vibrational amplitude Y_0 (i.e., the amplitude along the y axis, the sensing axis) is the largest when ω_d is equal to ω_y (the resonant frequency in the sensing direction y), and is $Y_0 = \frac{2X_0\Omega_z Q_y}{\omega_y}$, which should be maximized by making the driving-mode vibration X_0 and the quality factor Q_y large, as Ω_z is typically much smaller than ω_y. Consequently, the resonant frequencies of the driving mode ω_x and sensing mode ω_y should be as close as possible. If those are mismatched, the sensing-mode vibration amplitude Y_0 will be small; the larger the mismatch, the smaller the Y_0 will be.

6.4.2 Tuning Fork Gyroscope on Quartz

One of the early commercially successful micromachined gyroscopes is the one built on a quartz (that is micromachined with HF acid) using the quartz's piezoelectricity. A simple gyroscope on a quartz tuning fork with the drive and sense electrodes covering the tines is shown in Fig. 6.29.

When the drive electrodes are driven with the voltage as indicated in Fig. 6.29, there are electrical fields induced in the x- and z directions (i.e., E_x and E_z). But E_z does not produce any strain, according to the piezoelectric matrix for quartz (with piezoelectric coefficients d_{ik} and electric fields E_k and compliance coefficients s_{ij} and stress T_j) shown below, though E_x induces normal strains S_1 and S_2 as well as shear strain S_4. While S_1 and S_4 have relatively small effect on the bending of the tines, the normal strain S_2 causes the tine to bend in the x direction, and the two tines vibrate linearly in the two opposite x directions.

$$\begin{pmatrix} S_1 \\ S_2 \\ S_3 \\ S_4 \\ S_5 \\ S_6 \end{pmatrix} = \begin{pmatrix} d_{11} & 0 & 0 \\ -d_{11} & 0 & 0 \\ 0 & 0 & 0 \\ d_{14} & 0 & 0 \\ 0 & -d_{14} & 0 \\ 0 & -2d_{11} & 0 \end{pmatrix} \begin{pmatrix} E_x \\ E_y \\ E_z \end{pmatrix} + s_{ij}T_j \tag{6.64}$$

With the tines vibrating in the two opposite x directions, the tines experience Coriolis forces in the two opposite z directions when angular rotation is applied along the y axis. The z-directed Coriolis forces bend the tines in the z directions, and the bending produces electric fields in the two opposite x directions at the top half and bottom half of the each tine (of the two tines) along the z direction.

A company called BEI Technologies Inc. used to commercialize gyroscopes by etching quartz with HF and was very successful in commercial applications to vehicle stability control (VSC) for automobiles for a number of years [16]. Schneider Electric,

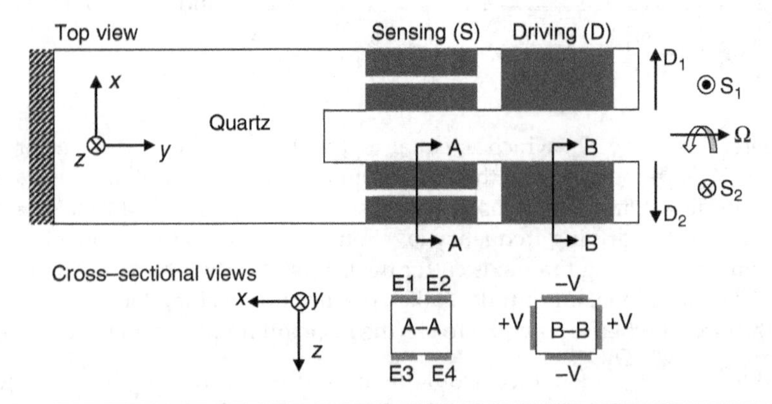

FIGURE 6.29 A simple gyroscope built on a quartz tuning fork with the drive and sense electrodes covering the tines as shown. With the drive voltage levels as indicated, the two tines vibrate linearly in the x direction (in opposite directions) and experiences Coriolis forces in the two opposite z directions when angular rotation is applied along the y axis. (Adapted from [15].)

a French manufacturer of equipment for electrical distribution and industrial control equipment, acquired BEI at about $560 M in 2005. In about 5 years, the competition from silicon MEMS gyroscopes forced Schneider Electric to abandon the quartz-based gyroscope.

6.4.3 Comb-Drive Tuning Fork MEMS Gyroscopes

Draper Lab reported the first gyroscope based on a comb drive that was driven in-plane, similar to what is illustrated in Fig. 6.30 [17]. For the applied angular rotation, the Coriolis force is in the out-of-plane direction and produces a larger displacement (and thus, a higher sensitivity) for a mechanically more compliant structure [18]. As the Coriolis force produces out-of-plane motion, the thickness of the combs needs to be quite thick (ideally > 10 μm) for a wide dynamic range. However, making such thick combs with chemical vapor deposited (CVD) polysilicon film is usually not cost effective due to the prohibitively long deposition time. Also, a very thick polysilicon film has a very rough surface. Consequently, many companies have stayed away from polysilicon comb drives for gyroscope applications except Analog Devices, Inc. which has been monolithically integrating polysilicon microstructures with circuits for inertial sensors. With the success of being able to detect extremely small change in capacitance (due to the fact that the sensor is integrated with circuits on a single chip), Analog Devices commercializes a gyroscope based on polysilicon comb drive [19].

Several companies have commercialized gyroscopes based on MEMS electrostatically actuated comb drives for smart phones, camera stability control, vehicle stability control, etc. Bosch is one of those and uses epitaxially deposited polysilicon as the comb's structural material [20]. With modified comb drives, the combs are driven in one of the in-plane axes, the x axis, and experience Coriolis force in the y axis (another in-plane axis) when there is an angular rotation around the z axis, the out-of-plane axis.

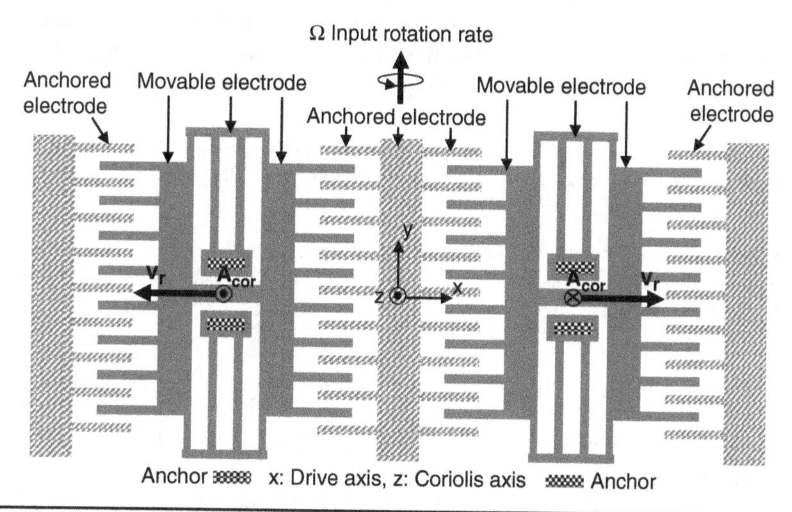

FIGURE 6.30 Conceptual diagram of a gyroscope based on an electrostatically actuated comb drive with the drive and sense (Coriolis) axes as well as the axis of the input rotation as indicated.

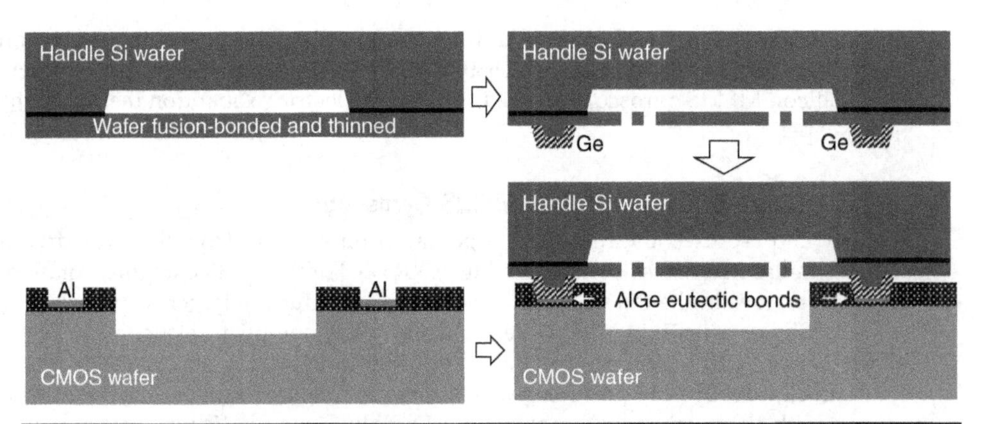

FIGURE 6.31 Brief fabrication process of InvenSense's gyroscope built on hybrid integration of the sensing elements and CMOS electronics through wafer bonding. (Adapted from [21].)

One of the major advantages of MEMS is low cost for manufacturing, but oftentimes the packaging cost for a transducer far exceeds the fabrication cost of MEMS device. One way to overcome this issue is to process multiple wafers: one for CMOS electronics, one for MEMS transducers and a few for capping/encapsulating (and/or packaging), and bond them, followed by dicing the bonded wafers for completely packaged, surface-mountable MEMS. InvenSense has come up with a relatively simple process (Fig. 6.31) to fabricate low-cost MEMS gyroscopes. In the process, there are basically two types of wafer bonding: silicon–silicon fusion bonding at a very high temperature (> 1,000°C) and Al–Ge eutectic bonding at a temperature low enough (< 450°C) not to damage Al in the CMOS. InvenSense integrates three gyroscopes on a single chip for sensing angular rotations in all three axes [21]. For x- and y-axis gyroscopes, two sets of combs are driven out-of-plane directions, so that the angular rotations around the in-plane axes (i.e., x- and y-axis) may produce in-plane motions, while a third set of combs are driven in-plane for z-axis gyroscope.

Question: Why may a gyroscope based on a comb-drive resonator (e.g., the one shown in Fig. 6.30) need a rather thick (e.g., 10–20 μm) film for the movable fingers (electrically driven in-plane) and fixed stators?

Answer: The movable fingers move out of the plane due to the Coriolis force induced by the applied angular velocity and that motion needs to be detected through capacitance variation between the movable and fixed fingers. If the heights of those fingers are short, the detectable range of the motion will be small.

With broad availability of MEMS accelerometers and gyroscopes at low cost and small size, MEMS inertial sensors offer unprecedented opportunity in performance enhancement and enablement in various systems because they can be placed exactly at the critical point of a system. For example, for an unmanned automatic vehicle (UAV) with multiple vision sensors, placing multiple MEMS inertial sensors right next to those many vision sensors will allow more accurate imaging through the vision sensors than placing one expensive and bulky (and more accurate) inertial sensor at a place that is far from many of the vision sensors spread out over UAV.

References

1. P. Osterberg, H. Yie, X. Cai, J. White, and S. Senturia, "Self-consistent simulation and modeling of electrostatically deformed diaphragms," Proceedings of MEMS 94, Oiso, Japan, January 1994, pp. 28–32.
2. Theory of Plates and Shells, McGraw-Hill Classic Textbook Reissue Series 2nd Edition, by S. Timoshenko, ISBN-13: 978-0070858206.
3. E.S. Kim, "IC processed piezoelectric microphone," M.S. Thesis, U.C. Berkeley, 1987.
4. Harris and Crede (Eds.), "Shock and Vibration Handbook," 2nd ed., McGraw-Hill, 1976.
5. H.N. Chu and G. Herrmann, "Influence of large amplitudes on free flexural vibrations of rectangular elastic plates," Journal of Applied Mechanics, 23, 1956, pp. 532–540.
6. K.H. Kim, J.S. Ko, Y.H. Cho, K. Lee, B.M. Kwak, and K. Park, "A skew-symmetric cantilever accelerometer for automotive airbag applications," Sensors and Actuators A: Physical, 50, August 1995, pp. 121–126.
7. S. Terry, "A miniature silicon accelerometer with built-in damping," Solid-State Sensor and Actuator Workshop, Hilton Head Island, SC, June 6–9, 1988, pp. 114–116.
8. M. E. Motamedi, "Acoustic accelerometers," IEEE Transactions on Ultrasonics, Ferroelectrics, and Frequency Control, 34(2), March 1987, pp. 237–242.
9. P.L. Chen, R.S. Muller, R.D. Jolly, G.L. Halac, R.M. White, A.P. Andrews, T.C. Lim, and M.E. Motamedi, "Integrated silicon microbeam PI-FET accelerometer," IEEE Transactions on Electron Devices, 29(1), January 1982, pp. 27–33.
10. P.L. Chen, R.S. Muller, R.M. White, and R. Jolly, "Thin film ZnO-MOS transducer with virtually dc response," Proceedings of IEEE Ultrasonics Symposium, Boston, MA, November 5–7, 1980.
11. Q. Zou, W. Tan, E.S. Kim, and G.E. Loeb, "Single-axis and tri-axis piezoelectric bimorph accelerometer," IEEE/ASME Journal of Microelectromechanical Systems, 17(1), 2008, pp. 45–57, 2008.
12. F. Rudolf, A. Jomod, and P. Bencze, "Silicon microaccelerometers," Technical Digest, 4th International Conference on Solid-State Sensors and Actuators, Transducers '87, Tokyo, Japan, June 2–5, 1987, pp. 395–398.
13. S. Suzuki, S. Tuchitani, K. Sato, S, Ueno, Y, Yokota, M. Sato, and M. Esashi, "Semiconductor capacitance-type accelerometer with PWM electrostatic servo technique," Sensors and Actuators, A21–A23, February 1990, pp. 316–319.
14. L.J. Spangler and C.J. Kemp, "ISAAC-integrated silicon automotive accelerometer," International Conference on Solid-State Sensors and Actuators, Transducers '95, 1, Stockholm, Sweden, June 1995, pp. 585–588.
15. J. Soderkvist, "Design of a solid-state gyroscopic sensor made of quartz," Sensors and Actuators, A21–A23, 1990, pp. 293–296.
16. A.M. Madni, L.E. Costlow, and S.J. Knowles, "Common design techniques for BEI GyroChip quartz rate sensors for both automotive and aerospace/defense markets," IEEE Sensors Journal, 3(5), October 2003.
17. J. Bernstein, S. Cho, A. T. King, A. Kourepenis, P. Maciel, and M. Weinberg, "A micromachined comb-drive tuning fork rate gyroscope," Proceedings of IEEE Micro Electro Mechanical Systems Workshop (MEMS '93), Fort Lauderdale, FL, February 1993, pp. 143–148.

18. Z.Y. Guo, Z.C. Yang, Q.C. Zhao, L.T. Lin, H.T. Ding, X.S. Liu, J. Cui, H. Xie, and G.Z. Yan, "A lateral-axis micromachined tuning fork gyroscope with torsional Z-sensing and electrostatic force-balanced driving," Journal of Micromechanics and Microengineering, 20, 2010, 025007.
19. H. Qu, "CMOS MEMS fabrication technologies and devices," Micromachines, 2016, 7, 14; doi:10.3390/mi7010014.
20. J. Liewald, B. Kuhlmann, T. Balslink, M. Trachtler, M. Dienger, and Y. Manoli, "100 kHz MEMS vibratory gyroscope," Journal of Microelectromechanical Systems, 22(5), October 2013, pp. 1115–1125.
21. J. Seeger, M. Lim, and S. Nasiri, "Development of high-performance, high-volume consumer MEMS gyroscopes," Solid-State Sensors, Actuators, and Microsystems Workshop, Hilton Head, SC, 2010, pp. 61–64.

Questions and Problems

Question 6.1 Using Rayleigh's energy method, we can calculate the fundamental resonant frequency of an elastic body by assuming a deflection curve. What would you do to estimate the resonant frequency as accurately as possible?

Question 6.2 When a beam (with a moment of inertia equal to 10^{-20} m^4) goes through a pure bending by a bending moment, the stress at a point 1 μm away (in the thickness direction) from the beam axis is measured to be 1,000 Pa. What is the stress at a point 0.5 μm away (in the same thickness direction) from the beam axis?

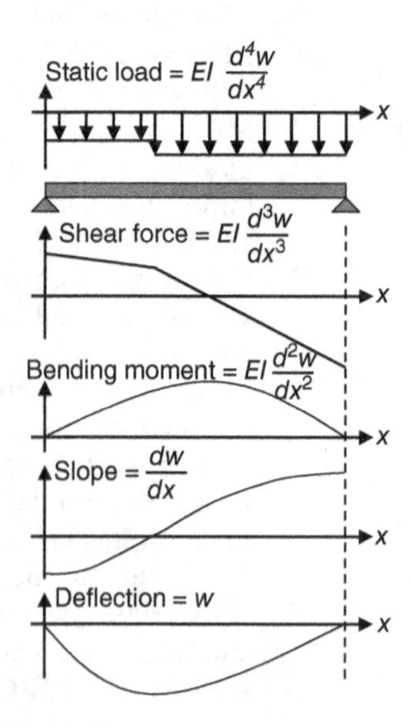

Question 6.3 A simply supported beam goes through a pure bending by a distributed load and is analyzed to have the shear force, bending moment, slope, and deflection as shown at right. If the beam has *clamped* edges (rather than simply supported edges), how the curves will look like? Sketch the new curves on the same figure at right.

Question 6.4 On the graph at right, sketch the shear force, bending moment, slope, and deflection when the beam (with its two ends being simply supported) is under a uniform distributed load (i.e., a load that is *not* varying, or constant in magnitude, along the beam axis, x). Note that the load that is shown at right has a step variation at a point along the beam axis and is not the load for this problem.

Question 6.5 A thin polysilicon beam (100 μm long, 2 μm thick, and 6 μm wide) is *simply supported* at its two edges. What *compressive* residual stress will cause the beam to be buckled? Assume the polysilicon has Young's modulus of 140 GPa.

Question 6.6 A thin polysilicon cantilever (100 μm long, 2 μm thick, and 6 μm wide) with one edge clamped is deflected by a static point force (1 dyne) as shown below. (a) Ignoring the cantilever weight, sketch the shear force, bending moment, and deflection as a function of the beam axis, *x*. Where is the maximum stress?

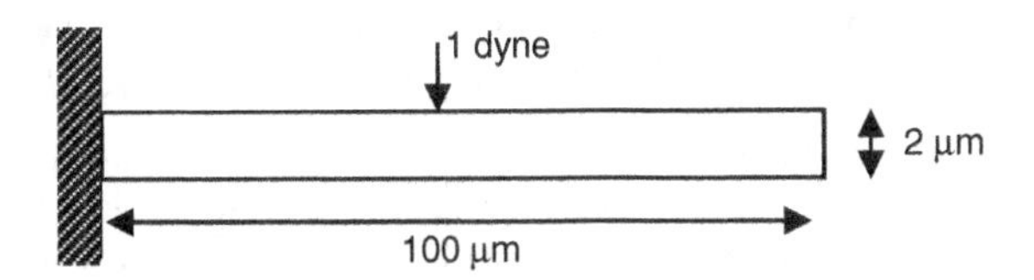

Question 6.7 A thin polysilicon bridge (100 μm long, 2 μm thick, and 6 μm wide) with two edges clamped is deflected by a static *point* force (1 dyne) as shown below. Sketch the shear force, bending moment, and deflection as a function of the beam axis, *x*, ignoring the cantilever weight.

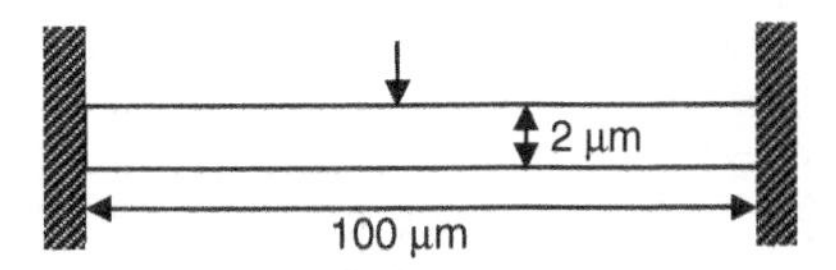

Question 6.8 A thin polysilicon beam (100 μm long, 2 μm thick, and 6 μm wide) is clamped at its two edges. (a) Sketch how the beam deflects under a static *uniform* loading (1 dyne/μm) shown below. In the same sketch, indicate the neutral plane where there is no strain and the point(s) where the bending stress is largest in tension. (b) If the fundamental resonant frequency of the beam shown below is measured to be 100 kHz, what is the expected resonant frequency for the second harmonic. (c) If the fundamental resonant frequency of the beam shown below is 100 kHz when the vibration amplitude is very small, what will be the resonant frequency when the vibration amplitude is large? Will be it less than 100 kHz, larger than 100 kHz, or 100 Hz?

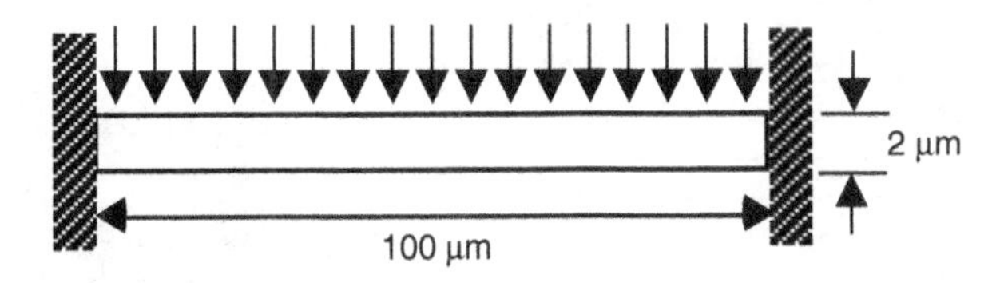

Question 6.9 A polysilicon bridge with its two edges clamped shows a frequency response of vibration amplitude as shown below. Sketch (on the figure below) how the frequency response will change, if the vibration amplitude is increased (say by increasing the actuating voltage).

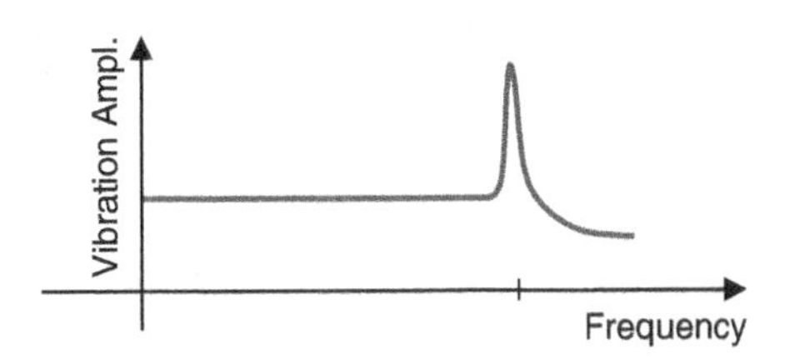

Question 6.10 For a square diaphragm with its four edges clamped, answer the following:

(a) How much larger will the center displacement be for a given uniform load, if the size of the diaphragm is increased to 1×1 mm^2 from 100×100 μm^2, while keeping everything else the same?

(b) How much smaller will the fundamental resonant frequency be, if the size of the diaphragm is increased to 1×1 mm^2 from 100×100 μm^2, while keeping everything else the same?

Question 6.11 An underdamped accelerometer of which the frequency response is shown at left below exhibits a time response to a step input as shown at right below. On the same figure at right below sketch a time response if the accelerometer is overdamped as indicated with B on the left figure below.

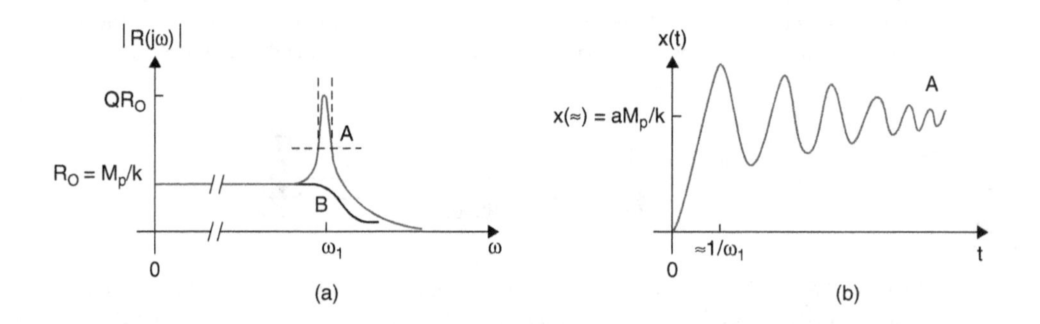

Question 6.12 For a piezoresistive accelerometer with a silicon proof mass that is suspended by four thin silicon beams on which eight piezoresistors are formed as shown below, indicate (on the table below with arrows) how each of the eight piezoresistors (named Z_1, Z_2, Z_3, Z_4, Y_1, Y_2, X_1, and X_2) will respond for *x*-, *y*-, and *z*-directed accelerations. The ones for Z_1 have been answered.

	ΔR with **x** acceleration	ΔR with **y** acceleration	ΔR with **z** acceleration
Z_1	↓	↔	↑
Z_2			
Z_3			
Z_4			
Y_1			
Y_2			
X_1			
X_2			

Question 6.13 Why do we define accelerometer's Figure of Merit (FOM) to be the product of DC response and square of fundamental resonant frequency (i.e., FOM $\equiv R_o\omega_o^2$), rather than $R_o\omega_o$?

Question 6.14 Shown below is a piezoelectrically actuated cantilever made for a projection display application. Assume that $d_{33} = 11 \times 10^{-12}$ C/N and $d_{31} = -5 \times 10^{-12}$ C/N for ZnO.

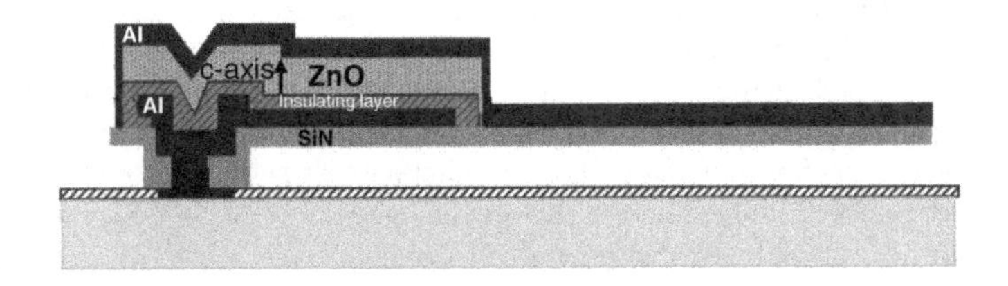

(a) What is the minimum thickness for the SiN layer, if we want the neutral plane (for cantilever bending) outside ZnO (i.e., not within ZnO)? Assume that the ZnO (0.3 μm thick) and the insulating layer (0.2 μm thick) have Young's modulus of 210 GPa and 70 GPa, respectively, while the top and bottom Al films (0.2 μm thick each) have Young's modulus of 70 GPa. Also, assume that Young's modulus of SiN is 270 GPa.

(b) If +10 V_{dc} is applied to the top Al with respect to the bottom Al, how will the cantilever be bent?

(c) If the fundamental resonant frequency of the cantilever is 1 kHz, *sketch* the magnitude of the cantilever tip displacement versus frequency (from 10 Hz to 10 kHz in log scale) for a sinusoidal voltage ($V_0\sin\omega t$) applied between the top and bottom Al, when the cantilever is under vacuum.

Problem 6.1 A thin polysilicon cantilever (100 μm long, 2 μm thick, and 6 μm wide) with one edge clamped is deflected by a static point force (1 dyne) that is applied at the free edge of the cantilever in the thickness direction, as shown below. Ignoring the weight of the cantilever, calculate and plot the shear force, bending moment. and deflection as a function of x. Assume that the Young's modulus for polysilicon is 150 GPa. Note that 1 Pa = 1 N/m^2 = 10 dyne/cm^2. Also, the moment of inertia (I) of a rectangular area around the centroidal axis is equal to $bh^3/12$. Where is the maximum stress? Calculate the maximum stress.

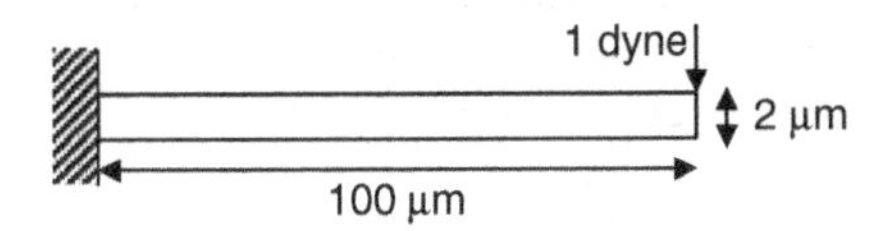

Problem 6.2 The figures below are the top- and cross-sectional view of a polysilicon cantilever (100 μm long, 2 μm thick, and 6 μm wide) with one edge clamped. The polysilicon cantilever is separated by 1 μm air gap from the underlying substrate and deflected by an electrostatic voltage applied between the cantilever and the substrate.

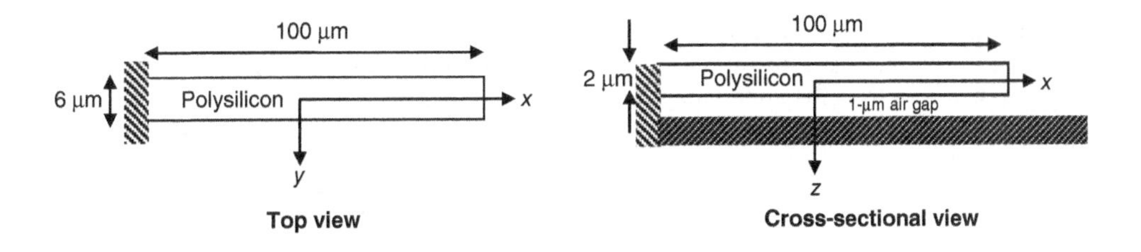

(a) At which points in the cantilever is the magnitude of σ_{xx} highest, when the cantilever is deflected by an electrostatic voltage applied between the cantilever and the substrate. Give the answers in values in x-, y-, z coordinates in μm such as (0, 0, 1), (50, 3, 0), etc. (b) Sketch σ_{xx} as a function of z from −1 μm to +1 μm at $x = 0$ and $y = 0$, when the cantilever is deflected by an electrostatic voltage applied between the cantilever and the substrate. (c) Sketch σ_{xx} as a function of y from −3 mm to +3 mm at $x = -50$ μm and $z = -1$ μm, when the cantilever is deflected by an electrostatic voltage applied between the cantilever and the substrate. (d) How much electrostatic force *per unit area* does the polysilicon cantilever experience, if a 1 GHz RF signal with a power level of 1 μW is applied between the cantilever and the substrate? (e) Calculate the cantilever displacement at its free end, if a 1 GHz RF signal with a power level of 1 μW is applied between the cantilever and the substrate. For this and (e), take the Young's modulus of the polysilicon to be 150 GPa. (e) Calculate the cantilever displacement at its free end, if 1 V DC voltage is applied between the cantilever and the substrate. (f) For a *small* AC voltage $V(t) = V_a \cos\omega t$ [V] applied between the polysilicon cantilever and the substrate, sketch the cantilever displacement at its free end as a function of frequency from 10 to 200 kHz in log-linear scale, assuming that the *static* displacement is 0.1 μm. Assume that the cantilever has *the fundamental resonant frequency of* 30 kHz with Q of 10.

Problem 6.3 The figures below are the top- and cross-sectional views of a polysilicon bridge (100 μm long, 2 μm thick, and 15 μm wide) with two edges clamped. The polysilicon bridge is separated by 1-μm air gap from the underlying substrate.

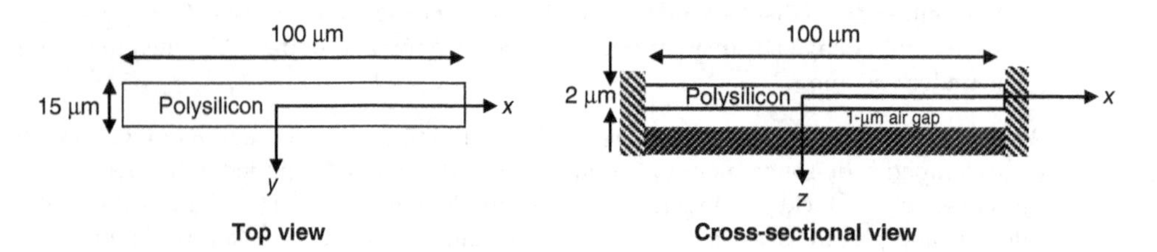

(a) At which points in the bridge is σ_{xx} highest in tensile stress, when the bridge is deflected by an electrostatic voltage applied between the bridge and the substrate. Give the answers in values in x-, y-, and z coordinates in μm such as (0, 0, 1), (50, 3, 0), etc. (b) Sketch σ_{xx} as a function of z from −1 μm to +1 μm at $x = 0$ and $y = 0$, when the bridge is deflected by an electrostatic voltage applied between the bridge and the substrate. (c) Sketch σ_{xx} as a function of y from −7.5 μm to +7.5 μm at $x = -50$ μm and $z = -1$ μm, when the bridge is deflected by an electrostatic voltage applied between the bridge and the substrate. (d) How much electrostatic force *per unit length* does the polysilicon bridge experience, if 1 V is applied between the bridge and the substrate? (e) For a *small* AC voltage $V(t) = V_a \cos\omega t$ [V] applied between the polysilicon bridge and the substrate, sketch the bridge displacements at its center as a function of frequency from 10 to 200 kHz in log-linear scale, for $V_a = 1$ V and $V_a = 5$ V, assuming that the *static* displacement is 0.1 μm for $V_a = 1$ V and that the cantilever has *the fundamental resonant frequency of* 160 kHz with Q of 10.

Problem 6.4 The figures below are the top- and cross-sectional views of a polysilicon plate (100 μm long, 2 μm thick, and 40 μm wide) with one edge clamped. The polysilicon plate is separated by 1-μm air gap from the underlying substrate.

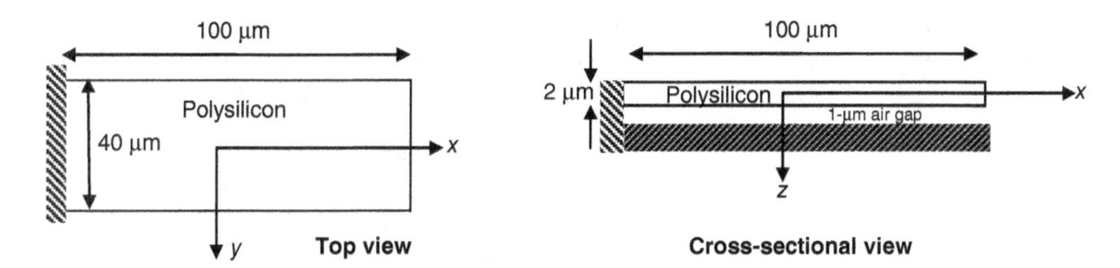

(a) At which points in the bridge is σ_{xx} highest in tensile stress, when the bridge is deflected by an electrostatic voltage applied between the bridge and the substrate. Give the answers in values in x-, y-, and z coordinates in μm such as (0,0,1), (50,20,0), etc. (b) Sketch σ_{xx} as a function of z from −1 μm to +1 μm at $x = -50$ and $y = 0$ when the bridge is deflected by an electrostatic voltage applied between the bridge and the substrate. (c) On a same plot, sketch both σ_{xx} and σ_{yy} as a function of y from −20 μm to +20 μm at $x = -50$ μm and $z = -1$ μm, when the bridge is deflected by an electrostatic voltage applied between the bridge and the substrate.

Problem 6.5 The figures below are the top- and cross-sectional view of a thin square diaphragm with the four edges simply supported (through electrostatic voltage applied between the free-plate diaphragm and the perforated backplate). For all the following problems assume that the square diaphragm (with $a = 1$ mm, the thickness of the free-plate diaphragm = 1 μm, mass density = 2.3 g/cc, and Poisson's ratio = 0.3) has the fundamental resonant frequency of 9 kHz with $Q = 40$. (a) Based on the fundamental resonant frequency, what is the Young's modulus of the free-plate diaphragm? (b) For a static uniform load of 1 Pa, what is the deflection at the center of the diaphragm (i.e., at $x = y = 0$)? (c) For a static uniform load of 1 Pa, what is $|\sigma_{xx}|$ at the center of the diaphragm (i.e., at $x = y = 0$) and at the top surface of the free-plate diaphragm? (d) For a static uniform load, sketch σ_{xx} as a function of z from the bottom surface to the top surface of the free-plate diaphragm at the center of the diaphragm (i.e., at $x = y = 0$). (e) For a static uniform load, sketch $|\sigma_{yy}|$ as a function of x from $-a/2$ to $+a/2$ at $y = 0$ and at the top surface of the free-plate diaphragm. (f) For a sinusoidal load $P(t) = \cos\omega t$ [Pa], what is the magnitude of the deflection at the center of the diaphragm (i.e., at $x = y = 0$) if $f = \omega/2\pi = 9$ kHz?

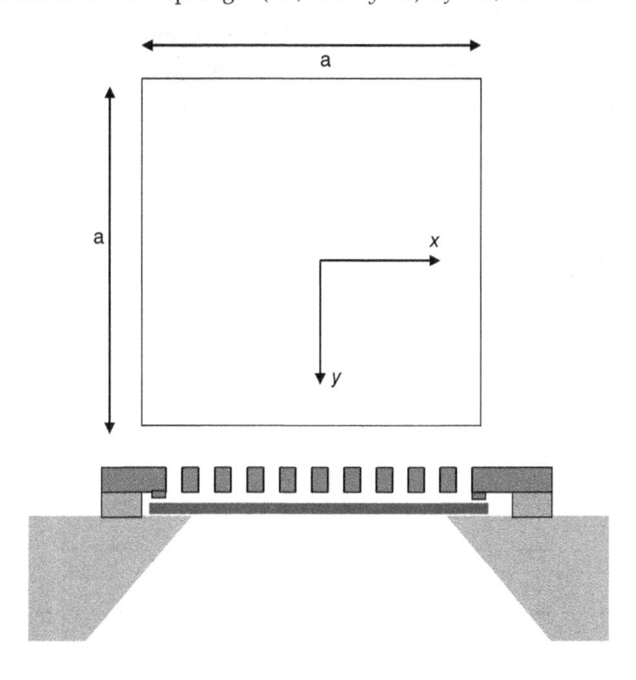

Problem 6.6 The figures below are the top- and cross-sectional view of a thin square plate with the four edges clamped. For all the following problems assume that the square plate with $a = 1$ mm and $h = 1$ μm has *the fundamental resonant frequency of* 36 kHz with $Q = 100$.

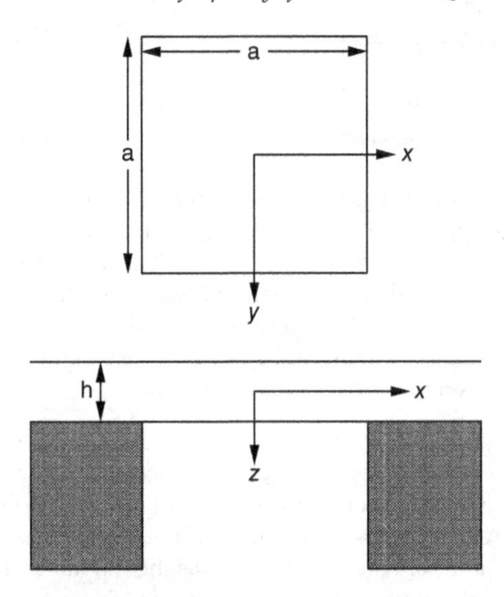

(a) For a small lateral sinusoidal load $P(t) = \cos\omega t$ [dyne/cm²], plot the amplitude of the center displacement as a function of frequency from 10 to 200 kHz in log-linear scale, assuming that the *static* center deflection is 0.1 μm. (b) At which points in the diaphragm is σ_{xx} highest for a static pressure applied uniformly over the diaphragm? And at which points in the diaphragm is σ_{yy} highest? Give the answers in values in x-, y-, and z coordinates such as $(0, 0, h/2)$, $(a/2, a/2, 0)$, etc. (c) Sketch σ_{xx} as a function of z from $-h/2$ to $+h/2$ at $x = 0$ and $y = 0$ for a static load applied uniformly over the diaphragm. Sketch σ_{yy} as a function of x from $-a/2$ to $+a/2$ at $y = +a/2$ and $z = -h/2$ for a static load applied uniformly over the diaphragm.

Problem 6.7 Using the figures shown in Problem 6.6 for the top- and cross-sectional view of a thin square plate with the four edges clamped, answer the following.

(a) For both *sinusoidally varying* and *step* pressures applied *uniformly* over the square diaphragm, sketch the center displacement in log-linear scale below. Assume that the *static* center deflection is 1 μm for 1 Pa lateral load and also that the square diaphragm has *the fundamental resonant frequency of* 10 kHz with $Q = 10$.

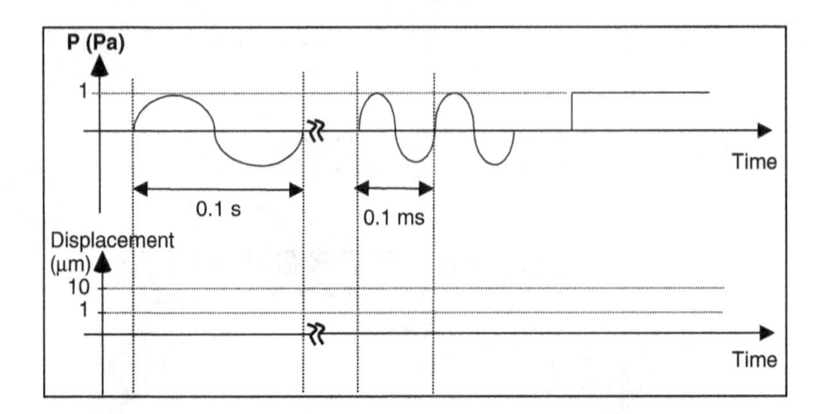

(b) If you want to make a pressure sensor through etching out silicon bulk by DRIE and forming a square diaphragm by a timed etch, where will you implant your piezoresistors? Give the answer in values in x-, y-, and z coordinates such as $(0, 0, h/2)$, $(a/2, a/2, 0)$, etc. [Hint: There are four excellent places.]

(c) What will be the resulting net stress, if we integrate the stress from $z = -h/2$ to $z = +h/2$ at a given area in the square diaphragm, when the diaphragm is deflected under a hydrostatic load applied uniformly over the diaphragm.

(d) Sketch σ_{xx} as a function of z from $-h/2$ to $+h/2$ at $x = a/2$ and $y = a/2$ for a hydrostatic load applied uniformly over the diaphragm.

Problem 6.8 Using the figures shown in Problem 6.6 for the top- and cross-sectional view of a thin square diaphragm with its four edges clamped, answer the following.

(a) Among the following four points (with their x, y, and z coordinate values according to the notation indicated in the figures), which is the best point to place a piezoresistor for a pressure sensing? In other words, where is the largest normal strain induced when a uniform pressure is applied over the diaphragm? Choose one from the following four: $(0, 0, 0)$, $(a/2, a/2, h/2)$, $(a/2, 0, h/2)$, $(0, 0, a/2)$.

(b) Calculate the largest normal stress induced, when a uniform pressure of 1 Pa is applied over the diaphragm, assuming that $a = 1$ mm, $h = 1$ μm, and $\nu = 0.3$.

(c) Which of the following are true, when the diaphragm is under a uniform distributed load.

$\sigma_y (x = 0, y = 0, z = h/2) = -\sigma_y (x = 0, y = 0, z = -h/2)$

$\sigma_x (x = a/2, y = 0, z = h/2) = \sigma_y (x = a/2, y = 0, z = h/2)$

$\sigma_x (x = a/2, y = 0, z = h/2) = \sigma_y (x = 0, y = a/2, z = h/2)$

$|\sigma_x (x = 0, y = 0, z = h/2)| > |\sigma_x (x = a/2, y = 0, z = h/2)|$

Problem 6.9 For a quartz tuning fork gyroscope shown in Fig. 6.29, (a) sketch (on the cross-section across B–B, repeated below) the electric field pattern produced by the drive electrodes.

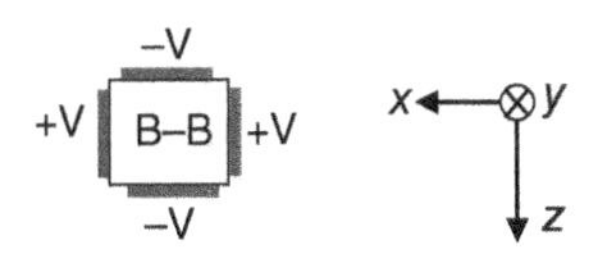

(b) What are nonzero strains when the voltages are applied on the drive electrodes as shown in Fig. 6.29? (c) How would you connect E1, E2, E3, and E4 to get an electrical voltage for the tuning fork responding to Coriolis force due to an angular velocity applied around the y axis?

Problem 6.10 For an accelerometer built on a simple spring-mass system shown in Fig. 6.17, answer the following, assuming that $k = 1$ N/m, $M_p = 1$ kg, and $D = 1$ Ns/m. (a) Calculate the resonant frequency ω_1 and quality factor Q. (b) Calculate the displacement magnitude of the proof mass, if a sinusoidal acceleration $a(t) = 9.8 \cos\omega t$ [m/s^2] is applied to the accelerometer with ω much less than ω_1. (c) Calculate the displacement magnitude of the proof mass, if a sinusoidal acceleration $a(t) = 9.8 \cos\omega_1 t$ [m/s^2] is applied to the accelerometer.

CHAPTER 7

Thin-Film Properties, SAW/BAW Sensors, Pressure Sensors, and Microphones

In this chapter, we will first study residual stresses in thin films and their impacts on MEMS performance and characteristics. Then we will look into material properties of polysilicon film and piezoelectric films, as they are often used in MEMS as structural or transducing materials. We will see that many material properties can be expressed as tensors, and piezoelectricity and piezoresistivity are characterized with a third- and fourth-rank tensor, respectively. As examples of piezoelectric and piezoresistive MEMS, we will study acoustic wave generation through piezoelectricity and mass resonant sensors based on such waves, and then silicon piezoresistive pressure sensors. Finally, we will study capacitive and piezoelectric microphones, which are basically sensitive pressure sensors with frequency response covering audio range.

7.1 Thin-Film Residual Stress

Materials used in MEMS are typically thin films, and we need to know their material properties, particularly static mechanical properties (such as residual stress, Young's modulus, and Poisson's ratio), dynamic mechanical properties (e.g., fatigue limits, internal friction related to mechanical Q), and other properties (such as static and dynamic friction, wear, and abrasion). Thin films (that are thermally grown or deposited/coated with various methods) are typically under residual stress, which can be broadly categorized into thermal stress and intrinsic stress.

7.1.1 Thermal Stress σ_t

If a layer is deposited as a stress-free thin film on a thick substrate at an elevated temperature (T_d), the film will be under stress at room temperature (T_r) due to the difference in the thermal expansion coefficients of the film (α_f) and substrate (α_{sub}), as the film experiences strain ε_t, which is equal to $(\alpha_f - \alpha_{sub})(T_d - T_r)$. If $\alpha_f > \alpha_{sub}$, the strain ε_t is greater than zero, meaning that the film is under tension, as the film has contracted

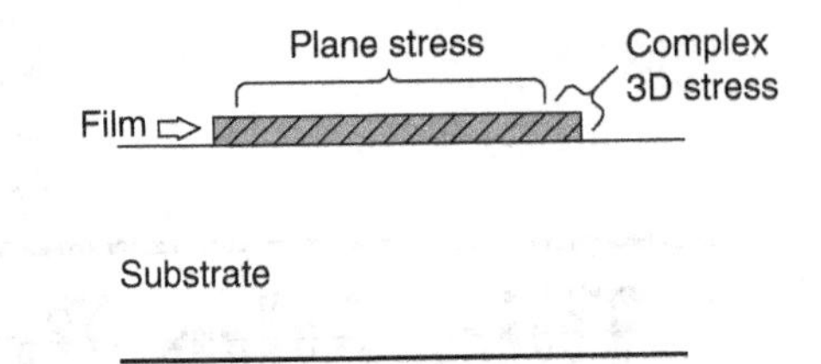

FIGURE 7.1 Cross-sectional view of a film under a plane stress with respect to substrate.

more than the substrate in the lateral direction. On the other hand, if $\alpha_f < \alpha_{sub}$, the film is under compression.

The strain ε_t can easily be converted to stress σ_t, assuming that the stress is "plane stress," not a complex three-dimensional stress, as illustrated in Fig. 7.1. In case of a thin film over a thick substrate, plane stress is a good approximation in most of the area except the patterned edges. For a plane stress, $\sigma_z = 0$, and we have

$$\varepsilon_x = \frac{1}{E}(\sigma_x - \nu\sigma_y) \quad \text{and} \quad \varepsilon_y = \frac{1}{E}(\sigma_y - \nu\sigma_x). \tag{7.1}$$

Consequently, we have the following equations for the stresses

$$\sigma_x = \frac{E}{1-\nu^2}(\varepsilon_x + \nu\varepsilon_y) \quad \text{and} \quad \sigma_y = \frac{E}{1-\nu^2}(\varepsilon_y + \nu\varepsilon_x). \tag{7.2}$$

If $\varepsilon_x = \varepsilon_y = \varepsilon, \sigma_x = \sigma_y = \frac{E}{1-\nu}\varepsilon = \frac{E}{1-\nu}(\alpha_f - \alpha_{sub})(T_d - T_r)$.

7.1.2 Intrinsic Stress σ_i

Intrinsic stress σ_i is usually larger than thermal stress σ_t, but is hard to pinpoint its sources, since there are many sources for intrinsic residual stress. Some are related to the film's nucleation and growth mechanisms, while others are due to structural mismatch between film and substrate. If it is due to dopants, the stress level depends on whether the dopants occupy substitutional sites (in which case the dopant's ionic radius in relation to the lattice size determines the stress type and level) or interstitial sites (in which case the doped layer would be under compression).

In case of sputter deposition, "atomic peening" effect causes compressive residual stress in the deposited film, as ion bombardment by sputtered atoms and working gas (e.g., argon) densifies the deposited film. Gas entrapment in the film also causes compressive residual stress. In case of sol–gel deposition, shrinkage of the coated layer during cure causes tensile residual stress. Polycrystalline films have grain boundaries, which affect residual stress, but the effects of grain boundaries on residual stress are complex.

7.1.3 Techniques to Control Residual Stress in Thin Films

As we saw with silicon-rich silicon nitride in Chap. 1, we can vary material composition during chemical vapor deposition (CVD) by varying the reactant gas ratio and/or deposition temperature in order to vary residual stress. In case of CVD silicon nitride

(Si_xN_y), the ratio between x and y can be varied from about 0.75 to about 1.0, resulting in a residual stress that is very tensile to compressive.

In case of sputter deposition, there are some parameters that can be adjusted to vary the residual stress, such as gas pressure, radio frequency (RF) power, substrate bias, substrate temperature, etc. However, the deposition condition that results in the lowest residual stress may not be the optimum condition for the film's property as a transducer.

Post-deposition annealing may induce recrystallization and grain growth, and reduce density of grain boundaries, resulting in change in the residual stress. One of the notable usages of post-deposition annealing is on making the residual stress uniform throughout the thickness direction for polysilicon in order to make the comb drives flat, as we studied in Chap. 3. If dopants are allowed, one can consider using dopants to compensate residual stress in a film.

7.1.4 Effects of Residual Stress

Effects of thin-film residual stress on released structures can be understood with the figures in Fig. 7.2, where four different cases are shown: (a) film under uniform tension, (b) film under uniform compression, (c) and (d) film with stress gradient along the thickness direction. If a film is stress-free when it is attached to a substrate, the lateral dimension of the film would be the same as that of the substrate, when it is detached from the substrate. However, if a film is under uniform tension (or compression) when it is attached to a substrate, its lateral dimension would be shorter (or longer) than that of the substrate, when it is detached from the substrate to form a cantilever, as illustrated in Fig. 7.2a and b. The cantilever will turn out to be flat, without any bending, since there is no stress gradient along the thickness direction. However, if the stress is not uniform along the thickness direction, the released cantilever will bend upward or downward, depending on how the stress varies along the thickness direction, as shown in Fig. 7.2c and d.

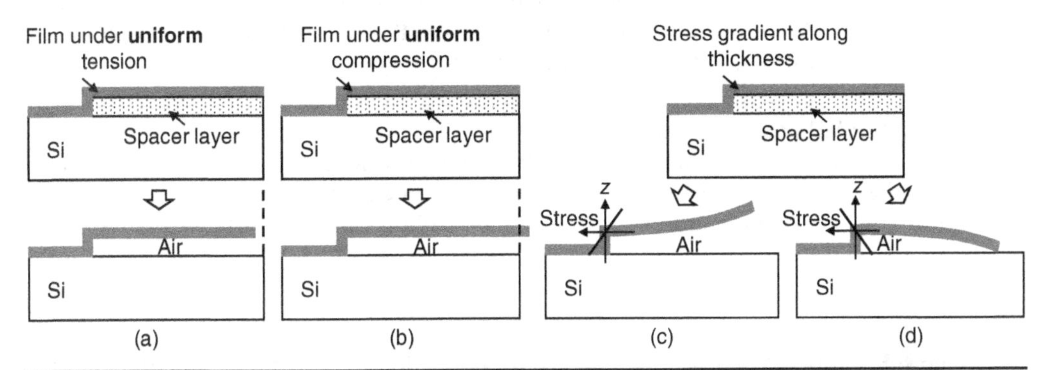

FIGURE 7.2 (Top row) Cross-sectional views of films attached to substrate with uniform tensile (a) or compressive (b) residual stress as well as with stress gradient along the thickness direction (c) and (d). (Bottom row) Cross-sectional views of released cantilevers with shorter (a) or longer (b) length than the length before being released due to the uniform residual stress; (c) and (d) if the residual stress varies along the thickness direction.

Question: If the residual stress varies along the thickness direction of a cantilever structural layer, the cantilever would be warped when it is released. If the residual stress is more tensile near the top than bottom of the cantilever layer before the structural layer is released, will the cantilever be bent upward (Fig. 7.2c) or downward (Fig. 7.2d)?

Answer: Bent upward.

7.1.5 Stress Measurement Techniques

Mechanical Deformation of Wafer

The mismatch in the lateral dimensions of a film and a substrate, when the film is released from the substrate (illustrated in Fig. 7.2a and b), can be viewed as the cause for bending of a whole wafer (both the film and the substrate), when the film is attached to the substrate [1]. Thus, flatness of a wafer can indicate global level of a film stress, since biaxial stress σ_f in a thin film on a thick substrate is proportional to the wafer curvature κ, and can be calculated from

$$\sigma_f = B_{sub} \frac{t_{sub}^2}{6t_f} \kappa, \tag{7.3}$$

where $B_{sub} = \frac{E_{sub}}{1-\nu_{sub}}$ is the biaxial modulus of the substrate, and t_{sub} and t_f are the thicknesses of the substrate and the film, respectively [1]. The classic Stoney formula [Eq. (7.3)] assumes that film thickness is much smaller than substrate thickness, i.e., the neutral plane is near mid-plane of substrate. In using the formula for measuring a film stress, a correction to the formula can be made with the following equation [2]:

$$\kappa' = \kappa \left[\frac{1+H}{1+Hb(4+6H+4H^2)+H^4 b} \right], \tag{7.4}$$

where $H = \frac{t_f}{t_{sub}}$ and $b = \frac{B_f}{B_{sub}}$.

This method, however, indicates only the average of a film stress over a whole wafer where the film thickness may vary substantially, and is usually inaccurate in estimating a film stress at a particular point or region on a wafer.

Membrane Deflection (Bulge Test)

One can fabricate a membrane with a thin film on a wafer and measure the membrane deflection as a function of applied differential pressure in order to measure a film stress. For circular or square membranes, neglecting bending stiffness, we have

$$p = \frac{ht}{a^2} \left(f(\nu) \frac{E}{1-\nu} \frac{h^2}{a^2} + c\sigma_o \right), \tag{7.5}$$

where p = applied pressure, h = measured deflection height, a = radius of a circular membrane or ½ side length of a square membrane, t = membrane thickness, E = Young's modulus, ν = Poisson's ratio, σ_o = tensile residual stress at $p = 0$, $f(\nu)$ = function of Poisson's ratio, and c = constant (which is about 4.0 and 3.4 for circular and square membrane, respectively). For a rectangular membrane that has a high aspect

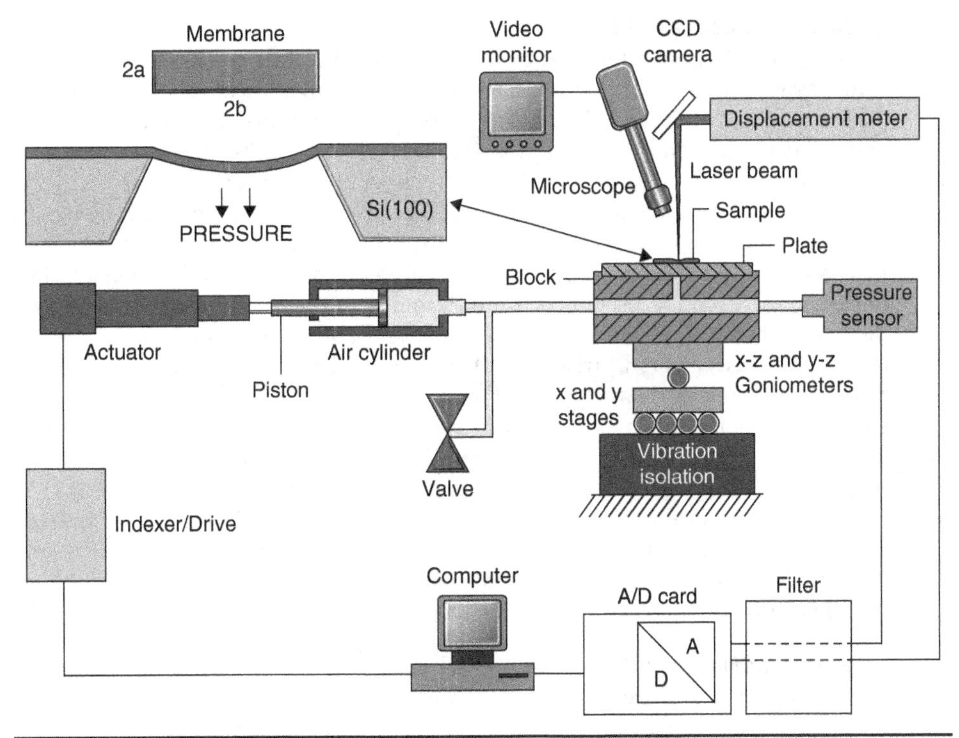

Figure 7.3 Advanced bulge test setup with a rectangular membrane micromachined in a silicon chip [4].

ratio (i.e., $b > 5a$ in Fig. 7.3), the $f(\nu)$ and c are $(0.75 + 0.75\,\nu)^{-1}$ and 2, respectively [3], and we have

$$p = \frac{ht}{a^2}\left(\frac{E}{0.75(1+\nu)(1-\nu)}\frac{h^2}{a^2} + c\sigma_o\right) = Ah\left(\frac{E}{1-\nu^2}h^2 + B\sigma_o\right) \tag{7.6}$$

Consequently, a measured curve of deflection h versus pressure p (that can easily be obtained with a setup similar to what is shown in Fig. 7.3) gives the residual stress σ_o and the material constant $E/(1-\nu^2)$ from the slope and the intercept of the curve, respectively. This method is commonly called a bulge test, and is the most accurate method to measure residual stress of a film. However, a deflectable membrane is needed, which is difficult to be fabricated with fragile materials.

For a circular or square membrane, Table 7.1 shows the function $f(\nu)$ and c that can be used in Eq. (7.5).

Membrane Type	$f(\nu)$	c	Method	Reference
Circular	$(0.385 + 0.0874\,\nu)^{-1}$	4.0	FEM	[5]
Square	$(1.98 - 0.585\,\nu)^{-1}$	3.4	FEM	[5]
Rectangular for $b > 5a$	$(0.75 + 0.75\,\nu)^{-1}$	2.0	Analytical	[3]

Table 7.1 Published Values for the Function $f(\nu)$ and c in Eq. (7.5)

Released Microstructures

Buckling threshold of a released bridge (with its two ends fixed) indicates the average compressive stress in a film [6]. Consequently, one can fabricate a series of bridges with different lengths and calculate a film's compressive stress from the bridge length at which buckling occurs. If the film's residual stress is tensile, one can microfabricate a released ring (with two anchors along one direction and a center beam along the direction orthogonal to the two-anchor direction), which allows a measurement of tensile residual stress that induces shortening or buckling of the center beam [7].

As can be seen in Fig. 7.2, a released cantilever has its lateral dimension increased or decreased from the patterned dimension before the release, depending on whether the residual stress is compressive or tensile, as the stress also is released with the cantilever's three edges being freed up. If the residual stress is uniform in the thickness direction, the released cantilever is flat, as shown in Figs. 7.4 and 7.5. However, if the stress varies in the thickness direction, as often is the case for a cantilever made of multiple layers (Fig. 7.6c), the released cantilever will be warped upward or downward (Fig. 7.6), not only in the length direction but also in the width direction.

A cantilever or plate that is warped in the width direction is harder to bend in its length direction, as can be visualized in Fig. 7.7a. Thus, the warping in the width direction affects not only the static deflection (in the length direction) but also the resonant frequency (of bending vibration in the length direction), as can be seen in Fig. 7.8.

X-Ray Diffraction for Lattice Strain

In case of polycrystalline or single-crystalline films (thicker than 0.5 μm), X-ray diffraction can be used to measure residual stress. As illustrated in Fig. 7.9, a normal scan of X-ray diffraction is done by scanning X-ray over 2θ while keeping the film at a fixed Ω (say, at 0°), and can be used to obtain the lattice constant d from the peak position and

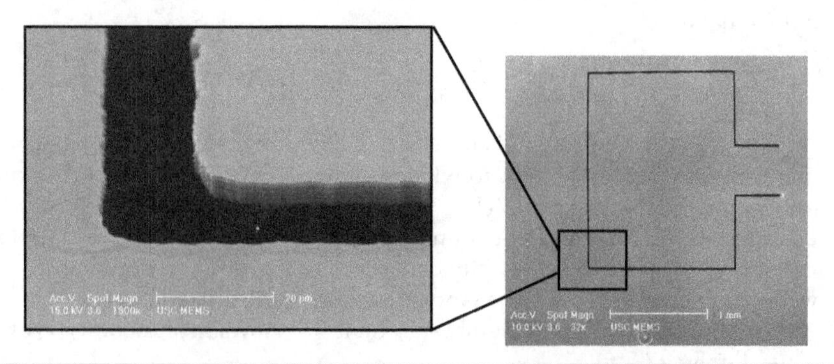

FIGURE 7.4 Photos of a flat cantilever made of the device layer of a silicon-on-insulator (SOI) wafer.

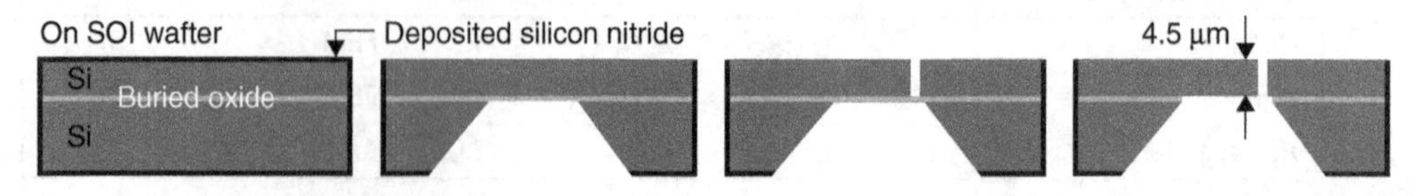

FIGURE 7.5 Fabrication process of a flat cantilever made of device silicon layer on a SOI wafer.

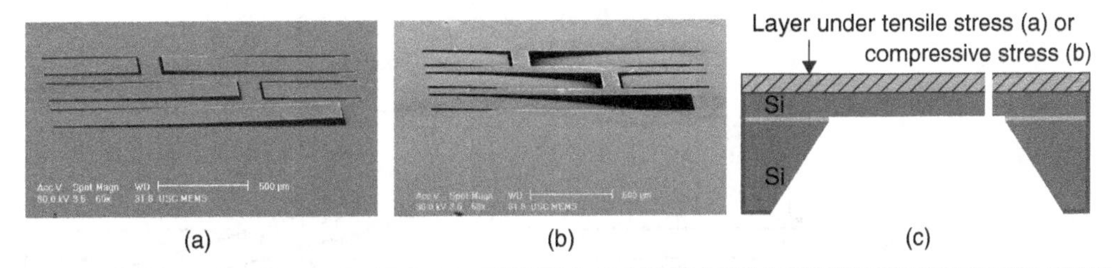

Figure 7.6 Photos of a multilayer cantilever warped upward (a) and downward (b) due to residual stress in the layer (above the silicon layer) that is tensile and compressive, respectively. (c) Cross-sectional view of the cantilevers shown in (a) and (b).

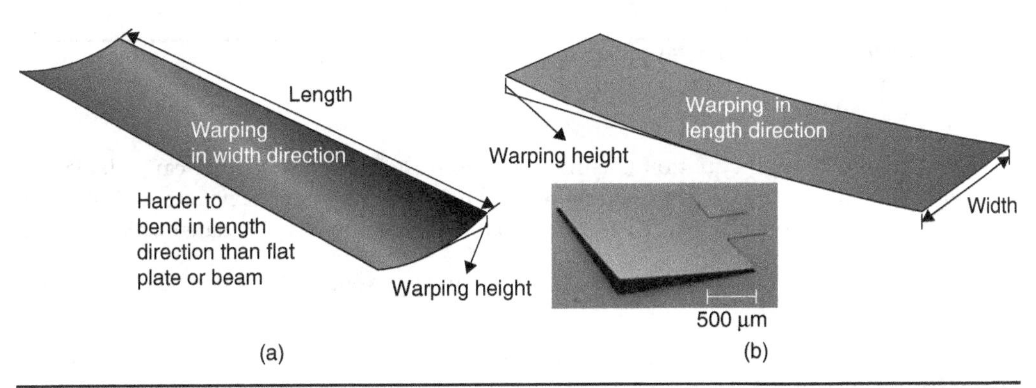

Figure 7.7 Warping in a beam due to residual stress or stress gradient: (a) in the width direction and (b) in the length direction. Also shown is a photo of a square plate (supported by a narrow beam) that is warped in both width and length directions due to residual stress gradient.

Bragg equation ($2d\sin\theta = \lambda$). A rocking curve, on the other hand, is obtained at a fixed 2θ while rocking the film over Ω, and provides information on grain-orientation distribution (obtained from full width at half maximum) and normality of c axis (obtained from off-normal angle of the peak).

For example, for ZnO film, a normal scan of X-ray diffraction gives a lattice constant c which is $2 \times (002)$ interplanar spacing d that can be obtained with (002) peak

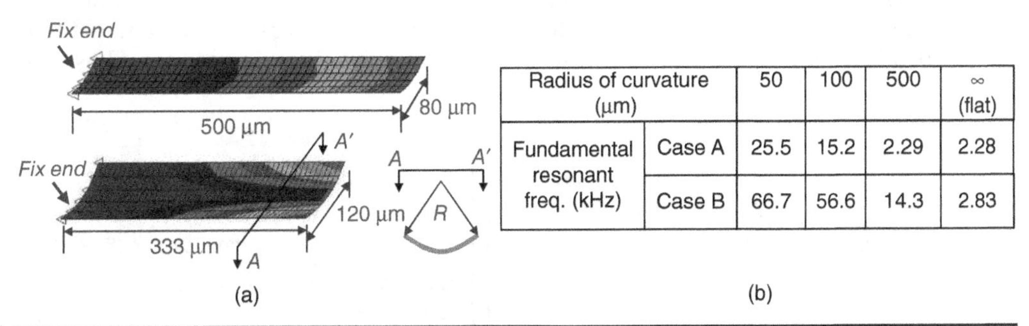

Radius of curvature (μm)		50	100	500	∞ (flat)
Fundamental resonant freq. (kHz)	Case A	25.5	15.2	2.29	2.28
	Case B	66.7	56.6	14.3	2.83

Figure 7.8 (a) Cantilevers warped in the width direction and (b) fundamental resonant frequencies of the cantilevers as a function of the radius of curvature.

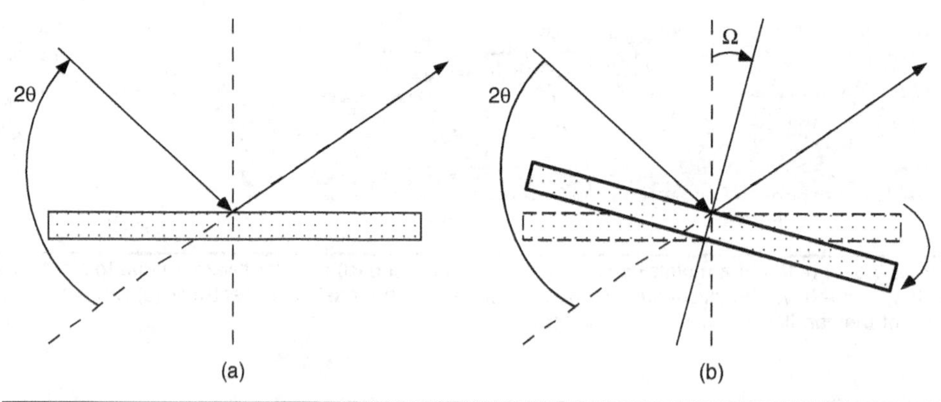

FIGURE 7.9 X-ray diffraction: (a) normal scan and (b) rocking curve.

position and Bragg equation ($2d\sin\theta = \lambda$). From the measured c, one can calculate a normal strain with $\varepsilon_z = (c - c_0)/c_0$, where c_0 and c are the lattice constants of bulk and thin-film ZnO, respectively. With the normal strain ε_z, one can calculate in-plane stress $\sigma = \varepsilon_z\left[2C_{13} - (C_{11} + C_{12})\left(\frac{C_{33}}{C_{13}}\right)\right]$ with C_{11}, C_{12}, C_{13}, and C_{33} being stiffness coefficients of ZnO.

7.2 Piezoelectric Films

7.2.1 Piezoelectric ZnO Film

Piezoelectric ZnO is generally known to have the strongest piezoelectric effect among non-ferroelectric materials, and ZnO film has been used for surface or bulk acoustic wave devices. Bulk ZnO material is used in paint, medical ointment, luminescent material for luminous screens, etc. As ZnO is a semiconductor with energy gap ≈ 3.2 eV at 300 K, it is typically n-type with excess Zn having a resistivity of 7–30 Ω·cm and electron mobility of 125 cm²/V·s. By doping ZnO with deep acceptors such as Li or Mn (at sub-percent level), the Fermi level can be pinned to at the middle of the energy gap, and the resistivity can be increased to 10^7 Ω·cm.

Piezoelectric ZnO films are commonly deposited in a RF magnetron sputtering system with deposition parameters listed in Table 7.2. The quality of a sputtered film depends on the gas pressure and substrate temperature. For piezoelectric films, one

Target	Hot pressed ZnO, 99.99% pure with 0.3% Mn	Target diameter	6–8 in
Target power	150–750 W	Sputter gas	Argon/oxygen (50/50)
Gas pressure	5–100 mTorr	Substrate temperature	200–300°C

TABLE 7.2 Parameters of ZnO Film Deposition in RF Magnetron Sputtering System

would want columnar grains with the grains preferentially oriented such that most of the grains have the *c* axis lined up perpendicular to the substrate surface [8].

7.2.2 Piezoelectric AlN Film

The quality of sputter-deposited AlN had been unreliable until a sputtering system was designed uniquely for AlN sputter deposition for film bulk acoustic resonators (FBARs) in late 90s. Now, sputter-deposited AlN film is the most commonly used piezoelectric thin film, as it has successfully been used in high-volume FBAR filters for smartphones. It can be deposited on Si, SiO_2, Al_2O_3 (basal plane), Kovar glass, quartz, Au, etc. Moreover, AlN is mechanically sturdy and does not need any support layer, when it is used for as air-backed diaphragm (e.g., for FBAR). Its high acoustic velocity and capability to withstand high temperature can be a plus.

However, AlN has a relatively low electromechanical coupling coefficient, and AlScN (Sc-doped AlN) has lately been explored. Various doping levels of Sc have been shown to increase the electromechanical coupling coefficient, more than double of that of AlN.

7.2.3 Ferroelectric $Pb(Zr,Ti)O_3$ (PZT) Film

Lead zirconate titanate (PZT) is ferroelectric, and thus, is piezoelectric, since all ferroelectric materials (having initial polarization after an excursion of electrical field) are piezoelectric. It has very high piezoelectric coefficients (and very high dielectric constants), and is ideally suited for actuator application. However, it has been difficult to deposit high-quality films of PZT with good repeatability and reliability as PZT is composed of four elements: Pb, Zr, Ti, and oxygen. The following three deposition techniques have been used for PZT depositions: (1) metal organic CVD (MOCVD), which requires accurate control of the flow rates of metal organic vapors, (2) RF sputtering, which also requires tight control of deposition parameters and the geometry between sputter target and substrate, as the individual incorporation rate of each element in PZT depends on the substrate material, and (3) sol–gel method, which involves spin cast of a chemical solution followed by sintering at 650°C, after which the deposited film is usually under great tension. Among the three, sol–gel method has been most widely used for MEMS, but requires additional buffer layers and interface metallization film for electrodes. All those layers have different thermal expansion coefficients, and thermal stress due to mismatch of thermal expansions/contractions produces localized microscopic cracks (during the firing cycle and/or thermal annealing) which propagate through the multilayers.

7.2.4 Brief Comparison of Piezoelectric Materials

Besides ZnO, AlN, and PZT, there are other piezoelectric or ferroelectric materials that have been explored for MEMS. These materials are briefly and roughly compared in Table 7.3.

7.3 Material Properties Expressed as Tensor

Scalar is commonly referred to a property having only magnitude, while vector has magnitude and direction. If a property has more than magnitude and direction, we now need to use a tensor, which has various ranks, as summarized in Table 7.4. For example,

		Piezoelectric Strain Const (10^{-12} C/N)		Coupling Coefficients		Density	Acoustic Velocity (km/s)		Melting Temp (°C)	Imped. (10^6 kg/m²s)	Relative Dielec. Const.	Loss dB/λs at 1 GHz	Resist. (Ωcm)
	Class	d_{33}	d_{31}	k_{33}	k_{31}	(kg/m³)	V_l	V_s	T_M	Z	ε_r	α	ρ
AlN	6 mm	5.6	−2.8	0.17		3,270	10.4		2,200	34.0	8.5	~5	10^{11}
ZnO	6 mm	10.6	−5.2	0.28	0.32	5,680	6.33	2.72	1,975	36.0	8.5	8.3	10^7
CdS	6 mm	10.3	−5.2	0.15	0.19	4,820	4.47	1.76	1,750	21.52	8.9	> 50	
GaN	6 mm	3.7	−1.9			6,150	6.9–8.2	5.0–4.0	2,500		8.7–9.7		
$LiNbO_3$	3 mm	6.0 ($d_{15} = 69$)	−0.9 ($d_{22} = 21$)	0.17	0.68	4,640	7.32	3.57	1,240	30.58	28–85	0.5–0.9	2×10^{10}
$LiTaO_3$	3 mm	8 ($d_{15} = 26$)	−2 ($d_{22} = 7$)	0.19	0.44	7,450	6.16	3.60	1,650	46.4	38–51	0.8	
									Curie Temp			Loss Tangent	
PZT4	6 mm	285	−122	0.70	−0.33	7,600			325°C				10^9
PZT8	6 mm	225	−97	0.64	−0.30	7,600			300°C				
PZT5A	6 mm	374	−171	0.71	−0.34	7,700			365°C				
PZT5H	6 mm	593	−274	0.75	−0.39	7,500			195°C				
PZT5J	6 mm	500	−220	0.69	−0.36	7,400			250°C				
Sol–gel PZT	6 mm	220	−89	0.49	−0.22						1,300	0.03	

Table 7.3 Material Properties of Piezoelectric and Ferroelectric Materials

Variable		Dependent Quantity		Connecting Property	
	Tensor rank		Tensor rank		Tensor rank
Temperature change ΔT	0th (scalar)	Polarization Change ΔP	1st (vector)	Pyroelectric coefficient	1st
		Strain S	2nd	Thermal expansivity	2nd
Electric current density J	1st	Electric field E	1st	Resistivity ρ	2nd
Electric field E	1st	Polarization P	1st	Electric susceptibility	2nd
Stress T	2nd	Polarization P	1st	Piezoelectric coefficient d	3rd
		Strain S	2nd	Elastic compliance s	4th
		Resistivity change $\Delta\rho$	2nd	Piezoresistivity π	4th
Strain S	2nd	Stress T	2nd	Elastic stiffness C	4th

TABLE 7.4 Variables and Dependent Quantities along with the Connecting Properties

a piezoelectric coefficient is a third rank tensor that connects a second rank tensor stress (or strain) and a first rank tensor polarization, while a piezoresistive coefficient is a fourth rank tensor that connects a second rank tensor stress (or strain) and a second rank tensor resistivity change.

7.3.1 Pyroelectricity as First Rank Tensor

A first rank tensor connects an independent scalar variable with a dependent vector quantity. An example is a pyroelectric effect connecting temperature variation with the electric polarization P. Polarization P of a specimen with volume V is defined as the dipole moment per unit volume, $P = \Sigma q_i r_i / V$, where r_i is the position vector of the charge q_i. Polarization P is a property of a specimen, and is not necessarily an intrinsic property of an extended crystal, since neutral specimens may be cut in such a way as to have different boundaries having different P, though in most cases the actual value of P is not observable directly because the specimen picks up sufficient charge from the surroundings to neutralize the polarization surface charge. Observable is the change of P, which is an intrinsic property of the crystal. A transfer of charge occurs in an external circuit connected to a pyroelectric crystal (or through the crystal), when temperature is changed. This kind of charge transfer is quite similar to a charge transfer in an open or short-circuit case of piezoelectric sensing (extensively studied in Chap. 3).

Ferroelectric crystal exhibits hysteresis when subjected to an electric field E of large amplitude variation, and has an electric dipole moment (called spontaneous or initial polarization) even when electrical field is zero, once the electrical field is raised high enough, as shown in Fig. 7.10. Ferroelectricity disappears above the transition temperature or the Curie point T_c, at which the crystal changes from the low-temperature polarized state to the high-temperature unpolarized state.

For example, barium titanate ($BaTiO_3$) has a crystal structure of perovskite; in a cubic structure Ba^{2+} ions are at the cube corners, O^{2-} ions at the face centers, and a Ti^{4+} ion at the body center, as shown in Fig. 7.10b. PZT (lead zirconate titanate or $Pb[Zr_xTi_{1-x}]O_3$ $(0 \leq x \leq 1)$) also has perovskite structure with Pb^{2+} ions being at the cube

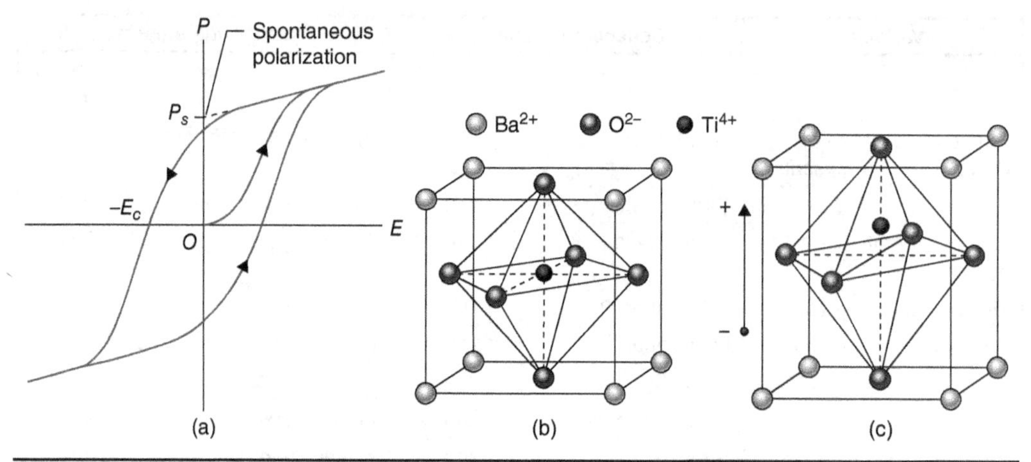

FIGURE 7.10 Ferroelectricity: (a) polarization versus electrical field, (b) $BaTiO_3$ before any electrical field is applied, (c) $BaTiO_3$ after being subjected to a sufficiently high electrical field, showing the spontaneous polarization due to displacement of atomic arrangement.

corners, O^{2-} ions at the face centers, and a Zr^{4+} or Ti^{4+} ion at the body center. Above the Curie temperature the ions are arranged as shown in Fig. 7.10b, and there is no net dipole. However, below the Curie temperature, the structure can permanently be deformed by subjecting $BaTiO_3$ to a large electrical field and then lowering the field down to zero, as shown in Fig. 7.10c, and the structure has a dipole or spontaneous polarization.

7.3.2 Second Rank Tensor

A second rank tensor connects an independent vector variable with a dependent vector quantity or an independent scalar variable with a dependent tensor quantity. For example, from Ohm's law, $J_i = \sigma_{ij} E_j$, or $E_i = \rho_{ij} J_j$, where the electrical conductivities σ_{ij} and the resistivities ρ_{ij} are second rank tensors. The equation $E_i = \rho_{ij} J_j$ can be written in matrix form as follows.

$$\begin{pmatrix} E_1 \\ E_2 \\ E_3 \end{pmatrix} = \begin{pmatrix} \rho_{11} & \rho_{12} & \rho_{31} \\ \rho_{12} & \rho_{22} & \rho_{23} \\ \rho_{31} & \rho_{23} & \rho_{33} \end{pmatrix} \begin{pmatrix} J_1 \\ J_2 \\ J_3 \end{pmatrix} \tag{7.7}$$

Question: Explain why mechanical stress and strain are second-ranked tensors, noting that pressure and force are scalar and vector quantities, respectively. What are typical units for strain, stress, Young's modulus, and Poisson's ratio?

Answer: Mechanical stress (or strain) needs two indexes to indicate (1) the direction of the force (or shape change) and (2) the plane where the force is acting on (or the shape change is happening). Strain and Poisson's ratio are unitless, while stress and Young's modulus have a unit of pressure (e.g., N/m^2 or Pa).

7.3.3 Piezoelectric Coefficients as Third Rank Tensor

A third rank tensor connects a second rank tensor to a first rank tensor or vice versa. Important examples are the piezoelectric effect d_{ijk} and its converse effect, for which there are two choices for the second rank tensor: stress T_{ij} and strain S_{ij}. Any two of three electric variables (D_i, E_i, and P_i) can be chosen for the vector variable, and we have the following equations for piezoelectric material:

$$D_k = d_{kij}T_{ij} + \varepsilon_{ki}E_i = P_k + \varepsilon_{ki}E_i \tag{7.8}$$

Stress tensor T_{ij} is the vector force F_i divided by scalar area with the unit vector normal to that area, n_j. If no torques are exerted on an element of a solid, $T_{ij} = T_{ji}$, as written in the first matrix below, and we have only six independent variables for stress tensor T_{ij}, which can now be written as T_k with $k = 1, 2, 3, 4, 5,$ and 6.

$$\begin{pmatrix} T_{11} & T_{12} & T_{31} \\ T_{12} & T_{22} & T_{23} \\ T_{31} & T_{23} & T_{33} \end{pmatrix} = \begin{pmatrix} T_1 & T_6 & T_5 \\ T_6 & T_2 & T_4 \\ T_5 & T_4 & T_3 \end{pmatrix} \tag{7.9}$$

Since it is more convenient to have only one index (rather than two indices), stress or stain is typically written with one index, and we have the following equation for $P_k = d_{kij}T_{ij}$ from Eq. (7.8)

$$\begin{aligned} P_1 &= d_{11}T_1 + d_{12}T_2 + d_{13}T_3 + d_{14}T_4 + d_{15}T_5 + d_{16}T_6 \\ P_2 &= d_{21}T_1 + d_{22}T_2 + d_{23}T_3 + d_{24}T_4 + d_{25}T_5 + d_{26}T_6 \\ P_3 &= d_{31}T_1 + d_{32}T_2 + d_{33}T_3 + d_{34}T_4 + d_{35}T_5 + d_{36}T_6 \end{aligned} \tag{7.10}$$

Wave Equations

For nonpiezoelectric and linear elastic solid, Hooke's law can be written as $T_{ij} = c_{ijkl}S_{kl}$ (i.e., stress = stiffness × strain), with strain $S_{kl} = \frac{1}{2}\left(\frac{\partial u_k}{\partial x_l} + \frac{\partial u_l}{\partial x_k}\right)$, where u_i and x_i are the displacement and coordinate variable, respectively. Then Newton's law can be written as $\rho\left(\frac{\partial^2 u_j}{\partial t^2}\right) = c_{ijkl}\left(\frac{\partial^2 u_k}{\partial x_i \partial x_l}\right)$, the solution of which for infinite medium is $u_i = A\exp[jk(l_i x_i - vt)]$, where $A = a_j x_j$, and the phase velocity v is measured along the propagation vector k whose direction cosines are the l_i. The shear wave velocities are usually lower than the longitudinal wave velocities by a factor of 2 or less.

For piezoelectric solid we replace the Hooke's law with the following piezoelectric equations of state:

$$T_{ij} = c_{ijkl}S_{kl} - e_{ijm}E_m \tag{7.11}$$

$$D_n = e_{nkl}S_{kl} + \varepsilon_{nm}E_m \tag{7.12}$$

where e_{ijm} = piezoelectric tensor (which brings in piezoelectric coupling), E_m = electric field, D_n = electric displacement, and ε_{nm} = dielectric permittivity tensor. In piezoelectric medium, elastic wave is accompanied by electric field wave, and wave velocity depends on elastic, piezoelectric, and dielectric properties. Piezoelectric coupling stiffens the medium, and increases wave velocity which is approximately equal to $\sqrt{\frac{\text{stiffness}}{\text{density}}}$. Effective elastic stiffness c is increased to $c_{\text{stiffened}} = c(1 + K^2)$ with $K^2 = e^2/c\varepsilon$, where K = electromechanical coupling coefficient; e, c, and ε depend on the propagation direction. The maximum value of K^2 is unity.

In case of bounded medium (unlike infinite medium), we need to satisfy stress and displacement matching conditions at boundaries, and solve for consistent velocity, assuming the following solutions for particle displacement components: $u_i = a_i \exp(jkl_3x_3)\exp[jk(l_1x_1 + l_2x_2 - vt)]$, where u_i = particle displacement component, x_3 = coordinate perpendicular to free space, x_1 and x_2 = coordinates in plane. Typical particle displacements of Rayleigh's surface acoustic wave (SAW) that is traveling in the horizontal direction x are shown in Fig. 7.11a as a function of the depth from the surface. Note the rapidly decreasing particle displacements (for both the vertical and horizontal particle displacements), as the location goes deeper into solid from its surface, indicating that SAW is mostly contained near the surface extending to about a half of the wavelength (λ_R) from the surface. The net particle motion (a combination of the vertical and horizontal particle displacements) at a point in the x direction is elliptical, as indicated in Fig. 7.11b, and the rotational direction changes from counterclockwise direction to clockwise direction at $0.192\lambda_R$ deep from the surface. These points are summarily illustrated in Fig. 7.12. The horizontal particle displacement (u_x) and vertical one (u_z) can be viewed as the ones belonging to the longitudinal P wave and shear vertical (SV) wave, respectively, which are parts of the Rayleigh–Lamb waves shown in Fig. 4.24 in Chap. 4.

Rayleigh's SAW which typically occurs over a homogenous, semi-infinite medium is one type of SAW, and the other type is called Love SAW which happens in a layered

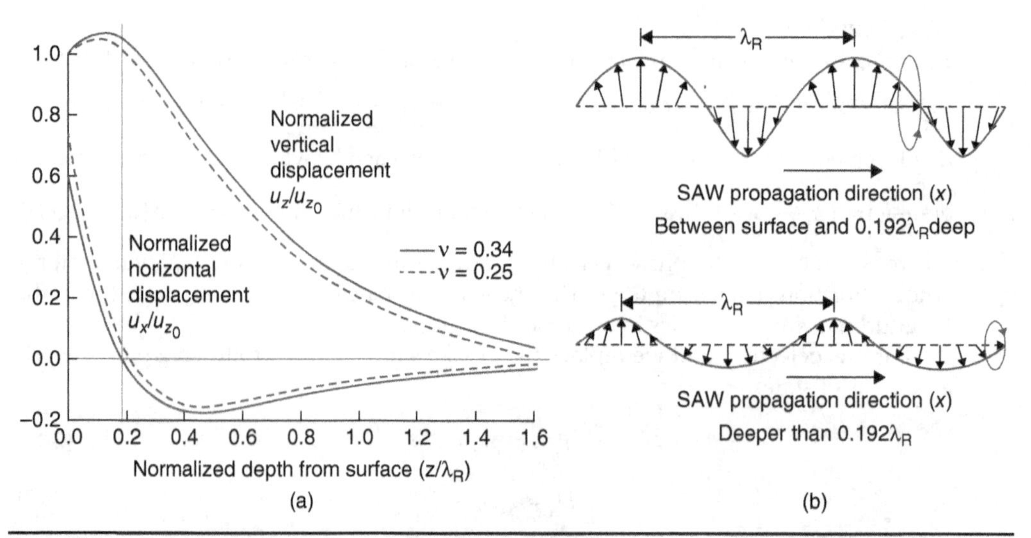

FIGURE 7.11 Typical particle displacements of Rayleigh's surface acoustic wave (SAW) that is traveling in the horizontal direction x: (a) normalized particle displacements (u_z/u_{z0} and u_x/u_{z0}) versus normalized depth from surface (z/λ_R) and (b) particle displacement vectors versus SAW propagation direction x at a particular point in time at two locations: one between the surface and $0.192\lambda_R$ (top) and the other at a location deeper than $0.192\lambda_R$ from the surface (bottom).

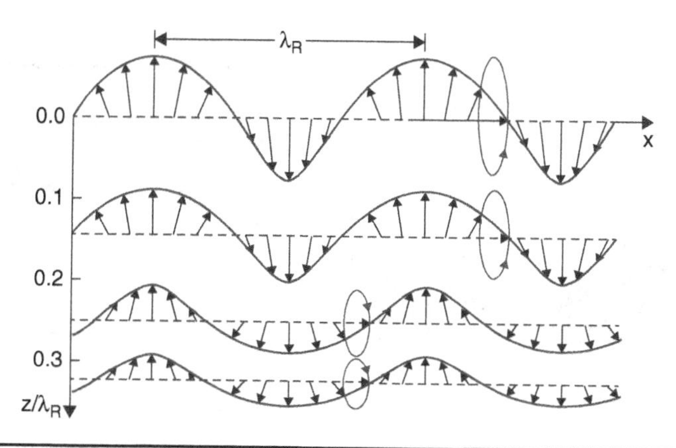

Figure 7.12 Particle displacement vectors versus SAW propagation direction *x* as a function of the depth from the surface which is deduced from Fig. 7.11a.

medium (having a layer of certain thickness with different material properties over a homogenous, semi-infinite medium). As waves go through multiple reflections and refractions in the layer in a layered medium, shear horizontal (SH) waves can be trapped in the layer, resulting in large horizontal particle displacements that are not present in Rayleigh's SAW. Such large horizontal particle displacements are present in the seismic SAW (associated with earthquake) and cause major damage on the surface.

Elastic waves that propagate in solid medium can be classified as bulk longitudinal wave (so-called P wave) where particles vibrate in the same direction as the wave propagation direction, bulk shear wave where particles vibrate in the direction perpendicular to the wave propagation direction (either SH and shear vertical [SV], depending on which of the two perpendicular directions that the particles vibrate along), and SAW where particles vibrate as shown in Fig. 7.11. If the medium carrying SAW is thin, the SAW becomes a so-called plate wave (or Rayleigh–Lamb wave), with the vibrational amplitude being either symmetric or antisymmetric (with respect to the plate's thickness), some details of which are shown in Fig. 4.24 in Chap. 4.

7.3.4 Surface Acoustic Wave Vapor Sensing

As shown in Fig. 7.13, an oscillator can be made with SAW in a feedback of an amplifier, with an oscillation condition of $2\pi fL/v_p + \phi_E = 2\pi n$, where v_p = SAW velocity ($\approx$ 3,000–4,000 m/s), L = distance between the two interdigitated transducers (IDTs), ϕ_E = phase shift in amplifiers, $n = 1, 2, 3\ldots$ With a typical SAW velocity, an IDT period of 2–400 μm result in an oscillation frequency f = 10–2,000 MHz.

A SAW-based oscillator can be used for sensing various measurands by measuring the oscillator's frequency. The sensing can be made to be selective through coating the section between the two IDTs with a selectively responsive layer, as shown in Fig. 7.13b. Other approaches for selective response in SAW sensors are to (1) use reference and active device (e.g., to reduce interference due to temperature change), (2) filter out unwanted particles (e.g., zeolite passing only the molecules smaller than zeolite's pore diameter), (3) separate vapors by rate of diffusion (e.g., molecules entering porous resonator coating), (4) use solubility parameter (e.g., to match absorber to vapor), (5) use

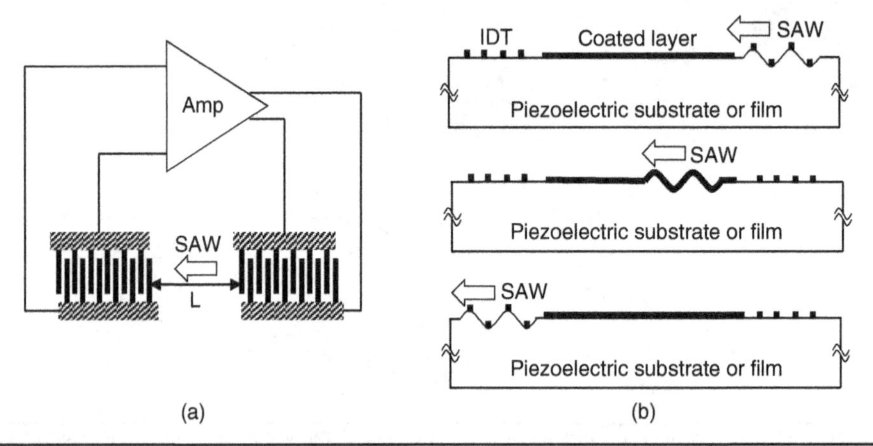

FIGURE 7.13 (a) Oscillator based on SAW transducers. (b) Cross-sectional views of SAW propagation from one interdigitated transducer (IDT) to another, with surface-coated layer between two IDTs for vapor sensing.

chemical reaction (e.g., substitution reaction) or biochemical reaction (e.g., antigen binding to immobilized antibody), (6) use spectroscopy (e.g., thermal desorption), and (7) use pattern recognition (i.e., seek pattern in responses of many nonselective absorbers), etc. Various coating techniques are listed in Table 7.5.

The oscillation frequency of an oscillator based on SAW (Fig. 7.13) is dependent on the SAW velocity in the coated layer between the two IDTs, which changes as the mass (m) and elastic stiffness of the layer change due to absorption of vapor. Effect of elastic stiffness change is often negligible, and the frequency change (Δf) is mostly due to mass change (Δm) and is $\Delta f \approx -kf_o^2\Delta m$, where k and f_o are a constant and the oscillation frequency, respectively.

7.3.5 Bulk Acoustic Wave Vapor Sensing

As mass loading reduces the resonant frequency (f_{res}) of a bulk acoustic wave resonator (BAR), the normalized resonant frequency change can be estimated with $\Delta f/f_{res} \approx -C_m\Delta m$, where C_m is proportional to f_{res}. For quartz crystal microbalance (QCM), $C_m \approx 2.3 \times 10^{-6} f_{res}$, where f_{res} is in Hz. Thus, mass sensitivity of BAR-based sensor is proportional to the resonant frequency. Because the fundamental resonant frequency of FBAR (film BAR) such as the one shown in Fig. 7.14 is usually hundred times that

Technique	Comments
Spin or spray coating	Thickness uniformity?
Inkjet printing	Good control and automated
Dipping	Langmuir–Blodgett monomolecular layers
Solvent casting	Thickness control?

TABLE 7.5 Various Polymer-Coating Techniques

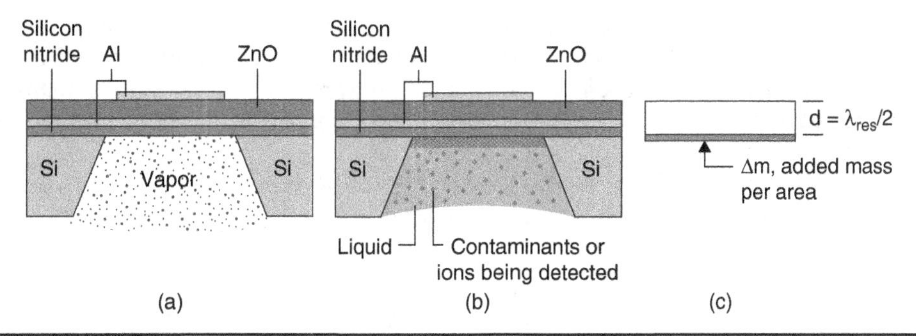

Figure 7.14 FBAR mass sensors: (a) vapor sensing, (b) mass sensing in liquid, and (c) effect of added mass on the surface of FBAR.

of a QCM, the resonant frequency shift in FBAR mass sensor (and thus the sensitivity) is hundred times that in QCM. However, a quartz crystal resonator has a Q hundred times that of an FBAR (and thus, hundred times lower noise floor), and both QCM and FBAR-based mass sensor have a comparable minimum detectable signal (MDS). Enhancement of Q in FBAR can potentially make FBAR mass sensor capable of smaller MDS than QCM. Even though MDS levels are similar for QCM and FBAR mass sensors, FBAR sensors offer size advantage over QCM at the cost of GHz electronics versus MHz electronics.

Question: For a resonant mass sensor, which one would you choose between quartz crystal microbalance (QCM) and FBAR-based sensor, if your only concern is the issue listed at the top row of the following table? Put check marks on the appropriate blanks.

	Size	Mass Sensitivity in Hz·cm²/g	Minimum Detectable mass in g/cm²
QCM operating at 4 MHz			
FBAR-based resonant Sensor operating at 1 GHz			

Answer:

	Size	Mass Sensitivity in Hz·cm²/g	Minimum Detectable mass in g/cm²
QCM operating at 4 MHz			√
FBAR-based resonant Sensor operating at 1 GHz	√	√	√

7.3.6 Piezoresistivity as Fourth Rank Tensor

The fourth rank tensor connects a second rank tensor variable with a similar dependent tensor. According to Onsager theorem, the second rank tensors have only six independent components. The fourth rank tensors, thus, have only 36 independent components (rather than 81) and can be represented by a 6×6 matrix.

The resistivity tensor ρ_{ij} can be written in a single subscript notation ρ_i, where $i = 1, 2, \ldots 6$, similar to the stresses T_i. Then, the change of the resistivity $(\Delta\rho)_i = \pi_{ij}T_j$, where π_{ij} are the piezoresistive coefficients. Though there are 36 components in π_{ij}, many of them are either zero or same due to crystal symmetry (e.g., silicon has only three independent constants, π_{11}, π_{12}, and π_{44}).

With single-indexed resistivity ρ_i, electrical field E_i is related to current i_i through resistivity as follows.

$$\begin{pmatrix} E_1 \\ E_1 \\ E_1 \end{pmatrix} = \begin{pmatrix} \rho_1 & \rho_6 & \rho_5 \\ \rho_6 & \rho_2 & \rho_4 \\ \rho_5 & \rho_4 & \rho_3 \end{pmatrix} \begin{pmatrix} i_1 \\ i_2 \\ i_3 \end{pmatrix} \tag{7.13}$$

In case of silicon, the off-diagonal elements of the resistivity are zero, while the diagonal elements all have the same value due to cubic crystal symmetry. Consequently, we have the following equation for the resistivity

$$\begin{pmatrix} \rho_1 \\ \rho_2 \\ \rho_3 \\ \rho_4 \\ \rho_5 \\ \rho_6 \end{pmatrix} = \begin{pmatrix} \rho \\ \rho \\ \rho \\ 0 \\ 0 \\ 0 \end{pmatrix} + \begin{pmatrix} \Delta\rho_1 \\ \Delta\rho_2 \\ \Delta\rho_3 \\ \Delta\rho_4 \\ \Delta\rho_5 \\ \Delta\rho_6 \end{pmatrix}, \tag{7.14}$$

where the resistivity change $\Delta\rho$ is related to stress σ_i through piezoresistive coefficients π_{ij}, as follows.

$$\frac{1}{\rho}\begin{pmatrix} \Delta\rho_1 \\ \Delta\rho_2 \\ \Delta\rho_3 \\ \Delta\rho_4 \\ \Delta\rho_5 \\ \Delta\rho_6 \end{pmatrix} = \begin{pmatrix} \pi_{11} & \pi_{12} & \pi_{12} & 0 & 0 & 0 \\ \pi_{12} & \pi_{11} & \pi_{12} & 0 & 0 & 0 \\ \pi_{12} & \pi_{12} & \pi_{11} & 0 & 0 & 0 \\ 0 & 0 & 0 & \pi_{44} & 0 & 0 \\ 0 & 0 & 0 & 0 & \pi_{44} & 0 \\ 0 & 0 & 0 & 0 & 0 & \pi_{44} \end{pmatrix} \begin{pmatrix} \sigma_1 \\ \sigma_2 \\ \sigma_3 \\ \tau_1 \\ \tau_2 \\ \tau_3 \end{pmatrix} \tag{7.15}$$

Thus, we have the following equation for silicon:

$$\begin{aligned} E_1 &= \rho i_1 + \rho\pi_{11}\sigma_1 i_1 + \rho\pi_{12}(\sigma_2 + \sigma_3)i_1 + \rho\pi_{44}(i_2\tau_3 + i_3\tau_2) \\ E_2 &= \rho i_2 + \rho\pi_{11}\sigma_2 i_2 + \rho\pi_{12}(\sigma_1 + \sigma_3)i_2 + \rho\pi_{44}(i_1\tau_3 + i_3\tau_1) \\ E_3 &= \rho i_3 + \rho\pi_{11}\sigma_3 i_3 + \rho\pi_{12}(\sigma_1 + \sigma_2)i_3 + \rho\pi_{44}(i_1\tau_2 + i_2\tau_1) \end{aligned} \tag{7.16}$$

7.3.7 Piezoresistive Silicon Pressure Sensor

A silicon pressure sensor is typically made by implanting dopants (to form resistors) on a bulk-micromachined silicon diaphragm, as shown in Fig. 7.15a. A differential pressure applied between top and bottom side of the diaphragm deflects the diaphragm,

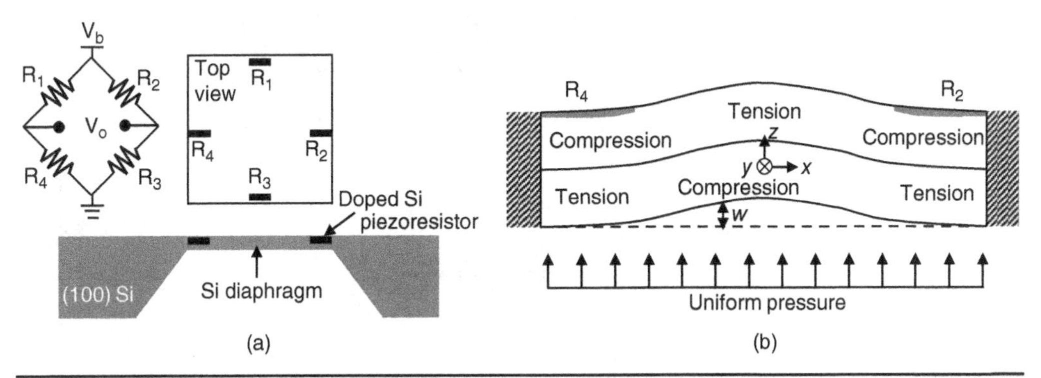

FIGURE 7.15 Piezoresistive pressure sensor made on a bulk-micromachined silicon diaphragm: (a) top view and cross-sectional view diagrams and (b) deflection of the diaphragm under a uniform pressure. Also shown at the left of the top view diagram in (a) is how the four piezoresistors (R_1, R_2, R_3, and R_4) are connected in Wheatstone bridge.

and stress is induced inside the diaphragm. The induced stress is not uniform, as we studied in Chap. 6, but varies as shown in Fig. 7.15b across the thickness direction. Under such induced stress the four resistors R_1–R_4 (formed at the top surface of the silicon diaphragm) experience uniaxial stress, which is in the same direction as the current (in case of R_2 and R_4) or in the direction perpendicular to the current direction (in case of R_1 and R_3). The normalized resistance change (i.e., $\Delta R/R$) is equal to $\pi_l\sigma_l + \pi_t\sigma_t$, where π_l and π_t are the longitudinal and transverse piezoresistive coefficients, respectively, while σ_l and σ_t are the stresses along the directions parallel and perpendicular to the current direction, respectively.

Longitudinal and Transverse Piezoresistive Coefficients

With π_{11}, π_{12}, and π_{44} measured in a coordinate system aligned to the <100> axes of the silicon crystal, all the piezoresistive properties can be calculated, as follows. In arbitrary Cartesian coordinates, the <100> axes can be transformed into a given coordinate system through the following:

$$\begin{aligned} e_1 &= l_1u_1 + l_2u_2 + l_3u_3 \\ e_2 &= m_1u_1 + m_2u_2 + m_3u_3 \\ e_3 &= n_1u_1 + n_2u_2 + n_3u_3 \end{aligned} \tag{7.17}$$

where l_i, m_i, and n_i are the direction cosines, while u_i are the coordinate system aligned to the <100> axes of the silicon crystal. When the uniaxial stress σ^*, electrical field E^*, and current i^* are all in the same direction (but not along a crystal axis), as shown in Fig. 7.16a, we can derive [9]

$$E^* = \rho i^* + \rho i^*[\pi_{11} + 2(\pi_{44} + \pi_{12} - \pi_{11})(l_1^2m_1^2 + l_1^2n_1^2 + m_1^2n_1^2)]\sigma^* \tag{7.18}$$

Thus, we have the following so-called longitudinal piezoresistive coefficient that relates $\Delta R/R$ to σ^*

$$\pi_l = \pi_{11} + 2(\pi_{44} + \pi_{12} - \pi_{11})(l_1^2m_1^2 + l_1^2n_1^2 + m_1^2n_1^2) \tag{7.19}$$

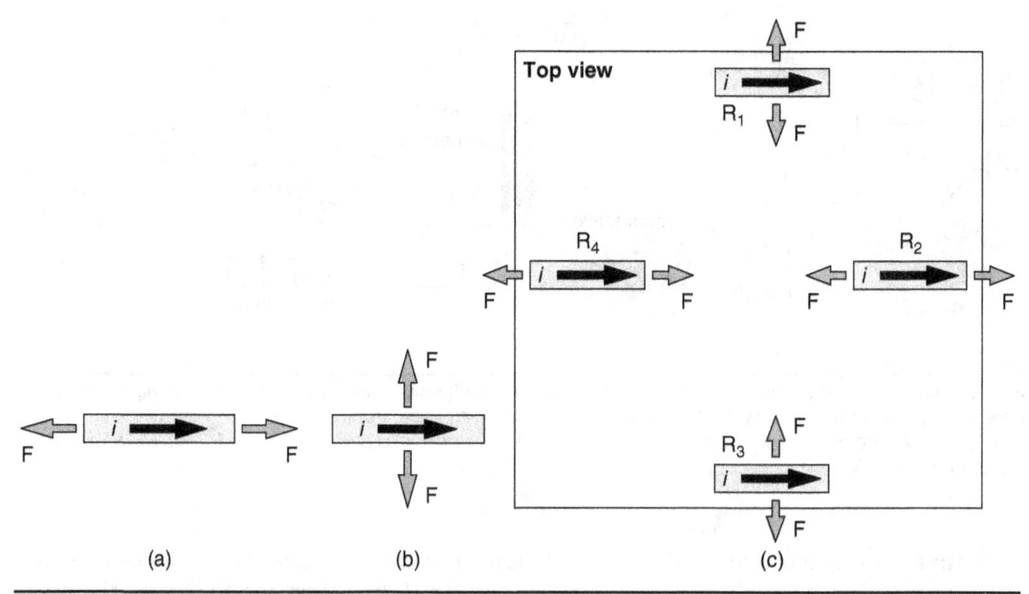

FIGURE 7.16 (a) A piezoresistive element with the uniaxial stress and electrical current all in the same direction, (b) a piezoresistive element with the uniaxial stress and electrical current perpendicular to each other, and (c) top view of the piezoresistive silicon pressure sensor shown in Fig. 7.15 with the electrical currents and dominant stress directions indicated.

In another case where uniaxial stress is perpendicular to both electrical field and current (Fig. 7.16b), we can derive the following transverse piezoresistive coefficient

$$\pi_t = \pi_{21} - (\pi_{44} + \pi_{12} - \pi_{11})(l_1^2 l_2^2 + m_1^2 m_2^2 + n_1^2 n_2^2) \tag{7.20}$$

In case of the pressure sensor shown in Fig. 7.15, its piezoresistive elements have the electrical currents and the dominant stress directions as indicated in Fig. 7.16c. Thus, piezoresistors R_2 and R_4 are dominated by the so-called *longitudinal* piezoresistive coefficient π_l, as the current direction is along the force direction, while the piezoresistors R_1 and R_3 are dominated by the so-called *transversal* piezoresistive coefficient π_t, as the current direction is perpendicular to the force direction.

Example: The longitudinal piezoresistive coefficient along the <111> direction can be obtained by noting that the direction cosines (for the coordinate transformation) are $l_1^2 = m_1^2 = n_1^2 = 1/3$. Using those in Eq. (7.19), we obtain

$$(\pi_l)_{<111>} = \frac{1}{3}(\pi_{11} + 2\pi_{12} + 2\pi_{44}) \tag{7.21}$$

The longitudinal and transverse piezoresistive coefficients for various crystal directions of silicon are listed in Table 7.6.

For a coordinate system aligned to the <100> axes of silicon crystal, the piezoresistive coefficients at room temperature are listed in Table 7.7. Silicon's piezoresistive coefficients decrease, as impurity concentration and temperature are increased [9].

	π_l	π_t
[100], [010], or [001]	π_{11}	π_{12}
[111], [$\underline{11}$1], or [11$\underline{1}$]	$(\pi_{11}+2\pi_{12}+2\pi_{44})/3$	$(\pi_{11}+2\pi_{12}-\pi_{44})/3$
[110], [101], or [011]	$(\pi_{11}+\pi_{12}+\pi_{44})/2$	$(\pi_{11}+2\pi_{12}-\pi_{44})/2$

Table 7.6 Longitudinal and Transverse Piezoresistive Coefficients for Various Crystal Directions of Silicon

	ρ (Ωcm)	$\pi_{11}(10^{-11}Pa^{-1})$	$\pi_{12}(10^{-11}Pa^{-1})$	$\pi_{44}(10^{-11}Pa^{-1})$
p-type	7.8	6.6	−1.1	138.1
n-type	11.7	−102.2	53.4	−13.6

Table 7.7 Piezoresistive Coefficients of Silicon for a Coordinate System Aligned to the <100> Axes of Silicon Crystal at Room Temperature

Question: A piezoresistive silicon pressure sensor shown Fig. 7.15 is made on an (100) n-type silicon having a resistivity of 11.7 Ωcm and has the following dimensions for the diaphragm: thickness = 20 μm (the Si membrane thickness), width = 1,000 μm, length = 1,000 μm. The normalized resistance change (i.e., $\Delta R/R$) is equal to $\pi_l\sigma_l + \pi_t\sigma_t$. Which product will dominate in R_1, R_2, R_3, and R_4 above? Is it $\pi_l\sigma_l$ or $\pi_t\sigma_t$?

Answer: $\pi_l\sigma_l$ for R_2 and R_4; $\pi_t\sigma_t$ for R_1 and R_3.

7.3.8 Capacitive Silicon Pressure Sensor

Instead of using silicon's piezoresistivity for pressure sensing, a capacitive pressure sensor can be made on a silicon. A capacitive pressure sensor with a 100 × 100 μm² polysilicon diaphragm was built on a silicon with MOSFET circuits integrated on a single chip and was reported to have a sensitivity of 0.93 mV/kPa and pressure resolution of 0.54 kPa (with the integrated MOSFET circuitry capable of detecting down to 30 attoFarad) [10]. If the pressure resolution were better by orders of magnitude, the pressure sensor could have been good as a microphone (that needs to detect a pressure level < 1 mPa).

7.3.9 Capacitive MEMS Microphone

One of the early MEMS capacitive microphones used surface micromachining for a metal backplate through electrodepositing metal over a sacrificial photoresist [11]. The backplate was 15-μm-thick copper, which was over an 8–9-μm-thick silicon diaphragm (1.8 × 1.8 mm²) that was bulk micromachined, similar to Fig. 7.17. The backplate is perforated with holes to allow air to pass through, when the silicon diaphragm deflects due to applied sound pressure. Without the holes, the diaphragm will be damped too much through squeeze film damping, and the sensitivity will start dropping as the frequency goes beyond tens or hundreds of Hz. Ideally, a microphone should be able to detect audio sound over 20 Hz–20 kHz with a uniform sensitivity. For cell phone applications, though, the needed bandwidth is narrower (down to 100 Hz–4 kHz), as most of speech information is contained over that range.

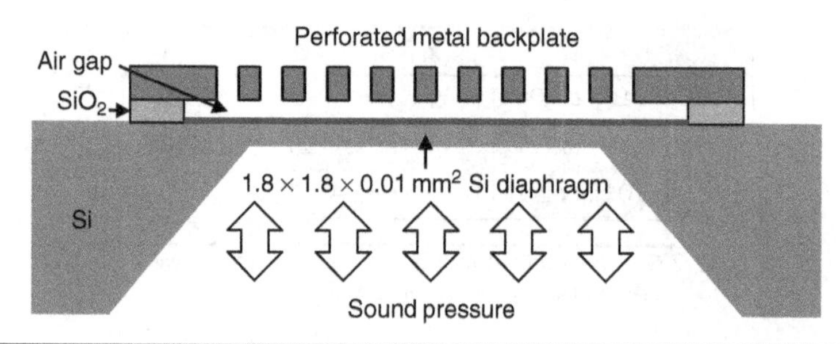

FIGURE 7.17 Cross-sectional view of a condenser microphone made with both bulk and surface micromachining cross-sectional view. (Adapted from [11].)

A simple way to convert the capacitance change (of a capacitive microphone) is to use a DC bias voltage, as shown in Fig. 7.18a. The bias voltage V_{bias} charges the microphone's air-gap capacitor C_m (formed by the diaphragm and the backplate in Fig. 7.17), and a static charge is developed across the capacitor. Now, when the capacitance changes due to a sound pressure (time varying with frequency 20 Hz–20 kHz) applied to the diaphragm, a time-varying charge change develops over the capacitor, and produces a voltage change when it passes through an amplifier. The capacitor C_c in the circuit is a so-called coupling (or decoupling) capacitor with a very large capacitance to prevent the DC bias voltage V_{bias} from affecting the amplifier without reducing the time-varying signal. This kind of microphone is often called a condenser microphone, and its sensitivity depends on the DC bias voltage; the higher the bias voltage, the higher the sensitivity (Fig. 7.18b).

Question: What is the main reason for almost three orders of magnitude difference in the sensitivities for the capacitive silicon pressure sensor and the capacitive MEMS microphone (Fig. 7.17)?

Answer: The size difference: 0.1×0.1 mm^2 versus 1.8×1.8 mm^2, a factor of 324 larger for Fig. 7.17.

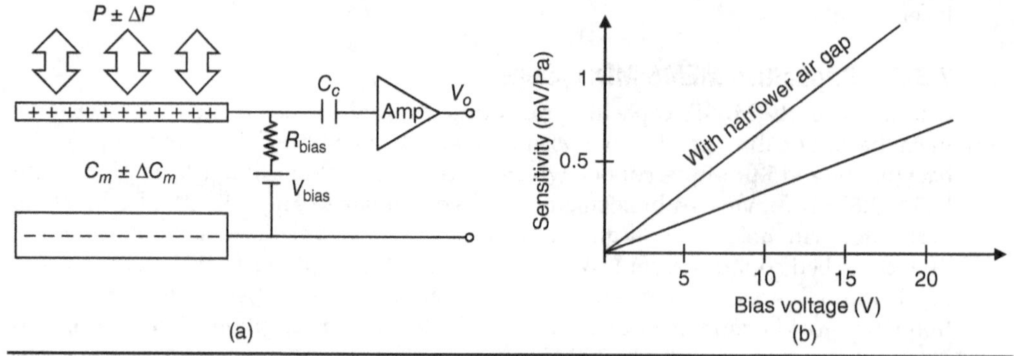

FIGURE 7.18 Condenser microphone: (a) a simple way to convert capacitance change into a voltage, using a DC bias voltage (V_{bias}) and (b) sensitivity versus the bias voltage (V_{bias}) as a function of air gap for a condenser microphone.

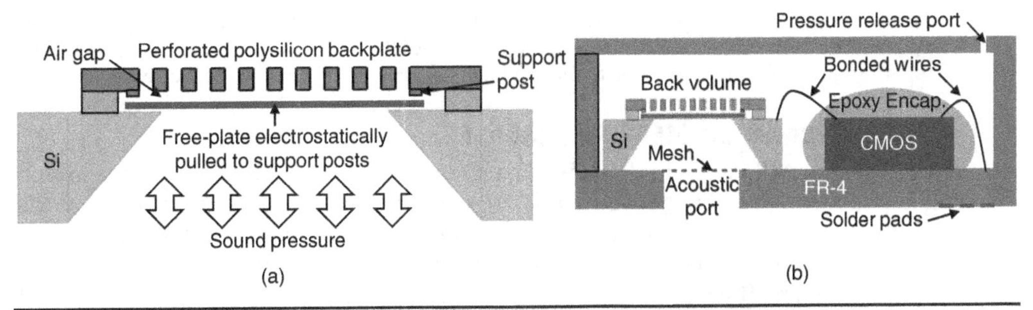

Figure 7.19 Cross-sectional views of Knowles' (a) condenser microphone and (b) package with hybrid integration of the microphone and a circuit chip. (Adapted from [12].)

Commercial MEMS Microphone

One of the commercialized MEMS microphones for high-volume cell phones is based on what is shown in Fig. 7.19. When it was first introduced around 2004, its price was substantially higher than the electret condenser microphone (ECM) that had been dominating the market. However, some cell phone manufacturers still went for the MEMS microphones, mainly because the MEMS microphone could withstand 250°C (a relatively high temperature needed for solder reflow), and can be mounted on a printed circuit board (PCB) along with other components through robotic pick-and-place. (Note: most ECMs use a polymeric membrane implanted with ionic charges, and lose the implanted charges when the temperature goes beyond 80°C. Consequently, those cannot be mounted on a PCB and require more cost in assembling them in cell phones.)

As shown in Fig. 7.19a, the diaphragm is completely released from the substrate and is free to move around within the confined region that is defined by the perforated backplate and substrate. The diaphragm is pulled to the support posts by an electrostatic voltage during operation as a microphone so that its circular edge becomes simply supported. This way any residual stress is released, and the diaphragm deflects more for a given pressure (than a diaphragm with built-in edge). However, there is no longer isolation of the front side (where there are electrodes) from the environment (to which only the backside could have been exposed, if the diaphragm was not completely released), and there are concerns of moisture condensation affecting long-term reliability or reliable operation of the microphone. The microphone is connected to a circuit chip through bonded wires, as shown in Fig. 7.19b, in a low-cost package with laminated FR-4, a commonly used material for PCB. Note that a "pressure release port" is needed in order to relieve static pressure that will build up inside the package due to temperature or ambient pressure. Also, there has to be sufficient "back volume" to prevent any squeeze-film damping from affecting microphone's frequency response, and for this reason, it may be advisable to make the "acoustic port" at the top lid rather than at the bottom. The size and performance of commercial MEMS microphones have vastly improved over years, once the microphone was adopted for cell phone applications.

Question: During the microphone operation, the edge of the free-plate diaphragm is simply supported to the support post in the microphone shown in Fig. 7.19a. How is the "free-plate diaphragm" made to contact with the support post? Referring to Fig. 7.19a, which of the following causes the free-plate diaphragm to be curled upward and downward after its fabrication: (1) residual stress, (2) residual stress gradient in the thickness direction, or (3) both?

Answer: The free-plate diaphragm is made to contact with the support post by applying electrostatic voltage to pull down the diaphragm to the support. Since the free-plate diaphragm is released from the substrate, it is warped only due to the residual stress gradient in the thickness direction.

7.3.10 Piezoelectric MEMS Microphone

A piezoelectric MEMS microphone was integrated with CMOS circuits on a single chip in late 80s to early 90s, as shown in Fig. 7.20, so that parasitic capacitance (as well as the connecting wire/electrode) between the microphone and the preamp might be minimized to increase the microphone sensitivity and to lower the noise from electromagnetic interference. The highest unamplified sensitivity of the microphone was reported to be 0.92 mV/Pa, while the A-weighted noise at the preamp's input was 13 μV, resulting in a signal-to-noise ratio (SNR) of 71 at 1 Pa, or 7.1 at 1 μbar, or a noise level of 57 dB SPL (sound pressure level, defined to be 74 dB SPL for SNR of 1 at 1 μbar), which needs to be reduced by about 20 dB (or about 10 times) for cell phone applications.

Compared to a condenser MEMS microphone, a piezoelectric MEMS microphone has advantages with no DC bias (or polarization) voltage needed, wide dynamic range of detectable pressure with good linearity, and simple fabrication process. However, the noise floor has been relatively poor, likely due to the residual stresses from the multiple layers and susceptibility to electromagnetic interference owing to the high impedance at the microphone and preamp's input.

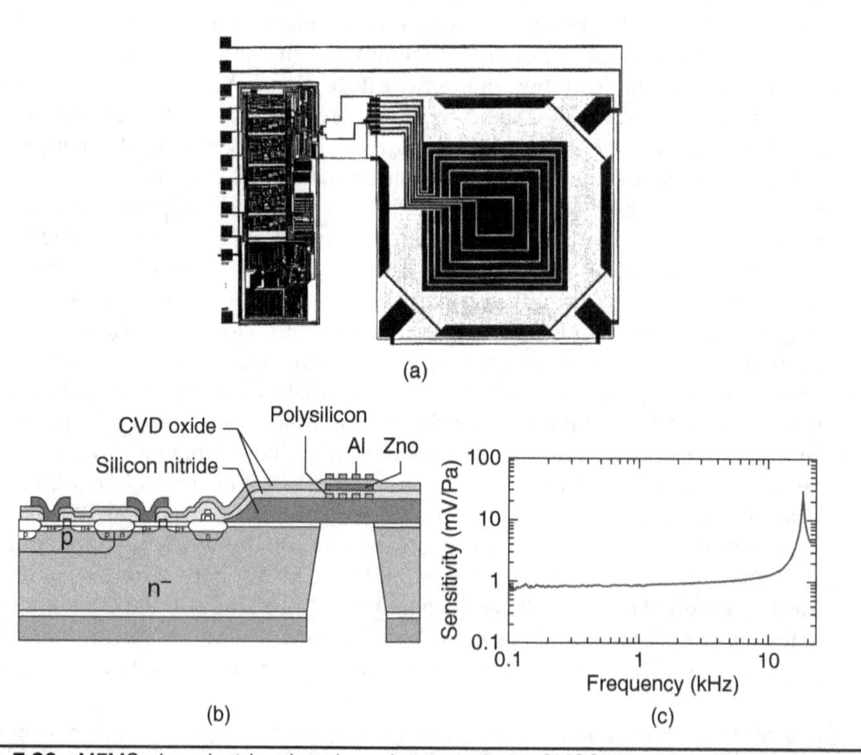

FIGURE 7.20 MEMS piezoelectric microphone integrated with CMOS circuits on a single chip: (a) top view of the CMOS circuits (on the left) and the piezoelectric microphone (on the right), (b) cross-sectional view of the circuits and microphone, and (c) unamplified sensitivity versus frequency [13].

References

1. W.D. Nix, "Mechanical properties of thin films," Metallurgical and Materials Transactions A., 1989, pp. 2217–2245.
2. O. Auciello and A. Krauss, eds, "In Situ Real-Time Characterization of Thin Films," John Wiley and Sons, New York, NY, 2001.
3. J.J. Vlassak and W.D. Nix, "A new bulge test technique for the determination of youngs modulus and poissons ratio of thin-films," Journal of Materials Research, 7, 1992, pp. 3242–3249.
4. H. Keiner, F. Preissig, H. Zeng, M. Nejhad, and E.S. Kim, "Advanced bulge test system," MRS Proceedings, 505, 1997, 229–234; doi:10.1557/PROC-505-229.
5. J.Y. Pan, P. Lin, F. Maseeh, and S. D. Senturia, "Verification of FEM analysis of load-deflection methods for measuring mechanical properties of thin films," IEEE 4th Technical Digest on Solid-State Sensor and Actuator Workshop, Hilton Head Island, SC, June 4–7, 1990, pp. 70–73.
6. H. Guckle, et al., "A simple technique for the determination of mechanical strain in thin films with applications to polysilicon," Journal of Applied Physics, 57, 1985, pp. 1671–1675.
7. H. Guckel, D.W. Burns, H.A.C. Tilmans, D.W. DeRoo, and C.R. Rutigliano, "Mechanical properties of fine grained polysilicon: The repeatability issue," 1988 Solid State Sensor and Actuator Workshop Technical Digest, Hilton Head Island, SC, June 6–9, 1988, pp. 96–99.
8. J A Thornton, "High rate thick film growth," Annual Review of Materials Science, 7(1), 1977, pp. 239–260.
9. S.M. Sze, ed., "Semiconductor Sensors," John Wiley and Sons, Inc., 1994.
10. J.T. Kung and H.S. Lee, "An integrated air-gap capacitor pressure sensor and digital readout with sub-100 attofarad resolution," IEEE Journal of Microelectromechanical Systems, 1(3), 1992, pp. 121–129.
11. J. Bergqvist and J. Gobet, "Capacitive microphone with a surface micromachined backplate using electroplating technology," Journal of MEMS, 3, 1994, pp. 69–75.
12. P.V. Leoppert and S.B. Lee, "SiSonicTM—The First Commercialized MEMS Microphone," Solid State Sensor and Actuator Workshop Technical Digest, Hilton Head Island, SC, June 4-8, 2006, pp. 27–30.
13. R.P. Ried, E.S. Kim, D.M. Hong, and R.S. Muller, "Piezoelectric microphone with on-chip CMOS circuits," IEEE/ASME Journal of Microelectromechanical Systems, 2, September 1993, pp. 111–120.

Questions and Problems

Question 7.1 Assume that a silicon nitride film (≈ 0.2 μm thick) on a silicon substrate (≈ 400 μm thick) is in great tension with *uniform* stress distribution along the thickness direction. If we make a cantilever out of the nitride film: (1) will the cantilever be flat, or curled upward, or curled downward? (2) Will the length of the fabricated cantilever be smaller, larger, or equal to the length patterned on the etch mask?

Question 7.2 Even after an annealing step at 1,050°C for 30 minutes, the polysilicon film (≈ 0.2 μm thick) on top of a 1-μm-thick SiO_2 film on a silicon substrate (≈ 400 μm thick) (shown below) is under *compressive* residual stress, though the stress is now *uniform* along the thickness direction. If we etch out the SiO_2 film to form a cantilever out of the polysilicon film, sketch the shape of the released cantilever on the figure below indicating whether the released cantilever is flat, curled upward, or curled downward and whether the length of the released cantilever is smaller, larger, or equal to the length before the release.

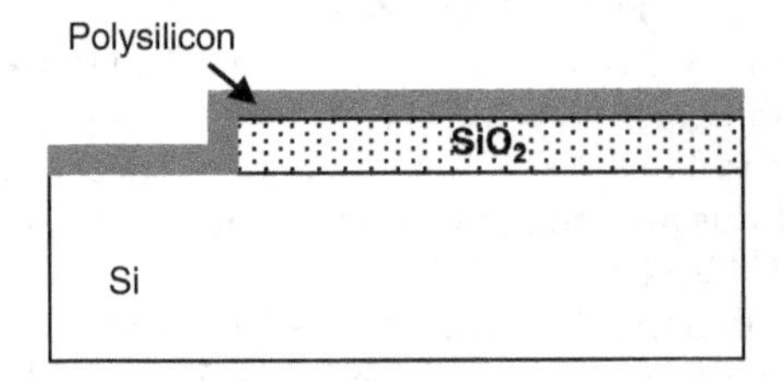

Question 7.3 A silicon wafer is deposited with a very thin (< 0.05 μm) silicon nitride film, and the top-side silicon nitride is etched away for a processed wafer of which the cross-section looks like what is shown below. If the wafer is now *thermally* oxidized to *grow* 0.6-μm-thick SiO_2 on the top side of the wafer, the wafer (at room temperature after the SiO_2 growth) will be deformed due to the residual stress from the SiO_2. Sketch the deformation shape of the whole structure noting that the SiO_2 growth will be mostly on the top side.

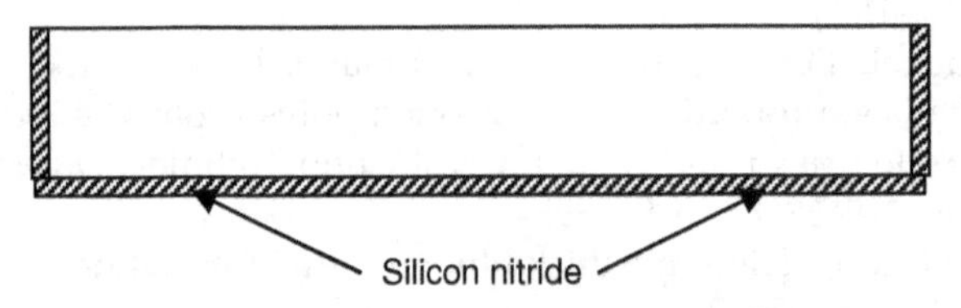

Question 7.4 A nickel thin film is deposited and patterned into a narrow and long serpentine line all over the top of silicon substrate (the cross-section of which is shown below) and is applied with an electrical current. Sketch the deformation shape of the whole structure, as the nickel film is heated due to the I^2R heating.

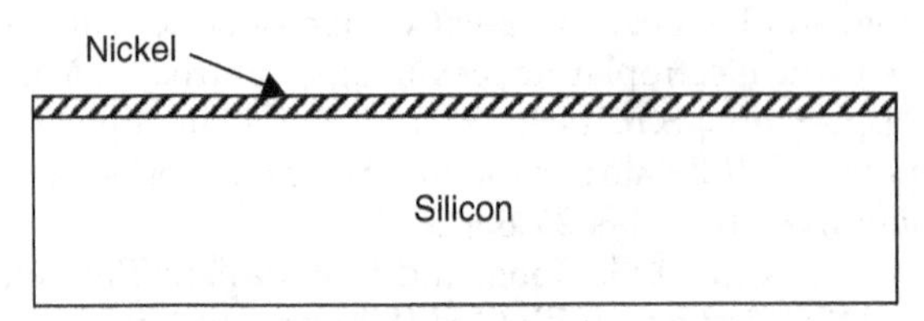

Question 7.5 Write out $S_i = s_{ij}T_j + d_{ik}E_k$ for $i = 1$ with $j = 1, 2, 3, 4, 5, 6$ and $k = 1, 2, 3$ in a usual equation form such as $S_1 = \dots + \dots + \dots + \dots$

Question 7.6 For two types of pressure sensors shown below, indicate in the table below which one of the two you would choose, if your primary concern is the issue shown in the first column, assuming that the diaphragm size (lateral dimension) is about the same for both types. Put a check mark on your choice.

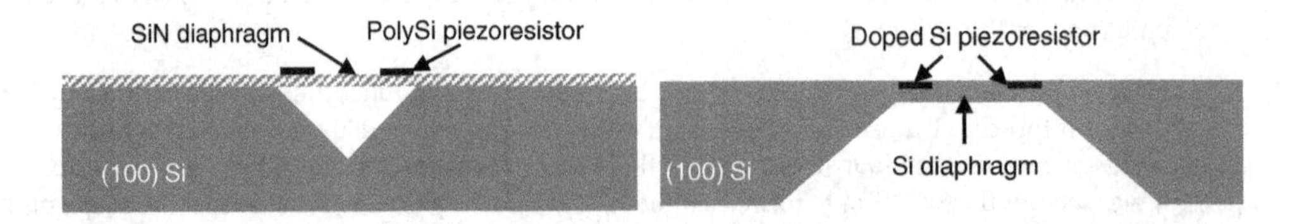

	Polysilicon Piezoresistor (left one)	Silicon Piezoresistor (right one)
Overall sensor size		
Sensitivity		
Easiness of sensor fabrication		
Long-term effects from environment		

Question 7.7 For a resonant mass sensor, which one would you choose between quartz crystal microbalance (QCM) and FBAR-based sensor (Fig. 7.14), if your only concern is the issue listed at the top row of the following table? Put check marks on the appropriate blanks.

	Sensing Electronics	Protection of Electrodes	Manufacturing Cost
QCM operating at 4 MHz			
FBAR-based resonant sensor operating at 1 GHz			

Question 7.8 For ZnO which has the following elasto-piezo-dielectric matrix, write out three equations for D_i $(= d_{ij}T_j + e_{ik}E_k)$, that is D_1, D_2, and D_3 in terms of d_{15}, d_{31}, d_{33}, ε_{11}, ε_{33}, $T_1 - T_6$, and $E_1 - E_3$.

Elasto-piezo-dielectric matrix								
C_{11}	C_{12}	C_{13}	0	0	0	0	0	e_{31}
C_{12}	C_{11}	C_{13}	0	0	0	0	0	e_{31}
C_{13}	C_{13}	C_{33}	0	0	0	0	0	e_{33}
0	0	0	C_{44}	0	0	0	e_{15}	0
0	0	0	0	C_{44}	0	e_{15}	0	0
0	0	0	0	0	$\frac{C_{11}-C_{12}}{2}$	0	0	0
0	0	0	0	d_{15}	0	ε_{11}	0	0
0	0	0	d_{15}	0	0	0	ε_{11}	0
d_{31}	d_{31}	d_{33}	0	0	0	0	0	ε_{33}

Question 7.9 The low stress silicon nitride (Si_xN_y) film (≈ 1.5 μm thick) on top of a 1 mm thick patterned SiO_2 film on a silicon substrate (≈ 400 μm thick) (shown below) is under residual stress along the thickness direction, as shown at left below, after LPCVD deposition. If we etch out the SiO_2 film to form a cantilever out of the Si_xN_y film, sketch the shape of the released cantilever, indicating whether the released cantilever is flat, curled upward, or curled downward; and whether the length of the released cantilever is smaller, larger or equal to the length before the release.

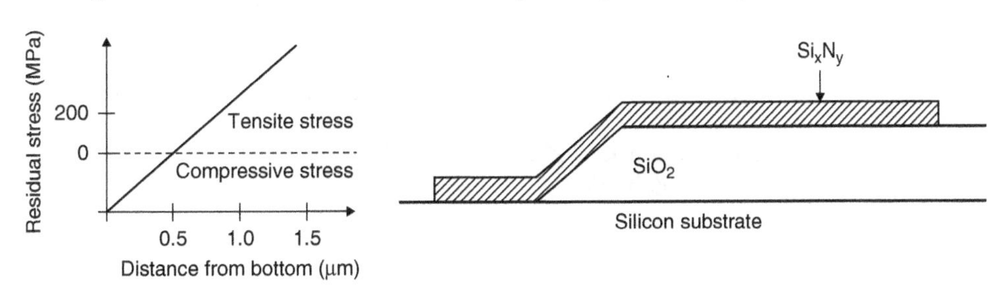

Problem 7.1 If the piezoresistive silicon pressure sensor shown at Fig. 7.15 is made on a (100) n-type silicon having a resistivity of 11.7 Ωcm, calculate the normalized resistance changes (i.e., $\Delta R/R$) for R_1, R_2, R_3, and R_4, assuming that $\sigma_l = 0.3$ kPa and $\sigma_t = 1$ kPa for R_1 and R_3, while $\sigma_l = 1$ kPa and $\sigma_t = 0.3$ kPa for R_2 and R_4.

Problem 7.2 Answer the following for a piezoresistive silicon pressure sensor made on (100) n-type silicon having 11.7 Ωcm resistivity with 20-μm-thick silicon membrane that is 1 × 1 mm in size, similar to what is shown in Fig. 7.15. (a) Calculate π_l and π_t for the piezoresistor. (b) The normalized resistance change (i.e., $\Delta R/R$) is equal to $\pi_l\sigma_l + \pi_t\sigma_t$. Which product will dominate in R_1, R_2, R_3, and R_4 above? Is it $\pi_l\sigma_l$ or $\pi_t\sigma_t$? (c) Calculate the normalized resistance change for R_2 (i.e., $\Delta R_2/R_2$) due to 1 kPa.

Problem 7.3 For an accelerometer shown in Fig. 6.20, the axial stress is maximum at the "base" (i.e., the clamped edge) on the top (or bottom) surface of the beam, where a piezoresistor is implanted to convert the stress (or strain) into electrical resistance change. For a piezoresistive accelerometer (Fig. 6.20) made on an (100) n-type silicon having a resistivity of 11.7 Ωcm, with the following dimensions: t = 10 μm (the beam thickness), w = 50 μm, l = 400 μm, M_p = 1 mg, l_p = 1,000 μm, answer the following. (a) The normalized resistance change (i.e., $\Delta R/R$) is equal to $\pi_l\sigma_l + \pi_t\sigma_t$. Which product will dominate in the accelerometer above? Is it $\pi_l\sigma_l$ or $\pi_t\sigma_t$? (b) Calculate π_l or π_t depending on your answer to the question (a). [Hint: The piezoresistor must have been aligned along <110> direction according to the shape of the structure that is made on (100) wafer.] (c) Calculate the normalized resistance changes (i.e., $\Delta R/R$) due to 1 g acceleration (i.e., 9.8 m/s^2) directed in the z direction. [Hint: Due to 1 g acceleration, there will be a force (i.e., F_{iz} directed in z direction as shown in Fig. 6.20) equal to 9.8×10^{-6} N. And the moment of inertia for a beam with a rectangular cross-section is $wt^3/12$, where w and t are the width and thickness of the beam, respectively.]

Problem 7.4 Answer the following for a piezoresistive silicon pressure below that is made on an (100) p-type silicon having a resistivity of 7.8 Ωcm, and is built on a 1 × 1 mm^2 square diaphragm with thickness of 20 μm.

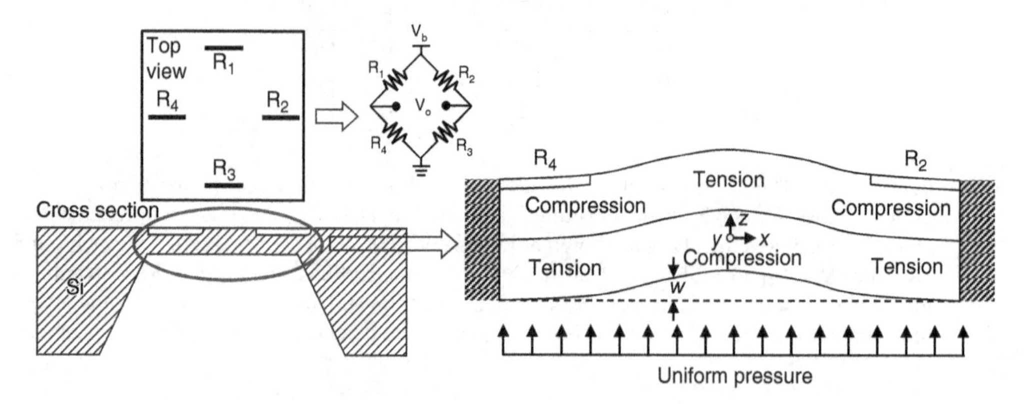

(a) For a uniform applied pressure of 1 kPa, calculate the normalized resistance changes (i.e., $\Delta R/R$) for R_1 and R_2.
(b) Calculate V_o for a uniform applied pressure of 1 kPa, if $R_1 - R_4$ are arranged in a bridge circuit as shown above with V_b = 10 V.

CHAPTER 8

Microfluidic Systems and Bio-MEMS

Miniaturizing fluidic devices such as channels, mixers, reservoirs, pumps, valves, sensors, etc. can potentially lead to lab-on-a-chip for chemical and biological analysis, environment monitoring, etc. The advantages of miniaturized fluidic devices include reduced size, reduced power consumption, reduced reagent, potential to make portable and disposable devices, potential to integrate a system into one single chip, etc. In this chapter, we will study microchannels, microvalves, micropumps, micromixers, and lab-on-chip. Specifically, we will learn fundaments on droplet formation through microchannels, electrowetting, wetting over structured surfaces, passive and active microvalves, electrically operated and capillary force–based micropumps, dielectrophoresis-based and acoustic wave micromixers (as well as passive micromixers), gene sequencing, and microelectromechanical systems (MEMS)-based polymerase chain reaction (PCR) systems including single-cell reverse transcriptase PCR system.

8.1 Microchannels and Droplet Formation

In gaseous flow, as explained in Chap. 1, the flow falls into one of several categories depending on the Knudsen number $K_n \equiv \lambda/d$, where d is the channel diameter, while λ is the mean free path of a molecule before colliding with another molecule and is equal to $\left(\pi n d_o^2 \sqrt{2}\right)^{-1}$ with d_o = molecular diameter and n = number of molecules per unit volume.

Fluid means gas and liquid that deform under stress, and two most important properties that characterize fluid are density ρ and viscosity μ. Viscosity is a measure of how resistant a fluid is to flow (analogous to friction), and is in unit of poise (= g/cm·s). In general, viscosity decreases rapidly with increasing temperature. In addition to the material properties, the dimension of the channel through which fluid flows is important in determining flow behavior, as the viscous force (or frictional force) of flow increases with decreasing channel diameter. Thus, the unitless Reynolds number (R_e) is used to determine whether flow is laminar (for $R_e < 2{,}000$) or turbulent (for $R_e > 4{,}000$): $R_e = \frac{\rho v d}{\mu}$, where v and d are the flow velocity and channel diameter, respectively.

In microchannels with such small channel diameters, liquid flow is laminar, and there is very little mixing between two flows over a log traveling distance, as illustrated in the middle channel in Fig. 8.1. Also shown in the figure are aqueous droplets formed at the nozzle end of the middle channel that are carried by immiscible oil, as the oil is pushed through the channel at the right end.

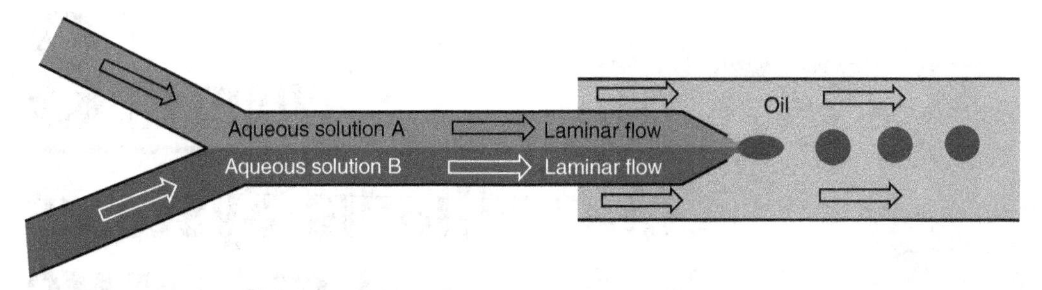

FIGURE 8.1 Flow through microchannels where the Reynolds number is low, and the flow is laminar: two channels carrying two different aqueous solutions (*left*), a channel where the two channels are merged (*middle*), and then another channel where immiscible oil is flowing with aqueous droplets formed at the nozzle at the end of the middle channel (*right*).

Based on the unique characteristics of the flow through microchannels, various techniques have been developed for generating and manipulating droplets in microchannels for cell manipulation and analysis, biochemical analysis, drug discovery, material synthesis, etc. [1]. For example, the same set of the microchannels shown in Fig. 8.1 can be used to produce bi-phase droplets of monomers [2], as illustrated in Fig. 8.2.

8.1.1 Electrowetting on Dielectrics

As illustrated in Fig. 8.3a, liquid droplet has a surface tension γ_{la} (in N/m or J/m^2), while the interface between solid and air has an interfacial force γ_{sa} (again in N/m or J/m^2) due to cohesive forces that are not balanced at the interfaces. When a liquid droplet is formed on a solid surface, the droplet's contact angle is equal to Young's equilibrium contact angle θ_Y, which comes from the balance of the three forces: the surface tension between liquid and air (γ_{la}), the interfacial force between solid and air (γ_{sa}), and that between solid and liquid (γ_{sl}), as follows [3].

$$\gamma_{la}\cos(180° - \theta_Y) = \gamma_{la}\cos\theta_Y = \gamma_{sa} - \gamma_{sl} \tag{8.1}$$

Now, if a voltage is applied between the liquid droplet and the electrode buried under an insulating layer on which the droplet sits, as illustrated in Fig. 8.4a, the contact angle of the droplet becomes smaller, or the droplet wets the solid surface more [4].

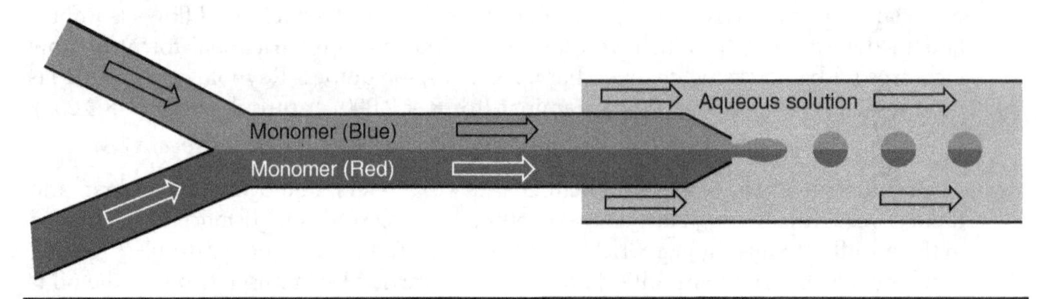

FIGURE 8.2 Formation of bi-phase droplets through the channels shown in Fig. 8.1.

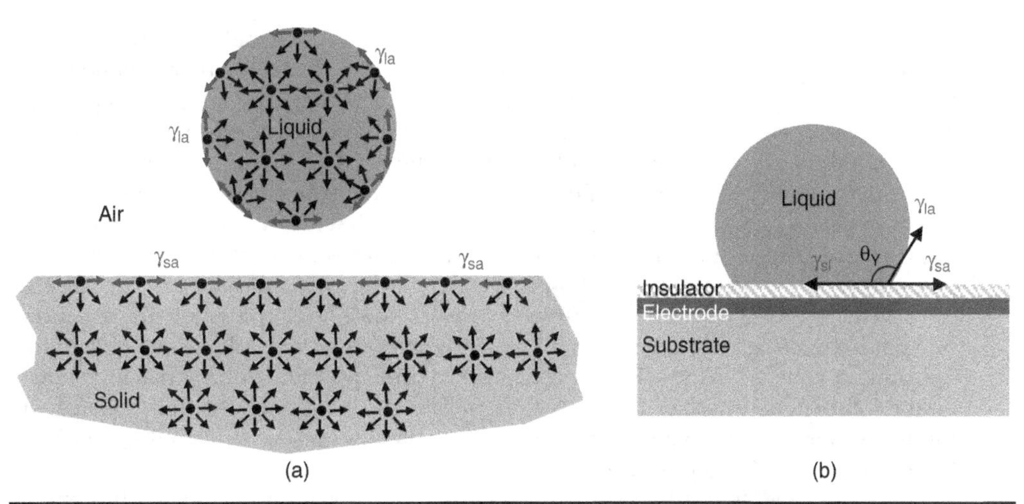

Figure 8.3 Surface and interfacial tensions between liquid and air (γ_{la}), solid and air (γ_{sa}), and solid and liquid (γ_{sl}) due to cohesive forces that are not balanced at the interfaces: (a) when liquid is freely suspended in air and (b) when liquid is in contact with solid (that is hydrophobic).

This electrowetting effect was first described by Gabriel Lippmann in 1875, and has been used in microfluidic channels for transporting, merging, or separating droplets as well as for liquid lens with electrical controllability of the focal length with an insulating layer between the droplet and electrode, thus, being called electrowetting on dielectric (EWOD). Generally accepted theory of electrowetting is that the interfacial force between solid and liquid (γ_{sl}) becomes a function of the applied voltage V_a as follows

$$\gamma_{sl}(V_a) = \gamma_{sl} - \frac{1}{2}\frac{\varepsilon}{t}V_a^2, \tag{8.2}$$

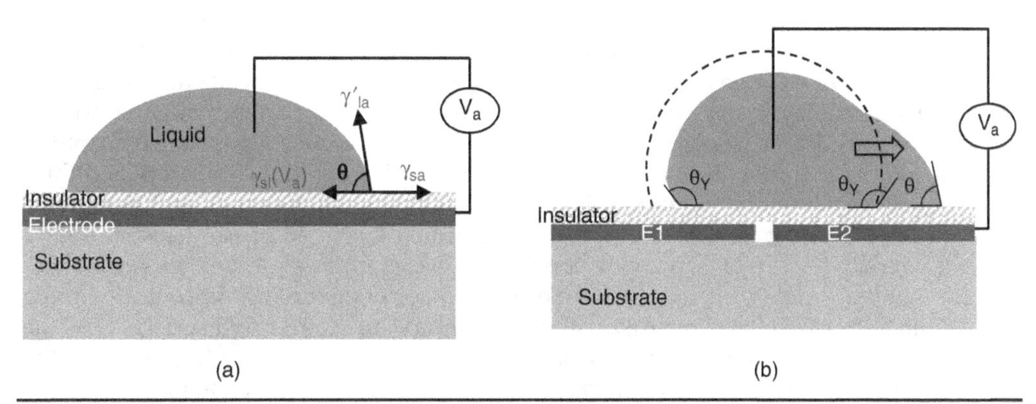

Figure 8.4 Electrowetting of a droplet: (a) under an applied voltage between conductive liquid droplet and electrode underneath an insulating layer in contact with the droplet, showing the contact angle of the droplet (θ) now being less than the contact angle (θ_y) without any applied voltage shown in the earlier figure, Fig. 8.3b, and (b) with the electrode patterned into two (E1 and E1) and a voltage applied only between E2 and the droplet (in which case, the contact angle is reduced on the right side, and the droplet moves to the right, if the applied voltage is greater than the threshold voltage).

where ε and t are the dielectric constant and thickness of the insulating layer, respectively. The applied voltage reduces the solid–liquid interfacial force, independent of the sign of the applied voltage, and amount of the reduction is equal to the capacitive energy (per unit area) stored across the insulation layer. Since the electrical field inside the conductive liquid is zero, resulting in all the applied voltage dropping across the insulating layer, the validity of Eq. (8.2) is viewed plausible but has not been backed by a universally accepted theory, though many papers have been published on it, some arguing that charges adhering on the solid–liquid interface exert radially outward in-plane force [5], while others pointing to the fringe electrical field around the droplet edge as the source of the radially outward in-plane force [6]. Experimentally, the equation has been measured to be reasonably accurate up to a certain voltage level called the "saturation voltage." Any voltage increase above the saturation voltage has little effect on the solid–liquid interfacial force, the reason for which has not been well established, though air ionization around droplet edge [5] or breakdown of the insulating layer [7] has been presented as a possible reason.

With the solid–liquid interfacial force reduced by the applied voltage, the contact angle θ is now governed by the following equation from the force balance, similar to Young's equilibrium contact angle θ_Y:

$$\gamma'_{la}\cos\theta = \gamma_{sa} - \gamma_{sl}(V_a), \tag{8.3}$$

where γ'_{la} is the surface tension of the droplet under the new contact angle θ. One can use the electrowetting effect to move a liquid droplet by segmenting the electrode and applying a voltage between one of the segmented electrode and the liquid droplet, as shown in Fig. 8.4b. As can be seen, the droplet experiences a reduced contact angle θ on the right end due to the applied voltage on that end, while the contact angle on the left end remains the Young's contact angle θ_Y due to no applied voltage on that end. This makes the droplet move from left to right.

8.1.2 Liquid Wetting over Structured Surface

Wetting of liquid over a rough surface is substantially different from that over a smooth surface because the interfacial tensions, both γ_{sl} and γ_{sa}, are increased by the "rough factor" c defined to be the ratio between ab and $a'b'$ in Fig. 8.5b, while the surface tension γ_{la} remains the same over a rough structured surface [8]. Comparing the two top views in Fig. 8.5, we see that a structured surface has a shorter length $a'b'$ (compared to the length ab in case of a smooth surface) over which a liquid droplet wets, and thus, the interfacial tensions between solid and liquid (γ_{sl}) and between solid and air (γ_{sa}) that are in N/m are larger by the factor of c compared to the interfacial tensions over a smooth surface where the wetting occurs over the length ab. However, the surface tension of the liquid (γ_{la}) is independent of the contact length or area, since it is between liquid and air and is the same over a smooth surface and a structured surface. How the changes in the interfacial tensions along with no change on surface tension affect the wetting behavior is illustrated in Fig. 8.6 and explained in the next paragraph.

The top row of Fig. 8.6 illustrates wetting over a smooth surface, when the surface is hydrophobic (b) and hydrophilic (c), while the bottom row shows wetting over a structured surface of Fig. 8.5b when the surface is hydrophobic (b) and hydrophilic (c). As can be seen in Fig. 8.6b, when the surface is hydrophobic, the contact angle of liquid droplet is larger over a structured surface than over a smooth surface because the net

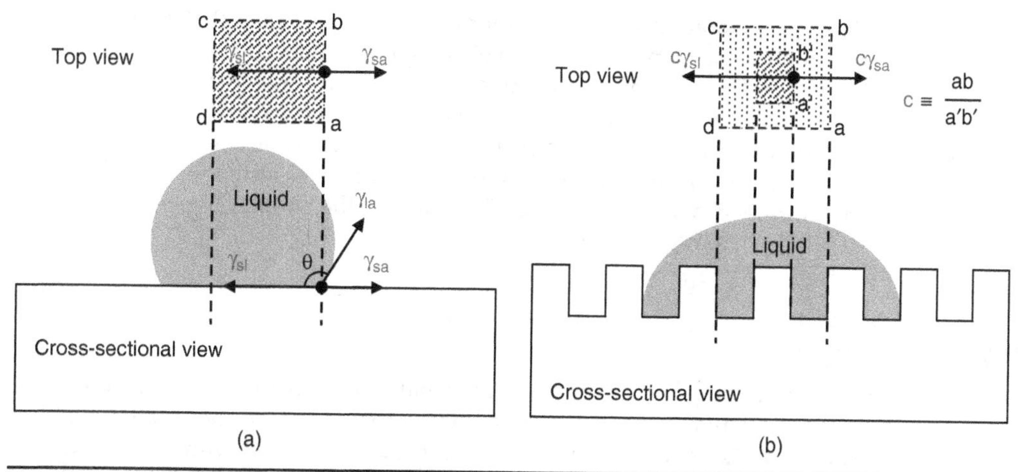

Figure 8.5 Wetting liquid droplet on (a) smooth surface with the contact angle determined by force balance of liquid and air (γ_{la}), solid and air (γ_{sa}), and solid and liquid (γ_{sl}) tensions and (b) structured surface with enhanced solid–liquid ($c\gamma_{sl}$) and solid–air ($c\gamma_{sa}$) tensions where $c \equiv \frac{ab}{a'b'}$. The effect of the structured surface is explained in Fig. 8.6 for hydrophobic and hydrophilic surfaces.

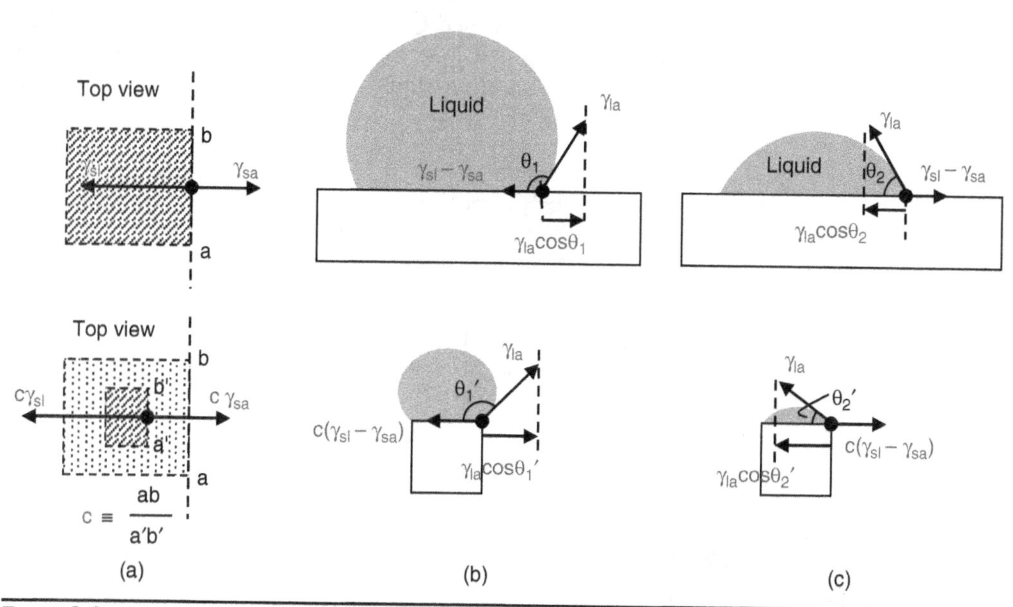

Figure 8.6 (a) Top views of the smooth and structured surfaces shown in Fig. 8.5, (b) cross-sectional views of liquid droplet wettings over the smooth (top) and structured (bottom) surfaces with the structured surface providing more hydrophobicity, if the smooth surface is already hydrophobic due to c ($\equiv ab/a'b'$) being greater than 1, and (c) cross-sectional views of liquid droplet wettings over the smooth (top) and structured (bottom) surfaces with the structured surface providing more hydrophilicity, if the smooth surface is already hydrophilic again due to c being greater than 1.

in-plane tension is equal to $c(\gamma_{sl} - \gamma_{sa})$, larger by a factor of c than that over a smooth surface, which is balanced by $\gamma_{la}\cos\theta_1'$ with θ_1' being larger than θ_1 (the contact angle for a smooth surface). Thus, a structured surface is more hydrophobic than a smooth surface if it is hydrophobic. If the smooth surface is hydrophilic, as illustrated in Fig. 8.6c, the contract angle becomes less over a structured surface, because the net in-plane tension is equal to $c(\gamma_{sl} - \gamma_{sa})$, larger by a factor of c than that over a smooth surface, which is balanced by $\gamma_{la}\cos\theta_2'$ with θ_2' being smaller than θ_2 (the contact angle for a smooth surface), resulting in a more hydrophilic surface.

8.2 Microvalves

There are two types of valves: (1) passive valve that is opened and closed by the energy from fluid and (2) active valve that is electrically or pneumatically actuated. A passive valve is simple in its operating principle, as can be understood with Fig. 8.7, which shows a passive valve made of Parylene on a plastic substrate with flow directions (along with the amount of the force carried with the flow) determining whether their valve is open or close.

Unlike a passive valve, an active valve requires an external energy for operation. Pneumatic pressure provides a powerful force for actuating a valve, but the input and output of air pressure requires proper fittings around which are sources of air pressure leak and failure. Compressed air is commonly used for the pneumatic pressure, but is not amenable for portable or handheld systems.

A popular valve for microfluidics (so-called Quake valve) is based on two stacked microchannels of polydimethylsiloxane (PDMS) made through steps involving soft lithography covered in Chap. 1. The fabrication requires two molding substrates for the

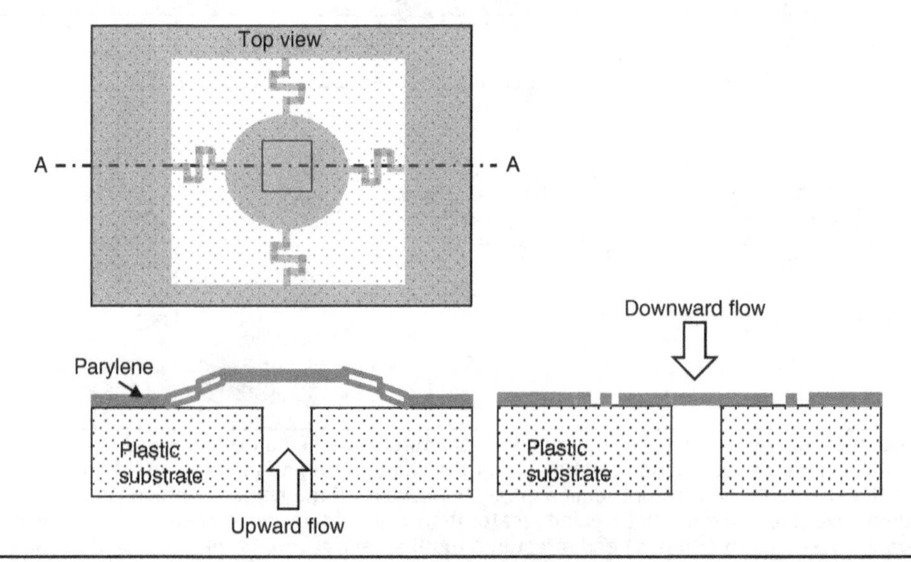

FIGURE 8.7 Top- and cross-sectional views of a passive valve with upward flow that opens up the valve (*left*) and downward flow that closes the valve (*right*).

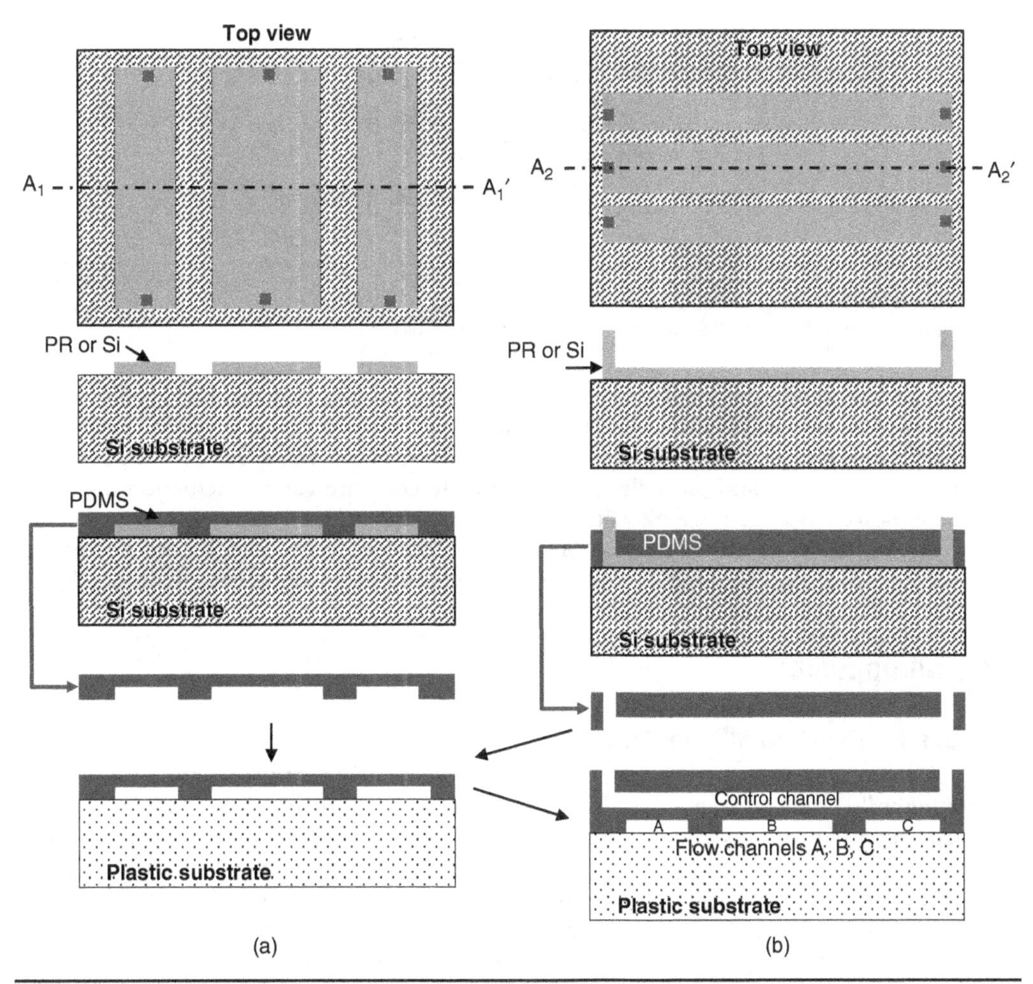

FIGURE 8.8 Top- and cross-sectional views (across A–A') of (a) the fabrication of three flow channels (into and out of the paper) with PDMS through soft lithography and (b) the fabrication of three control channels (from left to right) with PDMS followed by addition of the control channels onto the flow channels made in (a).

two layers of microchannels, as illustrated in Fig. 8.8, and two layers of PDMS formed from the molds are bonded together to form the two layers of microchannels. In Fig. 8.8, the microchannels in the lower layer are used to carry liquid, while the channels in the upper layer are used for pneumatic pressure to close the PDMS valve situated at the places where the lower and upper channels cross.

As illustrated in Fig. 8.9, the Quake valve relies on the air pressure deforming the elastomer (PDMS in this case) to close off the flow channels. According to the top-view designs of the two-layer microchannels shown in Fig. 8.8, each "control channel" with pneumatic pressure can control the open/close of the three flow channels at different location along the liquid flow direction. Interestingly, if the control channels are pneumatically actuated in a proper time sequence, a peristaltic pump can be obtained.

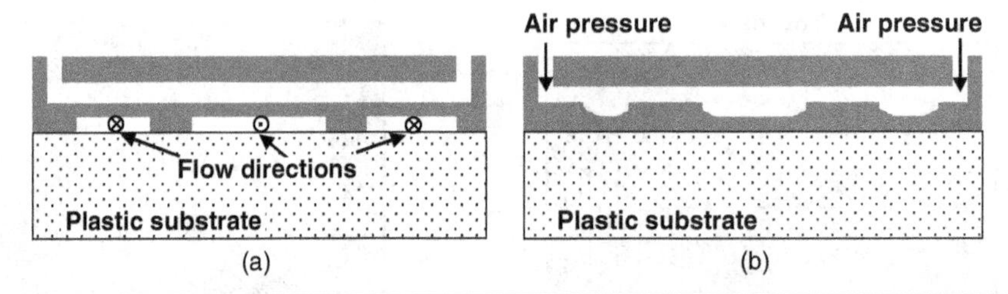

FIGURE 8.9 Operational principle of Quake valve shown in Fig. 8.8: (a) the three flow channels carrying liquid flows with no air pressure applied to the control channel and (b) the three flow channels closed by air pressure applied to the control channel.

In addition to pneumatic pressure, other actuation mechanisms have been used to open and close microvalves. Among them are thermally induced pneumatic pressure, electromagnetic actuator such as the one used in solenoid valve, electrostatic actuator, thermal bimorph, and piezoelectric bimorph. To compare various actuation methods, actuation energy density P may be a good figure of merit (FOM) as it is actuation energy (force F times distance z) per actuator volume V, that is, $P = Fz / V$ [J/m^3 or Pa]. According to such FOM, electrostatic actuation is inherently disadvantageous to others [9].

8.3 Micropumps

8.3.1 Valveless Micropumps

Micropumps are built with or without check valves, but valveless micropumps are inherently prone to mixing of downstream fluid with upstream. A valveless fluid pump can be based on flow rate difference through *passive* inlet and outlet flow channels (Fig. 8.10) or through electrothermally active channels (Fig. 8.11). In Fig. 8.10, when the chamber is expanded by the bending of a piezoelectric unimorph, fluid is pulled into the chamber through both the inlet and outlet channels with a higher flow rate through the inlet than through the outlet, since the fluid experiences increasingly larger channel diameter as it flows through the inlet channel, while it experiences increasingly smaller channel diameter through the outlet channel, resulting in more fluid getting into the chamber from the inlet than from the outlet. At the next cycle of the piezoelectric

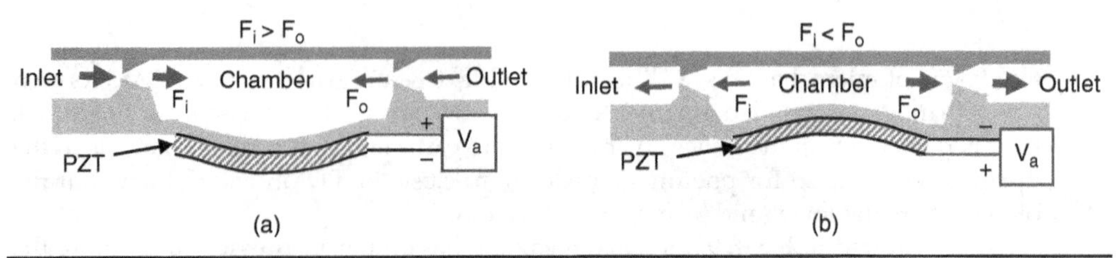

FIGURE 8.10 Operational principle of a valveless pump based on flow rate difference on *passive* flow channels: (a) when the chamber is expanded by the PZT unimorph bending, fluid is pulled into the chamber through both the inlet and outlet with a higher flow rate through the inlet than through the outlet due to the varying cross-sectional diameters of the flow channels. (b) When the chamber is contracted, the fluid is pushed out of the chamber but more through the outlet this time. When (a) and (b) are repeated sequentially, there is a net flow of fluid from the inlet to the outlet.

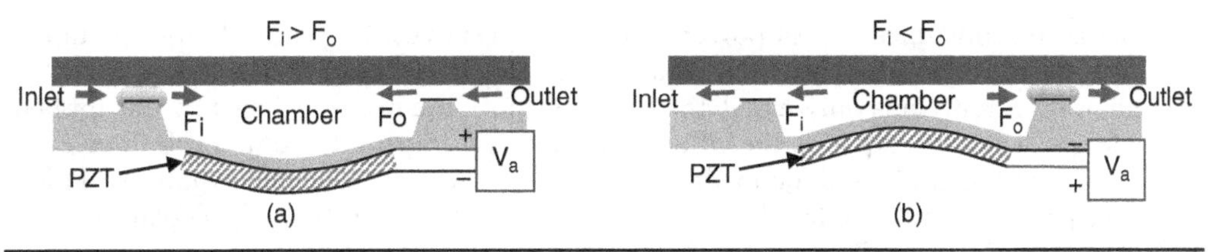

FIGURE 8.11 Operational principle of a valveless pump based on flow rate difference on *active* flow channels: (a) when the chamber is expanded by the PZT unimorph bending, fluid is pulled into the chamber through both the inlet and outlet with a higher flow rate through the inlet than through the outlet due to the electrical heating that lowers the fluid viscosity and thus increases the flow rate. (b) When the chamber is contracted, fluid is pushed out of the chamber but more through the outlet this time. When (a) and (b) are repeated sequentially, there is a net flow of fluid from the inlet to the outlet.

actuation that makes the chamber contracted, the fluid inside the chamber is pushed out of the chamber, but more through the outlet than through the inlet, resulting in more fluid getting out of the chamber through the outlet than from the inlet this time. When the chamber expansion and contraction are repeated sequentially, there is a net flow of fluid from the inlet to the outlet, and thus a pumping is achieved.

A similar flow difference happens through the inlet and outlet channels in Fig. 8.11, which present the same cross-sectional diameter but different temperature to the flow. As the chamber is expanded by the PZT unimorph bending, fluid is pulled into the chamber through both the inlet and outlet with a higher flow rate through the inlet than through the outlet due to the electrical heating that lowers the fluid viscosity and thus increases the flow rate. And when the chamber is contracted, fluid is pushed out of the chamber but more through the outlet this time, resulting in a net flow of fluid from the inlet to the outlet.

Active Diaphragm Bending through Electrothermal, Electrostatic, and Piezoelectric Actuation

Diaphragm bending that can produce the fluid motion can be obtained with electrothermal unimorph (or bimorph), electrostatic actuation, or piezoelectric unimorph (or bimorph), as illustrated in Fig. 8.12. Among the three approaches, the electrothermal

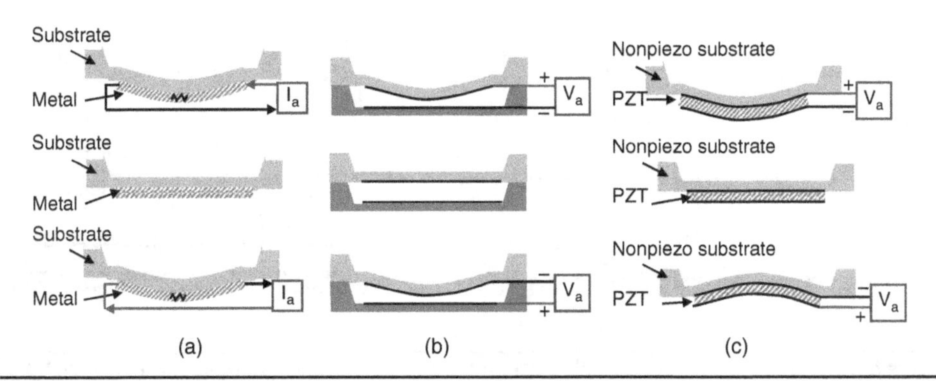

FIGURE 8.12 Three actuation methods for active microvalves under positive, no, and negative actuations: (a) electrothermal actuation with bimetal (e.g., Ni/Si), (b) electrostatic actuation, and (c) piezoelectric actuation with unimorph (e.g., PZT/Si).

actuation consumes largest power and responds slowest to electrical input (with the actuation frequency limited to hundreds of Hz). The bending displacement of the thermal actuation is only along one direction since the heat produced by the applied electrical current expands the volume of the metal causing the unimorph to bend as shown in Fig. 8.12a, independent of the current direction. The electrostatic actuation also produces unidirectional bending displacement independent of the polarity of the applied voltage, as illustrated in Fig. 8.12b. However, the piezoelectric actuation produces bidirectional bending displacement as the polarity of the applied voltage is changed (Fig. 8.12c) in addition to large actuation energy density P, low power consumption, and fast response (higher than hundreds of kHz).

8.3.2 Micropump Based on Electrostatic Actuation

Valveless micropumps (Figs. 8.10 and 8.11) are easy to fabricate, but have an inherent intermixing between the upstream and downstream. One-way valves, such as the ones shown in Fig. 8.7, cut down the intermixing but need to be flexible and ductile. Polymeric materials are more flexible and ductile than silicon or dielectric materials. Polyimide and PDMS have been used as valve materials, but tend to leak liquid/gas due to their relatively high permeabilities to liquid/gas. Parylene, on the other hand, has a very low permeability to liquid or gas.

A micropump made of two one-way Parylene valves and an electrostatically actuated diaphragm is illustrated in Fig. 8.13. Parylene may be a good valve material due to its low permeability to liquid and low Young's modulus (3 GPa, about 30 times less than that of silicon). The valve illustrated in Fig. 8.13 is bridge type as shown in Fig. 8.7, but can also be a cantilever type. The electrostatic actuation can be replaced with a PZT unimorph.

8.3.3 Micropump Based on Piezoelectric Unimorph

A piezoelectrically driven micropump can be built by stacking two one-way valve chips and a PZT unimorph chip, as shown in Fig. 8.14 [10]. The two valve chips provide the check valve function during pumping: one valve controls the flow in one direction,

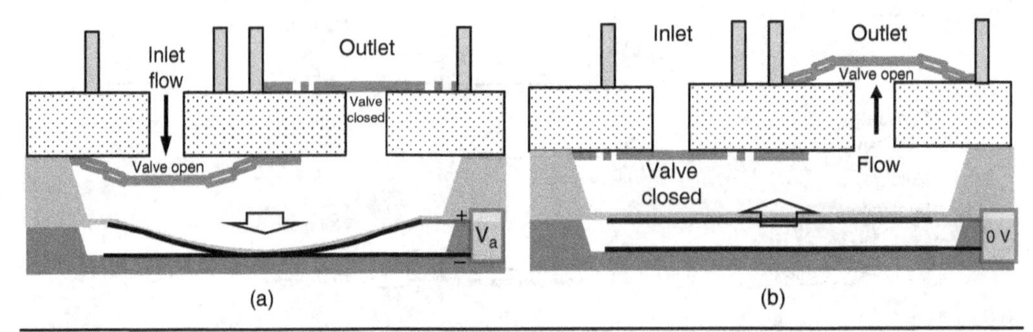

FIGURE 8.13 A micropump formed with a pair of the one-way valves illustrated in Fig. 8.7 that are being opened (a) and closed (b) by an electrostatically driven diaphragm. The pressure generating diaphragm can be driven by any of the three methods illustrated in Fig. 8.12. If (a) and (b) are repeated, there is a net flow of fluid from the inlet to the outlet.

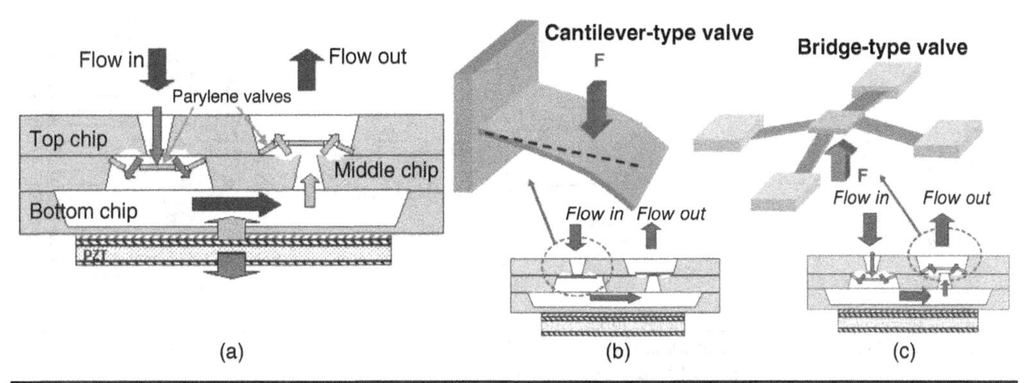

FIGURE 8.14 (a) Cross-sectional view of the piezoelectrically driven micropump based on PZT with arrows to illustrate its working principle and the micropump with two Parylene valves based on (b) cantilever type and (c) bridge type [10].

while the other controls the flow in the opposite direction. With the up and down motion of the PZT unimorph, liquid is brought in through the inlet valve and is pumped out through the outlet valve. Since each of the two valves blocks the unwanted leak to the inlet or outlet during each half of the pumping cycle, there is no intermixing between the upstream and downstream.

The piezoelectric unimorph is built by attaching a 200-μm-thick PZT substrate to an 80-μm-thick silicon diaphragm on a silicon substrate (Fig. 8.14a). Since the Young's modulus of silicon (162 Gpa) is about 2.5 times larger than that of PZT (62 Gpa), 80-μm-thick silicon diaphragm is chosen to match a 200-μm-thick PZT in the design so that the neutral plane may be located at the interface between the PZT and silicon. The silicon diaphragm is a part of a silicon chamber formed by silicon bulk micromachining, and is slightly larger than the PZT element.

The one-way valve can be a cantilever (Fig. 8.14b) or a bridge type (Fig. 8.14c). The cantilever-type valve has a rectangular shape with one of its shorter edge being fixed and the other three edges being free to move. Since the three edges are not anchored to the substrate, this cantilever type is not affected by residual stress (other than stress gradient in the thickness direction). The equivalent spring constant K (defined as applied force F divided by displacement y, in N/m) for a cantilever beam is

$$K \equiv \frac{F}{y} = \frac{2Eb}{3}\left(\frac{t}{l}\right)^3 \tag{8.4}$$

where b, t, l, and E are the beam width, thickness, length, and Young's modulus, respectively. The bridge-type valve has all of its four edges fixed and is greatly influenced by residual stress and diaphragm stretching effect. A beam with its two edges clamped has the following equivalent spring constant:

$$K = 32E \cdot b \cdot \left(\frac{t}{l}\right)^3 + 8\sigma \cdot (1-\nu) \cdot b \cdot \left(\frac{t}{l}\right) \tag{8.5}$$

where σ and ν are the residual stress and Poisson's ratio, respectively.

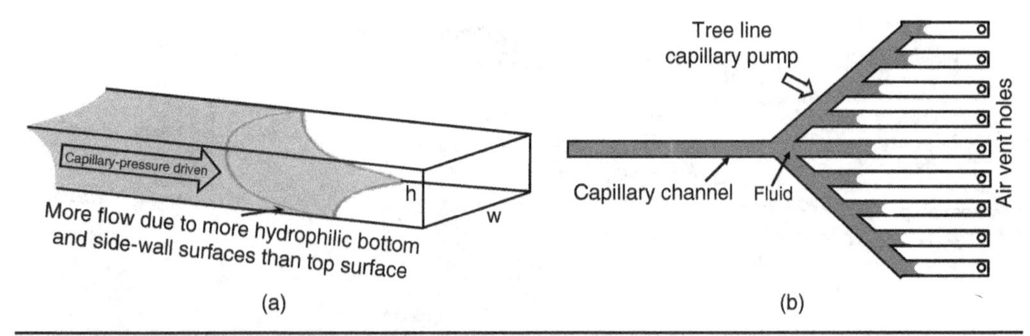

FIGURE 8.15 Passive capillary pumping: (a) capillary flow through a rectangular microchannel and (b) a tree-line capillary pump with air vent holes at its channel ends.

8.3.4 Passive Capillary Pumping

For a rectangular microchannel with height h and width w (Fig. 8.15a), a capillary pressure P capable of driving fluid can be obtained with the following equation [11]

$$P = -\gamma_{la}\left(\frac{\cos\theta_{top} + \cos\theta_{bottom}}{h} + \frac{\cos\theta_{left} + \cos\theta_{right}}{h}\right), \tag{8.6}$$

where $\theta_{top}, \theta_{bottom}, \theta_{left}$, and θ_{right} are the liquid contact angles with the top, bottom, left, and right surfaces of the microchannel, respectively. The capillary pressure can drive fluid and produce a pressure-driven, steady-state flow of fluid called Poiseuille flow. Solving the Navier–Stokes equation, one can derive an equation for the flow rate Q [12], which for $w >> h$, is

$$Q \approx \frac{h^3 w \Delta P}{12\mu L(t)}\left(1 - 0.63\frac{h}{w}\right), \tag{8.7}$$

where ΔP, $L(t)$, and μ are the capillary pressure difference, length of the fluid in microchannel, and fluid viscosity, respectively. The worst-case inaccuracy of Eq. (8.7) will be for a case of $w = h$ and is estimated to be 13%, which drops down to 0.2% for $w = 2h$ [12]. The flow rate per unit cross-sectional area (i.e., Q/wh) is equal to $dL(t)/dt$, and we have

$$L(t)\frac{dL(t)}{dt} \approx \frac{h^2 \Delta P}{12\mu}\left(1 - 0.63\frac{h}{w}\right), \text{ from which we obtain}$$

$$L(t) \approx h\sqrt{\frac{\Delta P}{6\mu}\left(1 - 0.63\frac{h}{w}\right)t} \tag{8.8}$$

For passive pumping based on capillary pressure, a capillary pump can be made out of tree-line capillary channels as illustrated in Fig. 8.15b. Note that air vent holes are purposely made at the ends of the capillary channels so that fluid may advance the channels by displacing air without backpressure from the air.

8.4 Micromixers

Due to a low Reynolds number in microchannels, mixing in microchannels requires external sources for turbulence or modified channel geometry (or built-in microstructures inside a microchannel). Electric, magnetic, or acoustic field can be used as an external source which produces turbulence in fluids. For nonconductive fluids, a nonuniform electric field can be used to produce dielectrophoresis (DEP) for chaotic mixing [13]. For nonmagnetic fluids, a magnetic stir bar or built-in microstructures can be embedded in a microchannel [14].

8.4.1 Dielectrophoresis

As noted in Fig. 3.18, any dielectric medium whether solid or liquid is polarized under an applied electrical field. If a dielectric solid sphere is immersed in a dielectric liquid and is under a uniform electrical field, both the solid and the liquid spheres are polarized as shown in Fig. 8.16a and b, if the solid is more and less polarizable than the liquid, respectively. Though the polarizations differ, the polarization difference between the solid and liquid is the same in magnitude and opposite in sign at the upper and lower hemispheres under a uniform electrical field. Thus, the sphere experiences equal and opposite electrostatic forces toward positive and negative potentials, thus zero net force, independent of whether the solid is more (or less) polarizable than the liquid. Note that the force directions are as indicated in Fig. 8.16 based on the net charges on the upper and lower hemispheres of the solid sphere. For example, the net charge at the upper hemisphere in Fig. 8.16a is negative (due to more polarization in the solid sphere than in the liquid), and the force from the upper electrode (with positive applied voltage) is attractive.

However, under a nonuniform field, the polarization difference between the solid and the liquid sphere is different in magnitude and opposite in sign at the upper and lower hemispheres, as illustrated in Fig. 8.17, due to the stronger field density at the lower hemisphere. Thus, the sphere is pulled toward the negative electrode (Fig. 8.17a), if the solid sphere is more polarizable than the liquid, but pushed away from the negative electrode (Fig. 8.17b), if the solid is less polarizable than the liquid. This phenomenon is called dielectrophoresis (DEP) and can be used for micromixing.

The classical time-averaged DEP force on a spherical particle (of radius a, permittivity ε_p, and conductivity σ_p) immersed in liquid (permittivity ε_m and conductivity σ_m) under an applied electrical field E can be calculated with the following equation [15]:

$$\langle F_{\mathrm{DEP}} \rangle = 2\pi a^3 \varepsilon_m \,\mathrm{Re}\left[f_{\mathrm{CM}} \right] \nabla E_{\mathrm{rms}}^2, \tag{8.9}$$

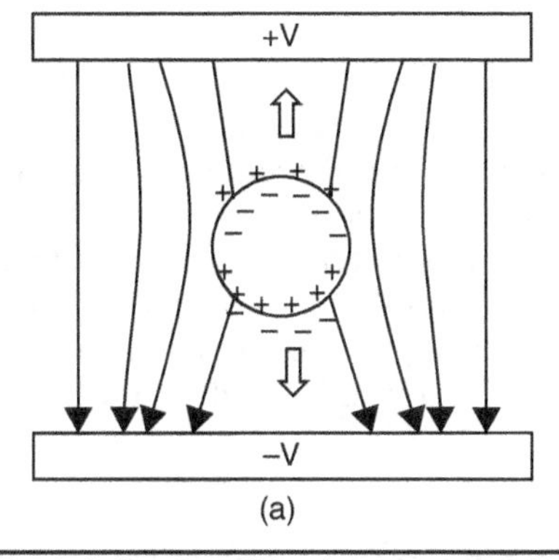

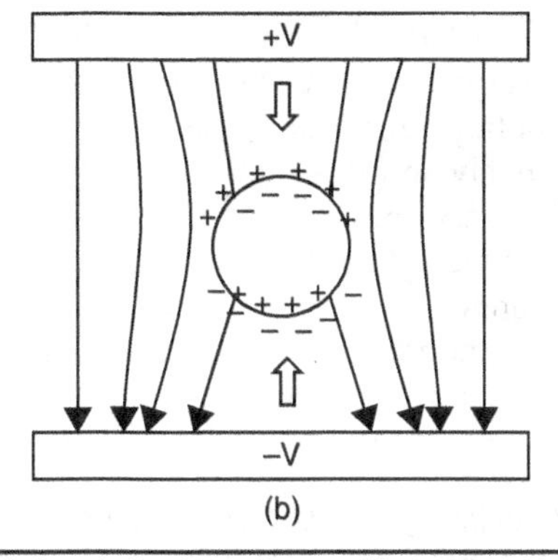

FIGURE 8.16 A dielectric solid sphere immersed in a dielectric liquid under a uniform electrical field: (a) when the solid is *more* polarizable than the liquid and (b) when the solid is *less* polarizable than the liquid. The electrostatic forces on the solid sphere are opposite in sign but the same in magnitude, and the sphere experiences zero net force in both (a) and (b).

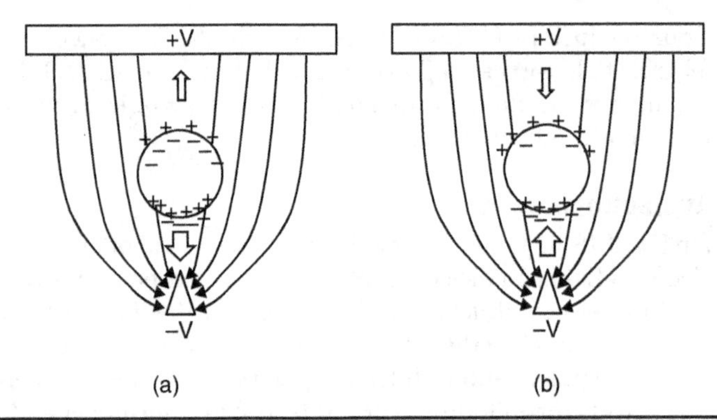

FIGURE 8.17 A dielectric solid sphere immersed in a dielectric liquid under a nonuniform electrical field: (a) when the solid is *more* polarizable than the liquid and (b) when the solid is *less* polarizable than the liquid. The electrostatic forces on the solid sphere are opposite in sign and different in magnitude, and the sphere experiences a downward and upward net force in (a) and (b), respectively.

where f_{CM} is a complex number called the Clausius–Mossotti factor, and is equal to

$$f_{\mathrm{CM}} = \frac{\varepsilon_p^* - \varepsilon_m^*}{\varepsilon_p^* + 2\varepsilon_m^*}, \tag{8.10}$$

where $\varepsilon_p^* = \varepsilon_p - j\frac{\sigma_p}{\omega}$ and $\varepsilon_m^* = \varepsilon_m - j\frac{\sigma_m}{\omega}$ are the complex permittivities of the sphere and the liquid, respectively, which depend on the frequency ω of the electrical field. The $\mathrm{Re}[f_{\mathrm{CM}}]$ can be any number between $-1/2$ and $+1$, depending on the permittivities, conductivities, and frequency. Thus, for a given set of material properties and setup of the electrodes, the DEP force direction can be changed by changing the frequency of the applied electrical field. Based on this particular characteristics, chaotic micromixing can be obtained with an array of microelectrodes integrated in a microchannel through actuating the microelectrodes with alternating frequencies (to generate alternating force directions) [13].

8.4.2 Acoustic Wave Micromixer

Acoustic waves of 10–900 MHz (generated by a piezoelectric film or substrate with its top and bottom electrodes patterned into Fresnel zone plate) interfere with each other constructively and destructively, as they propagate, and can produce strong fluid motion and/or mixing, as illustrated in Fig. 8.18a. A microfluidic mixer built on a PZT substrate can generate powerful fluidic mixing in a chamber above the mixer in Fig. 8.18b or can noninvasively be attached to a different substrate containing microfluidic system. Microsonic Systems has commercialized the mixing technology based on Fig. 8.18b to generate rotating lateral vortex in a microwell that is acoustically coupled to the PZT transducer through liquid medium [16].

Self-Focusing Acoustic Transducer

A self-focusing acoustic transducer (SFAT) built on a piezoelectric substrate (Fig. 8.19) effectively produces ultrasound when a sinusoidal voltage signal with its frequency matched to the thickness-mode resonant frequency of the piezoelectric substrate is

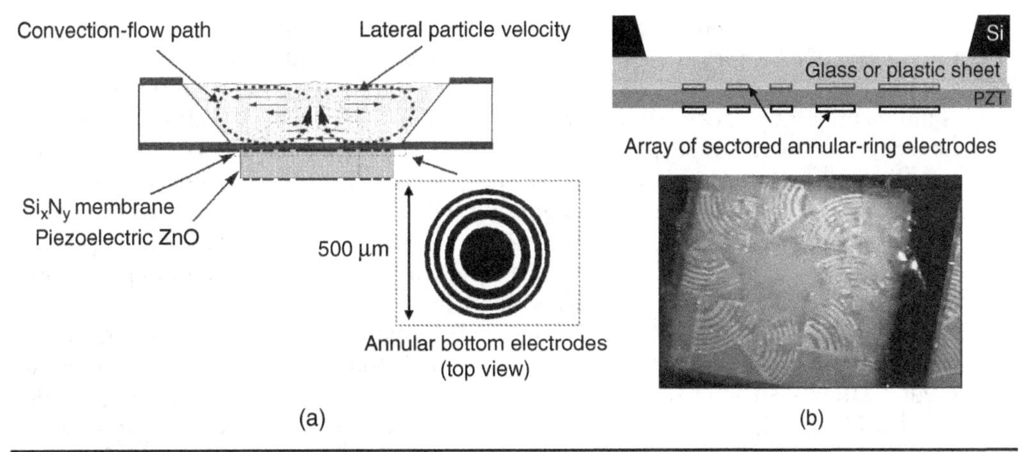

FIGURE 8.18 Acoustic micromixers: (a) cross-sectional view of a micromixer based on piezoelectric ZnO film with the top and bottom electrodes patterned into Fresnel lens pattern [17], (b) cross-sectional schematic and top-view photo of a micromixer made on a PZT substrate with an array of sectored electrodes [18].

applied onto the top and bottom electrodes sandwiching the piezoelectric substrate (e.g., PZT). There are basically two types of SFAT: one based on patterned electrodes and the other based on air cavities. In the first type, the electrodes are patterned into Fresnel half-wavelength band (FHWB) rings so that the waves are generated only on the electrode regions (Fig. 8.19a). In the second type, a Fresnel acoustic lens is formed through Parylene-sealed annular ring air cavities on top of a circularly patterned electrode (with conformal deposition of Parylene) (Fig. 8.19b). In either type, the annular rings are designed into FHWBs for a focal length F, whose boundary radius R_n is given by [19]

$$R_n = \sqrt{n\lambda \times \left(F + \frac{n\lambda}{4}\right)}, \tag{8.11}$$

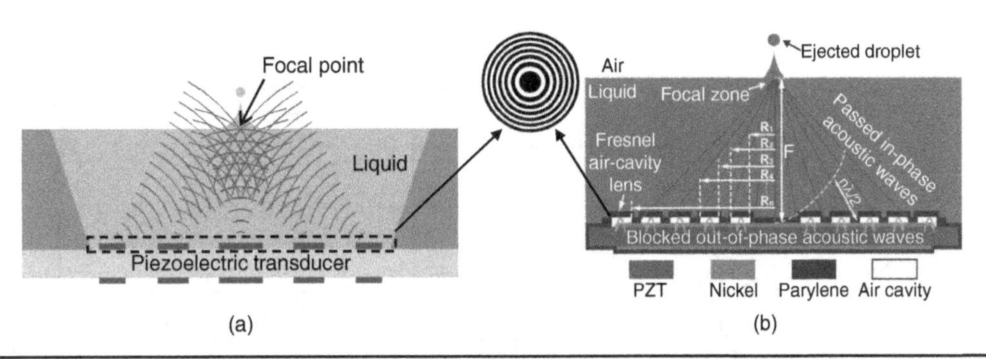

FIGURE 8.19 Cross-sectional views of self-focusing acoustic transducers (SFATs) based on (a) electrodes (sandwiching a piezoelectric substrate) that are patterned into Fresnel annular rings and (b) air cavities (above the top circular electrode) that are patterned into Fresnel annular rings, which are shown in between (a) and (b). The annular rings are designed such that the differences among the acoustic path lengths to the focal point (F) are integer multiples of the wavelength for the waves generated on the electrode areas (a) or for the "constructive waves" which are not reflected by the air cavities (b).

where λ is the wavelength in liquid. This way, the path length difference from two boundaries of a Fresnel ring band to the focal point equals half wavelength.

With SFAT shown in Fig. 8.19a, when RF power is applied between the electrodes (sandwiching the piezoelectric substrate) with its frequency corresponding to the thickness mode resonance of the piezoelectric substrate, strong acoustic waves are generated over the areas covered by the electrodes. The generated acoustic waves propagate in the liquid adjacent to the transducer, interfering with each other. Since the electrodes are patterned such that the difference among the acoustic path lengths to the focal point is a multiple integer of the wavelength, the waves arrive at the focal point in phase, and one can achieve wave-focusing effect without any acoustic lens.

Another version of SFAT uses Fresnel FHWBs made of air cavity rings (Fig. 8.19b). Because acoustic impedance mismatch between air (only 0.4 kRayl) and solid/liquid (over 1 MRayl), all acoustic waves leading to destructive interference (in rings where $R_n < R < R_{n+1}$, $n = 1, 3, 5, \ldots$) are blocked by air cavities, while constructively interfering acoustic waves (in rings where $R_n < R < R_{n+1}$, $n = 2, 4, \ldots$) propagate through Parylene layer, producing focused ultrasound at the focal point.

Micromixing Based on Acoustic Streaming Effects Produced by Sector SFATs

With a set of complete annular rings (Fig. 8.19), the acoustic field generated by SFAT is angularly symmetrical in the plane of the rings. However, when the rings are segmented into a pie shape, the symmetry is broken, and there is an in-plane body force in liquid which generates liquid flow. The body force is due to acoustic streaming effect caused by nonlinearity in the liquid.

For Fresnel rings sectored into a 90° pie on a PZT substrate (Fig. 8.20a), the electrode patterns for the top and bottom electrodes are designed to produce large acoustic pressure gradient (especially in the in-plane direction) near the focal points. With the coordinate system illustrated in Fig. 8.20b around the sectored SFAT, acoustic potential φ at any point in liquid due to the SFAT is given by

$$\varphi(r'', \psi'', z) = -\frac{u_0}{2\pi} \int_{r'=0}^{a} \int_{\psi'=0}^{\pi/2} \frac{e^{-jkR-\alpha R}}{R} r' d\psi' dr' \tag{8.12}$$

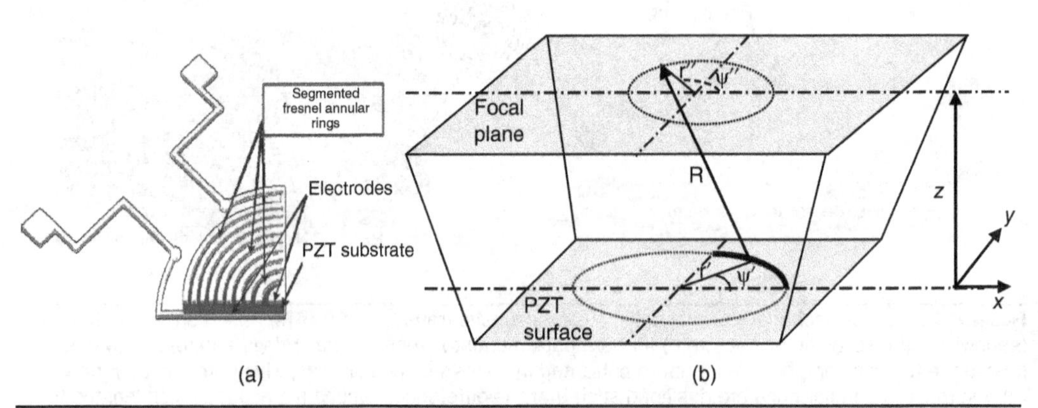

Figure 8.20 (a) Perspective view of a self-focusing acoustic transducer with Fresnel rings sectored into a 90° pie. (b) The notations and coordinates used for developing the equations.

where u_o and α are the particle displacement right above the transducer surface (i.e., at $z = 0^+$) and acoustic attenuation constant, respectively, with $R = \sqrt{z^2 + r'^2 + r''^2 - 2r'r''\cos(\psi'' - \psi')}$. As Eq. (4.6) ($Z_o \equiv -\frac{T}{v}$) indicates, stress T associated with acoustic wave is related to particle velocity v through acoustic impedance Z_o. And particle velocity v can be obtained from particle displacement u, since $v = \frac{du}{dt}$, which is equal to $-j\omega u$ for a harmonic particle displacement $u = u_A e^{-j\omega t}$. And particle displacements in the radial (r''), vertical (z), and circumferential (ψ'') directions at any point in the liquid over the 90° sectored SFAT are

$$u_{r''} = \frac{\partial}{\partial r''}\varphi(r'', \psi'', z)\,|_{\psi''=\text{const}, z=\text{const}}, \tag{8.13}$$

$$u_z = \frac{\partial}{\partial z}\varphi(r'', \psi'', z)\,|_{\psi''=\text{const}, r''=\text{const}}, \tag{8.14}$$

$$u_{\psi''} = \frac{\partial}{\partial \psi''}\phi(r'', \psi'', z)\,|_{z=\text{const}, r''=\text{const}}, \tag{8.15}$$

where acoustic potential φ can be obtained with Eq. (8.12).

With transformation of the cylindrical coordinates to the rectangular ones, we can get particle velocities (v_x, v_y, and v_z) from particle displacements (u_x, u_y, and u_z) as, $v_x = \frac{du_x}{dt}$, $v_y = \frac{du_y}{dt}$, and $v_z = \frac{du_z}{dt}$, leading to $\vec{v}_1 = v_x\hat{x} + v_y\hat{y} + v_z\hat{z}$ which can be used in calculating the steady body force due to acoustic streaming effect, as shown below.

For homogeneous isotropic fluid, the net dynamic force acting on a differential volume dv in the fluid is mostly due to surface stresses and can be obtained with

$$\vec{F} = \rho\left[\frac{\partial \vec{v}}{\partial t} + (\vec{v}\cdot\nabla)\vec{v}\right], \tag{8.16}$$

where $\rho(x,y,z,t)$ and $\vec{v}(x,y,z,t)$ are the mass density and particle velocity, respectively [20]. Using the continuity equation $\frac{\partial \rho}{\partial t} + \nabla\cdot\rho\vec{v} = 0$, we can write Eq. (8.16) as

$$\vec{F} = \frac{\partial(\rho\vec{v})}{\partial t} + \vec{F}' \tag{8.17}$$

where $\vec{F}' = \rho(\vec{v}\cdot\nabla)\vec{v} + \vec{v}\nabla\cdot\rho\vec{v}$.

To calculate steady body force $\langle\vec{F}\rangle$, we average Eq. (8.17) with respect to time. Since the temporal average of $\partial(\rho\vec{v})/\partial t$ is zero, we obtain

$$\langle\vec{F}\rangle = \langle\vec{F}'\rangle = \langle\rho(\vec{v}\cdot\nabla)\vec{v} + \vec{v}\nabla\cdot\rho\vec{v}\rangle \tag{8.18}$$

In addition, the Navier–Stokes equation of motions for Newtonian viscous fluid offers another equation for the net dynamic force

$$\vec{F} = -\nabla P + \mu\nabla^2\vec{v} + (\mu_B + 4\mu/3)\nabla\nabla\cdot\vec{v}, \tag{8.19}$$

where $P(x,y,z,t)$, μ_B, and μ are the pressure, dilatational viscosity coefficient, and shear viscosity coefficient, respectively. Time averaging of Eq. (8.19) yields another equation for steady body force as follows.

$$\langle \vec{F} \rangle = -\nabla \langle P \rangle + \mu \nabla^2 \langle \vec{v} \rangle \tag{8.20}$$

In obtaining Eq. (8.20), we take $\nabla \cdot \langle \vec{v} \rangle = 0$ which would be appropriate for incompressible fluid. The two equations above [Eqs. (8.18) and (8.20)] are the key equations for steady body force due to acoustic wave propagating in fluid where wave attenuation causes acoustic streaming effect.

To solve Eqs. (8.18) and (8.20), we use the method of successive approximations [20] with the following approximations for the pressure, mass density, velocity, and force:

$$P - P_0 = p_1 + p_2 + \cdots; \ \rho - \rho_0 = \rho_1 + \rho_2 + \cdots; \ \vec{v} = \vec{v}_1 + \vec{v}_2 + \cdots; \ \vec{F} = \vec{F}_1 + \vec{F}_2 + \cdots.$$

where the subscripts 0, 1, and 2 denote the steady-state values, the first-order, and second-order approximations to the steady-state values, respectively. The first-order approximations vary sinusoidally in time with frequency ω (corresponding to the frequency of the sinusoidal wave applied to SFAT), while the second-order approximations have time-independent terms as well as sinusoidal terms with frequency 2ω. From Eq. (8.20) we can see that the steady-state values and the first-order approximations produce zero body force, since the temporal averages $\langle P \rangle$ and $\langle \vec{v} \rangle$ are zero for those. However, the second-order approximations yield a steady body force equal to $\langle \vec{F}_2 \rangle = -\nabla p_2 + \mu \nabla^2 \vec{v}_2$, from Eq. (8.20). It can also be obtained from Eq. (8.18) as follows, noting that a term like $p_l(\vec{v}_m \cdot \nabla)\vec{v}_n$ is of the order of $(l + m + n)$, while a term like $\nabla \langle P_n \rangle$ is of the order of n.

$$\langle \vec{F}_2 \rangle = \langle \rho_0 (\vec{v}_1 \cdot \nabla) \vec{v}_1 + \vec{v}_1 \nabla \cdot \rho_0 \vec{v}_1 \rangle \tag{8.21}$$

Using $\vec{v}_1 = v_x \hat{x} + v_y \hat{y} + v_z \hat{z}$ obtainable from Eq. (8.12) as explained above, we can simulate the body force $\langle \vec{F}_2 \rangle$ in Matlab using Eq. (8.21), and show the body force at the focal plane of the 90° sectored SFAT in Fig. 8.21.

For a linear array of the 90° sectored SFATs shown in Fig. 8.22a, we can integrate the body force in the linear channel, and confirm that the body force will drive the fluid along the direction indicated in Fig. 8.22a. The steady body force, not the vibration from the PZT sheet, is indeed what is driving the in-plane fluid flow, as shown in [21] where the experimental observation of the in-plane fluid flow agrees well with the simulated flow trace. To experimentally verify the effectiveness of the sectored Fresnel rings, one can fabricate and test a similarly constructed transducer of which the electrodes are segmented into a pie but not into Fresnel rings as shown in Fig. 8.22b. In the case of Fig. 8.22b no in-plane liquid flow will be observed.

Acoustic Radiation Force

In addition to the body force due to acoustic streaming effect as studied in the previous section, a particle experiences a radiation force exerted by acoustic wave. For example, the acoustic radiation force (ARF) acting on a spherical particle (with radius R_s, mass

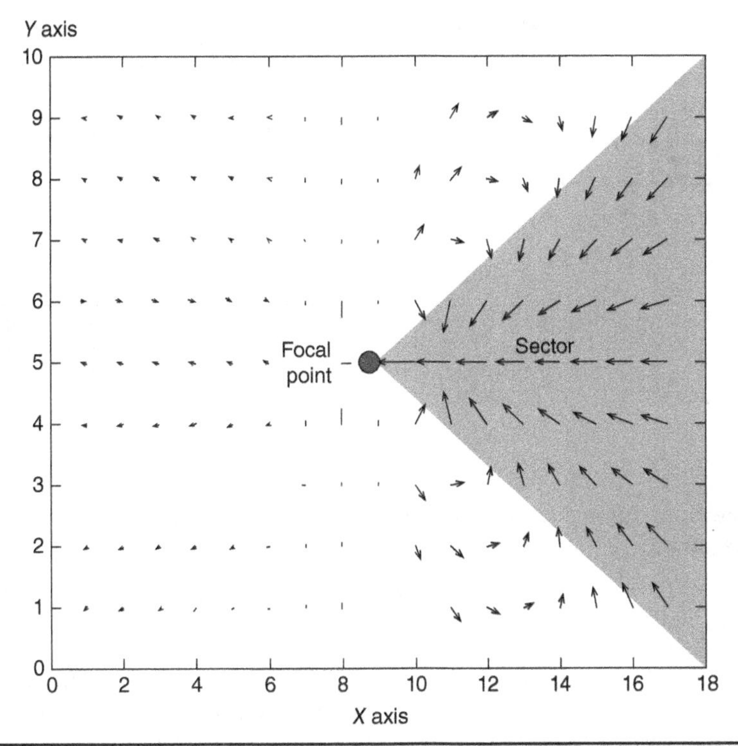

Figure 8.21 Top view of the simulated body force at the focal plane for the transducer shown in Fig. 8.20 (axis unit: 10 μm) [21].

density d_s, and sound velocity c_s) by a spherical acoustic wave in an ideal fluid (with mass density d_m and sound velocity c_m) is given by the following equation [22]

$$F_{\mathrm{ARF}} = \frac{4}{3}\pi R_s^3 \left(\frac{3d_s - 3d_m}{2d_s + d_m} \frac{\partial K}{\partial r} - \frac{d_s c_s^2 - d_m c_m^2}{d_s c_s^2} \frac{\partial P}{\partial r} \right) \tag{8.22}$$

where K and P are time-averaged kinetic and potential energy densities, respectively. For a spherical sound source (with radius R_p and radial velocity amplitude v_o

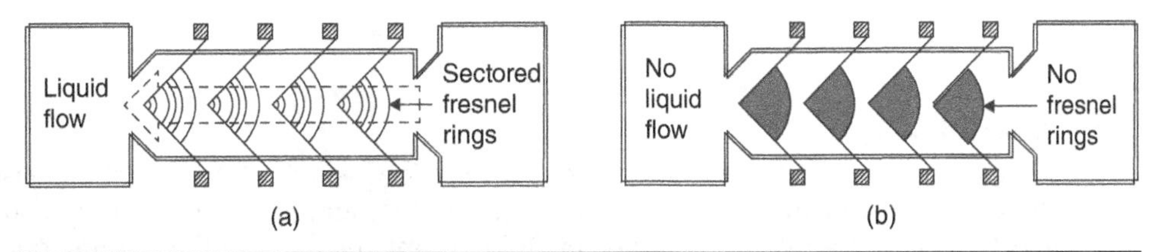

Figure 8.22 (a) Top-view schematic of the fluid driver composed of a linear array of acoustic transducers with sectored Fresnel annular rings. (b) Top view of the electrode pattern for the PZT transducer that shows no liquid flow due to little focusing effect by the pie electrode (that is not patterned into Fresnel rings) [21].

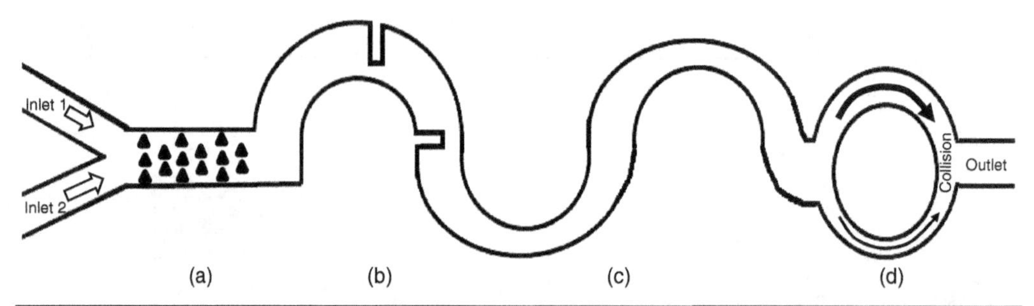

FIGURE 8.23 Four types of passive micromixing with (a) triangular barriers embedded in a channel, (b) radial barriers created by notching a channel wall, (c) convergence–divergence through a meandering channel with varying cross-section, and (d) unbalanced collision with differing flow rates through the top and bottom channel.

at the transducer surface), the radiation force at r from the source can be approximated with

$$F_{\text{ARF}} \approx \frac{4}{3}\pi R_s^3 d_m v_o^2 \frac{3d_s - 3d_m}{2d_s + d_m} R_p^4 r^{-5} \tag{8.23}$$

In general, one can first calculate acoustic radiation potential from simulated acoustic pressure and velocity fields along with the medium and particle properties, and then calculate ARF by taking the negative spatial derivatives of the radiation potential [23]. For a large particle whose diameter is comparable to or larger than the wavelength, ARF needs to be calculated through integrating the second-order momentum fluxes generated by the first-order pressure and velocity fields over the microsphere surface (e.g., using Equations 3–5 in [24]).

8.4.3 Passive Mixing in Microchannels

Passive micromixing is typically obtained through structural design of microchannels which produces fast molecular diffusion and chaotic advection (Fig. 8.23). For example, flow barriers of various shapes (such as triangular barriers) can be embedded in a straight or curved microchannel to produce disturbance in fluid flow for micromixing (Fig. 8.23a). Three other types of passive micromixing are illustrated in Fig. 8.23b–d and are reviewed in [25].

8.5 Lab-on-Chip

What we have learned so far can be implemented on a single chip of plastics, glass, or silicon for movement, metering, mixing, separation, amplification, heating, cooling, detection, and any other operations of fluids in microscale. In other words, one can build lab-on-a-chip or lab-on-chip (LOC) by integrating various discrete microcomponents (e.g., microchannels, microvalves, micromixers, etc.) on a single chip. Silicon is optically opaque, and more expensive than the alternatives such as plastics and glass.

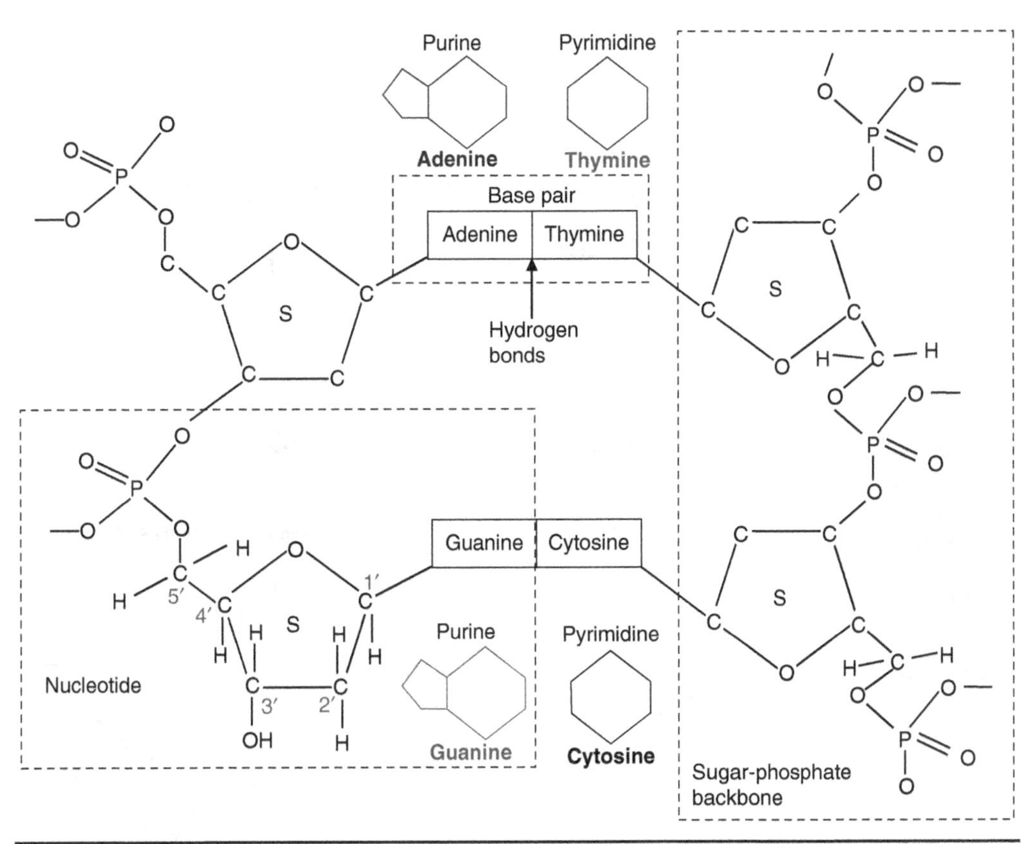

FIGURE 8.24 Basic DNA structure composed of sugar-phosphate backbone and base-pair ladders.

Thus, most of LOCs are made on plastics (and occasionally on glass which is more expensive than most plastics). Some recent examples of LOCs are microfluidic systems for high throughput cell screening [26] and exosome isolation/detection/analysis [27] based on droplets in microfluidic channels; single-cell, 2D and 3D microfluidic cell culture systems for studying cell-to-cell interactions or building tissue and organ [28, 29]; and microfluidic assays based on extracellular vesicles [30].

8.5.1 Gene Sequencing

Deoxyribonucleic Acid (DNA) and Ribonucleic Acid (RNA)

As illustrated in Fig. 8.24, nucleotide, a building block of deoxyribonucleic acid (DNA), is composed of a five-carbon sugar, one or more phosphate, and a nitrogen-containing base. In DNA, there are four different nucleotides: adenine (A), thymine (T), cytosine (C), and guanine (G). Adenine (A) always pairs with T, while G pairs with C. Each strand of double-stranded DNA (Fig. 8.25) is composed of a sugar-phosphate backbone and base-pair ladders. Sequence of the nucleotides uniquely codes living organism. In cell nucleus, DNA is transcribed to messenger ribonucleic acid (mRNA) by enzyme called

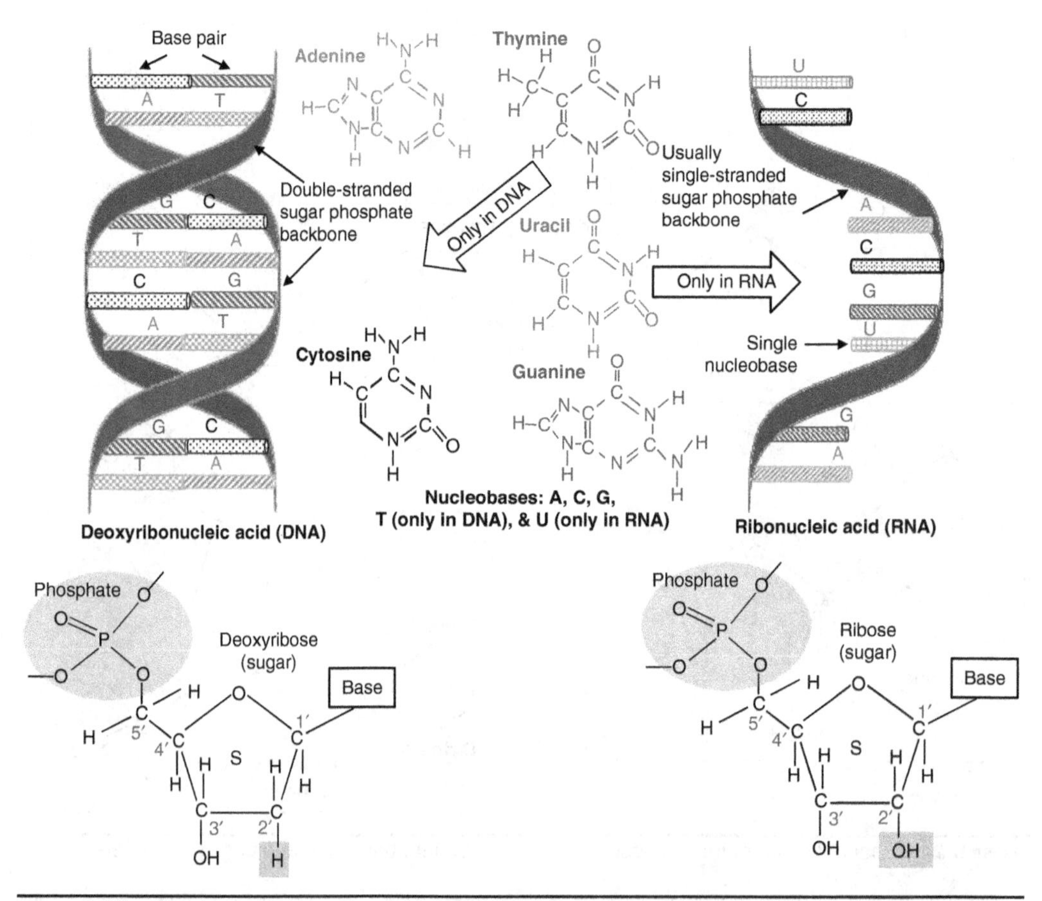

FIGURE 8.25 Structures of deoxyribonucleic acid (DNA) and ribonucleic acid (RNA). The sugars in DNA and RNA are *deoxy*ribose and ribose, respectively, which differ by one oxygen, as yellow-highlighted at the bottom. The lack of the hydroxyl group on the 2′ carbon makes DNA chemically more stable (or less reactive) than RNA.

"RNA polymerase II." There are also four different nucleotides in RNA, as illustrated in Fig. 8.25, three of which are the same as the ones for DNA, i.e., A, C, and G. However, in RNA there is a nucleotide called uracil (U) instead of thymine (T).

DNA → RNA → Protein

Proteins are synthesized through two sequential steps: transcription that occurs in the cell nucleus and translation that occurs in the cytoplasm. Through the transcription, DNA is converted to messenger RNA (mRNA) in the cell nucleus, and the mRNA is carried out to the cytoplasm where the translation occurs. Three-letter words in the mRNA carry the information on which amino acids, building blocks of proteins, will be produced through the translation. And the sequential order of the three-letter words in the mRNA determines the order of the amino acids to be joined together. Since RNA is composed of four nucleotides (A, C, G, and U), there are 64 ($=4^3$) three-letter RNA words (called codons). Sixty-one of the codons specify amino acids, and since there are only

20 different amino acids, some codons produce the same amino acid. For example, GUU, GUC, GUA, and GUG all produce amino acid valine, while GCU, GCC, GCA, and GCG all produce alanine.

Gene Sequencing

As the sequence of the nucleotides in DNA or RNA determines how proteins are synthesized, it is critically important to know the sequence. Typically, the sequence is analyzed through gene sequencing based on the fact that one single strand of DNA/RNA binds *selectively* with another strand of complementary nucleotide sequence through a process called hybridization.

Analysis of RNAs from cells offers dynamic picture of which genes are active in the cells, while DNA analysis provides static picture of what the cells may become. Analysis of the active genes in different cells through RNA analysis helps us to understand how gene expression changes in response to different stimuli.

RNA Extraction from Cell

For gene sequencing, RNAs (or DNAs) need to be extracted from cells through the procedure illustrated in Fig. 8.26. Cell lysis buffer (usually with RNase inhibitors that keep RNA intact) is added to a sample containing cells in a microcentrifuge tube, which is vibrated for mixing through pulse-vortexing and incubated at room temperature for cell lysis. The sample (containing lysed cells, proteins, and other contaminants) is then loaded in a microcentrifuge tube with a spin column composed of silica (Si-OH) matrix on which only RNA molecules bind under particular salt and pH conditions. The tube then goes through centrifugation to keep the RNA on the silica column, while pushing protein and other contaminants to the bottom of the tube. The silica column is then taken out of the microcentrifuge tube and placed into a clean collection tube, into which a wash buffer is added, followed by centrifuging of the tube to wash away any remaining impurities on the silica column so that only RNA is bound to the silica column. Afterward, the silica column is placed in a clean microcentrifuge tube, into which an elution buffer is added followed by centrifugation to make the elution buffer detach the RNA from the silica column, resulting in purified RNA.

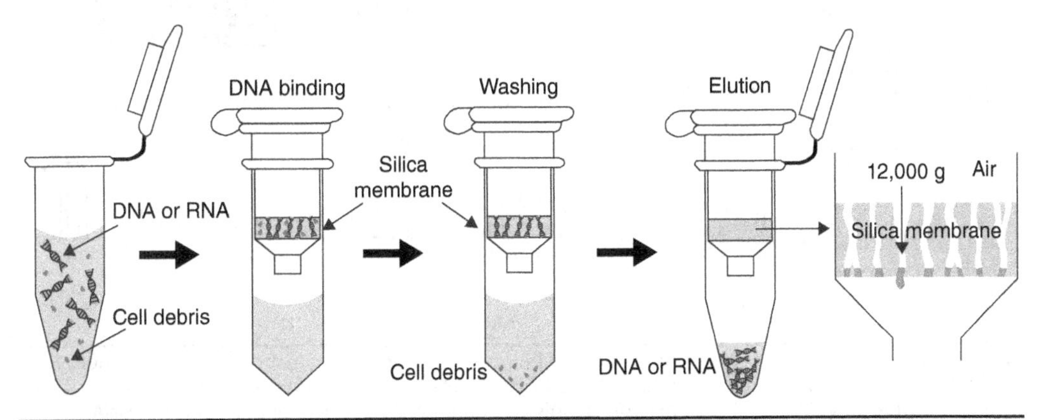

Figure 8.26 Purification process of DNA or RNA from lysed cells through DNA/RNA binding on silica membrane, washing of cell debris, and then ellution of DNA/RNA. (Adapted from [31].) Conventional process uses centrifugation force to drive the cell debris and DNA/RNA out of the silica membrane at different steps.

RNA → cDNA, Initial Denaturation, and Polymerase Chain Reaction

Sequences of RNA molecules need to be converted to their complementary DNA (cDNA) sequences through reverse transcriptase (RT) before polymerase chain reaction (PCR) that increases the number of DNA strands greatly for gene sequencing. The process of "RNA → cDNA → PCR" is called RT-PCR.

In RT-PCR process, the RNAs extracted and purified (and small in number) from cells are added to a mixture of RT enzyme, nucleotides (dNTPs), forward primer, reverse primer, TaqMan probe, DNA polymerase, and buffer in a microcentrifuge tube, which is vibrated by pulse-vortexing for mixing. At an appropriate temperature (e.g., 55°C) the RT will synthesize cDNA that is complementary to the RNA by synthesizing A for U, T for A, G for C, and C for G, resulting in RNA–cDNA double strands. Then the RNA–cDNA is denatured by raising the temperature to 95°C, while activating DNA polymerase and simultaneously inactivating RT. Single strands of the cDNA are multiplied in number through PCR.

As illustrated in Fig. 8.27, PCR is consisted of denaturation (at 95°C) step to denature double-stranded DNA, annealing (at 55°C) to anneal forward primer to its complementary part of the single stranded DNA, and extension step at 72°C to allow DNA polymerase to synthesize complementary DNA strand by adding nucleotides in the 5′ to 3′ direction. Afterward, the double-stranded DNA is denatured at 95°C, resulting in two single-stranded DNA molecules. After each thermal cycle, the number of the single DNA strands doubles; for example, after 40 cycles, which may take about 2 hours, the number of copies of the target DNA can reach about 550 billion.

In case of real-time PCR (not to be confused with reverse transcriptase PCR [RT-PCR]) or quantitative PCR (qPCR), after the initial denaturation at 95°C, the temperature is lowered to 55°C to allow annealing of the forward primer, reverse

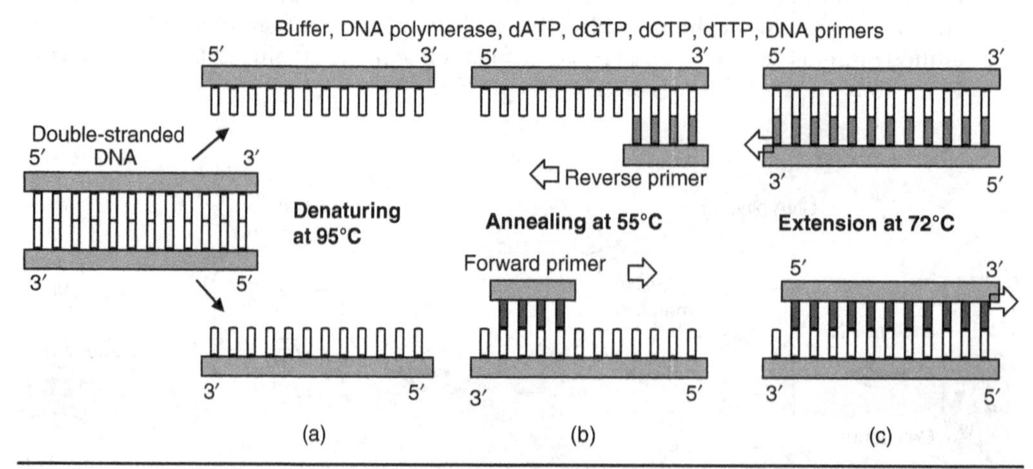

FIGURE 8.27 First thermal cycle of polymerase chain reaction (PCR) in mix buffer, DNA polymerase, dNTP, and DNA primers with one double-stranded DNA, producing two double-stranded DNA: (a) temperature raised to 95°C to split the double-stranded DNA into two single-stranded DNA, (b) temperature reduced to 55°C to hybridize DNA primers to the single-stranded DNA, and (c) temperature raised to 72°C to make DNA polymerase bind to the front end of the complex formed by primer and single-strained DNA to add matching nucleotides to the single-stranded DNA by incorporating nearby nucleotides.

primer, and TaqMan probe to each DNA strand. TaqMan probe is an oligonucleotide probe with a flurophore (or reporter of fluorescence) and a quencher of fluorescence attached to its 5′ and 3′ ends, respectively. The flurophore emits fluorescence when excited by a light with a proper wavelength. The presence of the quencher in close proximity of the reporter prevents detection of the fluorescence. In the extension step at 72°C, as DNA polymerase synthesizes new strands and reaches the TaqMan probe, it cleaves the probe separating the reporter from the quencher. The number of the released (or separated) reporters also increases as the thermal cycles increase resulting in fluorescence intensity proportional to the amount of DNA strands that have been synthesized.

Light from tungsten-halogen lamp goes through an excitation filter and reflects off from a mirror before going through a condensing lens which focuses the light onto a desired spot. The light reflected off from the spot goes through a mirror, followed by an emission filter, and collected or recorded by CCD camera which converts the image into digital data, allowing real-time PCR or qPCR.

Note: Quantitative PCR (qPCR) is also called real-time PCR, which is different form reverse transcriptase PCR (RT-PCR) used for RNA analysis. In qPCR the synthesis process of PCR is measured in real time during the synthesis process with fluorescent probes that are short sequences of nucleotides with fluorescence reporters and quenchers attached at the ends of the probes. Each probe has one reporter and one quencher at its two ends (in close proximity), and does not produce fluorescence light until the reporter is separated from the quencher by substantial distance. At the anneal step of PCR (i.e., at 55°C), the probes and the primers anneal to single-stranded DNA at different locations, and are degraded by DNA polymerase during the extension step (i.e., 72°C) at which the fluorescence reporters are released and separated from the fluorescence quenchers, and emit fluorescence (the intensity of which is directly proportional to the number of the released reporters).

8.5.2 MEMS-Based PCR and Single-Cell RT-PCR Systems

Scaling down PCR system with MEMS technology offers rapid thermal cycle due to the reduced thermal mass and high thermal resistance. With microfluidic chambers the thermal cycle time for PCR can be reduced by orders of magnitude, since only a very small volume of sample is thermally cycled, requiring short time to heat and cool due to low thermal mass of the sample. One of the earliest MEMS-based PCR systems was made with silicon bulk micromachining (Fig. 8.28), which has been commercialized [32, 33]. More recently, a prototype low-cost, handheld, battery-powered PCR system composed of a polyimide microchamber, heating module, and integrated complementary metal–oxide semiconductor (CMOS) optical detector was reported [34].

With a microfluidic channel passing through three different temperature zones (95°C for denaturization, 77°C for extension, and 60°C for annealing) on a chip, PCR can be performed continuously, as a sample and PCR reagents pass through the channel [35]. The sample and PCR reagents are pneumatically pumped through the long and meandering channel from inlets to an outlet, where multiplied DNA strands are collected.

Massively parallel gene sequencing on single cells (say, 100,000 cells) can offer information not measurable through a sequencing on the ensemble of the cells, but is

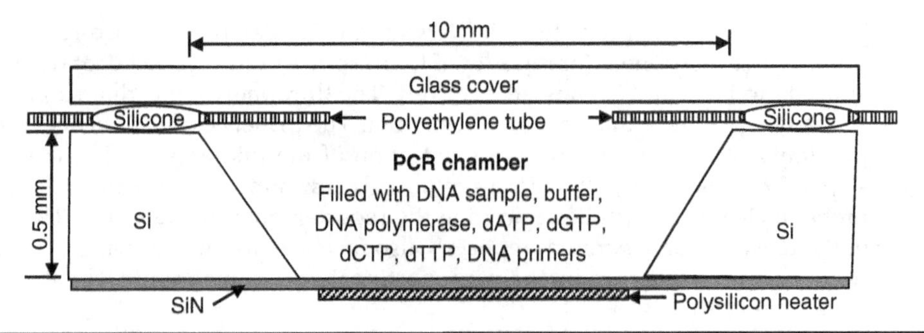

FIGURE 8.28 Conceptual cross-sectional view of MEMS PCR chip that appears to be the foundational technology of Cepheid's Xpert Xpress SARS-CoV-2 test cartridge. (Adapted from [32].)

labor intensive with conventional approaches. Thus, droplets formed through microchannels, similar to Fig. 8.29, were used for ~50,000 single-cell RT-PCR reactions in a single experiment [36]. In the setup, microfluidic channels were used (1) to produce droplets of a single cell and cell lysis reagents encapsulated with oil and (2) to add PCR reagents to droplets of lysed cells, while heating and optical detection were done outside the microfluidic chip.

In case of single-cell RT-PCR, the initial RNAs from a lysed cell are contained in a few picoliter, and would be diluted 10^6 times, if the RNAs are reverse transcribed into cDNA in a microtube of 10 μL (non-MEMS). Thus, a microfluidic chamber of 10 nL volume was used to perform the RT on 10 pg of RNA, and the result was compared with

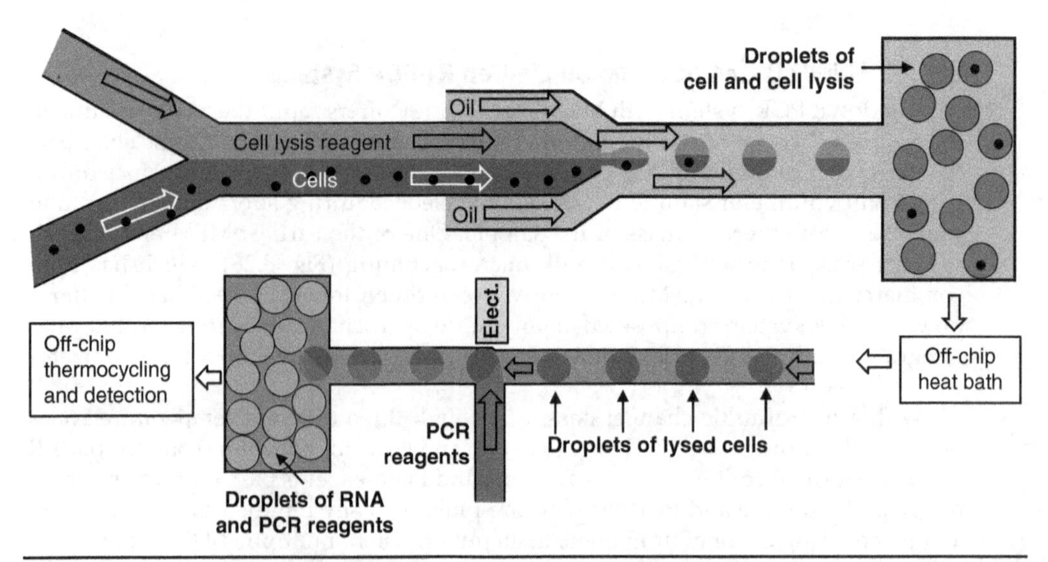

FIGURE 8.29 The droplet-based microfluidics for single-cell RT-PCR with off-chip heating for cell lysis, off-chip thermocycling, and optical detection for RT-PCR. (Adapted from [36].)

the one obtained with a 10 μL microtube, as the other concentrations of reagents were kept the same [37]. After RT-PCR, the number of genes that could be identified was 17 times more when the RT was performed in the 10 nl microfluidic chamber than when it was done in the 10 μL microtube.

References

1. L. Shang, Y. Cheng, and Y. Zhao, "Emerging droplet microfluidics," Chemical Reviews, 117, 2017, pp. 7964–8040.
2. T. Nisisako, T. Torii, T. Takahashi, and Y. Takizawa, "Synthesis of monodisperse bicolored Janus particles with electrical anisotropy using a microfluidic co-flow system," Advanced Materials, 18, 2006, 1152–1156.
3. F. Mugele and J.C. Baret, "Electrowetting: from basics to applications," Journal of Physics: Condensed Matter, 17, 2005, R705–R774.
4. S.K. Cho, H. Moon, and C.J. Kim, "Creating, transporting, cutting, and merging liquid droplets by electrowetting-based actuation for digital microfluidic circuits," Journal of Microelectromechanical Systems, 12(1), February 2003, pp. 70–80.
5. M. Vallet, M. Vallade, and B. Berge, "Limiting phenomena for the spreading of water on polymer films by electrowetting," European Physical Journal B, 11, 1999, pp. 583–591.
6. K.H. Kang, "How electrostatic fields change contact angle in electrowetting," Langmuir, 18 (26), 2002, pp. 10318–10322; doi: 10.1021/la0263615.
7. A.G. Papathanasiou and A.G. Boudouvisa, "Manifestation of the connection between dielectric breakdown strength and contact angle saturation in electrowetting," Applied Physics Letters, 86, 2005, p. 164102; doi: 10.1063/1.1905809.
8. R.N. Wenzel, "Resistance of solid surfaces to wetting by water," Industrial & Engineering Chemistry, 28 (8), 1936, pp. 988–994.
9. E.T. Carlen and C.H. Mastrangelo, "Electrothermally activated paraffin microactuators," Journal of Microelectromechanical Systems, 11(3), June 2002, pp. 165–174.
10. G.H. Feng and E.S. Kim, "Micropump based on PZT unimorph and one-way Parylene valves," Journal of Micromechanics and Microengineering, 14(4), 2004, pp. 429–435.
11. A. Olanrewaju, M. Beaugrand, M. Yafia, and D. Juncker, "Capillary microfluidics in microchannels: from microfluidic networks to capillaric circuits," Lab on a Chip, 18, August 2018, pp. 2315–2478.
12. H. Bruus, "Theoretical Microfluidics," Lecture Notes, Department of Micro and Nanotechnology, Technical University of Denmark, 3rd ed., 2006, p. 31.
13. J. Deval, P. Tabeling, and C.M. Ho, "A dielectrophoretic chaotic mixer," Proceedings of MEMS'02, 15th IEEE International Workshop Micro Electro Mechanical Systems, Las Vegas, NV, January 20–24, 2002, pp. 36–39.
14. K.S. Ryu, K. Shaikh, E. Goluch, Z. Fan, and C. Liu, "Micro magnetic stir-bar mixer integrated with Parylene microfluidic channels," Lab Chip, 4, 2004, pp. 608–613.
15. K. Khoshmanesh, S. Nahavandi, S. Baratchi, A. Mitchell, and K. Kalantar-zadeh, "Dielectrophoretic platforms for bio-microfluidic systems," Biosensors and Bioelectronics, 26, 2011, pp. 1800–1814.

16. http://www.microsonics.com/technology.html.
17. X. Zhu and E.S. Kim, "Microfluidic motion generation with acoustic waves," Sensors and Actuators A: Physical, 66(1-3), April 1998, pp. 355–360.
18. V. Vivek, Y. Zeng, and E.S. Kim, "Novel acoustic-wave micromixer," IEEE International Micro Electro Mechanical Systems Conference, Miyazaki, Japan, January 23–27, 2000, pp. 668–673.
19. J. C. Wiltse, "The Fresnel zone-plate lens," Proceedings of SPIE 0544, Millimeter Wave Technology III, 1985.
20. W.L. Nyborg, "Acoustic streaming," a chapter in "Nonlinear Acoustics," edited by M.F. Hamilton and D.T. Blackstock, Academic Press, 1988, pp. 207–229.
21. H. Yu, J.W. Kwon, and E.S. Kim, "Microfluidic mixer and transporter based on PZT self-focusing acoustic transducers," Journal of Microelectromechanical Systems, 15(4), Aug. 2006, pp. 1015–1024.
22. D.L. Miller, "Particle gathering and microstreaming near ultrasonically activated gas-filled micropores," Journal of the Acoustical Society of America, 84(4), 1988, pp. 1378–1387.
23. H. Bruus, "Acoustofluidics 7: The acoustic radiation force on small particles," Lab Chip, 12(6), 2012, pp. 1014–1021.
24. P. Glynne-Jones, P. Mishra, R. Boltryk, and M. Hill, "Efficient finite element modeling of radiation forces on elastic particles of arbitrary size and geometry," Journal of the Acoustical Society of America, 133(4), 2013, pp. 1885–1893.
25. G. Cai, L. Xue, H. Zhang, and J. Lin, "A review on micromixers," Micromachines, 8, 2017, 274.
26. M. Sesen, T. Alan, and A. Neild, "Droplet control technologies for microfluidic high throughput screening (μHTS)," Lab Chip, 17, 2017, pp. 2372–2394.
27. J.C. Contreras-Naranjo, H.J. Wu, and V.M. Ugaz, "Microfluidics for exosome isolation and analysis: enabling liquid biopsy for personalized medicine," Lab Chip, 17, 2017, pp. 3558–3577.
28. M. Rothbauer, H. Zirath, and P. Ertl, "Recent advances in microfluidic technologies for cell-to-cell interaction studies," Lab Chip, 18, 2018, pp. 249–270.
29. J. Kieninger, A. Weltin, H. Flamm, and G.A. Urban, "Microsensor systems for cell metabolism—from 2D culture to organ-on-chip," Lab Chip, 18, 2018, pp. 1274–1291.
30. Z. Zhao, J. Fan, Y.S. Hsu, C.J. Lyon, B. Ning, and T.Y. Hu., "Extracellular vesicles as cancer liquid biopsies: from discovery, validation, to clinical application," Lab Chip, 19, 2019, pp. 1114–1140.
31. H. Lee, W. Na, C. Park, K.H. Park, and S. Shin, "Centrifugation-free extraction of circulating nucleic acids using immiscible liquid under vacuum pressure," Scientific Reports, 8, 2018, p. 5467.
32. M.A. Northrup, M.T. Ching, R.M. White, and R.T. Watson, "DNA amplification in a microfabricated reaction chamber," Transducers 1993, Yokohama, Japan. pp. 924–926.
33. https://www.cepheid.com/coronavirus.
34. D.S. Lee, O.R. Choi, and Y.J. Seo, "A handheld and battery-powered realtime microfluidic PCR amplification device," IEEE Transducers Conference, Berlin, Germany, 2019, pp. 1063–1065.
35. Y. Zhang and P. Ozdemir, "Microfluidic DNA amplification—a review," Analytica Chimica Acta, 638, 2009, pp. 115–125.

36. D.J. Eastburn, A. Sciambi, and A.R. Abate, "Ultrahigh-throughput mammalian single-cell reverse-transcriptase polymerase chain reaction in microfluidic drops," Analytical Chemistry, 85, 2013, pp. 8016–8021.
37. N. Bontoux, L. Dauphinot, T. Vitalis, V. Studer, Y. Chen, J. Rossiera, and M.C. Potier, "Integrating whole transcriptome assays on a lab-on-a-chip for single cell gene profiling," Lab Chip, 8, 2008, pp. 443–450.

Questions

Question 8.1 Evaluate the three substrates on the leftmost column with respect to the issues listed on the topmost row.

Issues / Substrate	Substrate Cost? (high or low)	Fabrication Difficulty? (easy or difficult)	Lab-on-a-Chip Integration? (easy or difficult)	Need Treatment for Hydrophilic Surface? (yes or no)	Need Linker Layer? (yes or no)
Silicon					
Plastic					
Glass					

Question 8.2 For the Si check valve, the flow characteristics are shown below. On the same curve, roughly sketch a flow characteristic curve, if the valve is made of Parylene, assuming that Young's modulus for Si is 160 GPa, while that for Parylene is 3.2 GPa.

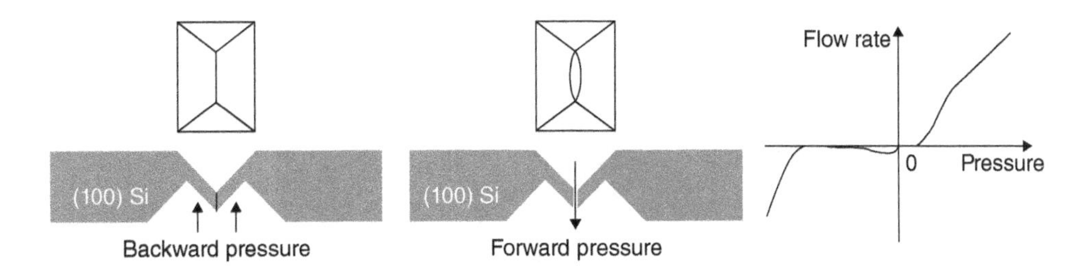

Question 8.3 For the valveless pump shown in Fig. 8.10, what are one key limitation and one key issue?

Question 8.4 Can the pump in Fig. 8.11 be used to pump liquid sample for PCR?

Question 8.5 Attachment of a PZT onto a silicon diaphragm (shown in Fig. 8.14) is not an easy process. So, can we consider using Parylene to replace the 80-μm-thick Si diaphragm?

Question 8.6 If an electrical field is applied between the top and bottom of a very long and narrow channel as shown below, will there be dielectrophoresis (DEP)?

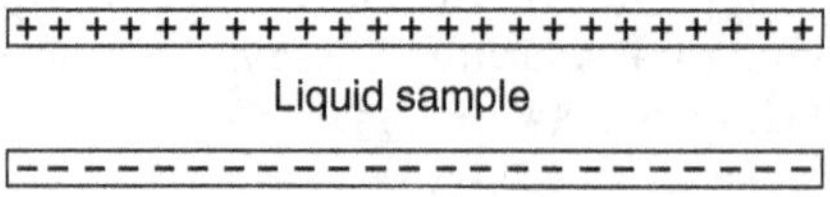

+ +

Liquid sample

\- -

CHAPTER 9
Power MEMS

There has been active research for energy harvesters for a compact, light, affordable, long-lasting power source to replace and/or supplement batteries. Among many renewable energy sources, vibrations are particularly attractive due to their ubiquity. There are three main transduction mechanisms for generating power from vibrating source: electrostatic, piezoelectric, and electromagnetic. Electrostatic energy harvester generates an electrical voltage or charge through varying capacitance. Although its fabrication process can be compatible with conventional integrated circuit (IC) technologies, the capacitor needs to be initially charged with an implanted charge or an external DC voltage source that limits practical application of electrostatic energy conversion. Piezoelectric materials can also be used for harvesting energy, as they are capable of generating a relatively high-voltage output due to mechanical strain caused by external motion. However, piezoelectric energy harvesters have high impedance, which mandates the load impedance to be high. Electromagnetic energy conversion, on the other hand, is advantageous due to its capability to drive a low impedance load and generate a high output current, and has commonly been used in macroscale power generation. However, microfabrication of electromagnetic energy harvester is quite difficult, as high-quality magnet and multiple-turn coil with low resistance do not come easily in planar microfabrication.

In this chapter, we will first study fundamental issues in vibration energy harvesting with electromagnetic and piezoelectric transduction. Then, we will study issues and potential approaches for power generation from vibration associated with human's walk without loading or limiting the person.

9.1 Electromagnetic Vibration Energy Harvesting

9.1.1 Mechanical Frequency Response of Vibration-Driven Energy Harvester

A vibration-driven power generator is consisted of a mass-spring-damper system with proof mass m suspended by a spring with spring constant k and moving within a frame (Fig. 9.1). A damper d incorporates any parasitic mechanical and aerodynamic damping losses. The absolute motions of the proof mass and the frame are $x(t)$ and $y(t)$, respectively, while the relative displacement between the proof mass and the frame is $z(t)$ $(= x(t) - y(t))$. Since the force on the proof mass m is equal to the force on the mass-spring-damper, the equation of motion for the system is given by

$$mz(t)'' + dz(t)' + kz(t) = -my(t)'' \tag{9.1}$$

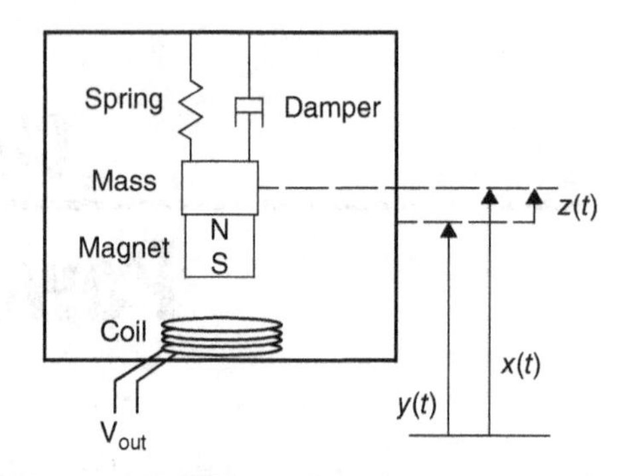

Figure 9.1 Model of a vibration-driven energy harvester with a proof mass (of a magnet) suspended by a spring and with a coil on the harvester frame which vibrates in response to an applied vibration [3].

For sinusoidal steady-state vibration $y(t) = Y_0 \cos\omega t = \mathrm{Re}\{Y_0 e^{j\omega t}\}$, $z(t) = Z_0 \cos(\omega t + \theta) = \mathrm{Re}\{Ze^{j\omega t}\}$, and we get $m\omega^2 Y_0 = -\omega^2 mZ + j\omega dZ + kZ$ from Eq. (9.1). Consequently,

$$Z = \frac{m\omega^2 Y_0}{-\omega^2 m + j\omega d + k} = \frac{(\omega/\omega_n)^2 Y_0}{1-(\omega/\omega_n)^2 + j2\zeta\omega/\omega_n} \tag{9.2}$$

where Y_0 is the frame's vibration amplitude or the input vibration amplitude; $\omega_n (= \sqrt{k/m})$ is the resonant frequency; ω is the vibration frequency; $\zeta (= \omega_n d/2k)$ is the damping ratio. Since the quality factor Q is equal to $k/\omega_n d$, the damping ratio ζ is equal to $1/2Q$. The magnitude of the relative motion Z_0 and the phase difference θ (between $y(t)$ and $z(t)$) are then

$$Z_0 = Y_0 \left(\frac{\omega}{\omega_n}\right)^2 / \sqrt{\left(1-\left(\frac{\omega}{\omega_n}\right)^2\right)^2 + \left(2\zeta\frac{\omega}{\omega_n}\right)^2} \tag{9.3}$$

$$\theta = -\tan^{-1}\frac{2\zeta\omega/\omega_n}{1-(\omega/\omega_n)^2} \tag{9.4}$$

Thus, the relationship between Z_0 and Y_0 is dependent on the frequency ω and damping ratio ζ, as shown in Fig. 9.2a. Near the resonant frequency ω_n, Z_0 increases as the damping ratio ζ decreases, while Z_0 is close to Y_0 at a frequency significantly higher than ω_n, as Eq. (9.3) simplifies to

$$Z_0 = \begin{cases} \dfrac{Y_0}{2\zeta} & (\omega = \omega_n) \\ Y_0 & (\omega \gg \omega_n) \end{cases} \tag{9.5}$$

9.1.2 Electromotive Force versus Frequency for Given Input Acceleration

In case of electromagnetic energy conversion with (1) a magnet being a portion of the proof mass and (2) a coil being a part of the frame as illustrated in Fig. 9.1, the relative

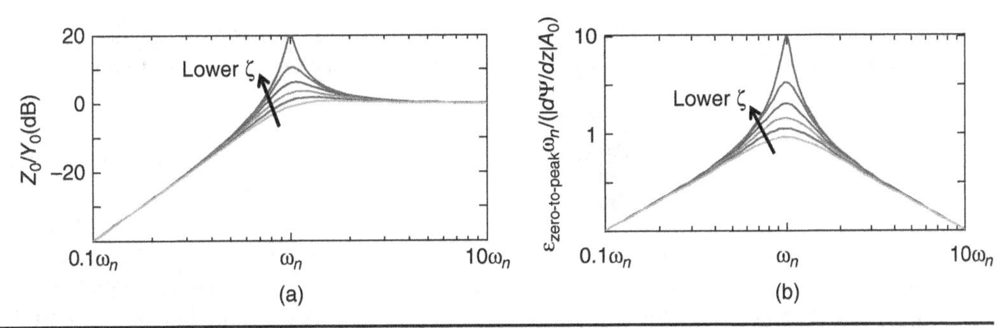

FIGURE 9.2 (a) Calculated relative displacement amplitude Z_0 versus vibration frequency ω as a function of damping ratio ζ for a given input vibration amplitude Y_0. Refer to Fig. 9.1 for the definition of z and y. (b) Calculated electromotive force (EMF) at a given input acceleration ($A_0 = \omega^2 Y_0$) versus vibration frequency ω as a function of damping ratio ζ.

motion of the proof mass, Z_0, (not Y_0, nor X_0) is what contributes to the power generation, and is the one that has to be maximized for a given input vibration amplitude Y_0. Interestingly, the relative motion of the proof mass, Z_0, is equal to the frame's vibration amplitude Y_0, when the vibration frequency is much higher than the resonant frequency, meaning that the absolute motion of the proof mass, X_0, is zero (i.e., only the frame moves in response to the vibration input at such a high frequency).

According to Faraday's law, the magnitude of electromotive force (EMF) ε is proportional to the time rate change of magnetic flux ψ through a coil, and is

$$\varepsilon = \left|\frac{d\psi}{dt}\right| = \left|\frac{d\psi}{dz}\frac{dz}{dt}\right| = \left|\frac{d\psi}{dz}\right|\omega Z_0 = \left|\frac{d\psi}{dz}\right| A_0 \frac{\omega/\omega_n^2}{\sqrt{\left[1-(\omega/\omega_n)^2\right]^2 + \left(2\zeta\frac{\omega}{\omega_n}\right)^2}} \tag{9.6}$$

where $A_0 = \omega^2 Y_0$ is the acceleration amplitude. [In deriving Eq. (9.6), Eq. (9.3) is used for Z_0 along with $A_0 = \omega^2 Y_0$ which indicates that at a fixed acceleration, vibration amplitude Y_0 is inversely proportional to the square of the vibration frequency ω.] Thus, the EMF depends on the vibration frequency, peaking at the resonant frequency with its magnitude dependent on the damping ratio ζ, as shown in Fig. 9.2b. Note that Fig. 9.2b is for a given input acceleration ($A_0 = \omega^2 Y_0$), not for a given vibration amplitude Y_0. At the resonant frequency, the EMF simplifies to

$$\begin{aligned}\varepsilon_0 &= \left|\frac{d\psi}{dz}\right|\frac{\omega_n Y_0}{2\zeta} \\ &= \left|\frac{d\psi}{dz}\right|\frac{\omega_n Y_0}{2(\omega_n d/2k)} = \left|\frac{d\psi}{dz}\right|\frac{kY_0}{d} = \left|\frac{d\psi}{dz}\right|\frac{m\omega_n^2 Y_0}{d} = \left|\frac{d\psi}{dz}\right|\frac{mA_0}{d}\end{aligned} \tag{9.7}$$

9.1.3 Mechanical Power Transfer Ratio

The instantaneous mechanical power transferred to the proof mass m from a vibrating surface is equal to the force on the mass times the velocity of the mass, i.e.,

$p_m(t) = -my(t)''[y(t)' + z(t)']$. Using $y(t) = Y_0\cos\omega t$ and $z(t) = Z_0\cos(\omega t + \theta)$ as well as Eq. (9.3) for the relationship between Y_0 and Z_0, we obtain the mechanical power to be

$$p_m(t) = -m\omega^3 Y_0^2 \cos\omega t\left[\sin\omega t + \frac{(\omega/\omega_n)^2}{\sqrt{\left[1-(\omega/\omega_n)^2\right]^2 + (2\zeta\omega/\omega_n)^2}}\sin(\omega t + \theta)\right] \tag{9.8}$$

This equation is rather complicated, and is probably the reason for lack of detailed analysis on energy conversion efficiency in literature. If the vibration frequency ω is equal to the resonant frequency ω_n, θ is equal to $-90°$ from Eq. (9.4) when $\omega = \omega_n$, and the power becomes

$$p_m(t) = -m\omega_n^3 Y_0^2[\cos\omega_n t \sin\omega_n t - \cos^2\omega_n t/2\zeta] \tag{9.9}$$

Even in this special case at the resonant frequency, the power transferred to the proof mas is not so straightforward. Depending on the damping ratio ζ, we have the following values for the *maximum* instantaneous power transferred to the proof mass m from a vibrating surface when $\omega = \omega_n$:

$$p_{m,\max} = \begin{cases} m\omega_n^3 Y_0^2\left(\dfrac{1}{4\zeta} + \dfrac{1}{2}\right) & \text{for } \zeta \geq 0.5 \ \text{(i.e., } Q \leq 1) \\ m\omega_n^3 Y_0^2 \dfrac{1}{2\zeta} & \text{for } \zeta \leq 0.5 \ \text{(i.e., } Q \geq 1) \end{cases} \tag{9.10}$$

9.1.4 Energy Conversion Efficiency

One measure of the energy conversion efficiency (ECE) for vibration energy harvesters has been a so-called harvesting efficiency (HE), which is defined as the ratio between the measured electrical power output and the maximum mechanical power transferred to an energy harvester [1]. Without detailed steps, a well-cited paper [2] shows the maximum mechanical power to be

$$P_{\max} = \frac{m\zeta Y_0^2\left(\dfrac{\omega}{\omega_n}\right)^3 \omega^3}{\left[1-\left(\dfrac{\omega}{\omega_n}\right)^2\right]^2 + \left(2\zeta\dfrac{\omega}{\omega_n}\right)^2} \tag{9.11}$$

Since Eq. (9.11) cannot be obtained from Eq. (9.8), I question the validity of Eq. (9.11). When $\omega = \omega_n$, Eq. (9.11) becomes

$$P_{\max} = \frac{mY_0^2\omega_n^3}{4\zeta} \tag{9.12}$$

which is different from the analysis leading to Eq. (9.10).

Yet, we must note that this HE accounts only the energy conversion once an energy harvester is driven into a vibration with vibration amplitude equal to Y_0. However, we know that there are efficiency losses (1) in the process of transferring mechanical energy from a vibrating body to an energy harvester and (2) due to the mass loading (of an energy harvester on a vibrating body) that reduces the vibration amplitude Y_0. The first loss mechanism is pronounced in case of a *light* energy harvester, while the second loss mechanism is prominent in case of a *heavy* energy harvester. Therefore, the so-called HE is not a good efficiency figure.

Interestingly, I note that literature is largely mute on the ECE for vibration energy harvesters, and present the following equations for the first time. The instantaneous mechanical power that a vibrating surface has, before an energy harvester is loaded, is

$$p_i(t) = -My_i(t)''y_i(t)' \tag{9.13}$$

where M and $y_i(t)$ are the mass of the vibrating surface and the vibration displacement of the vibrating body itself without an energy harvester, respectively. Since the kinetic momentum is $My_i(t)'$ before an energy harvester is loaded and because the harvester adds the mass of the proof mass m and other auxiliary mass m_h to M, we have $My_i(t)' = (M + m + m_h)y(t)'$ from conservation of kinetic momentum, and obtain

$$y_i(t) = \frac{M+m+m_h}{M}y(t) \tag{9.14}$$

Thus, the instantaneous mechanical power in Eq. (9.13) can be written in terms of Y_0 as follows.

$$p_i(t) = \frac{(M+m+m_h)^2}{M}\omega^3 Y_0^2 \cos\omega t \sin\omega t \tag{9.15}$$

And the maximum instantaneous mechanical power (that a vibrating surface has before an energy harvester is loaded) is

$$p_{i,\max} = \frac{(M+m+m_h)^2\omega^3 Y_0^2}{2M} = \frac{(M+m+m_h)}{2}\omega^3 Y_0^2\left(1+\frac{m+m_h}{M}\right) \tag{9.16}$$

With $p_{e,\max}$ defined to be the maximum instantaneous *electrical* power, the overall energy conversion efficiency is obtained by

$$\text{Energy conversion efficiency (ECE)} = \frac{p_{e,\max}}{p_{i,\max}} \tag{9.17}$$

In electromagnetic energy conversion, we obtain the maximum electrical power from Eq. (9.6) at the resonant frequency ω_n of an energy harvester as

$$p_{e,\max} = \frac{\omega_n^2 Y_0^2}{16\zeta^2 R}\left|\frac{d\psi}{dz}\right|^2 \tag{9.18}$$

where R is the load resistance that is matched to the energy harvester's source resistance. Thus, the maximum theoretical energy conversion efficiency of an electromagnetic energy harvester is

$$\text{ECE} = \frac{2M}{(M+m+m_h)^2}\frac{1}{16\zeta^2 R\omega_n}\left|\frac{d\psi}{dz}\right|^2 \tag{9.19}$$

For a given mass of an energy harvester ($m + m_h$), the ECE is maximized when $M = m + m_h$ and the maximum ECE is equal to

$$\mathrm{ECE}_{\max} = \frac{1}{2M}\frac{1}{16\zeta^2 R\omega_n}\left|\frac{d\psi}{dz}\right|^2 \tag{9.20}$$

For an energy harvester having a total mass of 90 g, a resonant frequency of 65 Hz, and ζ equal to 0.15, the ECE peaks at 18% when M is equal to 90 g. However, for a microfabricated energy harvester having a total mass of 0.5 g, a resonant frequency of 290 Hz and ζ equal to 0.03, the ECE is the highest when M is equal to 0.5 g, and is equal to 6%.

From the equation above, we see that to obtain the ECE, we will have to take the mass of a vibrating body into consideration. Consequently, the optimization of the efficiency will depend on specific vibrating energy sources. For example, if the mass of an energy harvester ($m + m_h$) is negligibly small compared to the mass of a vibration source (M), then one would just increase the number of coil turns, coil area, magnetic field gradient, and Q, while reducing the harvester's source resistance (which increases with increasing coil area and number of coil turns). However, if the mass of an energy harvester ($m + m_h$) is comparable to the mass of a vibration source (M), then the optimization of the efficiency will uniquely be different for different M's, involving the trade-offs between the harvester mass and the coil size/turns as well as the magnetic field strength.

Experimental or Measured ECE for Electromagnetic Energy Harvester

Experimentally, the ECE can be obtained through dividing measured maximum instantaneous *electrical* power $p_{e,\max}$ by the maximum instantaneous mechanical power $p_{i,\max}$ that a vibrating surface has before an energy harvester is loaded, and can be obtained from Eq. (9.16) using the masses of vibrating body and energy harvester, maximum vibration amplitude of the energy harvester's frame, and the frequency at which the maximum electrical power is obtained. The experimental ECE is expected to be close to the theoretical ECE, if the damping ratio ζ and $d\psi/dz$ are properly accounted for in calculating the theoretical ECE. Though it will be relatively easy to accurately estimate the damping ratio ζ and dB/dz for small vibration amplitude, those values will be increasingly difficult to estimate accurately as the vibration amplitude increases, since each of those will be a function of the vibration amplitude.

9.1.5 Increasing Efficiency in Electromagnetic Energy Harvesters

For energy harvesters, maximum power outputs are delivered into matched loads of which the resistances are equal to the coil resistances. Consequently, for a sinusoidal EMF, we get the power delivered to a matched load as follows, using Eq. (9.7) for the EMF magnitude ε_0,

$$P_L = \frac{\varepsilon_0^2}{8R} = \left|\frac{d\psi}{dz}\right|^2 \frac{m^2 A_0^2}{8Rd^2} \tag{9.21}$$

where R is the load resistance that is matched to the energy harvester's source resistance. It indicates that the power output (P_L) is proportional to the square of magnetic flux gradient $\left(\left|\frac{d\psi}{dz}\right|\right)$, proof mass ($m$), and acceleration amplitude (A_0).

Consequently, magnets with larger volume (which typically provide stronger magnetic field), more coil turns, larger coil area, or larger coil cross-sectional area result in a

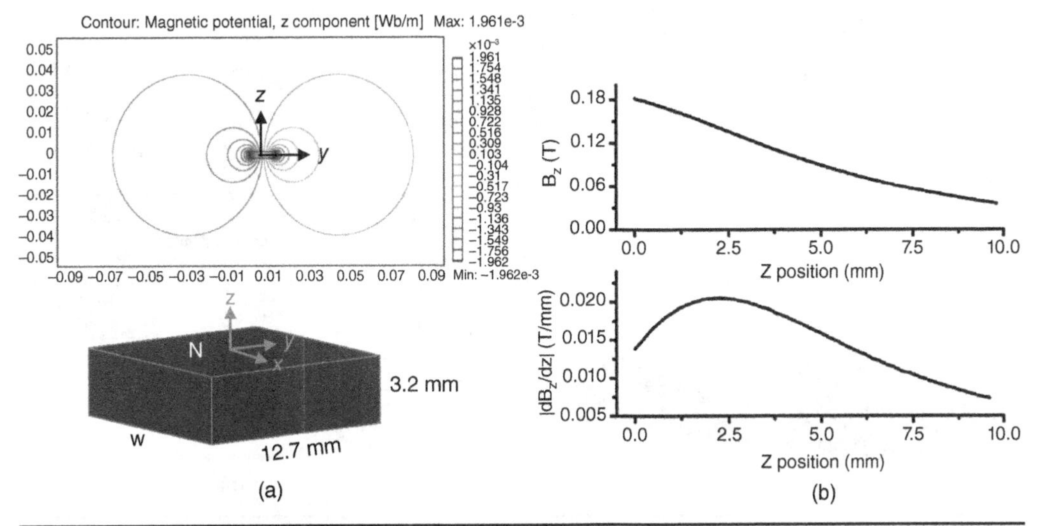

FIGURE 9.3 (a) One magnet and its coordinate system (bottom) and cross-sectional view of the magnetic field lines over *zy* plane produced by one magnet and (b) magnetic flux density (B_z) and its gradient ($|dB_z/dz|$) versus *z* position from one magnet [3].

higher power output, but increase the volume and weight. Thus, for electromagnetic energy harvesters, maximizing the spatial magnetic flux gradient $\left(\left|\frac{d\psi}{dz}\right|\right)$ is essential in increasing the power output, as the magnets or coils move in response to the environmental vibration.

For conventional electromagnetic energy harvester, the magnetic field is provided by a single magnet, and the magnetic flux change is caused by a distance change from the magnet. The magnetic field lines produced by one magnet, when simulated with COMSOL, are shown in Fig. 9.3, which also shows the calculated magnetic flux density (B_z) and its *z* gradient ($|dB_z/dz|$).

A rapidly changing field (for a high power output) is available over an array of magnets with alternating north- and south orientation arranged on a planar surface, as Fig. 9.4c shows that the change of magnetic flux density in the direction parallel to the planar surface ($|dB_z/dy|$, the *y* gradient, not the *z* gradient) peaks at the boundary between two magnets for different heights (d) over the magnet surface. The simulations show that the change of magnetic flux density ($|dB_z/dy|$) for the magnet array depends on the distance from the magnet surface, and can be more than hundred times higher than that of one magnet ($|dB_z/dz|$). Thus, one can use a magnet array (along with a coil array) and the in-plane (not out-of-plane) field gradient. In the simulations and experiments, Nd-Fe-B permanent magnet (Grade N52), which is one of the strongest commercial magnets, is used. Table 9.1 lists the parameters used in the simulations.

The magnetic flux (ψ) through a multi-turn coil (Fig. 9.5a) is obtained with

$$\psi = \sum_{i=1}^{n} \psi_i = \sum_{i=1}^{n}\left(\iint_{S_i} \vec{B}\cdot d\vec{S_i}\right) \tag{9.22}$$

where n is the number of coil turns; ψ_i is magnetic flux through the ith coil; B is magnetic flux density; and S_i is the area of the ith coil. The magnetic flux is a numerical integral of

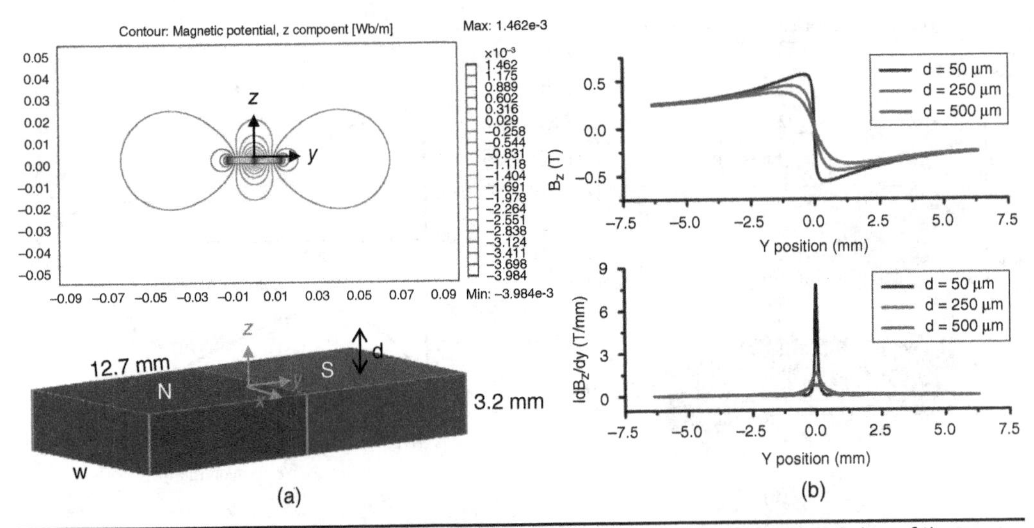

FIGURE 9.4 (a) Two abutting magnets and the coordinates (bottom) and cross-sectional view of the magnetic field lines over zy plane produced by two-magnet array (top) and (b) magnetic flux density (B_z) and its gradient ($|dB_z/dy|$) versus y position at a height of 50, 250, and 500 μm over two-magnet array [3].

Parameter	Value
Magnet length	12.7 mm
Magnet thickness	3.2 mm
Residual magnetic flux density	1.48 T
Relative permeability (in vacuum environment)	1
Coil turn*	100
Coil number	1
Lateral space between coils	60 μm
Outmost diameter of the coil*	12.7 mm
Distance between the coil and magnet surface†	250 μm

*For simulations when the coil is placed at different heights over the magnet array in Fig. 9.5c.
†For simulations when the coils with different outmost diameters are placed over the magnet array in Fig. 9.5d.

TABLE 9.1 Parameters Used in the Simulations

magnetic flux density across the entire coil area, and Fig. 9.5b shows the simulated magnetic flux (ψ_z) and its z gradient ($|d\psi_z/dz|$) versus the distance between the coil and the magnet for *a single magnet*. Similarly calculated magnetic flux (ψ_z) and its y gradient ($|d\psi_z/dy|$) for *an array of magnets*, when a coil is placed at a different position over the magnets, are shown in Fig. 9.5c. The $|d\psi_z/dy|$ for an array of magnets peaks when the coil center is located at the boundary between the magnets; increases as the distance (d)

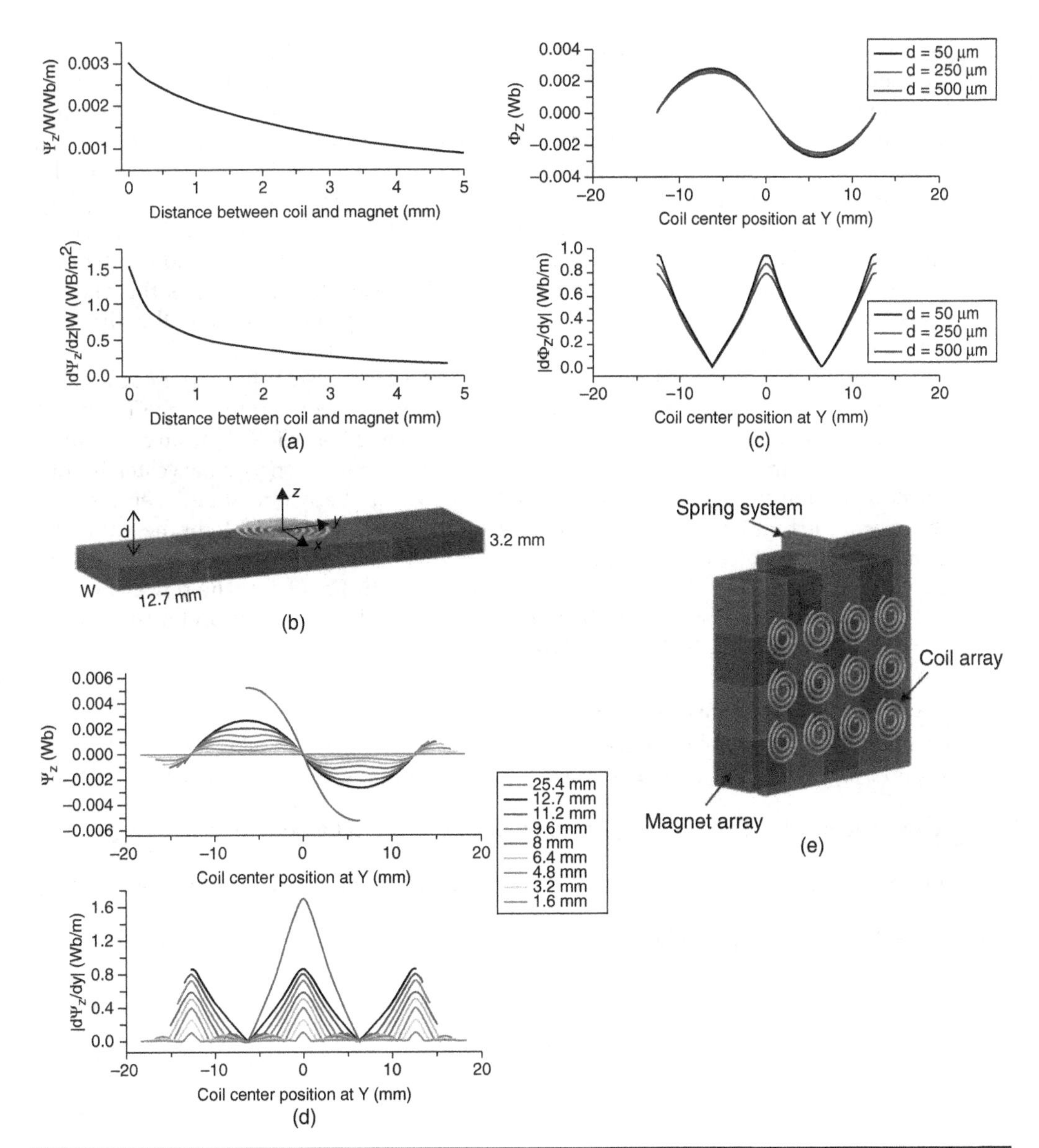

FIGURE 9.5 (a) Normalized magnetic flux (ψ_z) and its z gradient ($|d\psi_z/dz|$) versus the distance between a coil and a magnet for a single magnet, (b) a magnetic array and a multi-turn coil, (c) magnetic flux (ψ_z) and its y gradient ($|d\psi_z/dy|$) versus the coil center position when the coil is placed at a height of 50, 250, and 500 μm over the magnet array, (d) magnetic flux (ψ_z) and its y gradient ($|d\psi_z/dy|$) versus the coil center position when coils with different outmost diameters are placed at a height of 250 μm over a magnet array, and (e) schematic of a vibration energy harvester with alternating north- and south-orientation magnet array [3].

from the magnet surface is decreased; and is much higher than that for one magnet. The magnetic fluxes (ψ_z) and its y gradients ($|d\psi_z/dy|$) for coils with various diameters of the outmost coil (of a multi-turn coil) are shown in Fig. 9.5d, and we see that the magnetic flux change increases as the coil size increases. Thus, when an array of coils is placed over a magnet array, the optimal outmost diameter of the coils is equal to the side length of a magnet (if square magnets are used to form the array). One exemplary electromagnetic energy harvester based on magnet and coil arrays is illustrated in Fig. 9.5e with the magnets arranged on a planar surface such that north and south poles alternate. Coils are placed over the boundaries between the magnets, as the magnet array is suspended by a spring system providing a low spring constant in the direction parallel to the planar surface.

The one in Fig. 9.5e has been implemented to be a macroscale harvester of 51 × 51 × 10 mm (= 26 cc) and 90 g, and shown to produce an electromotive force (EMF) of $V_{p\text{-}p} = 28.8$ V with 270 mW power output (into a matched load of 96 Ω) from a vibration at 65 Hz with amplitude of 660 μm [3]. When the vibration energy harvester is connected to an incandescent light bulb *directly* (i.e., without any circuit between the harvester and the bulb), the harvested power level is so high that it can light the bulb. The measured power into a matched load (96 Ω) increases approximately in proportion to the square of the applied acceleration, as expected from Eq. (9.21). The power level of 270 mW is quite impressive, but the harvester size is relatively large and heavy, not to mention a relatively high vibrational frequency of 65 Hz.

9.1.6 Maximum Power Delivery

An energy harvester can equivalently be modeled with an ideal voltage source (V_S) and a source resistance (R_S), as shown in Fig. 9.6, and a voltage, $R_L V_S/(R_S + R_L)$, is developed across a load (R_L). Consequently, the power delivered to the load is $P_L = \dfrac{R_L V_S^2}{(R_S + R_L)^2}$. To find R_L for maximum delivery from the source, we differentiate P_L with respect to R_L and set the resulting equation zero as $\dfrac{\partial P_L}{\partial R_L} = \dfrac{V_S^2}{(R_S + R_L)^2} - \dfrac{2R_L V_S^2}{(R_S + R_L)^3} = 0$, from which we see that P_L is the maximum (equal to $\dfrac{V_S^2}{4R_L}$) when R_L is equal to R_S. In other words, a power delivered to a load from an energy harvester is the largest when the load

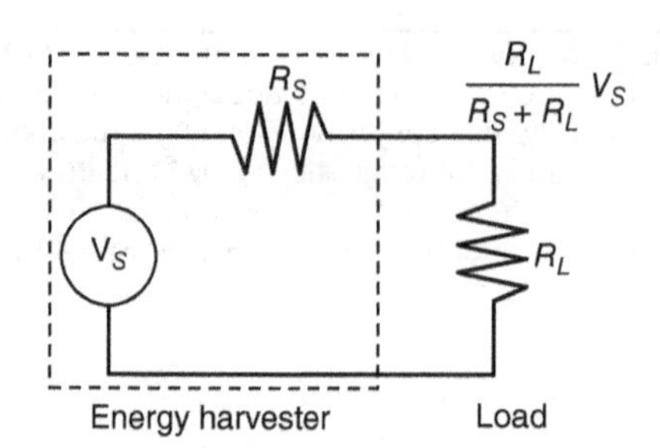

FIGURE 9.6 Equivalent circuit for an energy harvester with a load R_L, across which a voltage, $R_L V_S/(R_S + R_L)$, is delivered from the energy harvester.

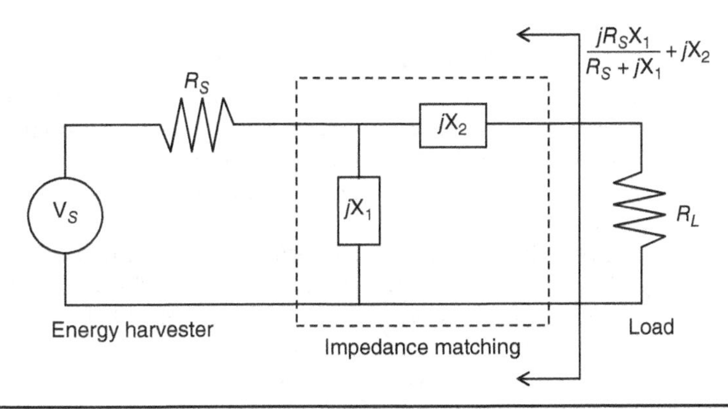

FIGURE 9.7 Impedance matching between a vibration energy harvester and a load, based on a capacitor and an inductor for the X_1 and X_2.

resistance (R_L) is matched to the source resistance (R_S) of the energy harvester. Thus, an impedance matching network may be needed between an energy harvester and a load.

For resonant energy harvesters, a simple matching network based on a capacitor and an inductor (shown in Fig. 9.7) would be the best in terms of size and weight as well as easiness to implement. For example, for $R_S > R_L$, one can choose $X_1 = -R_S\sqrt{\frac{R_L}{R_S - R_L}}$ (a capacitor with capacitance C) and $X_2 = R_L\sqrt{\frac{R_S - R_L}{R_L}}$ (an inductor with inductance L) [4], so that the impedance seen from the load is $\frac{jR_SX_1}{R_S + jX_1} + jX_2 = R_L$, which is matched to the load at a particular frequency (i.e., the resonant frequency of the energy harvester), and thus, the largest power is delivered from the energy harvester to the load. This approach, though, would not work well if R_L is too far from R_S (e.g., $R_S >> R_L$), since slight deviation from the ideal L and C values would cause a large error on the resulting impedance in such a case. Also, for energy harvesters (such as piezoelectric types) having a limited amount of current that they can deliver, the voltage on the load will be much smaller than a half of the open-circuit source voltage (V_s), even if one somehow manages to get the exact values for L and C for matching. However, most electromagnetic energy harvesters have already low source resistance (R_S) similar to most loads (R_L) as well as capability to deliver large current, and the LC-based impedance matching will be sufficient for resonant energy harvesters.

For non-resonant energy harvesters that cover broadband frequencies (1–5 Hz), impedance matching based on transformer would work, but it is not clear whether there exists a commercial-off-the-shelf transformer that covers such very low-frequency range and also whether, if there is such a transformer, it will be small and light enough for some applications where the energy harvesters and their accompanying electronics should be small and light. In the case of no good impedance matching for non-resonant energy harvesters, the harvesters can be designed to have its source resistance (through varying the number of turns for the coil) matched to that of the load (usually, several to tens ohm).

9.2 Piezoelectric Vibration Energy Harvesting

As we studied in Chap. 3, piezoelectric material can be used to pick up vibration energy and convert it to electrical energy. In one application, the vibration in a ski is converted into an electrical energy (with a piezoelectric energy harvester), which then is used to reduce the vibration through straining or stressing some key points on the ski (using a piezoelectric actuator).

A piezoelectric vibration energy harvester (VEH) can be built on a piezoelectric cantilever, as shown in Fig. 9.8a, which in response to vibrational motion produces a current that charges either a capacitor or a thin rechargeable battery. One commercial piezoelectric VEH is optimized for harvesting vibration at around 120 Hz in order to charge a capacitor or battery, weighing less than 5 g and occupying 1.35 cc, and is reported to generate a peak power of 0.1 mW at 1-g input acceleration at 120 Hz. It is not clear, though, how much power it can deliver into a load.

In case of the piezoelectric energy harvester based on a piezoelectric cantilever with a proof at the cantilever's free end (Fig. 9.8a), the outer frame is attached to a vibrating surface. The cantilever is typically composed of a piezoelectric layer and a support layer forming a unimorph. Alternatively, two piezoelectric layers having opposite piezoelectric polarities may be used to form a thin, flexible bimorph cantilever (or membrane across the frame) to suspend a proof mass. If random and unknown nature of vibration source prevents VEH's resonant frequency from being matched to the vibration source, VEH can be designed to have a very low resonant frequency (such as several Hz with a heavy proof mass), so that most vibrations cause the piezoelectric beam to bend.

To produce power in a usable form through piezoelectric transduction, energy will almost certainly have to be stored up, at least temporarily, because of the high resistance of a piezoelectric energy harvester. One can consider the storage of generated electrical energy in a capacitor C, as shown in Fig. 9.8b. The piezoelectric device is modeled by an equivalent voltage source V_{piezo} in series with an equivalent capacitance C_{piezo}. The two diodes (or similarly acting switches) allow the capacitor charge to build up monotonically.

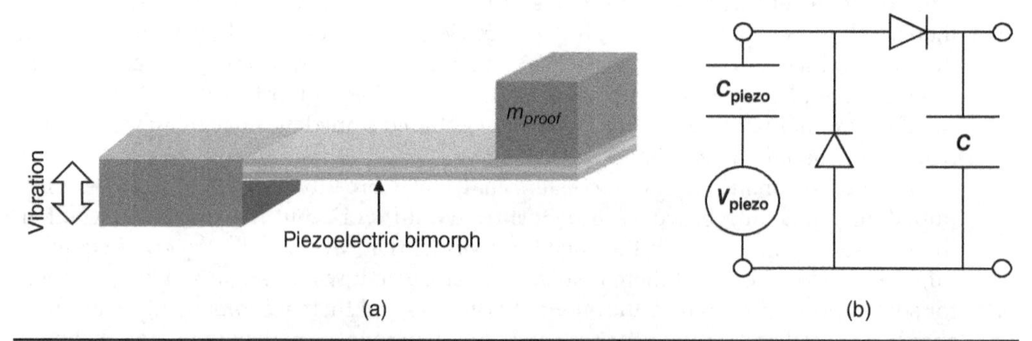

Figure 9.8 Piezoelectric vibration energy harvester: (a) conceptual diagram of a piezoelectric cantilever with a proof mass showing a vibration mode in response to applied acceleration and (b) equivalent circuit for a piezoelectric source along with a capacitor and two diodes for charging the capacitor with the piezoelectric current.

9.2.1 Example: PZT Bimorph-Based Energy Harvester

A vibration-energy harvester can be made with a simple cantilever based on a lead zirconate titanate (PZT) bimorph. A bimorph structure takes full advantage of opposite stress polarities at the clamped end of the cantilever. If a unimorph is used, a substrate material would have been necessary to support a PZT layer. The cross-sectional view of a commercially available PZT bimorph substrate is shown in Fig. 9.9a. If a proof mass (m_{proof}) is attached on the free end of a cantilever made of the PZT bimorph (with brass thickness t_b and PZT thickness t_{PZT}) as shown in Fig. 9.9b, the resonant frequency is

$$f_0 = \frac{1}{2\pi}\sqrt{\frac{3E_{\text{PZT}}I_{\text{eff}}}{m_{\text{eff}}l_{\text{beam}}^3}} \tag{9.23}$$

where E_{PZT} is the Young's modulus of PZT; I_{eff} is the effective moment of inertia of the rectangular beam cross-section, which is adjusted for the difference in Young's modulus between PZT and brass; m_{eff} is the effective mass of the cantilever beam; and l_{beam} is the length of the cantilever. The I_{eff} and m_{eff} are determined as follows:

$$I_{\text{eff}} = \frac{w_{\text{beam}}\left(t_b\dfrac{E_{\text{brass}}}{E_{\text{PZT}}} + t_{\text{PZT}}\right)^3}{12} \tag{9.24}$$

$$m_{\text{eff}} = 0.238m_{\text{beam}} + m_{\text{proof}} \tag{9.25}$$

If the weight of the energy harvester is not a critical parameter for a target application, one can increase the weight of the proof mass in order to reduce the total volume of the energy harvester or to increase the width. In both cases the power density of the energy harvester per volume is increased with a larger mass.

In an exemplary VEH, a commercial PZT bimorph (shown in Fig. 9.9a) is diced into a 5.5×4.0 cm^2 bimorph plate. One end of the rectangular plate is attached to a printed circuit board (PCB) with conductive silver epoxy, while the other end has a 31-g proof mass attached adhesively. The measured frequency response (Fig. 9.10a) indicates a resonant frequency of 27.2 Hz with quality factor of 13.9. Under a 0.1-g vibration source, the harvester delivers root-mean-square (RMS) power of 0.124 mW to a matched load of 32.3 kΩ at the resonant frequency, corresponding to an RMS voltage of 1.83 V. For a

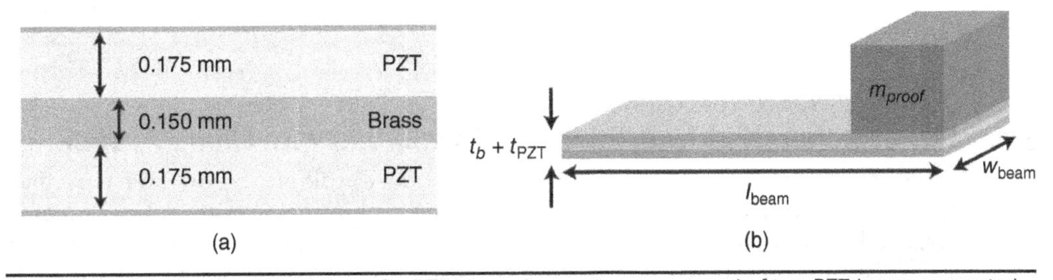

FIGURE 9.9 (a) Cross-sectional view of a PZT bimorph structure composed of two PZT layers separated by a brass layer. (b) Schematic of a PZT bimorph cantilever with a proof mass (m_{proof}) attached at the free end of the cantilever.

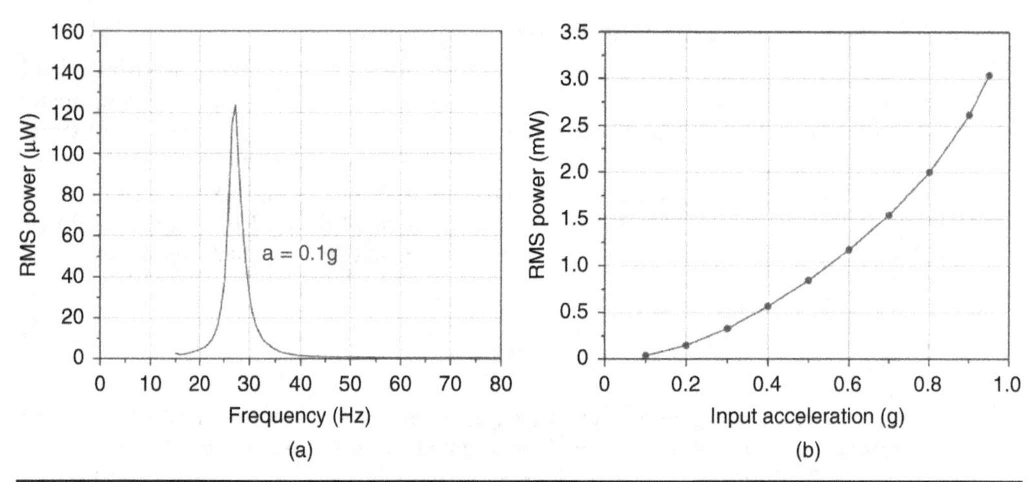

Figure 9.10 Plots of (a) the power delivered to a matched load of 32.3 kΩ by the PZT bimorph energy harvester versus frequency for 0.1-g input acceleration and (b) the root-mean-square (RMS) power harvested at varying vibration levels at the resonant frequency of 27 Hz.

given frequency such as the resonant frequency of 27.2 Hz, the power level increases in proportion to the applied acceleration (Fig. 9.10b), as expected from Eq. (9.21). Since the active volume of the energy harvester is 29.1 cc, the power density of the device is 4.26 μW/cc for a 0.1-g vibration source.

With multiple piezoelectric energy harvesters, connecting them in series will deliver more power to a matched load than a parallel connection. However, a serial connection increases the source resistance and requires the load resistance to be larger for maximum power delivery.

9.2.2 Piezoelectric versus Electromagnetic Energy Conversion

Piezoelectric vibration energy harvester is amenable to MEMS fabrication process and also is capable of producing relative high voltage (as long as a load is highly resistive such as > 100 kΩ), but is incapable to deliver large current to a load and to deliver high continuous power to low impedance load (< 1 kΩ). On the other hand, electromagnetic vibration energy harvester is typically incompatible with planar fabrication process (that is not accommodating for many-turn coil and permanent magnet) and tends to be bulky and heavy, but is capable of delivering large current and high continuous power to < 100 Ω load.

9.3 Power Generation from Vibration Associated with Human's Walk

As the internet of things (IoT) is expected to connect trillion sensors and now that smartphones and wearable devices (for fitness or health tracking) are getting ubiquitous, two key issues are how to power the devices and how to keep the devices working constantly and continuously without cumbersome battery recharging or replacement. Battery recharging or replacement means downtime and human intervention. One approach to avoid battery replacement is to minimize power consumption by operating a device intermittently, but mission-critical devices need to operate all the time not to miss any event. Thus, a battery-less, always-on device that can wirelessly send/receive information is what will be needed for IoT and wearable technology to have unprecedented impacts on human lives. There are plenty of places and objects that vibrate and provide sources for energy harvesting, such as bridges, building walls, automobiles, airplanes, ships, etc. Human body (especially when s/he walks) offers substantial amount of vibration energy, which can be converted into electrical power, particularly if the power generator can be made to be unobstructive and un-cumbersome to its wearer or carrier.

9.3.1 Power-Generating Shoe, Knee Cap, and Backpack

There have been efforts to generate a substantial amount of electrical power, as a person walks, with three different approaches: shoe, knee cap, and backpack. In case of the power-generating shoe, the electromagnetic version is bulky and inconveniences the wearer, while the piezoelectric version is incapable of delivering a large current to a load. In both cases, electrical wires must run from the shoe to deliver the power to the desired location, unless the location that needs the power is also in the shoe.

The power-generating knee caps generate electrical power as the wearer exerts force to move the knee cap [5, 6]. Those have been reported to produce a large power (5–15 W), though the weight of the knee cap is relatively light (0.9–1.6 kg). However, the knee cap inconveniences the wearer, as it limits the wearer's motion and must be individually tailored for each wearer.

In case of the power-generating backpack, the power output is proportional to the weight of the load that the backpack incorporates, and is very heavy (23 kg) for a large power (12–15 W) [7]. Also, there is limited space in the backpack for other things, due to the mechanical structure for the power generation. Consequently, the wearer is inconvenienced greatly due to the heavy weight and mechanical structure that affect the wearer's dynamic motion.

The challenge is on finding approaches to generate substantial power from human's walk without any of the disadvantages associated with the power-generating shoe, knee cap, and backpack. Specifically, a vibration energy harvester that does not load its wearer with weight or volume, nor inconveniences the wearer, nor limits the wearer's movability, nor requires individual tailoring or fitting will be highly desirable.

9.3.2 Challenges in Generating Power from Walking Motion without Loading the Person

Available Power from Human's Walking Motion

To estimate how much power can possibly be generated from human's walking motion, one can consider a potential energy for mass m, $E = mgh$, with g and h being the gravitational constant and height, respectively. A 36-kg mass on the back of a walking person requires 18 J of mechanical energy transfer for each step of 5 cm vertical displacement. If the person takes two steps per second, it means a potential power of 35 W [8]. However, if $m = 10$ g and $h = 5$ cm, then the energy $E = 0.005$ J and the power that can be generated is only 10 mW.

To check this simple calculation for sinusoidal vibration $y(t) = Y_0\cos(\omega t)$, the instantaneous mechanical power is $p_i(t) = -my(t)''y(t)' = m\omega^3 Y_0^2 \cos\omega t \sin\omega t$. Consequently, the maximum instantaneous mechanical power is $p_{i\max} = m\omega^3 Y_0^2/2$. For $m = 10$ g, $\omega = 2\pi \times$ 2 Hz, and $Y_0 = 2.5$ cm, the maximum power obtainable is $p_{i\max} = m\omega^3 Y_0^2/2 = 6.2$ mW, similar to the estimated power level with the simple calculation above. What the equation shows, though, is that the power level is proportional to the cube of the frequency. Thus, the extremely low frequency of subHz to several Hz associated with walking is what is limiting the maximum attainable power level.

Low Resonant Frequency and Nonperiodic Walking Motion

The extremely low frequency is not only limiting the maximum power attainable power level, but also presents challenge in constructing a mass-spring system for a power generator. With $\omega = \sqrt{k/m}$, 1-Hz resonant frequency requires very low spring constant and/or very large mass, and there is 25 cm initial displacement, due to gravity, for a spring-mass unit designed for 1-Hz resonant frequency, as can be seen in Fig. 9.11. This makes the harvester size to increase too much, unless there is some mechanism to reduce the initial displacement.

Moreover, human body's walking motion is not periodic, and various approaches will have to be explored to make the proof mass suspension be broadband. One can certainly take full advantages of gait analysis of human walking motion, as it provides information on acceleration direction at various parts of the body, the frequency spectra of the vibrations, and the locations on human body where largest vibration energy is available.

9.3.3 Magnetic Spring for Resonance at 2–4 Hz

Magnetic spring based on repulsive force from identical poles facing each other can be made to have a very low resonant frequency without a large volume. Consequently, to meet the challenging need for an extremely low resonant frequency without the large initial displacement, a magnetic spring can be used. Using a point charge (Q) model for

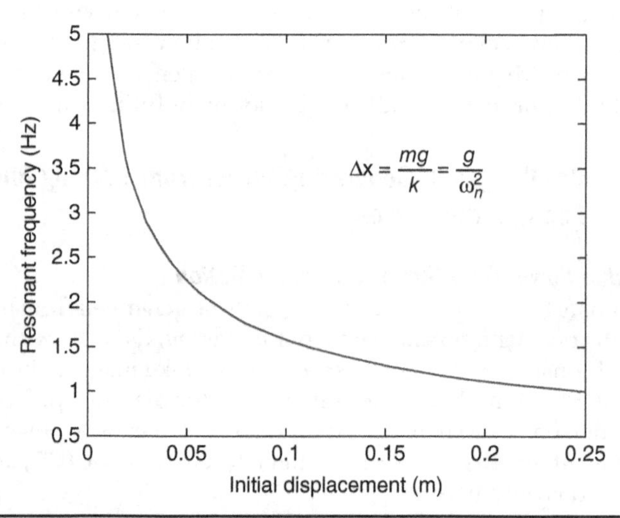

Figure 9.11 Resonant frequency versus initial displacement due to gravity for a spring-mass unit, showing 25 cm initial displacement (due to gravity) for 1-Hz resonant frequency.

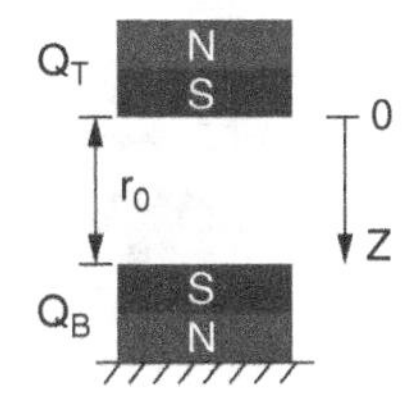

FIGURE 9.12 Magnetic spring, based on repulsive force between two magnets, which suspends a magnet with mass m in air.

the magnetic spring as shown in Fig. 9.12, we get the following equation for the motion, similar to Eq. (9.1) which we got with Fig. 9.1,

$$mz(t)'' + dz(t)' + (F_M - mg) = -my(t)'' \tag{9.26}$$

where m = mass of the suspended magnet, d = damping constant, g = gravitational constant, and $F_M = \frac{\mu_0 Q_T Q_B}{4\pi(r_0 - z)^2} = \frac{\mu_0 Q_T Q_B}{4\pi}\left(\frac{1}{r_0^2} + \frac{2}{r_0^3}z + \frac{3}{r_0^4}z^2 \cdots\right) \approx \frac{\mu_0 Q_T Q_B}{4\pi}\left(\frac{1}{r_0^2} + \frac{2}{r_0^3}z\right)$ (where r_0 = initial distance between the two magnets, and the approximation is for near the initial position).

Similar to Eq. (9.2), we get the following equation for sinusoidal vibration, noting $\frac{\mu_0 Q_T Q_B}{4\pi r_0^2} = mg$,

$$Z = \frac{m\omega^2 Y_0}{-\omega^2 m + j\omega d + \frac{\mu_0 Q_T Q_B}{2\pi r_0^3}} = \frac{m\omega^2 Y_0}{-\omega^2 m + j\omega d + \frac{2mg}{r_0}} = \frac{(\omega/\omega_n)^2 Y_0}{1 - (\omega/\omega_n)^2 + j2\zeta\omega/\omega_n} \tag{9.27}$$

And the resonant frequency near the initial position for a small vibration is

$$\omega_n = \sqrt{\frac{2g}{r_0}} = \sqrt{4g\sqrt{\frac{\pi mg}{\mu_0 Q_T Q_B}}}. \tag{9.28}$$

Thus, for a low resonant frequency, we would need strong magnets and light proof mass. A light proof mass, though, is not good for large power output, since similar to Eq. (9.8), the electromotive force (EMF) is

$$\varepsilon = \left|\frac{d\psi}{dt}\right| = \left|\frac{d\psi}{dz}\frac{dz}{dt}\right| = \left|\frac{d\psi}{dz}\right|\omega Z_0 = \left|\frac{d\psi}{dz}\right| A_0 \frac{\omega/\omega_n^2}{\sqrt{[1-(\omega/\omega_n)^2]^2 + \left(2\zeta\frac{\omega}{\omega_n}\right)^2}} \tag{9.29}$$

$$= \left|\frac{d\psi}{dz}\right| \frac{A_0}{2\zeta\omega_n} = \left|\frac{d\psi}{dz}\right| \frac{mA_0}{d} \text{ at the resonant frequency.}$$

The point-charge model above is simple, but is not very accurate when the vibrational amplitude is large. Consequently, to accurately calculate the force between two magnets, a *new* model (for a block magnet) based on a pair of parallel magnetically charged plates (Fig. 9.13) is needed [9].

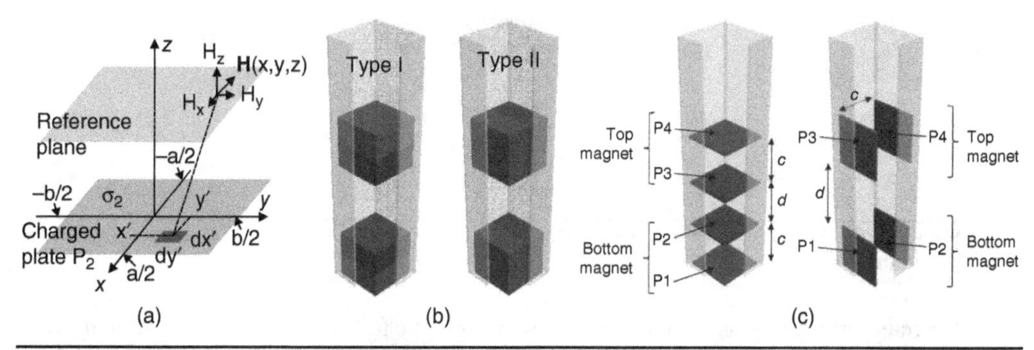

FIGURE 9.13 (a) Magnetic field established by uniformly charged rectangular plate P1. (b) Two types of magnetic spring. (c) In type I magnetic spring, two magnets are represented by four magnetically charged plates (P1–P4) with the resultant force between two magnets being composed of repulsive forces (between P1 and P4 and between P2 and P3) and attractive forces (between P1 and P3 and between P2 and P4) [9]. In type II magnetic spring, repulsive force comes from P1–P3 and P2–P4, while attractive force comes from P1–P4 and P2–P3.

For a rectangle of $a \times b$ with magnetic charge density σ_2 at $z = 0$ plane, as shown in Fig. 9.13a, the magnetic field $\vec{H}(\vec{r})$ in space is the superposition of the magnetic fields generated by infinitesimally small charges $\sigma_2 dx'dy'$ on the plane and its magnitude is

$$\left|\vec{H}(\sigma_2, x, y, z)\right| = \iint_{P2} dH = \iint_{P2} \frac{1}{4\pi} \frac{\sigma_2 dx'dy'}{(x-x')^2 + (y-y')^2 + z^2}. \tag{9.30}$$

In a magnetic spring with magnets aligned coaxially with the same poles facing against each other (Fig. 9.13c), the net lateral force perpendicular to the polarity axis is close to zero. Thus, the z component of the magnetic field is dominant, and is equal to

$$H_z(\sigma_2, x, y, z) = \frac{1}{4\pi} \iint_{P2} \frac{z\sigma_2 dx'dy'}{((x-x')^2 + (y-y')^2 + z^2)^{3/2}}, \tag{9.31}$$

which becomes

$$H_z = \frac{\sigma_2}{4\pi}\left[g\left(\frac{a}{2}+x, \frac{b}{2}+y, z\right) + g\left(\frac{a}{2}-x, \frac{b}{2}+y, z\right) + g\left(\frac{a}{2}+x, \frac{b}{2}-y, z\right) + g\left(\frac{a}{2}-x, \frac{b}{2}-y, z\right)\right], \tag{9.32}$$

with

$$g(x, y, z) \equiv \arctan\left(\frac{xy}{z\sqrt{x^2 + y^2 + z^2}}\right) \tag{9.33}$$

The force between two magnets can be modeled as the interaction of four charged plates (P1–P4 in Fig. 9.13c). The force between two charged plates can be obtained by integrating the infinitesimal force on the target plate, while treating the other plate merely as a field source. For two parallel plates with surface densities σ_2 and σ_3, separated by d, the force between them is

$$F_z(\sigma_2,\sigma_3,d)=\mu_0\iint_P \sigma_3\,dS\cdot H_z(\sigma_2,x,y,d)=\mu_0\sigma_3\int_{-a/2}^{a/2}dx\int_{-b/2}^{b/2}dyH_z(\sigma_2,x,y,d), \quad (9.34)$$

which is $F_z(\sigma_2,\sigma_3,d)=\frac{\mu_0}{\pi}\sigma_2\sigma_3\int_0^a\int_0^b g(x,y,d)\,dy\,dx$, due to $g(x,y,z)$ being symmetric with respect to x and y axis. Thus, we get

$$F_z(\sigma_1,\sigma_2,d)=\frac{\mu_0}{\pi}\sigma_2\sigma_3\cdot[f_1(d)+f_2(d)+f_3(d)], \quad (9.35)$$

with

$$\begin{cases} f_1(d)=ab\cdot\arctan\left[ab/\left(d\cdot\sqrt{a^2+b^2+d^2}\right)\right] \\ f_2(d)=\frac{1}{2}(a+b)d\cdot\left[\ln\left(\frac{a^2+b^2+d^2}{(a^2+d^2)(b^2+d^2)}\right)\right] \\ f_3(d)=(a+b)d\cdot\ln d \end{cases} \quad (9.36)$$

As F_z gives the force between two charged plates, we add up all the forces from four pairs of charged plates for a pair of block magnets as $F=F_{z,P1-P4}+F_{z,P2-P3}+F_{z,P1-P3}+F_{z,P2-P4}$, where P1, P2, P3, and P4 are the pole surfaces of two magnets. This surface charge model is compared with a dipole model and a point charge model, and is shown to give the best approximation to the experimental measurement and FEM simulation.

The newly developed equation provides some very valuable insights into how magnetic spring works. For example, according to the equation, the spring constant is dependent on the distance between the two block magnets, as the movable magnet moves closer to or further away from the stationary magnet (Fig. 9.14b), going from almost zero (when the magnets are separated by a large distance) to near infinity (as the two magnets are getting very close to each other). The spring constant depends also on the mass of the movable magnet (with a heavier magnet resulting in a larger spring constant), and the simulated natural vibrations (for a given initial velocity) are non-sinusoidal in time domain (skewed more in the "positive" direction), as shown in Fig. 9.15a. When the amplitude of natural vibration is plotted in time as a function of the initial velocity, we see that the dominant frequency component shifts lower, as the initial velocity increases (Fig. 9.15b), since a larger initial velocity means more time spent on the "positive" side of vibration displacement z (in Fig. 9.15b) where the spring constant is lower.

9.3.4 Magnet Levitation by Graphite with Resonance at 2–4 Hz

A magnetic spring based on the repelling force between the same pole is one way to suspend a proof mass (usually a magnet or magnet array) to obtain a low resonant frequency, but requires relatively heavy magnets for a resonant frequency as low as 1–5 Hz. Thus, it is worthy to study another suspension technique based on diamagnetic material

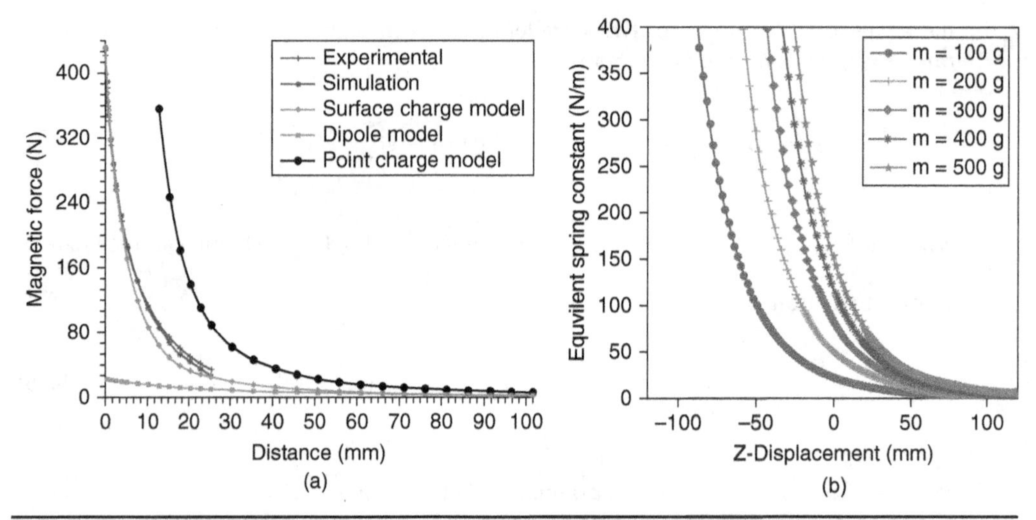

FIGURE 9.14 (a) Magnetic force versus distance obtained with experimental measurement, FEM simulation, and three different modeling methods: surface charge model (the newly developed model), dipole model, and point charge model. Curves are for a type I magnetic spring formed with two 2.56 × 2.56 × 2.56 cm^3 N52 magnets. (b) Equivalent spring constant versus separation distance between two magnets as a function of proof mass [9].

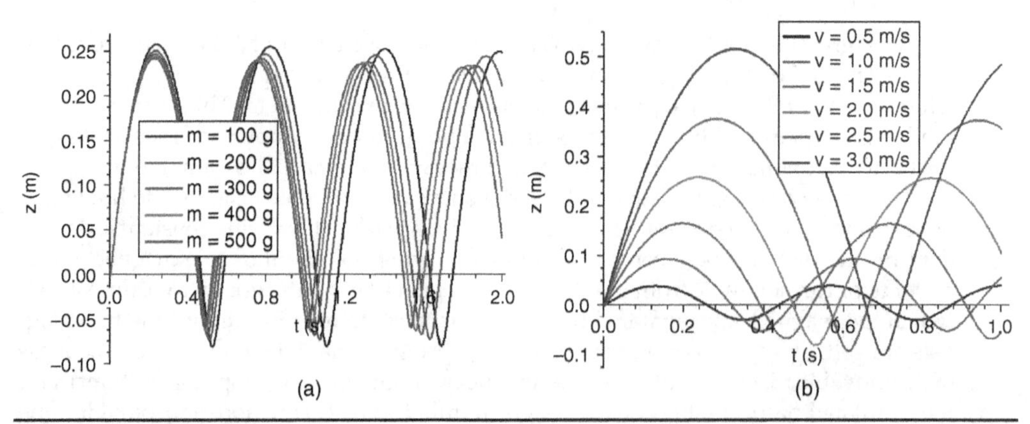

FIGURE 9.15 (a) Vibration amplitude of magnetic spring in time for different proof masses. (b) Vibration amplitude of magnetic spring in time for different initial velocities.

that can induce a magnetic field which is directed opposite to an externally applied magnetic field and repels the applied magnetic field.

In a VEH illustrated in Fig. 9.16, a pair of pyrolytic graphite substrates is used to provide the repelling forces so that a magnet may be floated and confined in air between the two diamagnetic substrates [10]. The floating magnet moves without contacting the graphite substrates in response to an applied vibration, and comes back to its static equilibrium position automatically when there is no applied vibration. The relative motion between the floating magnet and the nearby coil, as the magnet moves, induces

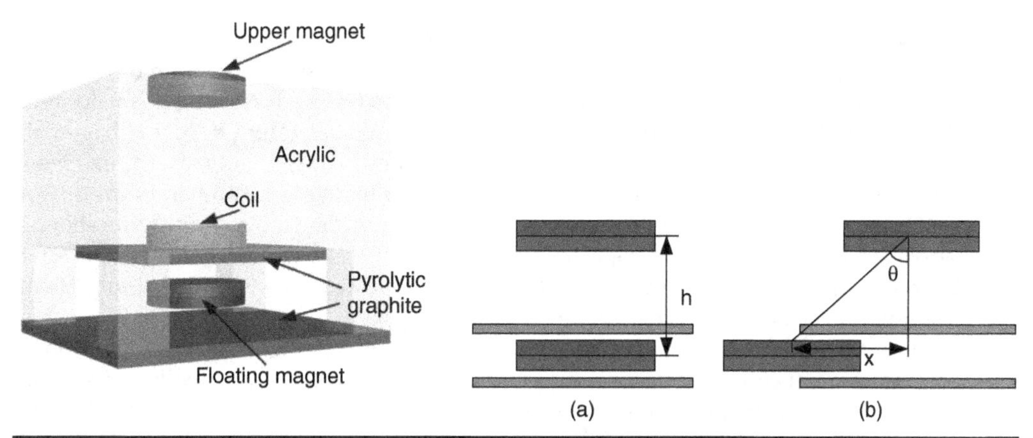

FIGURE 9.16 (Left) Schematic of a vibration energy harvester based on floating magnet through diamagnetic levitation. Though much of the gravity of the floating magnet is counteracted by an attractive force provided by the upper magnet, the floating magnet floats in air well balanced (without being completely pulled up to the upper magnet or down to the bottom graphite) because of the repelling force from the pair of pyrolytic graphite sheets surrounding the floating magnet. Relative movement between the floating magnet and coil due to in-plane vibration induces electromotive force. (Right) Operating principle: (a) the floating magnet in balanced position and (b) the floating magnet off from its balanced position due to applied vibration [10].

EMF. Since the repelling force from the bottom graphite is not strong enough to overcome the gravity of the floating magnet, an additional magnet (denoted as upper magnet in Fig. 9.16) is needed.

The upper magnet is a stack of multiple magnets, the number of which is chosen to provide enough attractive force to balance the gravity of the floating magnet. Though the attractive force may be large enough to make the floating magnet touch the top graphite, the floating magnet does not come in contact with the top graphite because of the repelling force provided by the top graphite. As the floating magnet gets closer to the bottom graphite, the attractive force from the upper magnet is reduced, while the repelling force from the bottom graphite is increased, and thus, the floating magnet does not come in contact with the bottom graphite, either. Instead, it stays stably between the top and bottom graphite sheets as a static equilibrium.

Using the illustration in Fig. 9.16, we can show that to the first order, the floating magnet vibrates in simple harmonic motion with the resonant frequency equal to $\sqrt{g/h}$, where g is gravitational acceleration constant. Consequently, a very low resonant frequency (e.g., 1–4 Hz) can be achieved with relatively small size for generating power from human's movement.

The idea of a floating magnet based on diamagnetic graphite with an additional magnet to counteract the gravity of the floating magnet has been implemented for electromagnetically generating electrical power from human's walking motion. A prototype energy harvester weighting 21.7 g produces a root-mean-square (RMS) voltage of 77.4 mV (in open circuit) and 68 μW power output (into 22 Ω load) at the harvester's resonant frequency of 3.4 Hz in response to a vibration amplitude of 43 mm (i.e., 2 g at 3.4 Hz). The low resonant frequency makes the energy harvester suitable for harvesting energy from human walking motion, and 11.7 μW is delivered to a 22 Ω load from 2 m/s walking speed from the harvester mounted on a human's back.

9.3.5 Liquid Spring for Resonance at 2–4 Hz

A rigid suspension is prone to breakage or failure under a strong vibration or in a long run. Thus, a sturdy suspension structure with low resonant frequency is highly desired. A ferrofluid-based liquid spring is a promising solution. Ferrofluid as spring has small volume and good robustness under strong vibration. Additionally, as the magnets which are also used as a proof mass are encapsulated by the ferrofluid and thus do not touch the frame, the friction between the magnets and the frame during vibration is greatly reduced. These advantages make the ferrofluid-spring-based energy harvester be more durable and efficient in producing electrical power from vibrating sources than a counterpart based on a rigid spring.

In a VEH based on liquid spring (Fig. 9.17), ferrofluid-based liquid suspension is formed in an enclosed chamber made by bonding a micromachined silicon (with an electroplated copper coil) and a laser-machined acrylic frame (Fig. 9.17a) or by bonding a laser-machined acrylic frame and a multilayer coil plate (Fig. 9.17b). Due to its unique magnetic property, the ferrofluid automatically suspends the magnet array right in the middle of the chamber.

As illustrated in Fig. 9.17c, the ferrofluid works as a mechanical spring, since ferrofluid is a liquid that becomes strongly magnetized in the presence of a magnetic field and is attracted by a magnet. The attractive forces counteract each other when the magnet array is in the middle with no applied acceleration. However, once the magnet array is displaced from its balanced position (due to applied acceleration), the part of the ferrofluid which has no symmetric counterpart will draw the magnets back with the amount of the unbalanced part of the ferrofluid (or the force to pull back the magnets) being proportional to the deviated distance.

A prototype harvester occupying $17 \times 11 \times 2.5\ \text{mm}^3$ and weighing 1 g was shown to a resonance around 15 Hz, at which point 176 nW was delivered into a load of 4.5 Ω from 3.5 g acceleration (corresponding to 3.9 mm vibrational amplitude) at 15 Hz [11].

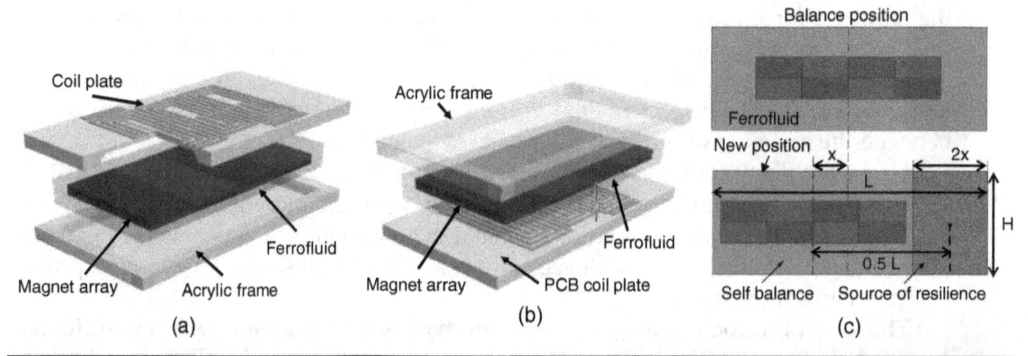

FIGURE 9.17 (a) Schematic of a power generator based on liquid spring. The magnet array is suspended by ferrofluid in a chamber formed by a micromachined silicon (with electroplated copper coil) and a laser-machined acrylic frame. (b) The magnet array is suspended by ferrofluid in a chamber formed by a print circuit board (PCB) with six-layer copper coil and a laser-machined acrylic frame. (c) Conceptual illustration for liquid spring: (Top) Attractive force between the magnet array and ferrofluid makes the magnet stay in the middle automatically. (Bottom) When the magnets are displaced from its balanced position due to applied acceleration, the part of the ferrofluid which has no symmetric counterpart to neutralize the attraction draws the magnets back with the force being proportional to the amount of the ferrofluid's unbalanced part [11].

The resonant frequency is dependent on the density of ferrofluid while the energy conversion efficiency is mainly dependent on the number of coil turns, which can be increased through a multilayer coil plate.

9.3.6 Non-resonant Suspension for Vibration Energy over Broad Frequency Range

Energy harvesters (or power generators) based on the resonance of a proof mass suspension system have their optimal performances only over a narrow frequency band. However, human's walking motion has vibration energy spread over broad frequency range below 4 Hz. Though a non-resonant energy harvester, in general, can respond to all the frequency components of a vibration over a wide frequency range, non-resonant harvesters typically suffer from low efficiency and figure of merit, as the relative motion between a proof mass (e.g., a magnet) and a frame (which houses, for example, a coil) is very small.

Liquid bearing allows the energy harvester to have a large relative motion between the magnet and coil with low friction and makes a non-resonant harvester highly efficient in producing electrical power. Thus, non-resonant structures based on liquid bearing would be worthy to consider for generating power from human's walking motion.

Let us study a non-resonant vibration energy harvester similar to what is shown in Fig. 9.18a [12]. The magnet array is built up with a number of magnets (that are arranged with alternating north and south poles for a large magnetic flux change when the magnets move laterally in response to applied vibration), and is suspended by liquid bearing (preferably of ferrofluid). A planar coil array is composed of coils that have the same shape and size as the magnets.

The magnet array is levitated by a liquid bearing over the coil array, and is laterally displaced from the underlying coil array (attached to the acrylic chamber that moves along with the vibration source) with almost no friction between the magnet and coil arrays. The magnetic field reaches its maximum value near the boundary between two abutting magnets. Consequently, ferrofluid would be the best for the liquid bearing, since ferrofluid concentrates along the boundary automatically (due to the fact that it is a liquid that is magnetized in the presence of a magnetic field), as can be seen in Fig. 9.18b, and will follow the array of the magnets during vibration [13]. As the self-assembled liquid

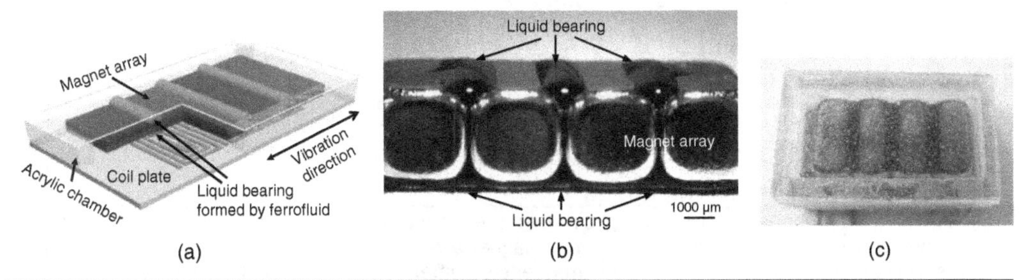

FIGURE 9.18 (a) Schematic of a non-resonant energy harvester with a magnet array levitated by liquid bearing formed by ferrofluid. (b) Photo of the self-assembled liquid bearing. The ferrofluid concentrates along the boundary of two abutting magnets automatically and forms the liquid bearing. Bearing on the bottom is flatter than the one at the top due to gravity pulling down the magnet. (c) Photo of a fabricated energy harvester based on liquid bearing with the acrylic frame cover appearing opaque due to superhydrophobic coating [12].

bearing allows a vibration source to produce a relative motion (between the magnet array and the coil array) with almost no friction, the relative motion induces electromotive force in the coil because of the magnetic flux gradient (with respect to the vibration direction), which is the largest near the boundary between two abutting magnets.

A prototype with an array of four NdFeB magnets and ferrofluid liquid bearing was made in a laser-cut acrylic frame that houses a coil array. The inside of the frame and the surface of the coil array were coated with a superhydrophobic coating, since hydrophobic surface reduces the friction of liquid bearing (Fig. 9.18c). The frame and the coil array were glued with epoxy resins. The fabricated energy harvester with ferrofluid liquid bearing was tested with in-plane vibration over 2–57 Hz on a linear actuator (Aerotech ACT115DL). Measured frequency responses of the energy harvester under various accelerations confirmed that the harvester indeed operated in non-resonant mode, as the harvested power increased in proportion to the frequency for a given acceleration, and harvested vibration energy over a broad frequency range (Fig. 9.19).

Beyond the peaking (not resonant) frequency (that is higher for a larger input acceleration) in Fig. 9.19a, the power output decreases as the frequency is increased for a given acceleration. What is physically happening is that before the peak, the power output is limited by the movable range of the magnet array, and increases with decreasing vibrational amplitude associated with increasing frequency for a given acceleration. However, once the frequency is higher than the peaking frequency, the magnet array touches the chamber at random points in time, and the relative velocity (the velocity between the magnet and coil arrays) is aperiodic and decreases as the frequency is increased for a given acceleration. According to a theoretical model [12], the critical vibration amplitude is determined only by the movable range of the magnet array, and is independent of the acceleration. The test results show the vibration amplitudes corresponding to the peaking frequencies at 0.5, 1, and 2 g accelerations are 0.55, 0.56, and 0.48 mm, respectively, which matches very closely to the critical vibration amplitude (0.55 mm) calculated with the model.

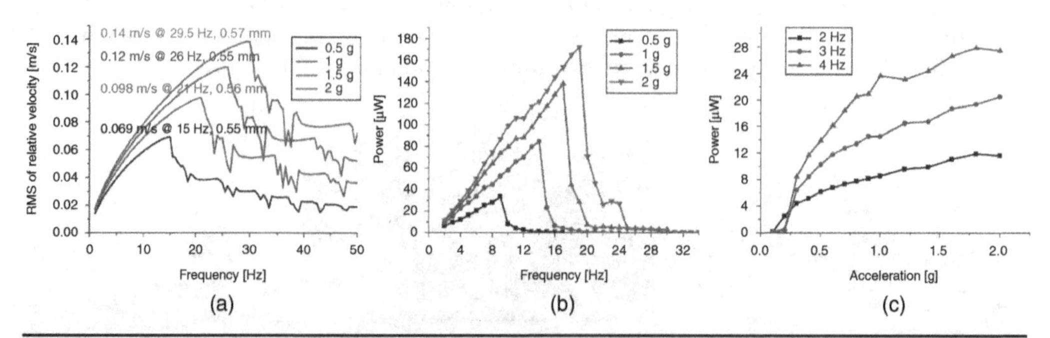

FIGURE 9.19 (a) Simulated relative velocity between the magnet array and the coil plate as a function of frequency under various accelerations, for the magnet array having a movable range of 2 mm. There is no resonance: The peaking (not resonant) frequencies differ at different accelerations and are 15, 21, 26, and 29 Hz under 0.5, 1, 1.5, and 2 g accelerations, respectively, while the vibration amplitudes at the peaking frequencies are 0.55, 0.56, 0.55, and 0.57 mm, respectively. (b) Measured frequency response of the non-resonant energy harvester with the movable range of the magnet array being 6 mm. The frequency where the power is the largest shifts as the applied acceleration increases, as simulated. (c) Measured power delivered to a matched load of 80 Ω as a function of applied acceleration at 2–4 Hz [12].

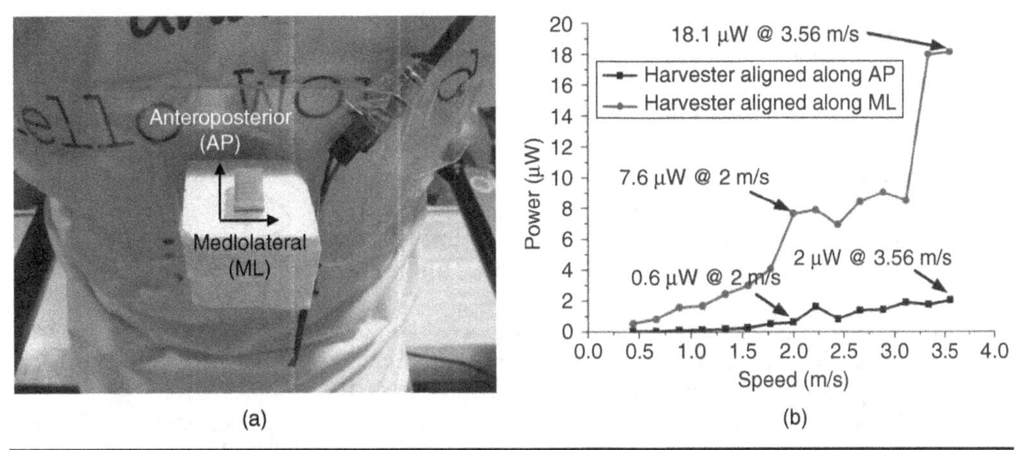

Figure 9.20 (a) A photo of the non-resonant electromagnetic energy harvester fixed on the back of a human body walking on a treadmill and (b) measured power output (to a matched load of 80 Ω) versus walking/running speed [12].

Note that for a resonant energy harvester, the output power is proportional to the square of the applied acceleration at a given frequency. But the output power of the non-resonant energy harvester based on liquid bearing is linearly proportional to the acceleration at a given frequency (Fig. 9.19c). The measurement with the fabricated harvester showed that the output power that could be delivered to a matched load of 80 Ω at the peaking frequency for acceleration of 0.5, 1, and 2 g were measured to be 5.58, 12.46, and 28.57 μW, respectively, as expected.

A non-resonant energy harvester (1.1 cc and 2.5 g) was placed on the back of a human walking on a treadmill, as shown in Fig. 9.20 [12], and was shown to deliver (to a matched load of 80 Ω) increasingly larger power as the walking/running speed was increased from 0.44 m/s (walking) to 3.56 m/s (running). The power level reached 18.1 μW at 3.56 m/s.

References

1. E. Sardini and M. Serpelloni, "An efficient electromagnetic power harvesting device for low-frequency applications," Sensors and Actuators A: Physical, 172, 2011, pp. 475–482.
2. C.B. Williams and R.B. Yates, "Analysis of a micro-electric generator for microsystems," Sensors and Actuators A: Physical, 52, 1996, pp. 8–11.
3. Q. Zhang and E.S. Kim, "Vibration energy harvesting based on magnet and coil arrays for watt-level handheld power source," Proceedings of the IEEE, 102(11), 2014, pp. 1747–1761.
4. W. Hayward, "Introduction to radio frequency design," American Radio Relay League, 1994.
5. J.M. Donelan, et al., "Biomechanical energy harvesting: Generating Electricity During Walking with Minimal User Effort," Science, **319**, 807 (2008).
6. Qingguo Li, Veronica Naing, and J Maxwell Donelan, "Development of a biomechanical energy harvester," Journal of NeuroEngineering and Rehabilitation, 6(22), 2009.

7. http://www.lightningpacks.com/.
8. L.C. Rome, et al., "Generating electricity while walking with loads," Science, 309, September 2005, pp. 1725–1728.
9. L. Zhao and E.S. Kim, "Analytical dual-charged-surfaces model for permanent magnet and its application in magnetic spring," IEEE Transactions on Magnetics, 56(9), September 2020, pp. 1–7.
10. Y. Wang, L. Zhao, A. Shkel, Y. Tang, and E.S. Kim, " Vibration energy harvester based on floating magnet for generating power from human movement," Solid-State Sensor and Actuator Workshop, Hilton Head Island, SC, June 5–9, 2016, pp. 404–407.
11. Y. Wang, Q. Zhang, L. Zhao, Y. Tang, A. Shkel, and E.S. Kim, "Vibration energy harvester with low resonant frequency based on flexible coil and liquid spring," Applied Physics Letter, 109, 203901, 2016; doi: 10.1063/1.4967498.
12. Y. Wang, Q. Zhang, L. Zhao, and E.S. Kim, "Non-resonant, electromagnetic broad-band vibration-energy harvester based on self-assembled ferrofluid liquid bearing," IEEE/ASME Journal of Microelectromechanical Systems, 26(4), 2017, pp. 809–819.
13. B. Assadsangabi, M. H. Tee, and K. Takahata, "Electromagnetic microactuator realized by ferrofluid-assisted levitation mechanism," IEEE/ASME Journal of Microelectromechanical Systems, 23(5), 2014, pp. 1112–1120.

Questions and Problems

Question 9.1 For an energy harvester having its source resistance two times a load resistance, how much more power can be delivered to the load, if a lossless impedance matching network is used to make the resistance seen by the load be the same as the load resistance?

Question 9.2 What will be the initial distance between the two magnets of the magnetic spring shown in Fig. 9.12, if we want 2-Hz resonant frequency?

Question 9.3 How much more power can be harvested, if one runs at 4 m/s, rather than walks at 2 m/s, with a vibrational energy harvester on her/his back? Assume that the travel distance by each step and the vertical height displacement on each step remain the same when the person changes her/his walking speed.

Question 9.4 How would Fig. 9.19b and c look like, if the device were a resonant vibration energy harvester rather than a non-resonant vibration energy harvester?

Problem 9.1 For $R_S = 50\ \Omega$ and $R_L = 8\ \Omega$, choose the inductance (L) and the capacitance (C) (of an impedance matching network shown below) that will make the maximum power delivered to R_L from a vibration energy harvester, when the vibrational energy is mostly at 2 Hz.

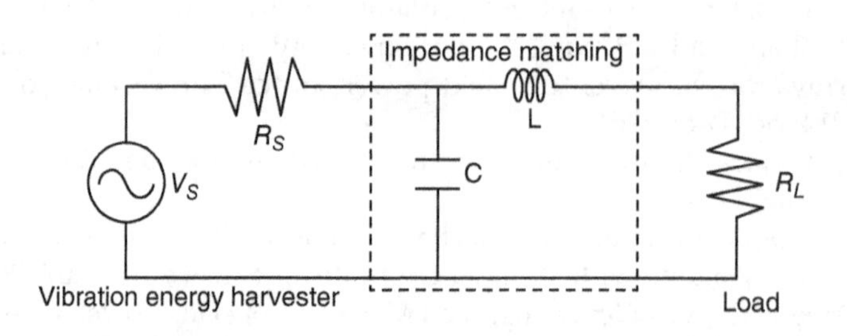

Problem 9.2 Knowing that the potential energy for a mass m is $E = mgh$, where g and h are the gravitational acceleration and height, respectively, how much mechanical power is there when a human carries a 1-kg mass on her/his back and walks two steps per second. Assume that each step of the walk accompanies 5 cm of net vertical displacement.

Problem 9.3 Derive the equation ($\sqrt{g/h}$) for the resonant frequency of the power generator (based on floating magnet through diamagnetic levitation) shown in Fig. 9.16.

Problem 9.4 Using Fig. 9.17c, show that the spring constant for the liquid spring is proportional to $\frac{HW\rho}{L^2}$, where H, W, L, and ρ are chamber height, chamber width, chamber length, and density of magnetic particles in ferrofluid, respectively.

CHAPTER 10

Electronic Noises, Interface Circuits, and Oscillators

In this chapter, we will study noise sources in electronic circuits and see how those may limit the noise floor or the minimum detectable signal for microelectromechanical systems (MEMS) sensors. Then we will study signal-to-noise ratio (SNR) and frequency response of voltage and charge amplifiers for piezoelectric sensors, particularly for piezoelectric microphone, and learn why the input resistance needs to be high and how to deal with electromagnetic interference at the high resistance node. We also study impacts of poles and zeros on frequency responses of linear systems, and use those to analyze and design oscillators based on bulk acoustic-wave resonators.

10.1 Input Referred Noise

10.1.1 Noise Sources

Shot noise of a diode or p-n junction is related to DC current (I_D) passing through a diode, and is independent of temperature and frequency. Mean square value of the shot noise current is $\overline{i_n^2} = 2qI_D\Delta f$, where q, I_D, and Δf are electronic charge (=1.6×10^{-19} C), diode's DC current, and observational bandwidth, respectively. For example, a diode carrying a DC current of 1 mA has $\overline{i_n^2} = 2 \times 1.6 \times 10^{-19} \times 10^{-3} \Delta f = 3.2 \times 10^{-16}$ A^2 for Δf = 1 MHz, and the root mean square (RMS) of the shot noise current is $\sqrt{\overline{i_n^2}} = \sqrt{2qI_D\Delta f} = 17$ nA.

In addition to shot noise, p-n junction has $1/f$ noise, mostly from surface states that provide energy levels between the conduction band edge (E_c) and the valence band edge (E_v) for electrons to jump from the conduction band to the valence band. This kind of noise is heavily dependent on fabrication technology that determines the surface states. The frequency where $1/f$ noise becomes larger than shot noise is usually lower than 10 kHz.

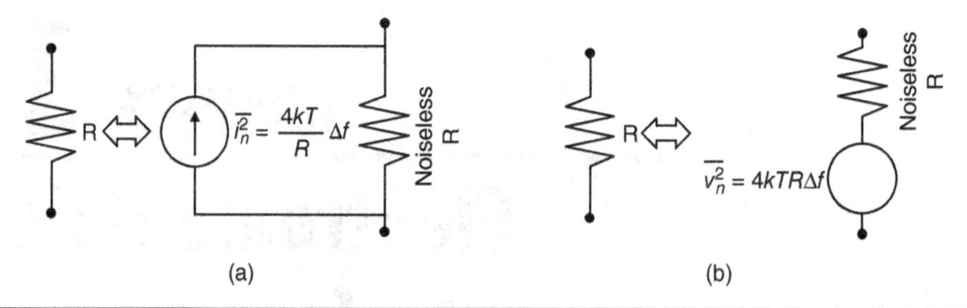

FIGURE 10.1 Thermal noise of a resistor modeled with an equivalent (a) current source and (b) voltage source.

Any physical resistor with resistance R has thermal noise that is due to molecules moving around to maintain thermal equilibrium, and is dependent on temperature but independent of frequency (i.e., it is white noise that is uniformly distributed over frequency). Thermal noise can be represented with an equivalent current or voltage source as shown in Fig. 10.1; $\overline{i_n^2} = \frac{4kT}{R}\Delta f$ or $\overline{v_n^2} = 4kTR\Delta f$, where k and T are Boltzmann constant (1.38×10^{-23} m^{2}kg^{-2}K^{-1} or J/K or VC/K) and temperature, respectively. Since 4 kT = 1.66×10^{-20} VC at 300 K, a 1-kΩ resistor at 300 K has $\overline{v_n^2} = (4\ \text{nV}/\sqrt{\text{Hz}})^2\Delta f$.

10.1.2 Equivalent Input-Referred Voltage and Current Noise Sources

For a linear, time-invariant system, the total noise power spectral density is the sum of the individual power spectral densities, as it is the case for linear resistors. For example, the net noise of two resistors can be modeled as shown in Fig. 10.2.

An amplifier with noise (Fig. 10.3a) can be represented with a noiseless amplifier with equivalent input noise voltage ($\overline{v_{eq}^2}$) and current ($\overline{i_{eq}^2}$) sources as shown in Fig. 10.3b.

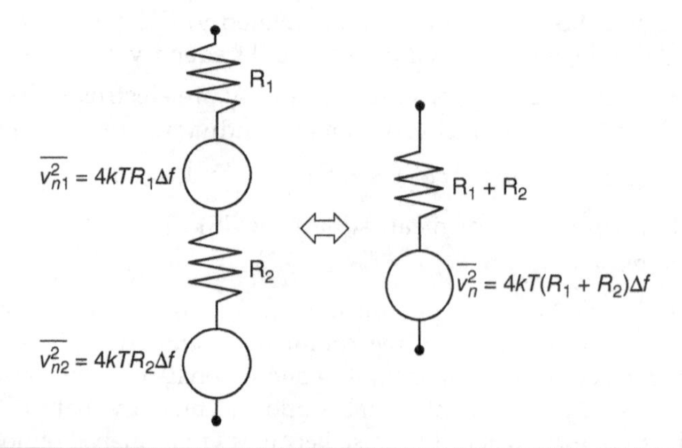

FIGURE 10.2 Equivalent representation of two resistors, each contributing thermal noise with one resistor and one equivalent noise source.

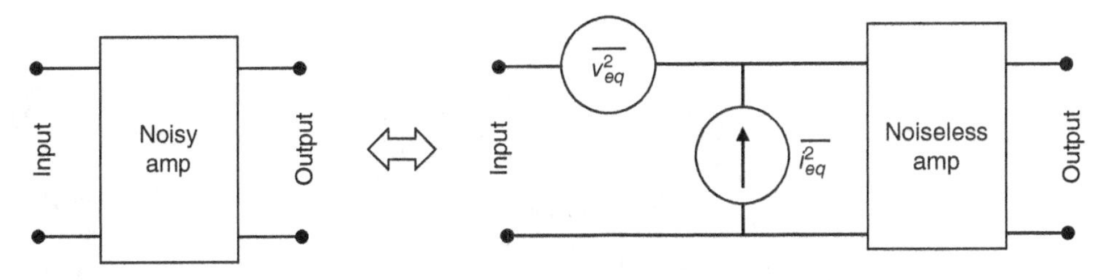

FIGURE 10.3 Equivalent input-referred voltage and current noise for an amplifier.

In general, $\overline{v_{eq}^2}$ and $\overline{i_{eq}^2}$ are correlated and dependent on frequency, and both $\overline{v_{eq}^2}$ and $\overline{i_{eq}^2}$ need to be considered. However, if the impedance of the signal source is much smaller than the input impedance of the amplifier, $\overline{v_{eq}^2}$ will dominate over $\overline{i_{eq}^2}$, since most of the current from $\overline{i_{eq}^2}$ will flow through the small source impedance. On the other hand, if the impedance of the signal source is much higher than the input impedance of the amplifier, $\overline{i_{eq}^2}$ will dominate over $\overline{v_{eq}^2}$, since most of the current from $\overline{i_{eq}^2}$ will flow through the amplifier's input impedance, while $\overline{v_{eq}^2}$ has very little effect on the voltage at the input of the amplifier. In most cases, source impedance is usually much smaller than input impedance of an amplifier, and only $\overline{v_{eq}^2}$ needs to be considered.

To obtain $\overline{v_{eq}^2}$ of an amplifier, one can take two cases in Fig. 10.3: an amplifier with noises (a real amplifier) and a noiseless amplifier with $\overline{v_{eq}^2}$ and $\overline{i_{eq}^2}$, and then short-circuit the inputs on both cases. Since the short-circuiting makes all the current from $\overline{i_{eq}^2}$ go through the short circuit, one can set the two outputs equal (i.e., $\overline{v_{o1}^2} = \overline{v_{o2}^2}$) and solve for $\overline{v_{eq}^2}$. Similarly, by open-circuiting the inputs for the two cases in Fig. 10.3, one can make the voltage from $\overline{v_{eq}^2}$ irrelevant to the output and can solve for $\overline{i_{eq}^2}$ by setting $\overline{v_{o1}^2} = \overline{v_{o2}^2}$ with the inputs open-circuited.

Minimum detectable signal (MDS) can be defined to be the signal level (at the source before an amplifier) at which the signal-to-noise ratio (SNR) at the output of the amplifier is equal to anywhere between 1 and 3. For example, if an amplifier has an equivalent input noise voltage ($\sqrt{\overline{v_{eq}^2}}$) as shown in Fig. 10.4a, the root-mean-square (RMS) input-referred noise voltage from 100 Hz to 10 kHz is equal to

$$\sqrt{\int_{100}^{10,000} (7 \times 10^{-9})^2 df} = 7 \times 10^{-9} \times \sqrt{10,000 - 100} = 0.7\ \mu\text{V}. \tag{10.1}$$

And the *peak-to-peak* input-referred noise voltage is $(2\sqrt{2}) \times 0.7\ \mu\text{V} = 2\ \mu\text{V}$. Thus, if the signal picked up by a sensor (with low source impedance) has 6 μV peak-to-peak voltage at any frequency between 100 Hz and 10 kHz, the amplified signal will look like as shown in Fig. 10.4b with SNR equal to 3, if the observation bandwidth is limited to 100 Hz–10 kHz.

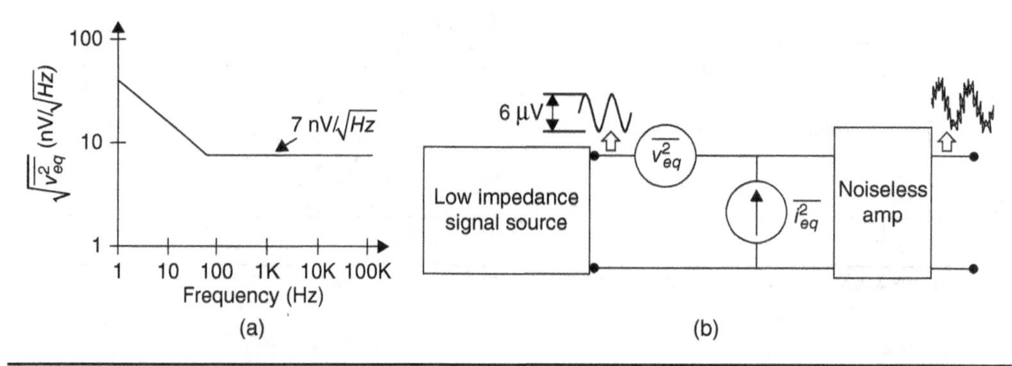

FIGURE 10.4 (a) Noise spectrum of an amplifier. (b) Signal at the output of the amplifier, if the signal input to the amplifier (presented by a low impedance signal source such as a sensor or an antenna) is sinusoidal with a peak-to-peak voltage of 6 μV at any frequency between 100 Hz and 10 kHz and if the observation bandwidth is limited to 100 Hz–10 kHz.

10.2 Voltage Amplifier versus Charge Amplifier for Piezoelectric Sensors

A piezoelectric microphone (based on a unimorph structure such as the one shown in Fig. 7.20) and a voltage amplifier A with a very high input resistance (> 10^{12} Ω) can equivalently be modeled with a circuit in Fig. 10.5a. A diode D (shown in Fig. 10.5a) is present only to provide a very high resistance (> 10^{10} Ω) DC path so that a small leakage path may be present to drain off stray charges that might otherwise accumulate on the exposed electrode and make the DC bias of the amplifier off. The capacitor C_e is comprised of parasitic capacitances due to the amplifier, the diode, and the conductor line connecting the microphone to the amplifier.

The capacitances on the equivalent circuit for the microphone are due to the ZnO layer (C_p) and the two insulating layers on the top and bottom of the ZnO layer (C_1 and C_2). The current source represents the piezoelectric activity of the ZnO that

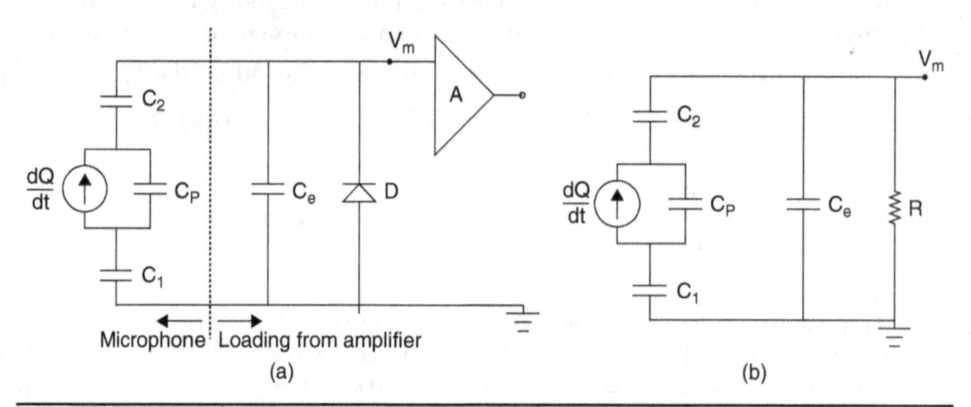

FIGURE 10.5 (a) Equivalent circuit for a piezoelectric microphone with a voltage amplifier. (b) Simplified circuit with R and C_e equivalently representing the loading from the amplifying circuitry.

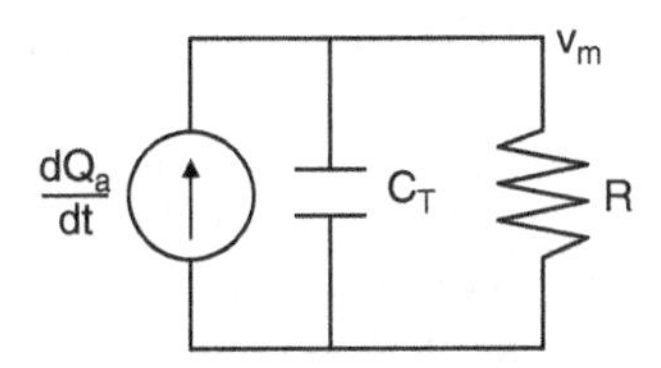

Figure 10.6 A further simplified equivalent circuit from Fig. 10.5b.

becomes strained when pressure deflects the diaphragm. For a sinusoidal pressure variation at a radial frequency ω, there is a piezoelectrically induced sinusoidal charge of $Q_o e^{j\omega t}$, where Q_o is proportional to the applied pressure. The current delivered by the current source is dQ/dt $(=j\omega Q_o e^{j\omega t})$.

The circuit shown in Fig. 10.5a can be simplified to Fig. 10.5b, which is further simplified to Fig. 10.6 with an equivalent current source (dQ_a/dt) and total capacitance C_T as follows:

$$\frac{dQ_a}{dt} = \frac{dQ/dt}{1 + C_p\left(\frac{1}{C_1} + \frac{1}{C_2}\right)} \tag{10.2}$$

$$C_T = C_e + \frac{C_p}{1 + C_p\left(\frac{1}{C_1} + \frac{1}{C_2}\right)} \tag{10.3}$$

Solving for v_m, the unamplified output of the microphone, we find

$$v_m = \frac{j\omega R}{1 + j\omega RC_T} \cdot \frac{Q_o e^{j\omega t}}{1 + C_p\left(\frac{1}{C_1} + \frac{1}{C_2}\right)} \tag{10.4}$$

In the frequency plot of v_m, shown in Fig. 10.7, v_m rolls off 20 dB per decade, as frequency is reduced, below the frequency $(f_{3\text{dB}})$ where v_m is 3 dB lower than that in the high-frequency range. The 3-dB frequency is $f_{3\text{dB}} = 1/(2\pi RC_T)$, and it is important to

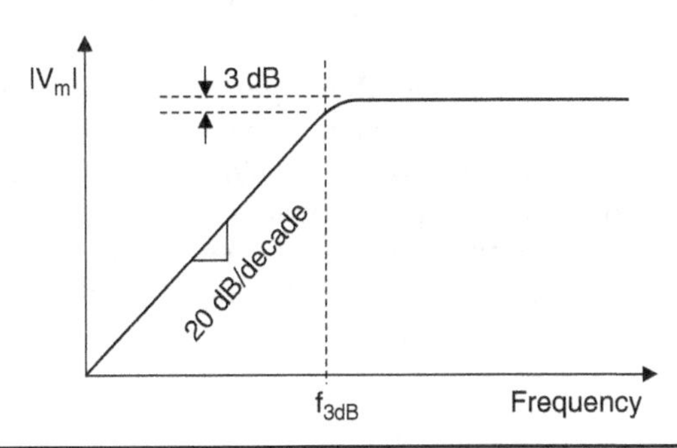

Figure 10.7 The magnitude of V_m (in Fig. 10.5) versus frequency showing a roll-off of 20 dB/decade below $f_{3\text{dB}}$.

keep f_{3dB} below the low end of the audio frequency range for uniform (and unreduced) sensitivity down to that frequency.

Consequently, the overall sensitivity depends on the input resistance (as well as the input capacitance) especially at low frequencies. For typical values of C_1, C_2, and C_e for a piezoelectric microphone, in order to have a good low frequency response down to 20 Hz, the input resistance R has to be greater than $10^9\ \Omega$, which limits the choice of the input transistors of the amplifier to field effect transistors (FETs). Also, to achieve $R > 10^9\ \Omega$, the semi-insulating ZnO layer has to be sandwiched with two insulating layers (C_1 and C_2).

For $\omega RC_T >> 1$, which is the condition for a good low frequency response, we have

$$v_m \approx \frac{1}{C_T} \cdot \frac{Q_o e^{j\omega t}}{1 + C_p(1/C_1 + 1/C_2)} = \frac{Q_a}{C_T} \tag{10.5}$$

With a *voltage* amplifier, the microphone sensitivity depends on the two insulating layers (C_1 and C_2), as can be seen in Eq. (10.5).

Now, to obtain an equation for the signal-to-noise (S/N) ratio, the most critical performance parameter, we use a voltage source (v_n) and a current source (i_n) to represent all the noise sources of the amplifier and connecting electrical lines. Since the voltage at the input node m induced by the noise–current source (i_n) is equal to $i_n/j\omega C_T$ for $R >> 1/j\omega C_T$, the S/N ratio is

$$\mathrm{S/N} \approx \frac{Q_a}{C_T v_n + \dfrac{i_n}{j\omega}} \tag{10.6}$$

10.2.1 Charge Amplifier

As shown in the previous section, with a voltage amplifier, the overall sensitivity depends on the input resistance (as well as the input capacitance), especially at low frequencies. In order to have a good low frequency response down to 20 Hz, the input resistance R has to be greater than $10^9\ \Omega$. To achieve $R > 10^9\ \Omega$, we have to sandwich the semi-insulating ZnO layer with two insulation layers (C_1 and C_2), which reduce the sensitivity through charge sharing, as can be seen in the following equation obtained from the previous section:

$$v_m \approx \frac{Q_a}{C_T} = \frac{Q_o e^{j\omega t}}{C_p + C_e C_p(1/C_1 + 1/C_2) + C_e} \tag{10.7}$$

According to the equation above, the larger the C_e is, the higher the sensitivity gets reduced by the two insulating layers (C_1 and C_2).

A charge amplifier, however, offers a distinct advantage in that the overall sensitivity is roughly independent of input impedance, as shown in the following derivations. When a piezoelectric microphone (Q_a and C_T being defined as in the previous section) is connected to a charge amplifier, as shown in Fig. 10.8, the output voltage of the charge amplifier due to the signal source is

$$v_o^a = \frac{-Q_a}{\left(1 + \dfrac{1}{A}\right)C_F + \dfrac{1}{A}\left(C_T + \dfrac{1}{j\omega R}\right)} \approx \frac{-Q_a}{C_F} \tag{10.8}$$

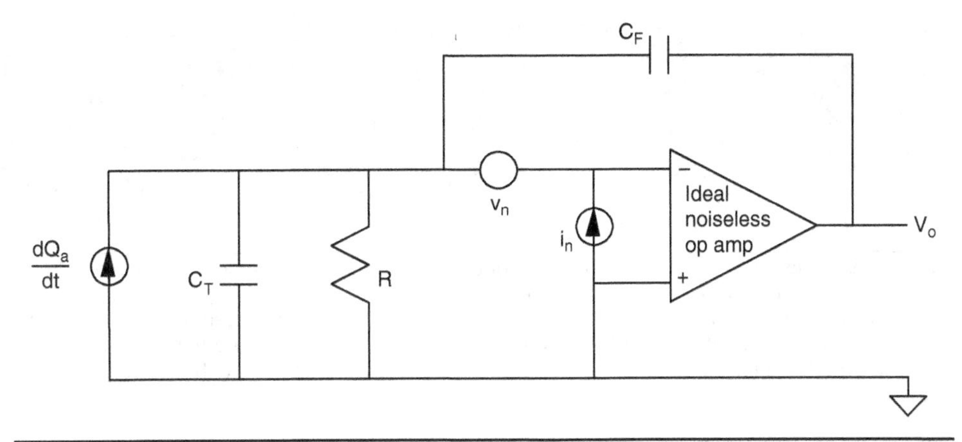

FIGURE 10.8 Simplified equivalent circuit for the piezoelectric microphone and a charge amplifier.

where A is an open-loop gain of the op amp. Thus, the overall sensitivity is independent of input impedance as long as A and R are large enough, and consequently, we can remove the two insulating layers (C_1 and C_2 shown in Fig. 10.5). Now, to obtain the signal-to-noise ratio, the output voltages due to the noise sources i_n and v_n are obtained as follows.

$$V_o^{i_n} = \frac{-i_n/j\omega}{(1+1/A)C_F + (C_T + 1/j\omega R)/A} \approx \frac{-i_n}{j\omega C_F}$$

$$V_o^{v_n} = \frac{[1 + j\omega R(C_F + C_T)]v_n}{j\omega RC_F + \frac{1}{A}[1 + j\omega R(C_F + C_T)]} \approx \left(1 + \frac{C_T}{C_F} + \frac{1}{j\omega RC_F}\right)v_n \tag{10.9}$$

Hence, the SNR with a charge amplifier becomes

$$S/N \approx \frac{Q_a}{\left(C_F + C_T + \frac{1}{j\omega R}\right)v_n + \frac{i_n}{j\omega}} \tag{10.10}$$

Consequently, the S/N ratio is typically worse with a charge amplifier than with a voltage amplifier due to C_F and R.

10.3 Electromagnetic Interference

For a sensor with high source impedance, if its amplifier has to have a high input impedance (not to lower the signal picked up by a sensor at a low frequency), electromagnetic interference (EMI) is a dominant noise source, since a high impedance node (between the sensor and the amplifier) is very susceptible to EMI. The noise picked up by such high impedance node directly competes with the sensor signal, and reduces SNR. Consequently, such high impedance node needs to be shielded from EMI by enclosing it in a metal box, as shown in Fig. 10.9.

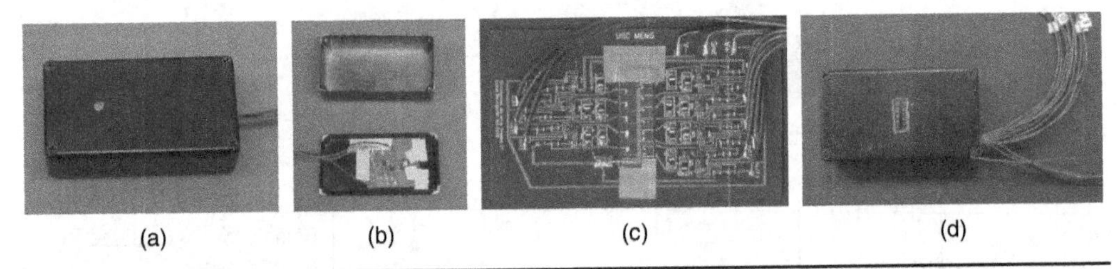

FIGURE 10.9 Photos of piezoelectric microphones (and their amplifying circuits) packaged in a metal box: (a) a metal box with a circular hole for sound input, (b) the box with its cover opened to show a microphone and a preamp, (c) an array of 13 microphones mounted and wire bonded to a preamplifier printed circuit board (PCB), and (d) a metal box containing the microphone array and PCB with a rectangular opening that exposes the diaphragms of the microphones to environment for sound input [1].

10.3.1 Noise Reduction through Low-Pass Filter

Noise at the output of an amplifier can be reduced with a low-pass filter that passes signals at low frequencies and rejects those at high frequencies, as shown in Fig. 10.10a, if the amplifier is to amplify low frequency signals such as audio signals. A low-pass filter can be implemented in an amplifier based on op amp, as shown in Fig. 10.10b and c, by adding a capacitor. The –3 dB frequency of a single-pole low-pass filter is the frequency at which the reactance of the capacitor equals the resistance of the resistor, and is determined by R_fC_f time constant.

10.4 Poles and Zeros

A transfer function $H(\omega)$ or $H(s)$ relates input $V_{in}(\omega)$ and output $V_{out}(\omega)$ as $H(\omega) = \frac{V_{out}(\omega)}{V_{in}(\omega)}$, where $V_{out}(\omega)$, $V_{in}(\omega)$, and $H(\omega)$ are phasors. For example, a simple RC circuit (Fig. 10.11a) has

$$H(\omega) = \frac{V_{out}(\omega)}{V_{in}(\omega)} = \frac{1/(j\omega C)}{R + 1/(j\omega C)} = \frac{1}{1 + j\omega RC}. \tag{10.11}$$

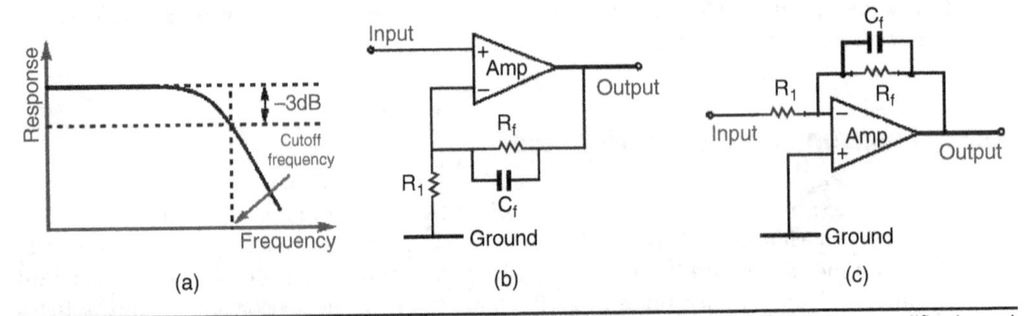

FIGURE 10.10 (a) Frequency response of a low-pass filter (LPF), (b) LPF in a noninverting amplifier based on an op amp, and (c) LPF in an inverting amplifier based on an op amp. The cutoff frequency is determined by R_fC_f time constant.

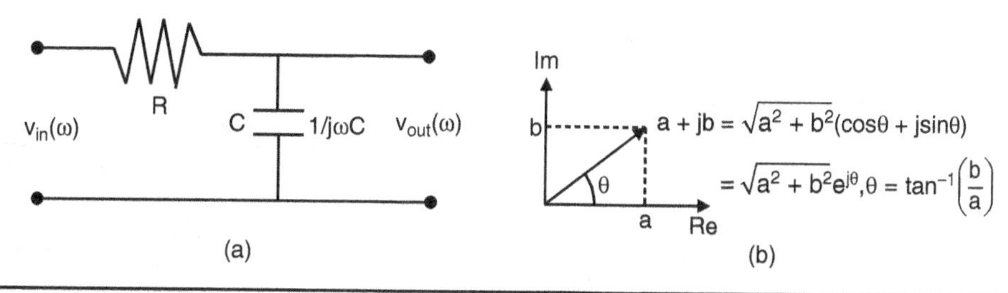

FIGURE 10.11 (a) A RC circuit resulting in a single pole. (b) A complex number $(a + jb)$ written with magnitude $\sqrt{a^2+b^2}$ and phase $\theta = \tan^{-1}(b/a)$.

Now, as shown in Fig. 10.11b, any complex number $(a+jb)$ can be converted into $Ae^{j\theta}$, where $A = \sqrt{a^2+b^2}$ and $\theta = \tan^{-1}(b/a)$ are the magnitude and phase, respectively. Consequently, any transfer function $H(\omega)$ can be converted to $|H(\omega)|e^{j\theta}$.

For the simple RC circuit (shown in Fig. 10.11a), we have the following for the magnitude $|H(\omega)|$ and phase θ of the transfer function $H(\omega)$ from Eq. (10.11):

$$|H(\omega)| = \sqrt{H(\omega)\cdot H^*(\omega)} = \sqrt{\frac{1}{1+j\omega RC}\cdot\frac{1}{1-j\omega RC}} = \sqrt{\frac{1}{1+(\omega RC)^2}} = \sqrt{\frac{1}{1+(\omega/\omega_{3dB})^2}} \quad (10.12)$$

where $\omega_{3dB} \equiv \dfrac{1}{RC}$,

and
$$\theta = \tan^{-1}\frac{\mathrm{Im}[H(\omega)]}{\mathrm{Re}[H(\omega)]} = \tan^{-1}\left(-\frac{\omega}{\omega_{3dB}}\right) = -\tan^{-1}\left(\frac{\omega}{\omega_{3dB}}\right). \quad (10.13)$$

In obtaining Eq. (10.13) for the phase θ, we need to convert Eq. (10.11) into $a+jb$ by multiplying the numerator and denominator with the complex conjugate as shown below:

$$H(\omega) = \frac{1}{1+j\omega/\omega_{3dB}}\cdot\frac{1-j\omega/\omega_{3dB}}{1-j\omega/\omega_{3dB}} = \frac{1-j\omega/\omega_{3dB}}{1+(\omega/\omega_{3dB})^2} \quad (10.14)$$

From Eq. (10.12), we see

$$|H(\omega)| = \sqrt{\frac{1}{1+(\omega/\omega_{3dB})^2}}$$
$$\begin{cases} \approx 1 & \text{for } \omega << \omega_{3dB} \\ \approx \sqrt{\dfrac{1}{(\omega/\omega_{3dB})^2}} = \dfrac{\omega_{3dB}}{\omega} & \text{for } \omega >> \omega_{3dB} \end{cases} \quad (10.15)$$

Thus, the plot of $|H(\omega)|_{dB} \equiv 20\log|H(\omega)|$ versus frequency in log scale looks like as shown in Fig. 10.12a, while the phase θ [Eq. (10.13)] versus frequency (in log scale) is as shown in Fig. 10.12b. In summary, the simple RC circuit shown in Fig. 10.11 has a transfer function with a single pole as in Eq. (10.11), whose magnitude and phase are as shown in Fig. 10.12.

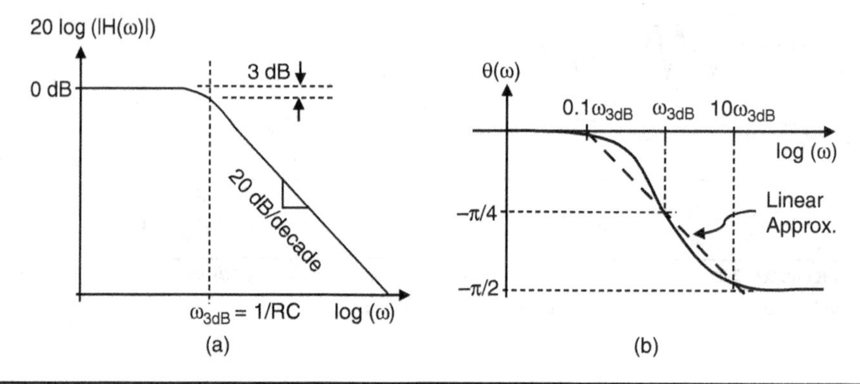

FIGURE 10.12 One-pole transfer function: (a) Bode plot of the magnitude $|H(\omega)|_{dB} \equiv 20\log|H(\omega)|$ versus frequency in log scale. (b) The phase $\theta(\omega)$ in linear scale versus frequency in log scale.

Now, let's consider a two-pole transfer function

$$H(\omega)=\left(\frac{1}{1+j\omega/\omega_{p1}}\right)\cdot\left(\frac{1}{1+j\omega/\omega_{p2}}\right)=H_{\omega_{p1}}(\omega)\cdot H_{\omega_{p2}}(\omega)\cdot \tag{10.16}$$

The magnitude $|H(\omega)|_{\text{dB}} \equiv 20\log|H(\omega)|$ is

$$20\log|H(\omega)|=20\log\left(\left|H_{\omega_{p1}}(\omega)\right|\cdot\left|H_{\omega_{p2}}(\omega)\right|\right)=20\log\left|H_{\omega_{p1}}(\omega)\right|+20\log\left|H_{\omega_{p2}}(\omega)\right| \tag{10.17}$$

The phase $\theta(\omega)$ can be obtained by noting

$$H(\omega)=\left(\left|H_{\omega_{p1}}\right|e^{j\theta_{\omega p1}}\right)\cdot\left(\left|H_{\omega_{p2}}\right|e^{j\theta_{\omega p2}}\right)=\left|H_{\omega_{p1}}\right|\cdot\left|H_{\omega_{p2}}\right|e^{j(\theta_{\omega p1}+\theta_{\omega p2})}=|H(\omega)|e^{j\theta(\omega)}, \tag{10.18}$$

where $\theta(\omega)=\theta_{\omega_{p1}}+\theta_{\omega_{p2}}$. Thus, the magnitude and phase of the two-pole transfer function are approximately as drawn in Fig. 10.13.

The impact of the phase θ in the transfer function can be seen in the input and output voltages in time, when they are compared to each other in time, as shown in

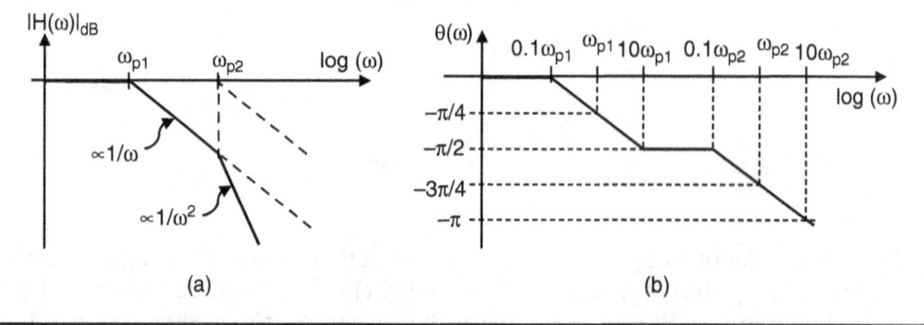

FIGURE 10.13 A two-pole transfer function: (a) Bode plot of the magnitude $|H(\omega)|_{dB} \equiv 20\log|H(\omega)|$ versus frequency in log scale. (b) The phase $\theta(\omega)$ in linear scale versus frequency in log scale.

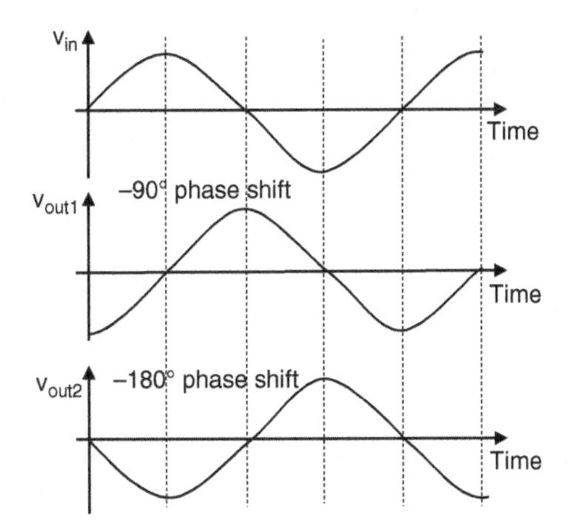

FIGURE 10.14 Phase lagging in the output voltage (compared to the input voltage) in time scale due to –90° and –180° phase shift in the transfer function.

Fig. 10.14. With a negative phase shift, the output voltage lags behind the input in time. However, a positive phase shift that occurs with a certain type of zero (i.e., a left-half-plane [LHP] zero, not a right-half-plane zero, as explained in the next paragraph) can make the output lead the input in time.

A system having a single zero at ω_z has a transfer function $H(\omega) = 1 \pm j\omega/\omega_z$, whose magnitudes and phase are shown in Fig. 10.15. If $H(\omega) = 1 + j\omega/\omega_z$, the zero is at the left half of the so-called *s*-plane, and the phase varies from 0 to 90° as a function of frequency. But if $H(\omega) = 1 - j\omega/\omega_z$, the zero is at the right half of the so-called *s*-plane, and the phase varies from 0 to –90°.

The circuit shown in Fig. 10.16a has one pole and one zero with the following transfer function

$$H(\omega) = \frac{V_{out}}{V_{in}} = \left(\frac{R_2}{R_1 + R_2}\right) \frac{1 + j\omega R_1 C}{1 + j\omega[R_1 R_2/(R_1 + R_2)]C} \tag{10.19}$$

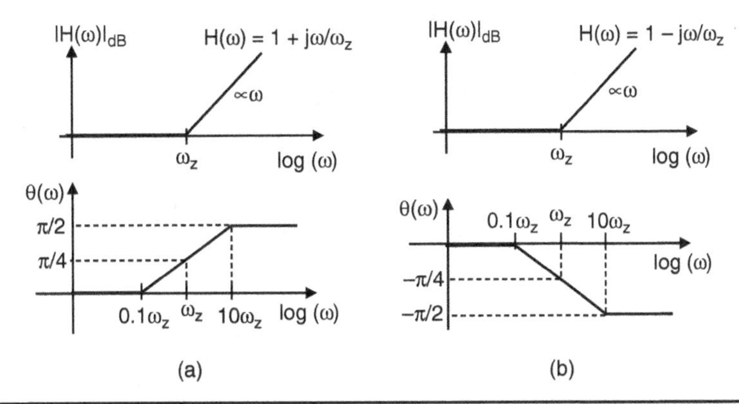

FIGURE 10.15 The magnitude and phase of the transfer function with a single zero versus frequency: (a) for $H(\omega) = 1 + j\omega/\omega_z$ and (b) for $H(\omega) = 1 - j\omega/\omega_z$.

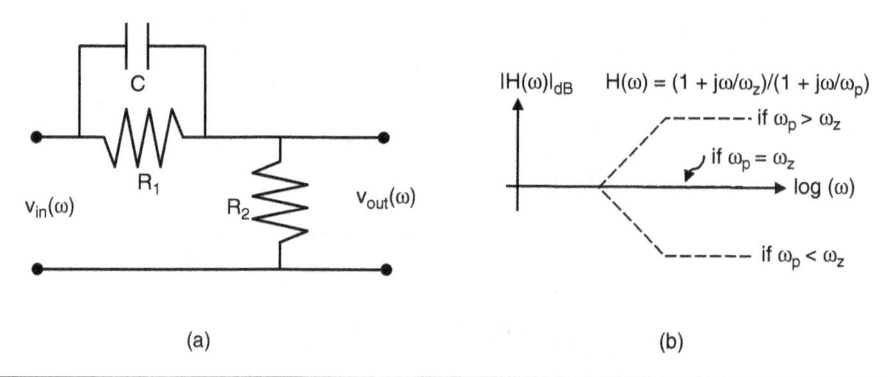

FIGURE 10.16 (a) A circuit with one pole and one zero and (b) the magnitude of the transfer function versus frequency as a function of the relative position of the pole with respect to the zero.

The magnitude of the transfer function versus frequency is shown in Fig. 10.16b as a function of the value of the pole ω_p (= $1/[(R_1\backslash\backslash R_2)C]$ for Fig. 10.16a) compared to that of the zero ω_z (= $1/[R_1C]$ for Fig. 10.16a). In case of Fig. 10.16a, though, ω_p will never be less than ω_z since $R_1\backslash\backslash R_2$ will never be greater than R_1.

10.5 Design and Analysis of Oscillators Based on Bulk Acoustic-Wave Resonators

Oscillators are used in wireless transceivers, waveform generators, clock signal generation, sensors, etc. Some oscillators are made into a voltage-controlled oscillator (VCO) so that the oscillation frequency can be controlled by an electrical voltage for phase lock loop (PLL), etc. In this section, we will develop techniques based on feedback to design and analyze oscillators built on bulk acoustic-wave resonators (BAR).

Referring to Fig. 10.17 where s is used instead of ω to indicate the frequency dependency, we have $v_o(s) = a(s)v_\varepsilon(s) = a(s)[v_i(s) - v_{fb}(s)] = a(s)[v_i(s) - f(s)v_o(s)]$, where $a(s)$ and $f(s)$ are the gain of the basic amplifier and the gain of the feedback, respectively. Consequently, the transfer function is

$$H(s) = \frac{v_o(s)}{v_i(s)} = \frac{a(s)}{1 + a(s)f(s)} \tag{10.20}$$

For a basic amplifier $a(s)$ with three poles and a feedback $f(s)$ with a real and positive amplification, the transfer function would look like

$$H(s) = \frac{a(s)}{1+a(s)f(s)} = \frac{A}{(1-s/s_1)(1-s/s_2)(1-s/s_3)} \tag{10.21}$$

Though all the three poles of $a(s)$ are in the left-half-plane (LHP) of s-plane, the closed-loop transfer function $H(s)$ may have poles in the right-half-plane (RHP) at a sufficiently

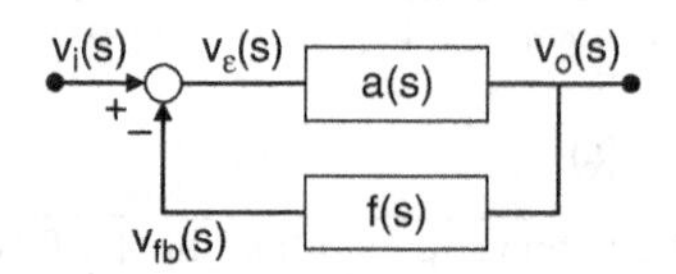

FIGURE 10.17 An amplifier with feedback $f(s)$ and open-loop gain $a(s)$.

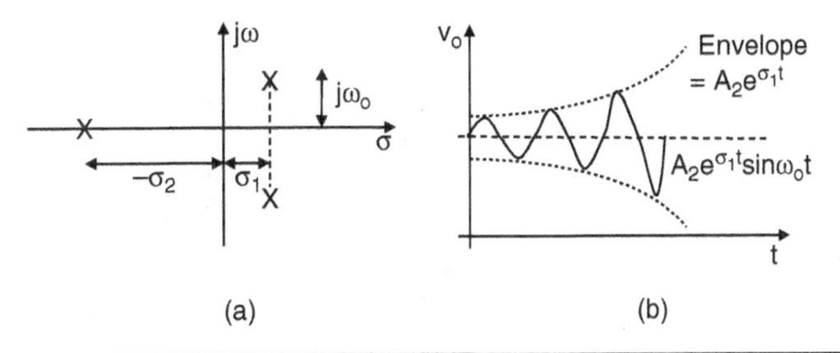

FIGURE 10.18 (a) Two right-half-plane (RHP) poles and one left-half-plane (LHP) pole in an s-plane and (b) transient response of the output voltage due to the two RHP poles.

large loop gain $a(s)f(s)$, such as the two shown in Fig. 10.18a. Consequently, the response of the circuit to any input is $v_o = \underbrace{f(\text{excitation})}_{\text{forced}} + \underbrace{A_1 e^{-\sigma_2 t} + A_2 e^{-\sigma_1 t} \sin \omega_0 t}_{\text{natural}}$, where A_1 and A_2 depend on initial conditions. The second term of the natural response is sketched in Fig. 10.18b.

As the voltage increases, the circuit's nonlinearities reduce the loop gain and limit the voltage level to a steady-state level, as indicated in Fig. 10.19. At the steady state of the oscillation, the loop gain (i.e., the so-called large-signal loop gain) is 1 (if not, the signal will continue to grow), while the total phase shift around the loop is 0°.

Pierce oscillator can be made by placing a bulk acoustic-wave resonator (BAR) between input and output of an inverting amplifier as shown in Fig. 10.20. The inverting amplifier is represented with an equivalent input resistance R_i (which is almost infinite for metal–oxide–semiconductor field-effect transistor [MOSFET]) and a dependent current source with small-signal transconductance g_m. The circuit oscillates at a frequency where L_{eff} of the BAR [Eq. (4.71)] resonates with $\frac{C_1 C_2}{C_1 + C_2}$, and the oscillation frequency is $\omega_0 = 1/\sqrt{L_{\text{eff}} \frac{C_1 C_2}{C_1 + C_2}}$.

10.5.1 Colpitts LC Oscillators

Before studying BAR-based Pierce oscillator with MOSFET for design and analysis, let us consider Colpitts oscillators based on L and C with MOSFET, shown in Fig. 10.21 without the DC bias circuit, in order to see necessary conditions for oscillation. In Fig. 10.21a, the small-signal input and output voltages (i.e., before the signal gets large

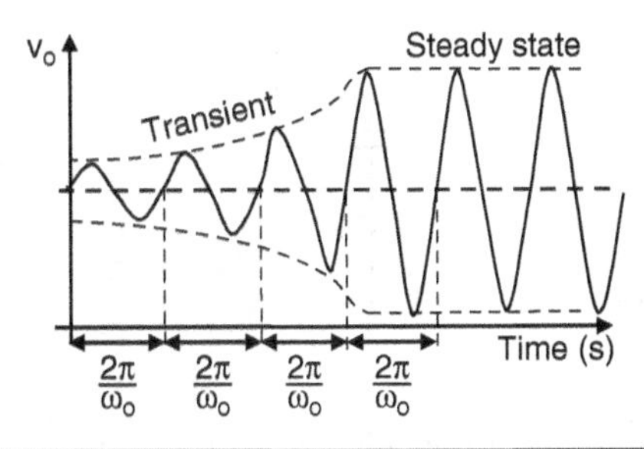

FIGURE 10.19 Oscillation voltage at steady state is set by the circuit nonlinearities as the voltage level increases.

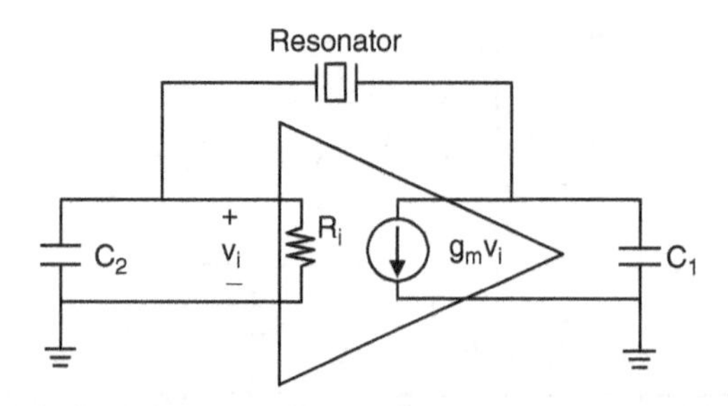

FIGURE 10.20 Pierce oscillator based on bulk acoustic wave resonator such as quartz crystal.

and reaches a steady-state oscillation) are related through the capacitor divider as $v_i = v_o \dfrac{C_1}{C_1 + C_2}$ and also through the voltage gain of the common gate amplifier $v_o = g_m R_L v_i$, where g_m is the small-signal transconductance of the MOSFET. Thus, the so-called initial gain A_e is $A_e = \dfrac{v_o}{v_i} = g_m R_L \dfrac{C_1}{C_1 + C_2}$, which needs to be greater than 1 for the circuit to oscillate and is typically designed to be 3–4. The oscillation frequency is determined by L and the serial combination of C_1 and C_2, and it is $\omega_0 = 1/\sqrt{L\dfrac{C_1 C_2}{C_1 + C_2}}$. Consequently, at the oscillation frequency, $\dfrac{1}{j\omega L} = -j\omega \dfrac{C_1 C_2}{C_1 + C_2}$, from which we see

$$j\omega L + \frac{1}{j\omega C_1} + \frac{1}{j\omega C_2} = 0, \tag{10.22}$$

the phase condition that needs to be satisfied for the circuit to oscillate.

In case of Fig. 10.21b (common-source Colpitts LC oscillator), there is a 180° phase shift from the input voltage to the output voltage through the MOSFET amplifier (which is configured to be a common-source amplifier which automatically carries 180° phase shift). Now, ignoring R_L, we see the reverse transmission from the output to the input to be $\dfrac{v_i}{v_o} \approx \dfrac{1/(j\omega C_2)}{j\omega L + 1/(j\omega C_2)} = \dfrac{1}{1 - \omega^2 L/C_2}$, which becomes $\dfrac{v_i}{v_o} \approx -\dfrac{C_1}{C_2}$, a negative number, at the oscillation frequency $\omega_0 = 1/\sqrt{L\dfrac{C_1 C_2}{C_1 + C_2}}$. Consequently, there is additional 180° phase shift from the output voltage to the input voltage, resulting in a total of 360 or 0° phase shift

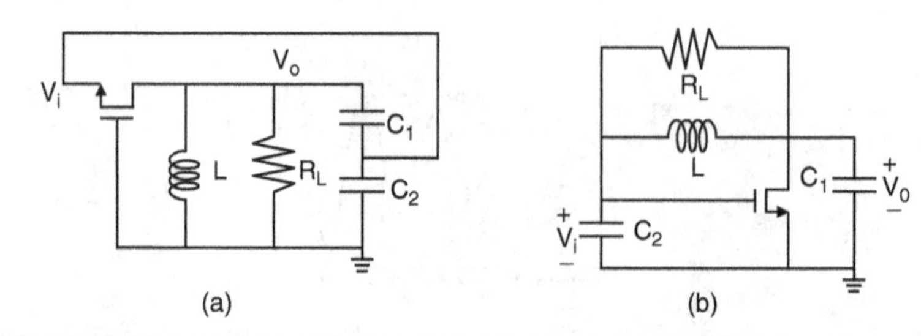

FIGURE 10.21 Colpitts LC oscillators with MOSFET: (a) common gate and (b) common source without DC bias circuit.

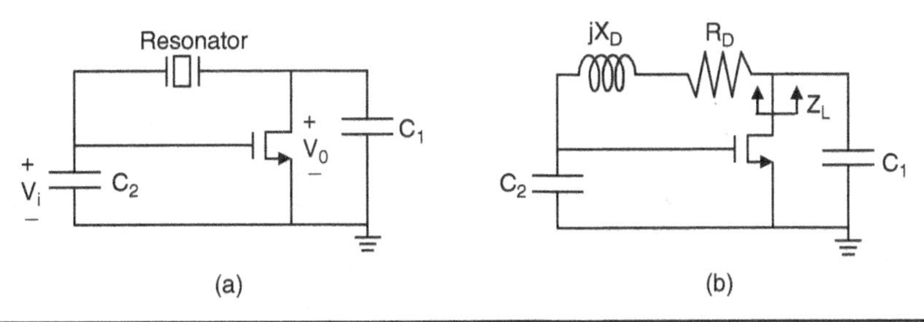

Figure 10.22 (a) AC circuit of a BAR-based Pierce oscillator and (b) the BAR represented with an equivalent inductor and a resistor at the oscillation frequency.

around the loop (i.e., from the input to the output, followed by from the output to the input). And if the initial loop gain is greater than 1, there will be oscillation.

10.5.2 Pierce BAR Oscillator

A BAR-based Pierce oscillator shown in Fig. 10.20 can be implemented with a common source MOSFET amplifier as shown in Fig. 10.22 where Fig. 10.22a is an AC circuit without the DC bias circuit, while Fig. 10.22b is the same circuit with the BAR replaced with an equivalent inductor (L_{eff}) and resistor (R_m). The drain of the MOSFET sees a load Z_L at the oscillation frequency $\omega_0 = 1/\sqrt{L_{\text{eff}} \dfrac{C_1 C_2}{C_1 + C_2}}$ as follows.

$$Z_L = \frac{\dfrac{1}{j\omega C_1}\left(R_D + j\omega L_{\text{eff}} + \dfrac{1}{j\omega C_2}\right)}{\dfrac{1}{j\omega C_1} + R_D + j\omega L_{\text{eff}} + \dfrac{1}{j\omega C_2}}$$

$$= \frac{\dfrac{1}{j\omega C_1}\left(R_D - \dfrac{1}{j\omega C_1}\right)}{R_D}, \text{ because } j\omega L_{\text{eff}} + \frac{1}{j\omega C_1} + \frac{1}{j\omega C_2} = 0 \text{ (Colpitts phase condition)}$$

$$\approx \frac{1}{\omega^2 C_1^2 R_D} \text{ for } \frac{1}{\omega C_1} >> R_D \text{ which is usually the designed case for a resistive load.}$$

Thus, the initial loop gain A_e is $A_e = g_m Z_L \dfrac{C_1}{C_2} \approx \dfrac{g_m}{\omega_o^2 R_D C_1 C_2}$, which is usually designed to be 3–4.

A complete BAR-based Pierce oscillator is shown in Fig. 10.23a, of which the AC small-signal equivalent circuit is shown in Fig. 10.23b. In designing the oscillator we will have to choose R_{B1}, R_{B2}, and R_{BS} to bias the MOSFET into its saturation region with a desired transconductance g_m for an initial gain A_e of 3–4, as will be explained in the next paragraph. The C_{Bypass} is chosen to be large enough to provide an electrical shorting at the oscillation frequency. The tunable capacitor (which is optional) is to eliminate the possibility of oscillation at the harmonic frequencies, while the radio frequency choke (inductor) is to bring V_{DD} to the MOSFET's drain in DC and to provide a high enough impedance at the oscillation frequency (so that it may not affect the loop gain).

The values for R_{B1}, R_{B2}, and R_{BS} are chosen such that the transistor is biased into its saturation region by ensuring that V_{GS} (the DC voltage between the transistor's gate and

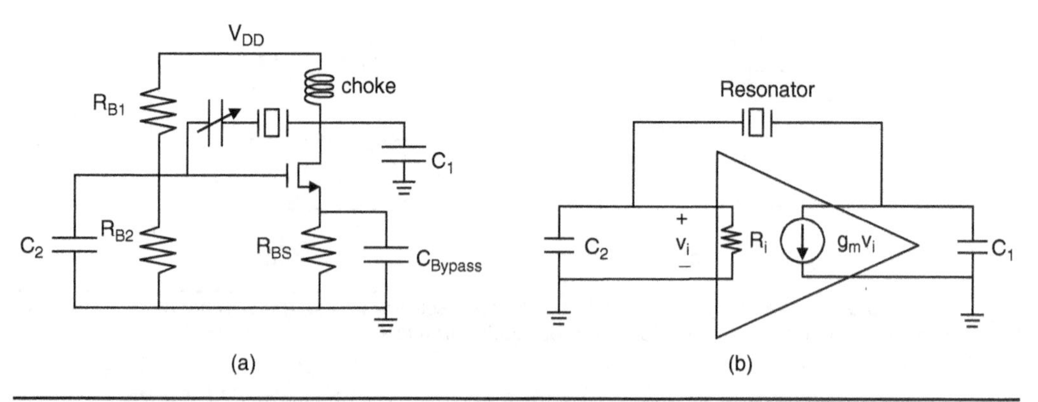

FIGURE 10.23 A BAR-based Pierce oscillator: (a) a complete circuit including the DC bias circuit (composed of R_{B1}, R_{B2}, R_{BS}, and C_{Bypass}) and frequency-trimming capacitor and radio frequency choke and (b) AC equivalent circuit of the oscillator where the MOSFET amplifier is equivalently represented with small-signal input resistance R_i and transconductance g_m.

source, which is V_{in} in Fig. 10.24a) is greater than the transistor's threshold voltage V_T, while the V_{DS} (the DC voltage between the drain and source, which is V_{out} in Fig. 10.24a) is greater than or equal to $(V_{GS} - V_T)$ or $(V_B - V_T)$, as shown in Fig. 10.24b. When the transistor is in saturation region, the current I_{DS} through the drain and source is

$$I_{DS} = \frac{k}{2}(V_{GS} - V_T)^2, \tag{10.23}$$

where k is a constant that depends on the transistor's dimension (width over length of the channel) and material properties, and $V_o = V_{DD} - I_{DS}R_D = V_{DD} - \frac{k}{2}(V_{GS} - V_T)^2 R_D$, which needs to be greater than or equal to $(V_{GS} - V_T)$ or $(V_B - V_T)$. For given V_{DD}, k, V_T, and R_D, we can calculate the maximum V_{GS} that will keep the transistor in saturation region. When the MOSFET is biased in the saturation region (through properly choosing

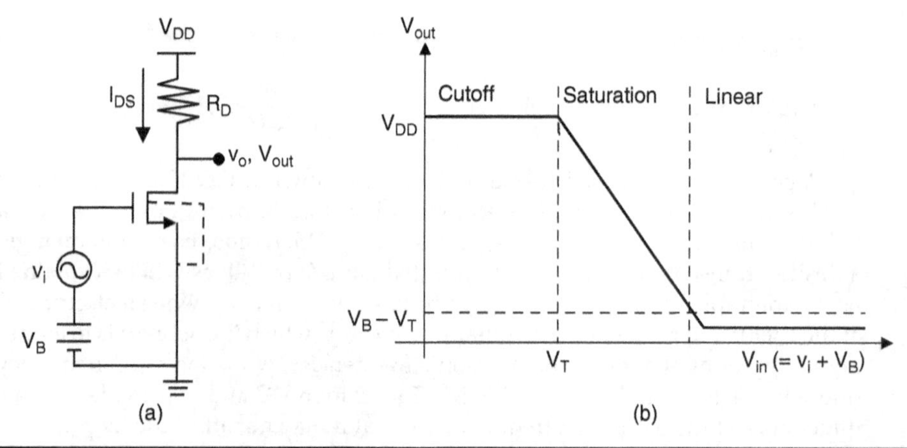

FIGURE 10.24 A simple common source MOSFET amplifier: (a) the circuit with the bias voltages and (b) V_{out} versus V_{in} with three distinct operational regions of the MOSFET.

R_{B1}, R_{B2}, and R_{BS} for given V_{DD} and V_T), there will be a bias current I_{DS} [Eq. (10.23), and the small-signal transconductance g_m is

$$g_m = \frac{\partial I_{DS}}{\partial V_{GS}} = \frac{\partial[k(V_{GS}-V_T)^2/2]}{\partial V_{GS}} = k(V_{GS}-V_T) = \sqrt{2kI_{DS}} \qquad (10.24)$$

10.5.3 HBAR-Based 3.6-GHz Oscillator

High-overtone BAR (HBAR) is typically built on a few hundred μm thick substrate with a metal/piezoelectric-film/metal transducing structure as illustrated in Fig. 10.25a. A piezoelectric ZnO film can be used as the piezoelectric material with its thickness ranging from submicron to a few microns, depending on the frequency of interest. The piezoelectric film generates acoustic wave, which propagates into the substrate. Then the whole unit including the transducer and the substrate becomes a resonant system. Since the substrate thickness is much larger than that of the piezoelectric film, most of the acoustic wave resides in the substrate. Thus, the quality (Q) factor is dominated by the acoustic property of the substrate. Since the sapphire has very small acoustic loss, an HBAR made on a sapphire substrate has a very high Q, which has been shown to be greater than 15,000 at 3–5 GHz [2].

A fabricated HBAR (Fig. 10.25b) for the 3.6-GHz oscillator (Fig. 10.25c) has a structure of 0.10 μm Al/0.88 μm ZnO/0.10 μm Al/400 μm Sapphire, and is measured to have a loaded Q of 15,000 and 19,000 at the series and parallel resonant frequencies around 3.67 GHz, respectively (Fig. 10.26b). The temperature coefficient of the resonant frequency is measured to be –28.5 ppm/°C. The 3.6-GHz oscillator (Fig. 10.25c) is a Pierce oscillator built on a printed circuit board (PCB) with the HBAR wire-bonded to a circuit composed of surface mounted bipolar junction transistor (BJT) and passive components. A 3.6-GHz Pierce oscillator shown in Fig. 10.26a is based on a HBAR, whose measured S_{11} (from which we can obtain the HBAR's impedance as a function of frequency) is shown in Fig. 10.26b. This design uses a bipolar junction transistor (BJT) rather than a MOSFET, but the differences between the transistor types in oscillator design are rather minor: (1) BJT has base, emitter, and collector that act like gate, source, and drain of MOSFET, respectively, (2) BJT has $V_{BE(on)}$ (≈0.7 V) instead of V_T for MOSFET for BJT to be on, (3) V_{CE} has to be greater than 0.2 V for BJT to be in a region similar to MOSFET's saturation region,

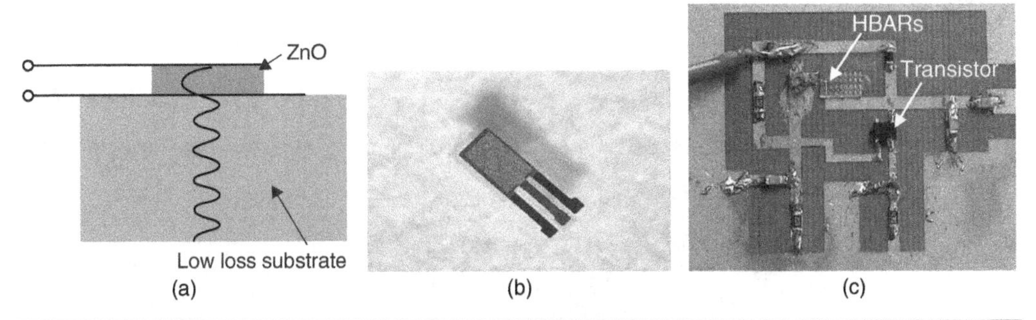

FIGURE 10.25 Cross-sectional schematic (a) and top-view photo (b) of HBAR [2]. (c) Photo of a PCB containing the discrete HBAR-based Pierce oscillator [2].

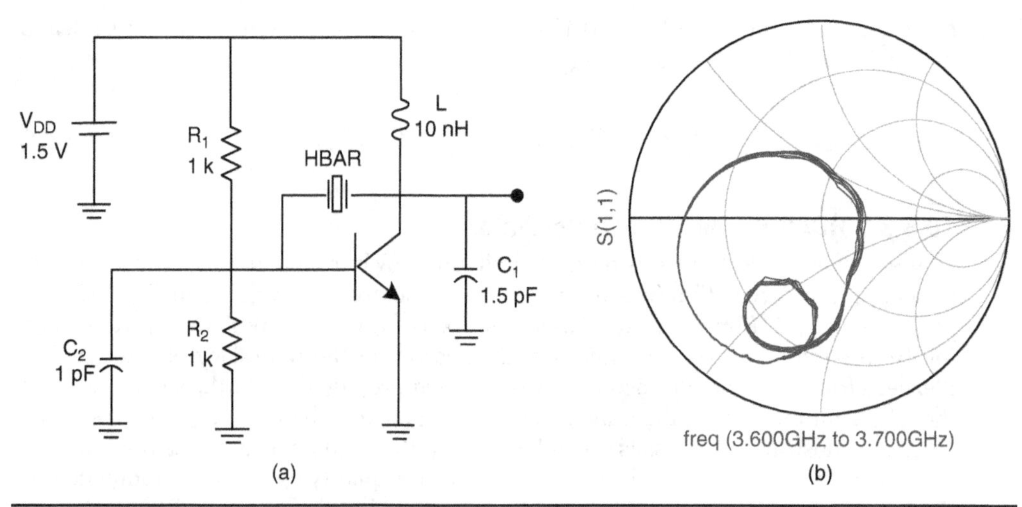

FIGURE 10.26 (a) 3.6-GHz Pierce oscillator based on HBAR and (b) measured S_{11} of the 3.6-GHz HBAR.

and (4) there is nonnegligible bias current into the base of BJT, though there is virtually zero bias current into the gate of MOSFET, etc.

To meet the oscillation condition, the initial small-signal loop gain must be larger than one at the oscillation frequency where the phase shift around the loop reaches 360°. Near the oscillation frequency of 3.6 GHz, the HBAR has impedance $Z_{\text{HBAR}} = R_c + jX_c$, where R_c = 10–100 Ω and X_c = 0–50 Ω. Two HBARs (considered for the oscillator) have $Z_{\text{HBAR}} = 8 + j24$ Ω and $26 + j23$ Ω near 3.6 GHz, and the bonding wire for the HBAR has about 0.6 nH/mm inductance which provides tens of ohm (e.g., 77 Ω at 3.6 GHz if the length is 7 mm). Impedances jX_1 and jX_2 of C_1 (1.5 pF) and C_2 (1 pF) are $-j29$ Ω and $-j44$ Ω, respectively. Consequently, R_1 (1 kΩ), R_2 (1 kΩ), and $j\omega L$ (= $j226$ Ω) can be ignored in calculating the initial loop gain.

With $Z_{\text{HBAR}} = R_c + jX_c$, $jX_1 = 1/(j\omega C_1)$, and $jX_2 = 1/(j\omega C_2)$, the small-signal gains from the base to the collector (V_c/V_b) and from the collector to the base (V_b/V_c) are

$$\frac{V_c}{V_b} = -g_m Z_L, \text{ where } Z_L = \frac{jX_1(R_c + jX_c + jX_2)}{R_c + jX_c + jX_1 + jX_2}$$
$$\frac{V_b}{V_c} = \frac{jX_2}{jX_c + R_c + jX_2} \tag{10.25}$$

The initial loop gain is $A_e = \dfrac{g_m X_1 X_2}{R_c} = \dfrac{g_m}{\omega^2 C_1 C_2 R_c}$ when the phase condition $jX_c + jX_1 + jX_2 = 0$ is satisfied, as shown below, independent of how large R_c is compared to X_c.

Case 1: neglible R_c

$$\frac{V_b}{V_c} = -\frac{jX_2}{jX_1} = -\frac{X_2}{X_1} = -\frac{C_1}{C_2}$$

$$Z_L = \frac{jX_1(-jX_1)}{R_c} = \frac{X_1^2}{R_c}$$

Case 2: R_c dominant

$$\frac{V_b}{V_c} = \frac{jX_2}{R_c}$$

$$Z_L = \frac{jX_1 R_c}{R_c} = jX_1$$

$$\frac{V_c}{V_b} = -g_m \frac{X_1^2}{R_c} = -g_m \frac{1}{\omega^2 C_1^2 R_c} \qquad \frac{V_c}{V_b} = -jg_m X_1$$

$$A_e = \frac{g_m X_1 X_2}{R_c} = \frac{g_m}{\omega^2 C_1 C_2 R_c} \qquad A_e = \frac{g_m X_1 X_2}{R_c} = \frac{g_m}{\omega^2 C_1 C_2 R_c}$$

For given or chosen C_1, C_2, R_c, and oscillation frequency ω_o, the initial loop gain A_e needs to be designed to be greater than 1, ideally 3–4, by biasing the transistor such that its $g_m \left(= \frac{I_c}{26 \text{ mV}}\right)$ for BJT at room temperature is large enough. Agilent ADS' harmonic balance tool can be used to design the oscillation to happen near the HBAR's parallel resonance where the Q is the highest, though it was not used for this design.

The completed oscillator is measured (with Agilent 4448 spectrum analyzer) to oscillate at 3.677 GHz with an output signal power of −15 dBm. The oscillation frequency is located exactly in the inductive region of the HBAR near its parallel resonant frequency. Though the phase noise of the oscillator can be measured with a spectrum analyzer, the oscillator was evaluated by a group at National Institute of Science and Technology (NIST) for accurate measurement of the phase noise as well as Allan deviation [3].

The phase noise and Allan deviation of a free-running oscillator are measured with a Timing Solutions TSC 5120A phase noise test set, after down-converting the signal to the required range of 1 and 30 MHz. The reference frequency comes from a frequency synthesizer that is externally referenced to a hydrogen maser ensemble. Fig. 10.27a shows the measured phase noise of the oscillator running with 1.24 V power supply, demonstrating the oscillator's outstanding performance at sub-Hz–100 kHz offset. The spikes are due to 60-Hz power and harmonics, monitors, fluorescent lights, etc. The measured phase noise of the oscillator consuming mere 3.2 mW is −54 and −102 dBc/Hz

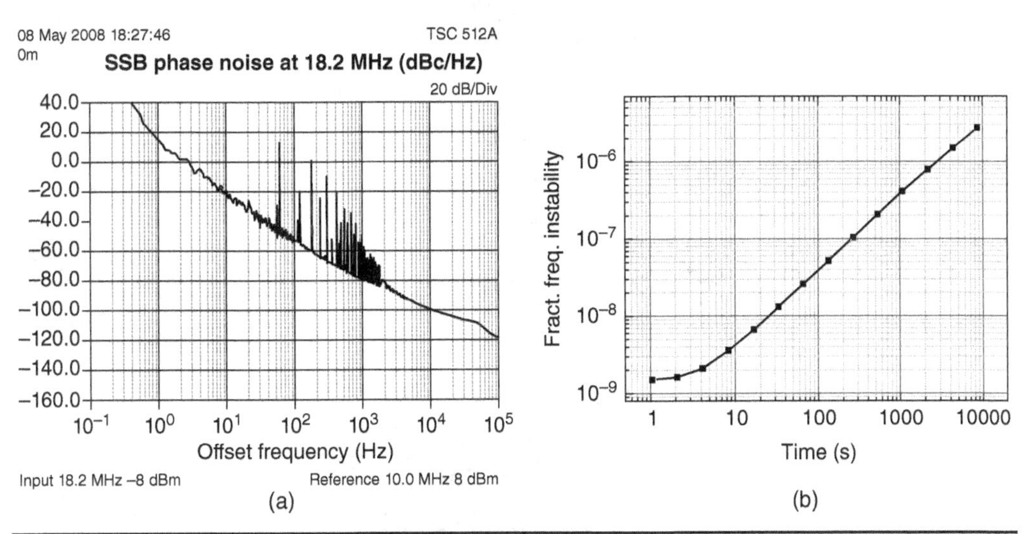

FIGURE 10.27 (a) Phase noise of the 3.6-GHz oscillator measured with a Timing Solutions TSC 5120A phase noise test set. (b) Measured Allan deviation of the 3.6-GHz oscillator that consumes low power (about 3 mW), free-running oscillator [3].

at 100 Hz and 10 kHz offset, respectively (Fig. 10.27a). To measure the Allan deviation, the oscillator was put in a well-shielded box (to prevent electrical interference), which then was put in a foam box to somewhat protect against ambient temperature fluctuation. The measurement was long enough that the drift does not seem to show up on the measurement until after the 10-second mark (Fig. 10.27b).

References

1. L. Baumgartel, A. Vafanejad, S.J. Chen, and E.S. Kim, "Resonance enhanced piezoelectric microphone array for broadband or pre-filtered acoustic sensing," IEEE/ASME Journal of Microelectromechanical Systems, 22, 2013, pp. 107–114.
2. H. Yu, C. Lee, W. Pang, H. Zhang, and E.S. Kim, "Low Phase Noise, Low Power Consuming 3.7 GHz Oscillator Based on High-overtone Bulk Acoustic Resonator," IEEE International Ultrasonics Symposium, New York, NY, October 29–31, 2007, pp. 1160–1163.
3. H. Yu, C. Lee, W. Pang, H. Zhang, A. Brannon, J. Kitching, and E.S. Kim, "HBAR-based 3.6 GHz oscillator with low power consumption and low phase noise," IEEE Transactions on Ultrasonics, Ferroelectrics, and Frequency, 56(2), 2009, pp. 400–403.

Questions and Problems

Question 10.1 An amplifier with a low-frequency voltage gain of about +10 has a ***pole*** (for the voltage gain) at 100 kHz. Assuming that the amplifier has the ***zero*** (right-half-plane zero) at 100 MHz and the ***second pole*** at 10 GHz, sketch the magnitude (in dB) and phase (in degree) of the voltage gain as a function of frequency.

Question 10.2 Calculate the shift of the oscillation frequency ω_o, if there is a spurious phase shift of –15° in the oscillation loop for an oscillator based on a quartz crystal with Q of 10,000.

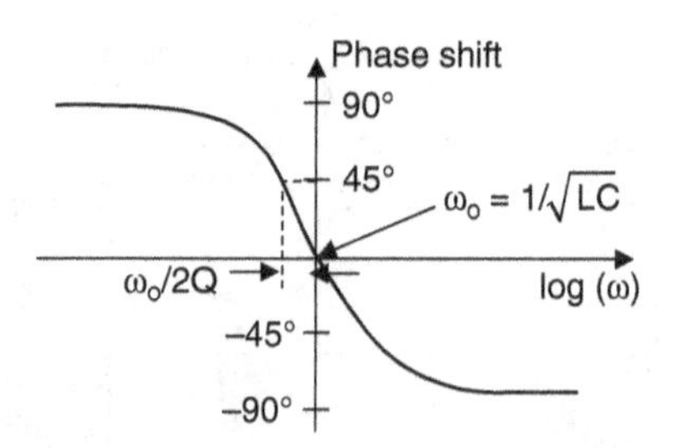

Question 10.3 Show that the reactive current through the inductor L in Colpitts oscillator (Fig. 10.21) is Q times the current in R_L.

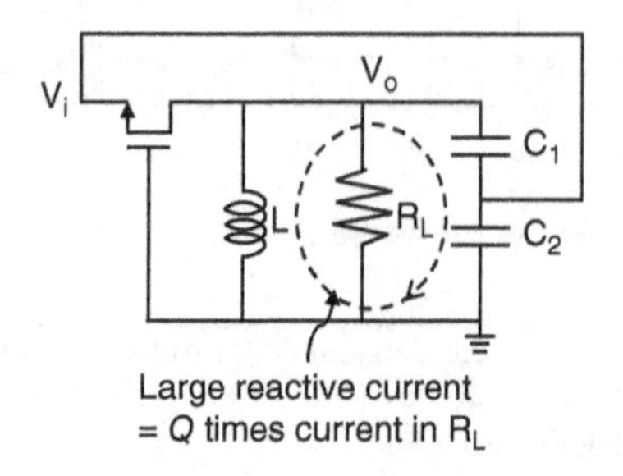

Problem 10.1 Answer the following for an op amp that has a low-frequency voltage gain a_o of 40,000, with the dominant pole at 2.5 Rad/sec and another pole at 100 kRad/sec, while all the other poles and zeros are at far above 1 MRad/sec, so that the voltage gain a may be $a(j\omega) \approx \dfrac{a_o}{(1+j\omega/2.5)(1+j\omega/10^5)}$

(a) If the op amp is connected into a negative feedback circuit as shown below so that the closed-loop gain A_{CL} (i.e., v_{out}/v_{in}) at low frequency is equal to 10 (with feedback factor $f = \dfrac{1\text{ k}\Omega}{1\text{ k}\Omega + 9\text{ k}\Omega} = 0.1$), what is the magnitude of the closed-loop gain $|A_{CL}(j\omega)| = \left|\dfrac{a(j\omega)}{1+a(j\omega)f}\right|$ at the frequency where the magnitude of the loop gain ($T \equiv a(j\omega)f$) is 1 and sketch $|A_{CL}(j\omega)|$ versus frequency.

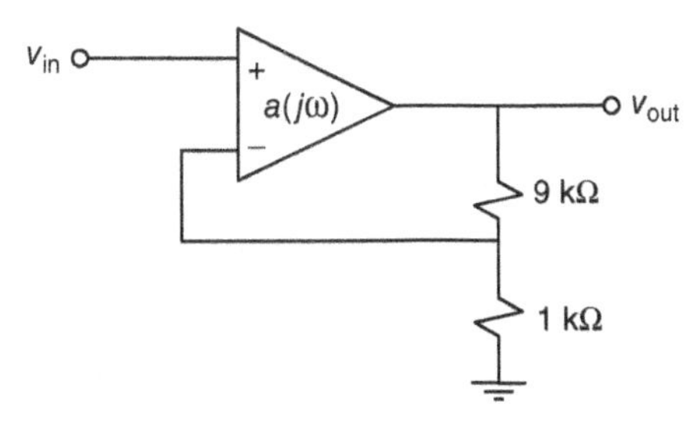

(b) If the op amp is connected into a unity-gain feedback as shown below so that the closed-loop gain A_{CL} (i.e., v_{out}/v_{in}) at low frequency is equal to 1, what is the magnitude of the closed-loop gain $|A_{CL}(j\omega)| = \left|\dfrac{a(j\omega)}{1+a(j\omega)f}\right|$ at the frequency where the magnitude of the loop gain ($T \equiv a(j\omega)f$) is 1 and sketch $|A_{CL}(j\omega)|$ versus frequency.

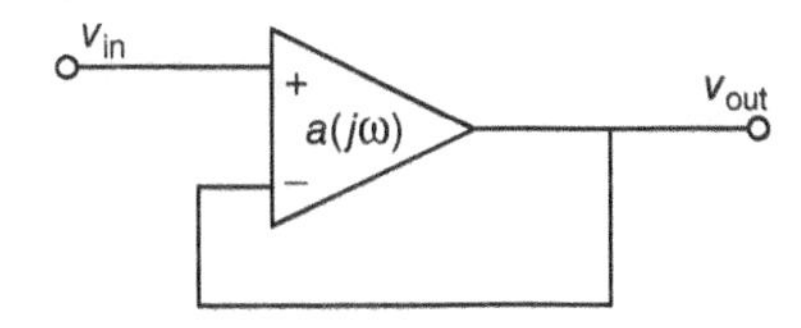

Problem 10.2

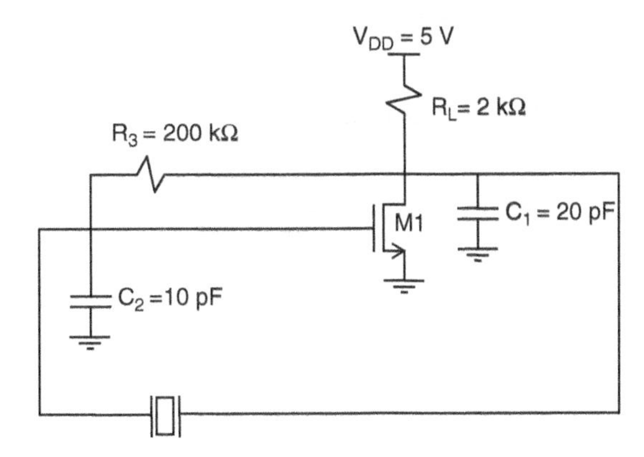

Crystal Data: $L_m = 1$ H, $C_m = 0.01$ pF, $R_m = 500\ \Omega$, Negligible C_o.

(a) Calculate the oscillation frequency in Hz to *seven* significant digits. (b) What is the loop gain at the *steady state* when the circuit is oscillating? (c) Ignoring the effects of R_3 and R_m, write down an equation for the reverse transmission ratio (i.e., v_{gs1}/v_{ds1}) in terms of ω, L_{eff}, and C_2 with $L_{eff} = L_m\left(1 - \frac{1}{\omega^2 L_m C_m}\right)$. Then calculate the reverse transmission ratio at $\omega = \frac{1}{\sqrt{L_{eff}\frac{C_1C_2}{C_1+C_2}}}$.

(d) *Sketch* the drain-source current (i_{ds1}), drain-source voltage (v_{ds1}), and gate-source voltage (v_{gs1}) of M1 at the *steady state* (when the circuit is oscillating) as a function of time.

Index

T